THE CENTRALIA *Wizard*

The Chance Company Story

Volume I

1896 thru 1936

BY DONALD R. DARST

THE
DONNING COMPANY
PUBLISHERS

The Donning Company Publishers
184 Business Park Drive, Suite 206
Virginia Beach, VA 23462

Steve Mull, General Manager
Barbara B. Buchanan, Office Manager
Richard A. Horwege, Senior Editor
Scott Rule, Graphic Designer & Imaging Artist
Scott Rule, Director of Marketing
Tonya Hannink, Marketing Coordinator

Ed Williams and Steve Mull, Project Directors

Library of Congress Cataloging-in-Publication Data

Darst, Don R.
The Centralia wizard : the Chance Company story / Don R. Darst.
v. <1> ; cm.
Includes bibliographical references and index.
Contents: In the beginning (up to 1899) — The birth of Chance Company (1900/1910) — The formative years (1911/1916) — Growth and expansion (1917/1924) — Success and hardships (1925/1930) — Reorganization and survival (1931/1933) — A depression rebirth (1934/1936)
ISBN 978-1-57864-406-3 (hard cover : alk. paper)
1. A. B. Chance Company—History. 2. Telephone companies—Missouri—History. 3. Telephone—Missouri—History. I. Title.
HE8846.A23D37 2007
338.7'62415—dc22

2007009338

Printed in the USA by Walsworth Publishing Company

Dedication

To the men and women who labored for Chance Company in fulfilling their dreams and the dreams and visions of the founder, A. Bishop Chance, and his son, F. Gano Chance. Their combined efforts positively impacted the world. The historical record of these facts is now officially recorded.

—D.R.D.

Chance Company

POLE LINE SPECIALTIES

GENERAL OFFICE AND WORKS

Centralia, Missouri

A. BISHOP CHANCE, PRESIDENT.

OFFICES
BOSTON
CHICAGO
ATLANTA
PITTSBURGH
KANSAS CITY
SAN FRANCISCO

World's Largest Anchor Manufacturers

Table of Contents

Volume I:
The Centralia Wizard

The Centralia Wizard

Foreword

It was mid-April of 1907 in Centralia, Missouri, when the A. B. Chance Company was born. This newly formed company was located in a small frame building that was once a church on the south side of Sneed Street, just across from the City Square.

A. Bishop Chance began what was to become a worldwide electrical manufacturing company and its stock would be traded on the New York Stock Exchange some seventy years later. In the early days, Bishop was the only employee. Bishop's grandfather, Garrison Chance, settled in Boone County in 1840, only nineteen years after Missouri had achieved statehood. Prior to the founding of A. B. Chance Company, Bishop and his father, John, owned and operated Centralia's first telephone exchange. During the Great Depression, Chance struggled, but was able to provide work for the one hundred employees. A huge loss for A. Bishop was the death of his oldest son, Buford, in an automobile accident. In 1939 at the age of thirty-four, Bishop's younger son, F. Gano Chance, assumed the presidency of the company and "Boss" Chance, as he was affectionately called by his employees, became chairman of the Board. Gano's brother-in-law, Nathan Toalson, some years later, successfully guided the company.

It was 1978, a few years after Chance Company merged with Emerson Electric of St. Louis, Missouri, that Gano, my father, decided it would be a good idea to have a book written about the history of the company. The Chance Foundation was chosen to be the vehicle to accomplish this project and Roy Hansen assumed the responsibility of interviewing longtime employees. Who better to tell the story of Chance than those who had worked there? Hansen himself was a longtime employee of Chance and had recently retired as president. Roy took tape recorder in hand and talked with just about everyone he could find that had retired from the company. At the time, we did not know how valuable these interviews would become. Helen Winn spent countless hours transcribing these conversations from tape to paper. The next task Gano had was to find a suitable author. Through the years of being in the electrical business Dad had become acquainted with Paul Galvin, founder of Motorola, Inc. A book had been written about Galvin's company by Mark Petrakis and Dad was so impressed with his writings that he contacted him about doing a book on Chance Company. Mark made a number of trips to Centralia and to Scottsdale, Arizona, where Mother and Dad spent their winters. But in early 1981, Dad became ill and a couple of months later passed away while in Scottsdale. The writing of the history of A. B. Chance Company was put on hold. Time passed, and Mark Petrakis became ill and died.

Not until approximately ten years ago had there been much thought about the book, until Don Darst talked to Art Allen about the possibility of writing a book on the history of Chance Company. Don showed us some of the books he had written previously and we were quite impressed. It was discussed with members of the Foundation Board, and a decision was made to continue the work on Chance history. Don spent long hours gathering information on the company, going to libraries, reading publications, and visiting with longtime employees. He made countless phone calls, and traveled many miles in his desire to make sure the written word was absolutely correct and accurate. Don's wife, Valerie, was a great help working right along with him. Don had been an employee of Chance for forty years and he personally knew many longtime employees and their families.

I said earlier that the interviews Roy Hansen made were valuable, well, they were priceless because by this time most of the people interviewed had passed away. Art Houston, who was a key man in Chance Company's advertising and marketing program, added much to the efforts of accumulating company historical materials in the earlier days. Bill Miller and Janet Holdeman of The Chance Foundation added their support.

As I look at what has been accomplished by the recording and writing of the A. B. Chance history, there is one thing that is saddening to me. My Grandfather, A. B., my Father, Gano, and my brother, Phil, are not here today to enjoy this book. Neither are the many loyal employees that have gone on before us. I know they would be proud of the work done here.

In recording the history of Chance Company, it is impossible not to write about the history of Centralia and the many wonderful people that have made this community the greatest city in America. Don Darst has accomplished just that. In closing, I would like to pass on the ideas that have guided the A. B. Chance Company through decades of time.

I quote:

"Those are the ideas that guided us."

This is the story about a man, his company, the employees, the people of the community, and the town they called home.

1. We believe in and work to preserve the God-given rights of freedom of choice and freedom of action expressed in the Declaration of Independence and guaranteed to Americans in the Bill or Rights. We believe in representative government of, by, and for the people. . . . government whose powers are limited to those willingly given by the people through the Constitution. We actively support and encourage individuals and organizations who hold similar beliefs.

We strive to help make our nation, and particularly each plant community, a better place in which to live and work by assuming our civic and political responsibilities as a corporate citizen. We encourage our employees to do likewise as individuals.

2. We believe in vigorous, ethical competition within the framework of our free, competitive economy where there are both profits and losses. We prosper by providing useful, high-quality products and reliable service to the electric power and communications industries and other customers. Through these products and services, we fill a need of our society. By performing our functions effectively and efficiently, we make a profit and help those with whom we do business to profit.

Our customers are providing vital services necessary to the growth and progress of our nation and world. We have faith in their future and ours. We believe in sharing our progress in better products and service with our customers to improve their operations and reduce their costs.

3. We seek to provide useful, stimulating work, plus the opportunity and training for those who have the desire and ability to attain positions of increased responsibility. We believe in sharing our profits with our employees as an incentive in addition to regular wages and salaries.

4. We recognize our responsibility to manage the business efficiently and prudently so that it will be strong financially and capable of providing stability and opportunities for growth.

5. Because we believe in the dignity of man, it is our practice to respect the individual rights of others and to be honest, fair, and courteous in our relationships with individuals and organizations.

Jack Chance
(John Hardin Chance)
November 2006
Centralia, Missouri

SKY-ROCKET

"NEVER CREEP" ANCHORS

Chance Company

MANUFACTURERS OF

ELECTRICAL SPECIALTIES

CENTRALIA MO. U.S.A

"SKY ROCKET" BRAN
LIGHTNING ARRESTERS
TERMINAL BRACKETS
400 COMPANION
LIGHTNING ARRESTERS
BACK TRUSSES
ANCHORS, ETC., ETC.

THE ANCHOR THAT IS SUPERIOR TO THE "DEAD-MAN"

Introduction

From the day I went to work for A. B. Chance Company in 1961, there has continually been a thirst for knowing more about the history of the company. Like F. Gano Chance, it became my feeling that the Chance Company story was too rich and important to be left undone. Some twenty-five years ago, I had a discussion with Public Relations Director Wally Schramm about such a venture. Little did I realize that it would become my destiny to fulfill this wish.

After publishing two International Harvester books, the Chance book actually began by writing short Chance Company stories. My I.H. experiences had helped prepare me, but it had also revealed that a quality hard cover book, befitting to Chance Company, would be well beyond my financial means. In addition, besides my own collection of Chance material, hardly any other official company documents appeared to have survived. Realizing that the Chance Foundation might be able to provide support, I approached them. Just like in earlier times, it was more of a handshake and a verbal OK from Jack Chance and Art Allen to proceed. At that time I was unaware that in the mid-1970s Gano had orchestrated several initiatives in an attempt to have a Chance book published. One result was the involvement of Art Houston, Roy Hansen, and his son Jack in gathering historical material. When Jack and Art revealed and released to me the Hansen/Chance and Petrakis interview material that was gathered in the 1978–81 period, it gave my project a strong boost. It represented interviews with sixteen key veteran employees and officials of the company, plus family friend R. L. (Linwood) Hope. In addition to personal remembrances of thousands of discussions over forty years, I also began to interview and talk with veteran employees in earnest. Hundreds of contacts have been made for Volume I, especially valuable as many employees have since passed on such as Tom and Anna Marshall, Keith Claxton, and Perry Roberts. Thanks goes out to those who provided additional stories, company details, and support.

Years passed and slowly the book began to take shape. Those who knew of the project surely reasoned it had dead-ended again, and even Jack and Art became concerned as I did as the one-hundredth anniversary grew closer. It became clear that the book was not going to be completed by 2007 and it had become too large for one publication. It was decided and agreed that it would become two volumes and work to complete Volume One would be expedited. This worked out beautifully as Volume One would focus on the founder, A. Bishop Chance, and Volume Two on his son, F. Gano. Volume One ends in 1936, as it is prior to the Tips Hot

Line Tool era that is associated with F. Gano Chance's reign of company leadership.

There are many ways the Chance story could have been composed. Had Mr. Petrakis successfully completed his Chance volume, I'm convinced it would have been different. One goal was an honest portrayal of Chance Company without being unduly influenced by any one source, although the *Centralia Fireside Guard* and employee interviews were heavily relied upon. It was purposely written in a "story book" fashion in an attempt to entertain the reader exposed to a multitude of historical information. This would hopefully enable the reader to be transported back to relive those times and events in a vividly realistic manner. Period local, national, and world events were also utilized, so the reader could have a sense of the times and the events that helped shape and mold the Chance story. It was also written in a personal way, as that is reflective of A. Bishop Chance and the Chance Company in its early years. There is considerable background material presented, especially involving the telephone. This is historically significant from a Chance perspective, as it honed Mr. A. Bishop's business and technical skills and was the industry that inspired the birth of Chance Manufacturing Company.

Two of the most vital elements in our modern world are the telephone and electricity. One purpose of this book was to record the important and frequent innovative contributions made by Chance Company to the telephone and electrical utility industries. Another purpose was to capture the distinctive operating style of this unique firm. A policy statement encompassing those principles is splendidly displayed in "Those are the ideas that guided us," in Jack's Foreword. In addition, an important goal was to educate the reader on the benefits of America's industrial might of the past, and how difficult it was to amass those capabilities. In the future when the reader is reminded of the term *Businessman*, hopefully it will bring a positive connotation.

To complete the task of a lifetime, in my later years, is a dream come true, especially when it is something I was not supposed to do. Maintaining the course, develops the suggestion for each of us to not give in, but to give out. This is what Mr. A. Bishop Chance believed as you read the adversities he and his employees faced and overcame. This book would not have been undertaken and completed without Valerie and my Christian faith that speaks of service in return for the blessings of living through difficult, but truly the best of times. The daily cost in time and personal sacrifices has been substantial for Valerie J. (Macrander) Darst. Over the last ten years she has maintained a computer file and processed constant handwritten rewrites. Her professional librarian background and love of reading has generated advice and suggestions that are immeasurably important to this book. To date, some ten-thousand-plus hours have been expended by Valerie and me on the two volumes. As Volume I neared completion, Ralph Gerhard and Andy Kropf provided assistance in helping refine format and content. Through the years, Art Allen and Jack Chance of the Chance Foundation continued to provide additional materials and support that has proven valuable, especially private Chance family and company resources.

This book is not a perfect product, but it has become so much more than initially anticipated. Sometimes only small bits of information were available, but it was enough to develop a subject such as the Taylor Anchor saga, the beginning of the Chance Expanding Anchor story. It is a hope that school and college students, former Chance employees and families, later generation utility people, and the general public will be exposed to this information. Maybe even a movie could be made of the exploits of "The Centralia Wizard," as it is truly an All-American story!

November 2006 Don Darst

Chapter 1
In the Beginning
(Up to 1899)

A New Birth

In the beginning God created the heaven and earth. And the earth was without form and void; and darkness was upon the face of the deep. And the spirit of God moved upon the face of the waters. And God said, Let there be light; and there was light. And God saw the light, that it was good; and God divided the light from the darkness. And God called the light Day and the darkness he called Night. (Genesis 1:1–5, KJV)

The earth grew and prospered, but it also suffered setbacks. A huge meteor struck the surface of the earth in a fiery crash, generating a cloud cover that snuffed out the sunlight, which provided life to dinosaurs and other early life forms. Warmth turned to cold, water to ice, light again to darkness, and the great ice age began. Ice formations in the northern polar regions grew to immense proportions.

As the atmosphere began to clear, traces of sunlight shot through breaks in the cloud cover and the warmth of the light began to slow the growth of further ice formations. As the sun began to penetrate the atmosphere more powerfully, the great northern ice cap began to melt and glaciers formed. These huge immeasurable layers of ice inched their way southward over North America, crushing and grinding up everything in their path as they pressed southward. The territory over which they passed was dramatically changed forever. As the Biblical great flood had retreated, so the ice began its slow melt down and receded back northward. A few drops began a trickle that became a stream, branch, creek, and river as the torrent of melting ice intensified in its quest to flow to our grand inland seas and the oceans. In an isolated area between two great rivers this water divided, some flowing easterly and some flowing to the south. In this plateau of higher elevated ground, stretching from the east to west and then back north, like a dogleg for a couple of hundred miles, a rich prairie would develop. So began the slow regeneration of a huge mid-continent area, rich in diversity and raw material resources.

Local historian Ralph Gerhard described in volume one of his book *The Early Day History of Moberly, Missouri* the prairie that became modern-day central Missouri.

> *On this flat plateau covered with loamy soil, the main vegetation for thousands upon thousands of years was the tall, native*

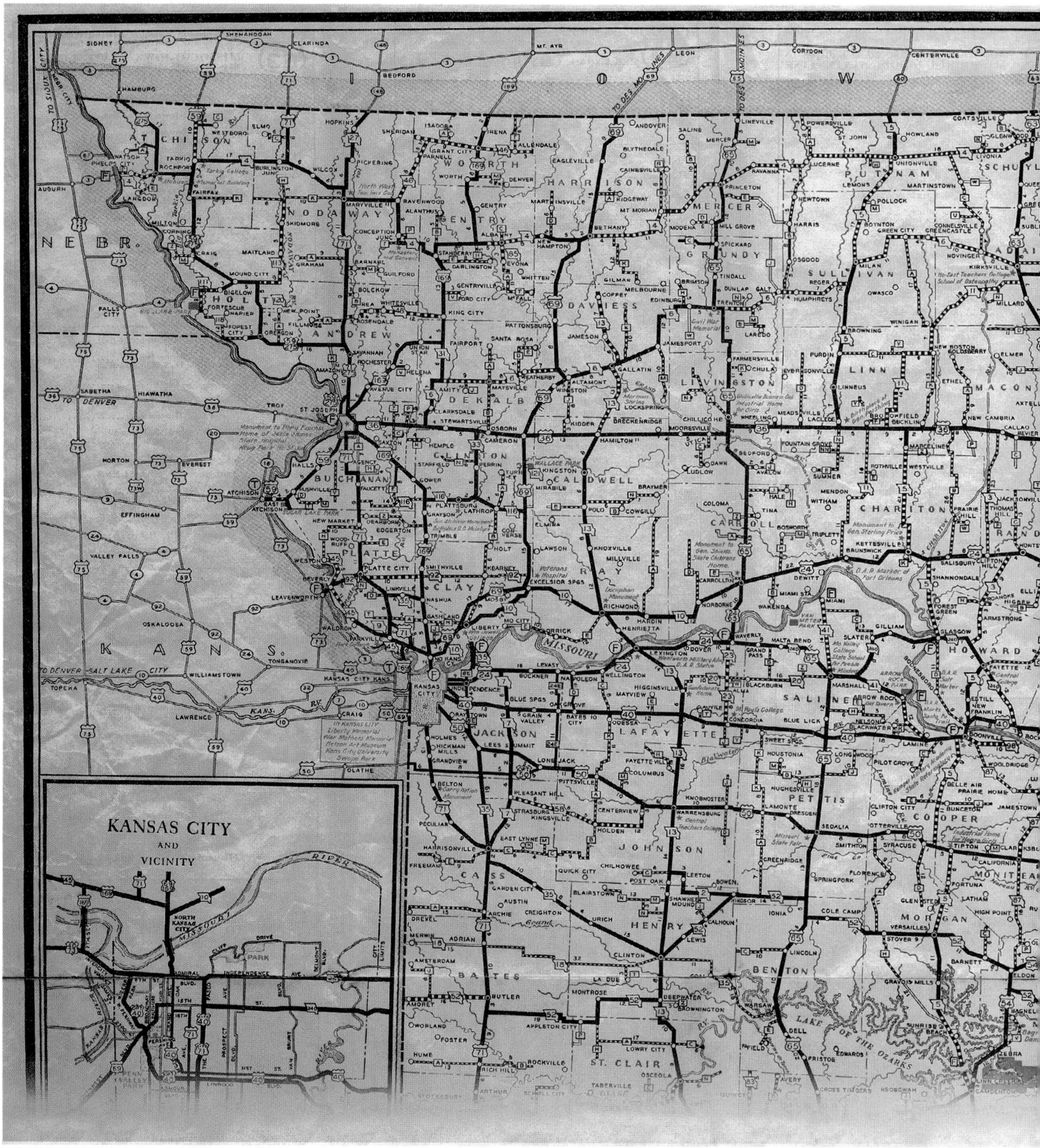

1936 Missouri Map. (Darst Collection)

prairie grass which flourished everywhere. Surveyors of the North Missouri Railroad, in the Civil War period reported that the grass grew taller than the head of a man on horseback. Natural fires kept this upper prairie virtually free of trees or brush but the natural grasses of bluestem, speargrass, dropseed and blooming forbs, such as silphium flourished. Buffalo, elk, antelope, deer, wolf, fox, bear, and other smaller mammals as well as many types of birds, reptiles, rodents and insects were here in abundance. (Gerhard, p. 5)

The prairie was also described during a visit by Harvey Hulen, a frontier schoolteacher in the area of what is now Centralia, Missouri, who was in a privileged position of seeing it in its natural state. The *Centralia Fireside Guard* of October 20, 1922, reported his comments about this area in the 1850 period. "Not so many years ago, where we now stand (First Christian Church in Centralia) was a vast wilderness of grass, seemingly unbounded, extending for miles and miles. Did you ever see this great prairie in its pristine beauty? It was a grand sight." He told of the first road across it from Paris to Columbia, almost as a crow flies. It carried the commerce and produce of the early pioneers. "The prairie grass stood so high that a man on horseback was almost invisible from a short distance." But before those from the Old World settled the North American continent there were already inhabitants which today are referred to as Native North Americans, American Indians, or more properly by their tribe or nation (i.e., Osage). The use of *Redman* is a misnomer as this appearance was reflective of a light to dark-skinned people who used a red solution to color their faces. An archaeological source suggests, "Some researchers hold that the Americas were discovered and inhabited as long as fifty thousand years ago, while others propose that the first settlers did not arrive until about twelve thousand years ago." (Chapman, p. 23)

Cahokia Mounds near St. Louis, established by Indians around A.D. 700, was, according to the Illinois Historic Preservation Agency, "the most sophisticated prehistoric Indian civilization north of Mexico . . . and Monks Mound is the largest Indian mound north of Mexico and the largest prehistoric earthen construction in the New World." Current (2000) Cahokia Mounds material suggests it had a population of ten to twenty thousand. One of the earliest European excursions into North America is traditionally attributed to the Spaniards. De Soto's movement into southern Missouri and Arkansas around 1540, not only brought disease, but his army's cruelty toward the Indians in killing many village leaders is credited with changing the "townspeople" Indian culture! Thus Native Americans were struggling against unsurmountable odds as disease alone almost wiped out their whole civilization and according to *Indians and Archaeology of Missouri*, "many tribes became extinct." (Chapman, p. 10) It did more effectively what guns, the slaughter of the buffalo, and reservations would later complete according to another source. "By the late 1600s many Mississippians and Missouri Indians had died, and the rest had fled their villages to escape illness." (La Doux, p. 20) This was prior to the settlement of the Europeans in mass, especially in the mid and west continent areas, and indicates why Indians appeared nomadic. They had found epidemics could be controlled, or its overall affects reduced, by living in small groups. Disease, therefore, played a significant role in the strength, organization, culture, and characteristics of later Indians that early settlers confronted on the American frontier.

Indian tribes including the Osage, Missouri, Foxes, Pottawatomies, Delaware, Shawnee, Piankashaw, Peoria, Kaskaskia, Cheokee, Miami, Wea, Iowas, Kickapoos, Sac (Sauk), and the Otoe settled in Missouri at one time or another. According to the book *Stories of Missouri*, Missouri was an Indian name of a tribe of Native Americans living along the Missouri River upstream before the Big Muddy flowed into the Mississippi. "That great, dark stream which flowed into the Mississippi was called by Father Marquette 'Pekitanoui' (pek-i-tan-o-wee), meaning Muddy Water. The river was known as Muddy Water until 1712, when it was named Missouri from a tribe of Indians who inhabited the country at its mouth." (Musick, p. 12). The Missouri and Osage were the dominant Indian tribes in Missouri before other tribes were pushed westward into Missouri. The Missouri tribe met the French on the Pinnacles in about 1714. "The tribe was numerous and powerful, the first and strongest to be met when traveling up the Missouri River. They gave their name to the river, then the territory, and eventually the state. The village site where the French first encountered them has been preserved in Van Meter State Park by the Missouri State Park Board." (Chapman, p. 101) The French were well received by the Missouri and Osage and, "in 1723 Bourgmont established Fort Orleans, essentially a trading post, on the north side of the Missouri River." (Chapman, p. 99) Note that this predates Ste. Genevieve (1735) near St. Louis, traditionally considered the oldest settlement west of the Mississippi. Bourgmont persuaded the Little Osage (*Village des Petits Osages*) to settle close to the Missouri village (*Village des Missouris*) so they could also trade. Missouri State Park officials suggest the first French contact took place in the 1680 period, resulting in the spread of an epidemic which reduced the Missouri Tribe from several thousand to some six hundred. This sapped their strength, making them powerless against stronger tribes and they eventually would cease to exist. But in spite of all the problems that the Europeans brought, the Missouri and Osage like other tribes, continued to aggressively pursue trade relations.

But change would be the future rule as even the suggestion of Indians adjusting to these "New Americans" would be challenge enough, especially when mixed with the uncertainty brought, in short order, by three different official governing bodies in the Missouri area: the French, then the Spaniards, and lastly the Americans. The French were highly regarded in their treatment of the Indians. Problems, especially with the Osage, were often ignored by the French in the 1700s, because of the substantial revenues being generated from the rich fur trade with the Osage. However, trading was not something that can be credited to Europeans as during the Mississippi period a thousand years ago, an elaborate Indian trade network was established, stretching hundreds of miles from the Cahokia center. "An almost continuous stream of people from the hinterlands brought salt, meat, paint, skins, bow wood, nuts, and other items to barter for goods of equal value or to exchange as presents on a friendly basis." (Chapman, p. 71)

In the 1600s and 1700s fur traders and those searching for gold and silver were traveling the great rivers including an exploration team of Lewis and Clark (1804–06) after the Louisiana Purchase in 1803. Control of the inland waterways, especially the Mississippi and a possible connection to the Pacific which many thought the Missouri provided, seemed more important than the land mass of the Louisiana Territory. Missouri truly was the Gateway to the West, at the heart of America's great waterway system, a landscape separating the great eastern woodlands with the beauty and vastness of endless prairies and the stepping off point of the great trails that would provide access and routes to the Southwest, West, and Northwest. With the Louisiana Purchase, America had finally gained control of the Mississippi River, owing to the foresight of President Thomas Jefferson. However, France and Spain were so financially strapped by European wars and American Indian problems, and were plagued with such difficulties in protecting their business interests against intrusions by unlicensed dealers, especially remotely from across the Atlantic, that they in essence were pleased to dispose of it.

Early frontiersmen such as Daniel Boone visited Missouri and Boone County. Confirmation of this appears in the February 8, 1873 *Our Fireside Guard*, the local Centralia newspaper referred to as the *Guard*, reporting the death of seventy-nine-year-old Mrs. Susan (Patton) Pace on January 29, 1873, in Boone County. Mrs. Pace, an early area pioneer, came from Allen County, Kentucky, in 1820. "She was intimately acquainted with Daniel Boone who frequently made her home his stopping place." Daniel Boone made his home in 1797 or soon after near St. Louis (Femme Osage). His two sons, plus other parties, established a successful salt mine operation some forty-five miles west of Centralia, Missouri. Boone County and Boonesboro were named after the legendary Boone family. A Missouri State Historical Site, Boone's Lick is in the area that is now Howard County. Daniel Boone's exploits were widely known, and reportedly when an Indian chief heard Boone was inside a fort they planned to attack, they withdrew instead. Such confrontations were also against their war tactics of making lightning-like attacks to gain the advantage of surprise. The Indians' respect of Boone apparently was also afforded to his two sons, known for their courage and enterprise. A similar lick (Freemore's Lick) was developed on the Salt River in Ralls County in 1809, but all members of that party, except one, were soon killed by Indians who were considered more warlike north of the Missouri River. The Salt (*Auhahah* or *Ochahah*) and its tributaries drained the northeast corner of the Northeast Prairie area of Missouri to the Mississippi River, with Centralia located on its southwestern edge.

"The Pottawatomies, who were the great horse thieves of the frontier, made frequent raids upon the Boone's Lick County. In addition to this, the more warlike Iowas, Foxes, and Kickapoos threatened the lives of the settlers." (Musick, p. 137) The reference to the Pottawatomies' attacks were near Franklin and Boone's Lick where some of the first settlements in the interior of Missouri were located and where the Santa Fe Trail would originate in the rapidly growing town of Franklin prior to 1825. The Santa Fe Trail would later move up river and originate officially at Fort Osage before moving on to Independence. "From its exposed position, vast territory, and sparse population, north Missouri suffered more from Indian depredations in the early history of the territory than any other part of the country. With few exceptions, the Indians on the Missouri River were peaceable, even during the War of 1812. The Northern Indians including the Sacs, Pottawatomies, and Iowas made frequent incursions into the inhabited portions of the Territory. They would murder and plunder, and then, as soon as they were pursued by rangers they would flee to the north." (Musick, p. 114–115)

The Missouri Rangers were a military-style organization of volunteers developed to protect settlers on the Missouri frontier from Indians. According to remembrances of a member of the Rangers, Richard Chitwood, recorded in *Stories of Missouri*, "the first regiment of rangers was organized and commanded by Colonel David Musick of St. Louis, who was one of the first representatives from that county in the Territorial Legislature."(Musick, p. 119) No official record apparently exists of this group or the record may have been "destroyed when the State Capitol burned (1837), about twenty-five years later." (Musick, p. 118) This would put the Missouri Ranger formation in the 1810 time frame or a decade before the much-touted Texas Rangers were formed in 1823. Nathan Boone, a key figure in the salt operation at Boone's Lick, was also a prominent officer in the Rangers.

But just as there was hostility between the early settlers and Indians there was also fighting amongst the Indians, especially as they were pushed westward. It wasn't necessarily a question of understanding the European settlements on the East Coast and their westward expansion, but instead Native Americans were reacting in a territorial sense, being concerned about survival, resulting in their attacking, kidnapping, and killing each other. It is also evident that after Spanish control of the territory, "In the 1790s the Delaware and Shawnee were two of the several tribes making raids on the Osage at the instigation of the Spanish who, until 1794, intended to destroy the Osage." (Chapman, p. 120) The Sauk/Fox who later claimed much of northeast Missouri also attacked the Osage/Missouri tribes and settlers. Tribes such as the Missouri who had become small in number had to combine with other tribes and some even lived within the white man's settlements for safety and survival in later years. The early settlers thus upset the whole order of the Indian nations, generating chaotic and violent conditions. The influx of settlers was described by one Osage elder: "They just keep coming like ants." (Wolferman, p. 86) Although for the Osage, not unlike other tribes, "warfare dominated Osage life, and all Osage men wanted to become warriors." (Wolfermen, p. 15) It was an honor to die fighting and the Osage became revered for their fighting spirit. Thus the Iowas and other migrating tribes fought against the Osage (who were favored by the French/Americans) whereas, other rifle-bearing tribes were supported by the British/Spanish, sometimes for disruptive purposes. The athletic and physical attributes of some Indians suggest they were on the order of

Olympian athletes. "Many Osage men were over six feet (1.8 meters) tall. They shaved their eyebrows and heads, leaving only a small patch of hair on top of their heads." (LaDoux, p. 9) Washington Irving, who traveled around the country and met many native tribes, said the Osage were "the finest looking Indians in the west. . . . The men were very tall, usually six feet or more, and very well proportioned." (Wolferman, p. 12) But the Indians' resistance would fall short as the onslaught of European settlers for a land rich beyond comprehension, swept declining remnants of Native North Americans westward, either threatening or resulting in future extinction of their purity of race and tribal connections.

The movement of settlers inland from the Missouri River was slow, as the woodlands around the rivers and in other areas were preferred over the prairies. The prairie was hard to cultivate and had no wood for fuel and building materials or protection from prairie fires and cold winter winds. Swampy areas, common around what would become Centralia, during the summer were infested with mosquitoes, plus rattlesnakes and other undesirables including thick vegetational growth producing a hostile environment that some simply considered uninhabitable for humans. Its remote location also brought transportation limitations as the great rivers were the main channels through which pioneers traveled and commerce flowed as it had with the prior riverine culture of the Mississippi period of Native Americans.

From River to Rail (1822–1879)

Missouri and its people are and have been one of the most diverse in the nation. *Stories of Missouri*, an 1897 textbook's view of Missourians, helps explain:

> *Nearly all the American settlers in Missouri were the descendants of pioneers who had grown accustomed to the trying scenes of forest life. They were a rigidly honest people. No locks were on their doors, and they had no need of any. . . . As they were dependent on each other for protection, what belonged to one was freely loaned to his neighbor. If a pioneer wanted to borrow his neighbor's ax, and did not find the man at home, he took the ax without asking, and returned it when done with it. . . . The early Missourians were rude and uncultured, but unselfish and brave. (Musick, p. 5)*

The Chance family and the Chance Company family of native Missouri employees, were influenced by those early traits. Albert Bishop Chance (1873–1949), was the founder of Chance Manufacturing Company, later becoming the A. B. Chance Company, and also known simply as Chance Company, the primary subject of this book.

In 1822, William Sexton was one of the earliest settlers in the southwestern section of what would become Centralia Township. This settlement was along a small branch of Silver's Fork. Like many other early settlers in this part of Missouri, they were originally from Madison County, Kentucky, by way of Howard County, Missouri. Madison, Missouri, is said to have gotten its name from those transplanted from that Kentucky county. An early Boone Countian and schoolteacher, Harvey Hulen, recalled some of the family names of early area settlers in a speech to the Centralia First Christian Church, recorded in the October 20, 1922 *Guard*: "Silvers, Jennings, Roberts, Barnes, Newmans, Leaches, Boyds, Wainscotts, Mastersons, Sexton, Hunter, Fountains, Hulett, Bryson, Turner, Williams, Pool, and Daniels." Several relatives of these early pioneers were names that would later grace the employee list of A. B. Chance Company. In addition, at least four of these families would become relatives of the Chances through marriage. In the May 18, 1923 *Guard* Harvey Hulen listed some of the over one hundred students he had taught at Locust Grove School, beginning in 1867. These included "M. S. Bush, John A. Chance and his sister Lizzie Chance." All three will be identified later.

The word *chance* could be viewed as a "game of chance," which will become surprisingly appropriate from a Chance Company product perspective. It also could be construed to mean taking a chance referring to one who is confronting that which has poor odds. The origin of the surname *Chance* isn't exactly clear; however, the name *Chance* can have positive connotations "Deo Non Fortuna" which is associated with good luck and fortune. On March 20, 1823, Garrison Hardcastle Chance (1823–1922) was born in Delaware near Dover. He was the son of Thomas Chance of New Jersey and Delaware, whose father had Welsh origins. According to Garrison's obituary in the *Centralia Fireside Guard* of February 24, 1922, he was born in 1822 and passed away one month shy of his one hundredth birthday anniversary. Even the year 1822 is in question as remarks that Mr. Chance made at his ninety-fourth birthday celebration, recorded in a letter by J. T. Roberts to the *Guard* in the March 24, 1916 issue, suggested that according to his count it was his ninety-fifth, which would have meant that he was born in 1821. He did possess a fabulous recall of dates and events for his age, as noted in the many pieces of correspondence that Mr. Roberts wrote from Tulsa back to their Centralia friends and relatives. But 1823 established by later family research, will be carried forth in this book.

In 1827 the Thomas Chance family moved to Ohio per a Boone County history source and in the 1830–33 period, they moved further west to Scott County, Illinois. Information from a celebration of G. H. Chance's birthday in 1921, per an article in the April 8, 1921 *Guard*, concurs with more detail. "Mr. Chance came to Ohio with his parents in an ox-wagon when he was eight years old. A few years later they again loaded their household effects and family into the ox-wagon and lumbered across the prairie to Illinois." Garrison's obituary said his parents died in 1831 in Oxville, Illinois, but later information suggest his mother Francis (Anderson) Chance passed away in 1838 at Oxville.

In 1840, Garrison H. Chance and brother Albert, born in 1820, moved to Boone County Missouri. On November 21, 1844, Garrison H. Chance married Miss Martha Ann Roberts (1827–1871), daughter of Mr. and Mrs. John (Big John) Roberts. There was reported to be six sons and two daughters of this union. According to Francis Gano Chance (1905–1981), the second son of Albert Bishop Chance, he didn't remember his great-grandfather, Garrison Chance, because he moved away when Gano was about three. Garrison had lived in Oklahoma fourteen years before his death in 1922, or when Gano was eighteen. In 1909, a year after his departure, Garrison visited Centralia, but at four Gano may have been too young to remember. Since there is no recorded visits found between the Albert Bishop family and Garrison, apparently Gano never personally knew him. This is surprising because A. Bishop traveled much and he was so outgoing that one would expect there would have been contact.

The repeated use of the same first name in succeeding generations, as is explained in the next section concerning the Sexton family with Chance family ties, was also practiced by the

Roberts family. The January 13, 1905 *Guard* covered the death of John Roberts, eighty-nine, who was known as Left-Handed John. The reason for the nickname was because there were four Johns in the Roberts family, Little John, Big John, John Louis, and Left-Handed John. The John M. Roberts who passed away in 1875 appears to be Garrison's father-in-law who was also known as Big John because as Garrison's obituary suggested, "so called because of his enormous size." He was also likely the source of Garrison's and Martha's selection for the name of their son, J. A. (John) Chance.

On July 2, 1849, John Albert Chance (1849–1934), according to his obituary, was born in Illinois, to Garrison Hardcastle and Martha Ann (Roberts) Chance. Mr. G. H. Chance was originally from Scott County, Illinois, and Mrs. G. H. Chance was a native Boone Countian. John lived in Centralia on a regular basis from 1891 to his death in 1934. His home was eventually west of Allen on Singleton along the Wabash branch line to Columbia. It later became the home of Ollie Chandler until it was moved to allow the street to be straightened.

The Chance move back to Illinois is discussed in a Harvey Hulen letter in the February 18, 1921 *Guard*. Garrison's brother Albert and son Weeden moved back to Illinois prior to the beginning of the Civil War in 1861, which is logical for those who held positive sentiments toward the Union in an area predominantly settled by those with Southern sympathies. Weeden's enlistment into the Union Army, prior to or during the Civil War, supports such a premise. At the time this story is told in 1921, Weeden was said to be living in Springfield, but Missouri not Illinois. Another reason for such a move may have been to follow in his brother Garrison's footsteps. But the *History of Boone County* said Garrison and family did that much earlier in 1847 and did not return to Missouri until 1865. This statement was made in reference to John, who was said to have gone with his parents to Illinois. This is impossible as J. A. Chance wasn't born until 1849! But that wasn't the initial reason that raised suspicions; instead it was Garrison's obituary in the February 24, 1922 *Guard* that stated: "he made Missouri his home from that time until he moved to Oklahoma 14 years ago." Because of some marriage irregularities, this statement wasn't immediately accepted, but supporting information about his farms in the Hallsville/Centralia area explained the significance of the years 1847 and 1865. As the obituary continued: "Mr. Chance obtained a patent from the government and secured a quarter section of land in 1847 in northern Boone County. He cleared the tract and built a home here, where he lived until 1865. He sold it and bought a farm southwest of Centralia that year and had practically lived in and around Centralia from that time until he moved to Oklahoma." One explanation could be that the Garrison Chance family may have spent a short time in Illinois about the time of John's birth. If not, the question arises if John was born in Missouri instead of Illinois. An October 20, 1916 *Guard* article about the Chance family said John was living with his parents "on the old Dunn place west of town, then the Chance farm." All of this information strongly suggests that Garrison Chance and growing family resided before and during the Civil War period in Boone County, Missouri. This information was all related, but the wrong Chance was noted and the years suggested did indeed relate to Garrison, but only in his in-county movements. He thus established the strong roots that "anchored" the Chance family to this locale for five generations.

A state charter was granted in March 1851 to the North Missouri Railroad to build a rail line from St. Louis to the Iowa state border. This was one of three railroads envisioned by St. Louis interests to help develop the nation to the west of St. Louis. As its base, St. Louis would reap the benefits of such an arrangement and continue to be the Gateway to the West. Construction began in May 1854 and by

August 20, 1855, trains were operating to a ferry landing across the river from St. Charles. In a letter to the *Guard* of January 21, 1921, a former resident John M. Daniel, who came in the spring of 1856, saw the first lots sold in Centralia. He also heard Mr. Southgate, who surveyed the route of the North Missouri Railroad, say Centralia was the highest point in elevation between St. Louis and Kansas City!

By 1856 the railroad survey crew had reached the point chosen by the original city fathers of Centralia and headed west toward Sturgeon, which was established in 1856 a year earlier than Centralia. Following the dogleg of higher elevation on north to the Iowa border, generated a fairly level and easy to construct roadbed for the rail line. "On February 1, 1858, North Missouri rails reached Coatesville (Iowa border) and for the first time a railroad linked St. Louis with St. Joseph and Hannibal." (*Railroad History* No. 133) As the name implies, the Hannibal and St. Joseph Railroad ran from the Mississippi on the east to the Missouri at St. Joseph to the west. The North Missouri Railroad crossed over this line at Macon, which allowed the transfer of passengers and freight from either line. If one notes the homes of Civil War soldiers killed during the Centralia Massacre in 1864, Maryville and points west are mentioned which suggests the advantages of such a transfer point.

The railroad's 1856 passage through the area where Centralia now lies is further confirmed by former resident Ben H. Ewens of Searcy, Arkansas. In a letter to the *Guard* of August 1, 1919, he told of having ridden on one of the first work trains that proceeded west out of where Centralia was to be located.

> *He came from south of where Centralia now stands, with his brothers, to see the railroad they had heard was being built, and got their first sight of the railroad. They found that the construction train was out west of this place, where the Tinsley, Ellston & Co. store was the only building. So they went out to the old Jim Bryson farm and got their first sight of the iron horse and the train. When the men were ready to move the train over the newly laid rails, to the westward, Conductor Johnson put the smaller boys up on a flat car and told the rest of the boys to climb up on the car. Before starting the train Mr. Johnson said: "You boys can say you rode on the first train that ever went over the North Missouri Railroad." When the train stopped again about a mile or so down the track, Mr. Ewens' grandfather rode up on a horse and when he found that he and his brothers had been riding on the cars, he came near giving them a whaling. "You might have been killed," he said. Mr. Ewens says that was his first ride and his last ride was over the same railroad a week ago when he came to Centralia. He says he likes to think of the old days as he is now so thoroly [sic] modern that he seldom ever looks up to see an airplane go over, as Tom Foreaker's farm in Searcy, Arkansas, where Mr. Ewens stays, is on the route taken by the army aviators and there are machines going back and forth most every day.*

An area pioneer who had moved away forty-seven years earlier continued to recall remembrances of the early days. Harvey Hulen penned another letter, reproduced in the January 21, 1921 *Guard,* which told of the railroad's construction through the Centralia area in the 1856–57 period. The route chosen had been discussed in the February 14, 1919 issue as being influenced by a prominent central Missourian, Major Rollins. Mr. Hulen suggested that Columbia was too close to the Missouri River and owners wanted to divide the state between the two rivers to avoid riverboat competition. The higher, flatter ground through Mexico, Centralia, Moberly (Old Allen), and Kirksville, on to Coatsville and Moulton was apparently chosen because of the ease of construction and operation. It also fulfilled the riverboat issue. Some of the towns mentioned didn't even exist prior to the railroad's passage. The track was constructed, according to Mr. Hulen, almost entirely by Irish laborers with shovels and wheelbarrows. Scoops and dump wagons were used for long hauls. He was impressed at how smoothly the steam locomotive on the construction train ran. There were hundreds of people from surrounding areas who came to view the construction daily, including Mr. Hulen.

By May of 1857 the town of Centralia had been laid out. Its location was earlier determined simply by where the Paris and Jefferson City (Columbia) highway and stagecoach route crossed the grade stakes of the approaching North Missouri Railroad from the east. All along the rail line small railroad stations were being built, which like Centralia marked the railroad origin of many towns and cities in the Midwest and West. This lent impetus to the construction of Centralia, which reportedly contained some thirty structures by 1860.

In May 1916 the *Guard* was continuing articles about the early years and remembrances of early settlers. A Mrs. Mary (Crutchers) Robinson said the prairie grass was taller than a man's head when she first saw the Centralia area prairie in 1856. This is different than the earlier "horseback" version. A couple of weeks later a piece entitled "Boone County Facts" suggested Centralia's name derives from two sources; first the central mileage location on the North Missouri Railroad between St. Louis and Ottumwa (Coatsville?), Iowa, and second because it was located near the center of the vast prairie between Mexico and Huntsville and between Columbia and Paris. The rail line did eventually reach Ottumwa and another line splitting off at Moulton later entered Des Moines.

The year 1861 would be the beginning of a long period of difficulties for the small prairie hamlet of Centralia in which it would face virtual destruction. How, in the face of such continuing adversity, would destroyed businesses and homes be rebuilt and its citizens continue to endure such hardships, requires further analysis. Although Centralia was located in a very fertile farming area there is the question of how much prairie had been conquered at this point. A very interesting twist on this was covered in the February 12, 1915 *Guard*. It suggested that north of Centralia lay a vast prairie for ten or more miles that was used primarily by cattle herders. Most of the town trade came from the south where the terrain became hillier with creeks and streams. This rougher, wooded area was chosen over the richer prairie because it reminded early settlers from Kentucky of their former home. This was confirmed by local pioneer Harvey Hulen who said that the first settlers concentrated their homesteads along creeks. Of course timber was a source of building material and fuel, which was a vital commodity ensuring early settlers' survival and success. The *Guard* suggested on March 20, 1880, "Where a few years ago was to be seen nothing but raw prairie now is all under cultivation." Another factor that bound local residents to this area was the commonality of the people having similar roots; the majority appeared to have come from Kentucky. According to *Stories of Missouri*, the early settlers were primarily from Kentucky, Tennessee, and Virginia. This resulted in a strong Confederate connection, enough in fact that the surrounding area came to be labeled Little Dixie. This was an important element that bonded these people together, but there was also another factor that made the area attractive, the railroad.

In America, the Land of Opportunity, progress was heavily driven by financial opportunities. This was as simple as an immigrant's quest for their own homestead to provide a living for their families or other business ventures. Nothing symbolized more perfectly this westward advancement of the white man's form of civilization and the progress of the industrial revolution than the railroad. It not only provided people with an efficient mode of transportation that linked them with the outside world, but it was also the conveyor belt that allowed for the development and sustained the expansion of our great commercial and industrial might.

Slavery was a way of life on many local farms, homes, and even businesses in this area. This had financial implications that generated bitter opposition to the Union. Some of the decisions made by political, business, and military leaders and the harsh actions of certain military forces drove the intensity of the war and strongly influenced many to choose sides. One of the more prominent examples of this was the James family in northwest Missouri, which also reflects the consequences of what Union forces saw as local support of opposing forces. Even those who chose not to take part were often caught in the middle of the conflict and were treated miserably by both sides. This was true of the Little Dixie area where Northern or Union forces fighting the "Bushwhackers, Guerrillas, and Rebels" generated a continuous war zone type of atmosphere. Troops on both sides stole, robbed, killed, and destroyed even unto those who were supporting them!

Ironically the railroad sustained Centralia, but during the Civil War it also brought conditions that resulted in death and destruction. As it opened the door to the outside world for the forces of good, it in the same way provided the rapid mobility and concentration of warring factions. Missouri was a hotbed of Civil War activity. In fact the number of the various confrontations between Union and Confederate forces in Missouri was said to be second only to Virginia. Because the state was split between the warring factions, the way in which the war was waged generated much destruction and suffering. Inflamed by border state conflicts, it reached very low levels at times with guerrilla warfare, which can no better be represented than around Centralia in 1864!

The importance of the North Missouri Railroad to Union forces is clearly reflected in the many destructive attacks by Confederate troops during the Civil War period. Confederate General Sterling Price issued an order in July 1861 calling for the destruction of the North Missouri Railroad. Under Captain Bill Anderson every bridge in over one hundred miles of rail line was destroyed, plus extensive damage to track, rolling stock, and burned fuel supplies and ties all along the railroad.

The railroad depot in Centralia was burned by the Anderson Gang in 1864, according to the *Guard* of March 14, 1919. The first depot, around 1861, was either built by the railroad or by the Centralia agent, T. S. Sneed. The railroad replaced it with a larger one in 1862 that was later destroyed during the Centralia Massacre.

In one of the more notorious and controversial battles in late September of 1864, which displayed the coldness and violence of guerilla activities, the railroad and the struggling community of Centralia was involved. The "Centralia Massacre" as it came to be labeled, led by Captain (Bloody Bill) Anderson, born in nearby Huntsville, included the robbing of Centralia, a stagecoach, and a train. The massacre of 23 Union soldiers on the train was followed by the some 350-man Confederate force destroying a Union contingent of 123 recruits under Major Johnston. Jack Chance, grandson of the founder, suggests there were 79 buried along the railroad east of town. They were later in 1873 reburied at the National Cemetery in Jefferson City. The other 44 were buried elsewhere. Some of the soldiers on the train had been on Sherman's March to the Sea. Postdated accounts of witnesses to the Centralia Massacre became almost routine over several decades. The actual number of casualties, as reported on various occasions and from numerous sources, frequently vary and should be viewed as suspect. Even the *Guard* wasn't immune from confusion on this issue. This event gave Centralia notoriety and insured its place in history.

Thirty-three summers had come and gone as Centralia residents prepared for winter in mid-September 1897 as they had in September 1864. There was one important difference in 1864, however, as their town was suddenly attacked, partially burned and all businesses and homes burglarized and ransacked by a band of Confederate guerillas who terrorized the community. But the unmerciful killings of unarmed soldiers and those trying to escape, the scalping incidents, and throat-slashings had only begun. It was reported that as many as 123 Union soldiers were eventually killed, but only three Confederates. Some would suggest it was Centralia's location on the North Missouri Railroad that attracted these marauders and it was the railroad in September 1897 that transported one of the participants back into Centralia. It was a common occurrence for Columbia-bound rail passengers to have temporary layovers in Centralia, waiting for the Columbia branch train to arrive. This particular visitor was said to be taking his son to Columbia to enroll him in college. The editor of the September 17, 1897 issue of the local paper didn't miss much around Centralia. As the word spread that one of the James boys was back in town, or by then en route to Columbia, those residents who witnessed the 1864 massacre surely had mixed emotions, regardless that the governor had pardoned Frank. A few days later on Sunday, fifty-four-year-old Frank James (1843–1916) returned, but the *Guard* editor, Adam Rodemyre, witnessed that notoriety had changed how he was received. A portion of a September 24, 1897 editorial stated: "How time changes public sentiment. In 1884 men were eager to capture or kill Frank James; in 1897 they fall over each other in order to shake his hand." The October 11 *Guard* described this now famous outlaw as follows: "Frank James is a raw boned, sharp featured six footer, of 54." Frank was escorted by townspeople out to the battlefield on the old J. D. Gavard (Garrard) farm, southeast of town where he pointed out where forces were and what occurred. He was "reticent" about the mutilations and said he

Adam Rodemyre, editor of the *Centralia Fireside Guard* in the late 1800s. (Centralia Historical Society)

didn't witness any such scenes. It was mentioned that Jesse (1847–1882), Frank's younger and wilder brother was also in Centralia during the battle. There was then as well as today a lot of controversy surrounding the James family. One of the last members of the James Gang, Cole Younger, died in March 1916, following the death of Frank James a few months earlier. But that wasn't the end as a hundred years later they are still talked about and immortalized and their fame doesn't seem to have subsided. Factions from several states have tried to prove, even in current times (2003), that the final resting place of Jesse is somewhere other than near the James farm where he was buried. The James and their exploits could be blamed on the sharp division between Union and Confederate sympathies and forces in Missouri.

Another by-product of the Civil War period was the putting down of roots of families that are traceable today to many current residents. Some were here before the war while others came later. The post–Civil War period brought an influx of people to this area, which would complete its settlement rather quickly. However, what attracted early settlers to an area has no magic in retaining later generations and like leaves in the winter wind, families eventually became dispersed again. This is no better displayed than examining a Centralia phone book at the end of the twentieth century and seeing no Chance name recorded there.

Wars tend to drive technological advances but the horrible consequences they generate can never justify them. The importance of railroads and even their vulnerability to enemy forces had clearly been shown. Many local business and community leaders believed that their town's access to a rail line was vital to its future success. Towns such as Columbia were close to a major river but they found themselves without rail service in the Reconstruction period after the war. Such nearby upstarts like Centralia, which was some forty years Columbia's junior, found themselves in an advantageous position. Soon after the war there were several proposals for rail lines by Columbians, but in May 1866 the approved route to the North Missouri Line via Centralia was chosen. This not only holds some significance to Centralia, but eventually to the Chance Company as well.

The transition from the upheavals of war back to normalcy again brought a focus on life in general. Making a living, socializing, attending church and school, arguing about politics, traveling, and marriage became the important issues of the day. John Albert Chance married Miss Adaline (Addie) Sexton (1855–1945), the daughter of Solomon and Sarah (Sallie) Ann (Hunter) Sexton. The year of the marriage is in question as Addie's obituary in the August 6, 1945 *Moberly Monitor Index* indicated 1870. John's obituary in the *Guard* recorded the marriage year as 1871 (supported by later sources), and a Boone County historical source suggested 1873! The Sextons were an old and prominent Boone County family.

The year 1871 not only brought *Our Fireside Guard* (*Guard*), a newspaper, to a thriving Centralia community, but also the completion of the massive North Missouri Railroad Bridge over the Missouri River at St. Charles. Also the contract for a new Christian Church was given to J. B. Lofland who had the low bid of $15,37 (Note: comma after second digit as recorded). The church was given three lots on the south side of the city square by Major Rollins, said to be some of the finest lots in Centralia. A local farm family by the name of Chance was or would become members of this congregation. Three decades later when the church had been outgrown and had become a storage facility, a family member of the same name would rent it as a parts depot and later would begin a small manufacturing operation.

In 1872 a fire destroyed the business district of Centralia for the fourth time, as it had in 1860, during the Civil War, and again in 1868. To fight the destructive results of fires, the evolution of buildings constructed of brick and the development of a modern water system would be embraced by a future Centralia citizen, A. B. Chance. On May 26, 1873, Albert Bishop (also known as A. B., A. Bishop, Bishop, Boss, The Centralia Wizard, or Founder) Chance (1873–1949) was born in Centralia to John A. and Adaline (Sexton) Chance. He had one sister Sallie (1879–1959), who later married W. T. (Willis) Fountain. They had one child, a daughter Lucille, Mrs. Gilmore Stone (1911–1992).

Bishop mirrored certain characteristics of his mother Adaline (Sexton) Chance according to her grandson Gano Chance: "She had a lot of pep and I think that my father got that more from her than he did from his father. My grandfather was a very quiet, slow moving sort of fellow who didn't get worked up about anything. On the other hand, my grandmother had a lot of pep and she got into many things." It would appear that Bishop got his energetic and inquisitive nature from his mother and his business head

A rush to the train station was a familiar scene in Columbia when classes were out. The Wabash railroad put on extra coaches and baggage cars to handle the sudden surge of extra passengers. (University of Missouri Yearbook of 1926)

and strong convictions from both of them. A splendid example of Mrs. Adaline's strong and enduring spirit is displayed in the September 21, 1923 *Guard*. It recorded her participation in the local fair where she displayed fifty-four containers of fruits, vegetables, and meats when she was nearly eighty years old!

Gano continued to point out other similarities. "She was quite an experimenter (where my father got it) and she considered going into the cosmetic business. She developed a lot of old-time remedies, etc., and she cooked them up and sold them." Gano felt that John Chance, the father of Bishop, was a stabilizing influence who controlled things in a more conservative manner. Bishop's association with long-time Chance employee Miss Ida Lou Traughber mirrored certain qualities found in the successful relationship of his mother and father. Gano further described his grandfather as being a large, quiet sort of fellow that did not get disturbed about anything but he wasn't mechanically minded. Granny Ryman described John Chance as having "a great long beard." Paul Sturgeon said, "he was a tall slender man. As I recall, he had a short-cropped beard and walked very erect." (1979)

An article in the March 28, 1919 *Guard* told of the construction of the long-planned Columbia branch line. It was first organized on January 30, 1857, as the Boone County and Jefferson City Railroad. It was envisioned that it would connect with the Pacific line at Jefferson City, but the Civil War halted any further progress on the proposed rail line, until construction began in 1866. Switching operations in Centralia was always a tricky proposition as the Branch trains got on the main line and several wrecks and injuries resulted. The Wabash had long attempted to obtain land to build a wye to turn around the trains and a separate siding track for the Columbia train. A separate depot on the south side of the main line or switch track for boarding/disembarking Columbia passengers was even discussed for passenger safety. The local paper of May 12, 1922, told of renewed Wabash activities to buy land where the branch line connected with the main line, for the long-awaited wye. The August 18 issue told of the beginning of construction, which was later completed on October 27, 1922. The wye required a lot of fill along with a long switch track to the Goodwater Creek a mile west of town. This track then or later was run to the Centralia Wabash yard in the Jefferson Street area to the east.

The railroad had a powerful influence on the Centralia area and Chance Company. Its impact parallels that of an interstate highway running through town. Noting news items in the local paper from towns and cities along the original North Missouri Railroad reveals it also helped carry the news. Being on the same rail line had a way of connecting those small towns and cities together as it did later for a young Bishop Chance. This influence was increased by having the Columbia branch line originating at Centralia. Unlike Kansas City, St. Louis, or Iowa bound trains, which may have only stopped briefly in Centralia, Columbia passengers usually had to disembark and wait to board the Columbia branch train, resulting in numerous significant layovers.

One long tradition that was alive and well, even as early as 1873, was the movement of Columbia college students by rail through Centralia. This peaked at least four times during the school year: the beginning of classes in the fall, leaving and returning from Christmas holiday break, and the closing of classes in the spring. A week after a sleet storm blanketed the area in early January of 1873 another winter storm brought blizzard-like conditions with blowing, drifting snow. Ice doesn't affect trains as it does automobiles, because the first set of wheels of the locomotive squeezes it off the rail. Snow is another matter, especially in right-of-way cuts through hills that can become clogged with drifting snow. Such conditions overwhelmed the Columbia Branch train loaded with Stephens College and University of Missouri students returning from the Christmas holiday. The train and its passengers remained stuck for twenty-four hours. However, instead of presenting a life-threatening situation, there were reports of much laughter, singing, and socializing. Of course in such a setting there had to be some mischief, in this case directed either intentionally or unintentionally toward

Stephens College Students. (*Follow the Flag*, H. Roger Grant)

a drummer who unfortunately found himself amongst all these young people. A young man decided he could sleep better by placing his legs and stocking feet up over the facing seat in the general area of the drummer's face. When the drummer awoke the next morning he realized it wasn't a bad dream of fearing he had spent the night in a pigpen.

College students coming into Columbia prior to completion of the Columbia branch came by stagecoach, carriage, horseback, or even on foot. Columbia, with its origin traceable back to 1817, became the Boone County Seat but more importantly it became a college town. Columbia has a rich history as a base of higher education that reaches worldwide. Its crowning achievement was officially becoming the site of the state University of Missouri in 1839. Stephens College was officially established in 1870, although its roots go back to the Columbia Female Academy of 1833, which later became the Baptist Female College of 1856–57. Christian College (Columbia College) opened in 1851 with both it and Stephens College resulting from Baptist and Christian Church differences from which its original name came. It was very impressive for Missouri to have a college for women, even predating the actual opening of the University of Missouri. How appropriate as Chance Company likewise mirrored this acceptance of women.

The active involvement of the Christian Church in the organization of a women's college is also reflective of its strength in the early 1870s in Centralia. The Christian Church also had a strong influence on A. Bishop Chance who would also involve a lady in the management of his company. His strong Christian faith helped develop characteristics and inner strength that were crucial to Chance Company's success. This drove the continuing effort to produce a quality product and to develop a sincere relationship with the customer, which generated customer allegiance and friendship toward the company. His son Gano also mirrored these church-instilled influences. This reveals the specialness of a young, struggling America, especially within its heartland, the Bible belt and specifically in the Centralia and Columbia areas, truly a historical region that inspired and produced giants, not only in personalities but also in institutions and forward directed movements.

On February 13, 1875, there were two legal notices in the *Guard*, which may have first introduced the Chance name to its readers. The first was the "Dissolution of Partnership" between C. E. Chance (1846–1921) and W. H. Roberts. C. E. was the son of Garrison and a brother to John, who was the father of A. Bishop Chance. The reason cited by C. E. was "retiring from the firm." This indicates that C. E. had some sort of business arrangement other than farming. This is telegraphic of what would follow when C. E. entered another partnership in the 1880s involving a Centralia hardware store. The second document was an "Administrators Notice" served by G. H. (Garrison) Chance. Garrison married Martha Roberts who was the daughter of Mr. and Mrs. John (Big John) Roberts. The estate that Garrison is administering is that of John M. Roberts, who appears to have been his former father-in-law. Also note that the dissolution of the partnership involving his son C. E. included another Roberts (W. H.). The significance of these documents and the activities they point to, are early signs of the Chances holding responsible positions and showing certain signs of prowess as businessmen. It is significant that the first notice of the Chances in the local paper involved business dealings.

By early 1878 the final survey was made for the route of a new railroad through the northern section of Boone County and the construction began in the spring of 1878. Another *Guard* source from February 1919 suggests it was surveyed in 1878 and built in 1879. The Chicago and Alton (C&A) became known simply as the Alton. The Alton became famous later for its red passenger trains and the introduction of Pullman (sleeping) car service to the world. For a town to be served by the main line of two railroads was quite a distinction especially when many communities were fighting and struggling just to get one. There was, however, a difference between this new line and the one a few yards to the south that had already been in operation for some twenty years. The C&A was a Chicago line whereas the North Missouri (Wabash) Railroad was a St. Louis based line. There was an early attempt by Illinois interests to establish Alton (Illinois side of Mississippi River) instead of St. Louis as the metropolitan center of that region. As we now know this was unsuccessful, but there was a continuing struggle between the Chicago and St. Louis based interests, in having their railroads serving the Kansas City and western areas of our expanding nation. St. Louis would be headquarters for such rail lines as the Missouri Pacific, Wabash, Katy, and Frisco, whereas other lines such as the Alton, Burlington, Rock Island, Chicago and North Western, and Santa Fe would have Chicago connections. The Alton and the several railroads to the north of the Missouri River were Chicago based, whereas

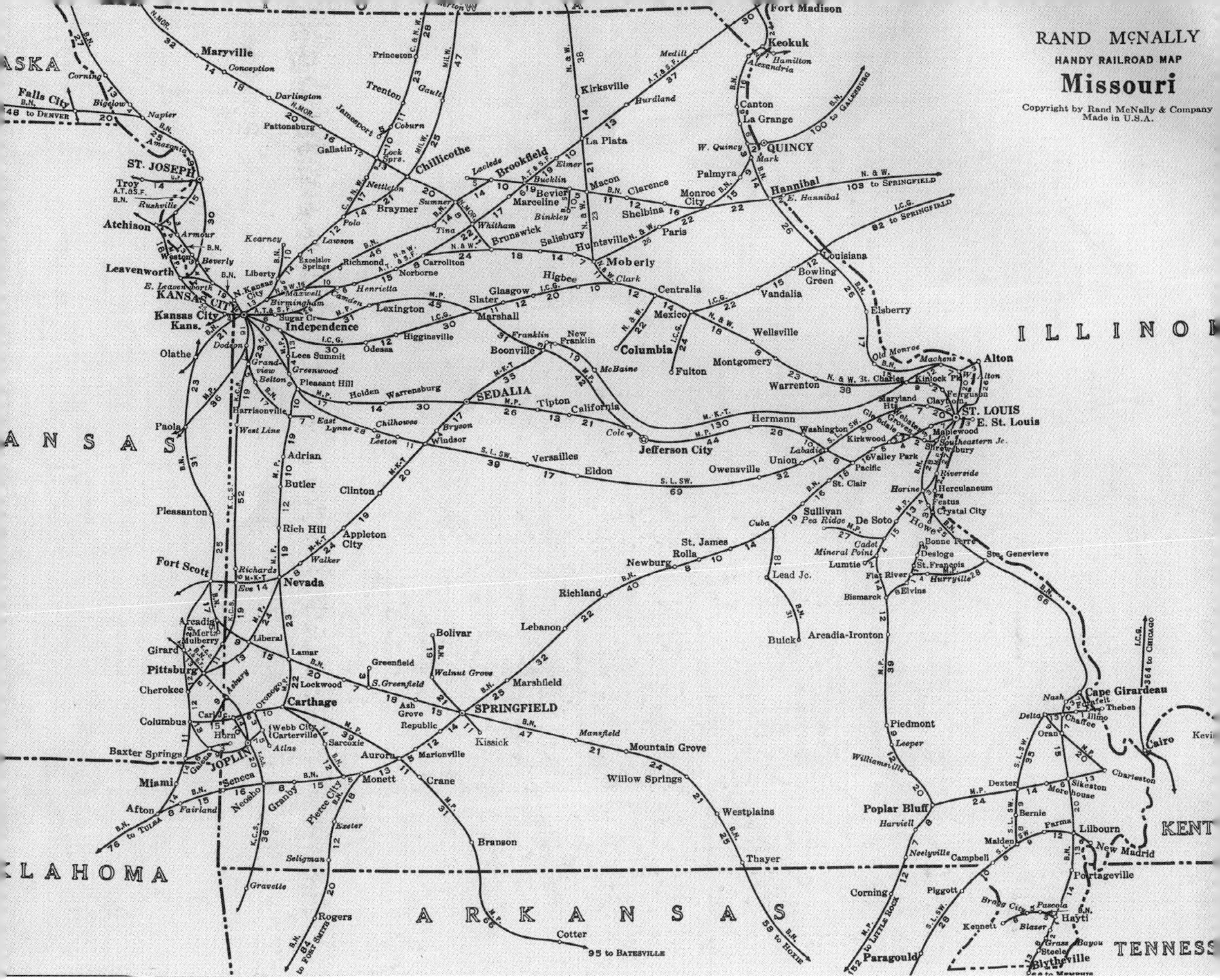

Railroads of Missouri. N&W was formerly the Wabash. (Rand McNally)

the North Missouri and lines to the south of the Missouri River were St. Louis based. The C&A split in western Illinois with the Kansas City branch crossing the Mississippi River at Louisiana and the Missouri at Glasgow. The C&A would have both a passenger (original, twenty-eight feet by forty-two feet) and freight depot on the north side of Centralia. After fire destroyed the passenger depot in late 1906, a new one with a slate roof and concrete platform was completed in early 1907. It was located on Allen Street to the east alongside the future Chance Main Office or factory building at 210 North Allen. That area became a Chance Company parking lot.

As its name depicts, the North Missouri line shot northward and stayed north of the Missouri River from St. Charles to its western terminus of either Council Bluff (Omaha) or North Kansas City, Missouri. Of course its name originally derived or described the original rail line stretching north to Macon, Kirksville, and on into Iowa where it later split and proceeded either to Ottumwa or Des Moines. The birth of Moberly is directly related to the later decision to build the North Missouri (Wabash) westward toward Kansas City, as this is where the north and west lines split. The North Missouri line, however, was the south line from Mexico to Centralia, where it and the C&A ran parallel to each other. The C&A veered northwest of

Centralia bypassing Sturgeon to eliminate the curve and extra distance required before the two railroads crossed each other at the Clark switch, which was to become the new town of Clark.

Centralia was 121 miles from St. Louis, 153 miles from Kansas City, and 312 miles from Chicago. Comparison of the three major rail lines and their distances between St. Louis and Kansas City were as follows: C&A, 305 miles; Missouri Pacific, 282 miles; and St. Louis, Kansas City and Northern, 277 miles. It was suggested in the *Guard* article that lighter grades on the new C&A would make for quicker times, but there was intense competition between the lines, especially the two through Centralia. The C&A line into Centralia was completed a few days before the May 22, 1878 issue of the *Guard*. The line into Glasgow was completed in late August and rails reached Kansas City in March of 1879. But trains couldn't get across the Missouri River at Glasgow, where reportedly there were two to three thousand men working at times on the bridge. The May 3, 1879 *Guard* reported that rail traffic had begun crossing the new Glasgow bridge, the eighth across the Missouri River, and one that was considered a high bridge because it was some fifty feet above the high water mark. It was also noteworthy because it was the first large all-steel bridge of its type in the world. The Alton stations at Glasgow and Kansas City were said to be special. The Glasgow station was recently restored, which is near the current bridge that replaced the original around 1900. In the upper superstructure of the Glasgow railroad bridge, the bold letters displaying C&A proudly reflects its original owner and builder. It is reassuring to see proof of the past when most figures and material being discussed have long ago slipped away from us. The flood of 1909 raised questions concerning the wisdom of locating the railroad bridge at Glasgow with its western approach built along side the Missouri River. Serious problems would develop again in the flood of 1993 when severe damage again threatened the line's abandonment, saved only by the insertion of federal funding. U.S. Senator "Kit" Bond from Mexico (along rail route) was instrumental in getting approval for this funding.

A Developing Business Culture (1880–1899)

The year 1880 would bring the Chance name into the public spotlight with their involvement in a local hardware store, Chance and Sexton Brothers. Coverage in the *Guard* late that year would initiate nearly a century of business news associated with the Chance family, but first, a matrimonial event that surprisingly may have connections to the before mentioned business activity involving the Sexton and Chance families. On September 16, 1880, the local paper reported the marriage of Miss Ettie F. Chance (daughter of G. H. [Garrison] Chance and sister of John A. Chance) to Mr. Thomas D. Sexton. All were Boone Countians. Some of the earliest settlers in the Centralia Township in the 1820s were Thomas and William Sexton. They were likely neighbors later to a prominent local farmer, Garrison Chance. Randolph County historian Ralph Gerhard suggested that it was common at that time for a family to repeat a first name over and over again. Mrs. Ettie F. (Chance) Sexton's new husband Thomas D. was likely a grandson of pioneer Thomas, although he could have been a son as childbirth deaths of women in those times found many men remarrying, sometimes resulting in wide spreads in ages of children. However, the Sexton family was also represented by at least two others: physician and surgeon in Centralia (1880) a M. P. Sexton and the prominent Perchetown Sextons, George (in the mail and stagecoach business) and his son Charles of Everett, involved in a plow factory and other ventures. The Chance/Sexton family connection gets even more interesting looking back to 1872 when John Chance married Miss Adaline Sexton!

The December 18, 1880 *Guard* included an advertisement for the "Chance and Sexton Bros." hardware store. It offered groceries, stoves, tinware, hardware, and Queenware. They were also agents for Smith Wagons. Early general/hardware stores commonly became farm implement dealers before motorized farm machinery required separate repair facilities and generated more specialized and separated businesses. The wagon was big business in those days with everything from carriages to stagecoaches being required for the transportation of people and materials. Boone County historical information pointed to Sexton Brothers in reference to another business, but it isn't known if they were one and the same. One of the brothers involved in the Centralia hardware store was said to be Thomas. Was this John Chance's brother or father-in-law, or another Sexton family branch? The bottom of the ad more officially listed the names of this new hardware store as "Chance & Sexton Bros." In 1880 the Sextons were a prominent family name in Boone County, more so than the Chance family, but this was going to change, suggesting the constant cycling of the importance of a family name. Centralia was certainly the place to be in 1880 as it was outgrowing all other Boone County towns. Small hamlets such as Perchetown, which failed to get a railroad, slowly evaporated. Adam Rodemyre even suggested (to enhance circulation of his newspaper), that certain Sturgeon businessmen would like to find buyers for their firms so they could start over again in Centralia!

The year 1881, from the richly illuminated perspective of the *Guard*, reflected a "Boom Town" kind of atmosphere in and around Centralia, which had quickly gained a population of some nine hundred residents. All of the hustle and bustle taking place reflects the image of a prospering prairie town as depicted in western movies. But the rawness of the environment of a semi-frontier town wasn't yet sophisticated enough to honestly be crowned the Prairie Queen. Growth from the continuing settlement of that area's land along with the wholesale capacity for the movement of materials and people brought by two railroads, point to seemingly unlimited business opportunity. But it also represented a quagmire that could spell sudden doom to the unwary. Frequent business start-ups, changes in partners, or resale to some wide-eyed opportunist were commonplace. Several such examples occurred in the heavily competitive grocery and hardware business during 1880–81, including the Chance family entrance into the local, nonagricultural economy. Contemplating the advantages in the successful operation of a local business for someone from the surrounding area with fifteen or twenty years of residency versus those with days or even hours to their credit, is an intriguing question. But business savvy and capital, plus good eastern supply sources may have more decisively impacted the survival of a business. Thus the volatile nature of these times had a liquidating influence on instability and the rapid movements of people in and out intensified credit pressures on proprietors. Those too compassionate quickly found themselves with a high level of bad debt. So this appears to be the kind of atmosphere John Chance became involved in with his seven-year-old son Albert Bishop, establishing an early inquisitiveness as he looked over his father's shoulders. Later references to John Chance point to one hardened by these fiery times and conservative in his business activities. It is now clearer that some of the tough early influences helped mold him.

The Chance and Sexton Brothers association appears to have lasted some six months. In May 1881 the large "Fire Sale" type ads in the *Guard* that began in December 1880, ceased. No such ads appeared again later in 1881, but almost two months after the last Chance and Sexton Brothers ad, on June 25, 1881, the Chance Brothers were featured in the "Local Matters." This featured local news, gossip, and short spots promoting local merchants and their wares. They were offering cookstoves at cost, among other items. It's not clear what broke off the earlier Sexton relationship, but the *History of Boone County* reported in 1881 that John A. Chance purchased a controlling interest in a Centralia hardware store from Thomas Sexton and brother C. E. Chance became his partner. They called their business partnership "Chance Brothers." During this time John became a member of the Board of Deacons of the Centralia Christian Church, which was founded in 1867 by J. A. Berry.

In 1882, materials were being accumulated to build the Centralia Baptist Church. The Chance Brothers brought out a large ad in the paper on February 4, 1882, entitled: "Still Alive!" For ten weeks running, the public understood that this ad confirmed solvency. Having to restate your continued existence is reflective of how competitive things were at that time and how fragile long business life was in Centralia. Besides the financial advantage of running an ad repeatedly to spread out set-up costs, there was another reason driving the Chances to advertise. In the March 25 *Guard* was an announcement by the Chance's in-laws-to-be and former partners the Sexton Brothers, who were opening a new grocery and dry goods store. This was just what the other three or four Centralia grocery businesses didn't need. The ad did reflect the early resolve of the Chances, especially when a statement appeared in "Local Matters" in the paper on March 4 of how they practiced business in an aggressive and fair manner.

> *Three things we want you to remember. First—We allow no man to sell cheaper than we do. Second—We allow no man to sell better goods than we do. Third—We will allow [no] man to treat you better than we will. Come in and see us, whether you want to buy or not. Chance Bros.*

With all the charm and hospitality of the perfect host, the Chances opened their doors wide, but not unlike other local businesses, the delinquent credit problem also had to be faced. On July 15, the *Guard* reported: "Special Notice—All persons knowing themselves indebted on open accounts are paying 10 per cent every 30 days. Chance Bros." Although other businesses frequently made requests in the paper for debtors to pay off their overdue accounts, this was a somewhat different approach that is even used today. One other proprietor must have thought the Chance way was worthy as he repeated it not long after. The intensity of these types of notices grew at harvest time when cash crops were being sold.

Change continued to be the rule in 1883, as John Chance withdrew from the Chance Brothers retail business. The *Guard* carried the following notice on September 22, 1883:

> *DISSOLUTION NOTICE—Notice is here by given that the firm of Chance Bros. has been dissolved by mutual consent. John A. Chance withdrawing. The business will be continued by C. E. Chance and A. E. Lenier who will assume liabilities and receive all money due the late firm. All persons indebted to the firm, either by note or account must come forward and settle at once. J. A. Chance and C. E. Chance.*

What might have been expected was for John, a direct link to A. Bishop and his business success, to continue until he got into the hotel business and then moved on into the local phone exchange. After almost three years it may have been clear to John that this kind of business was a dead end, and looking back to 1875 and C. E.'s earlier business involvement, John would appear to be the newcomer who may still have been trying to find his own niche. John Chance seemed to have disappeared from the local spotlight from September 1883 through 1884.

The year 1884 continued to see C. E. Chance participating in the original Chance Brothers retail business. He was apparently establishing himself within Centralia; as in the Centralia Public School announcement for fall registration, C. E. is the district clerk (Board of Education) with whom students are to register. The Chance family had a long history of local support for the Centralia Schools, which may have begun here. Centralia was said to have one of the best schools in the county as it did later during the twentieth century when A. B. Chance Company became successful and contributions above and beyond required taxes were liberally given to the local schools. The father of C. E. and John also makes the September 13 paper as he was the road overseer south of town. Garrison Chance's farm was south of Centralia near Hallsville.

The December 26, 1885 *Guard* reported the death, from consumption of Mrs. G. H. Chance age forty-eight years on December 5. Although Garrison's obituary states his second wife was the former Miss Lucinda (Lurinda) Wainscott (Winscot?), who passed away in 1883, the local paper suggests it was 1885 instead. So this appears to be Mrs. Lurinda (Winscot) Chance (1837–1885), the daughter of the Inyard Winscots. This second marriage of Garrison produced four sons, Aubrey being deceased prior to 1922. The September 15, 1893 *Guard* seems to clear this up by reporting G. H. Chance's marriage on September 13 to Miss M. S. Neal. Miss Mattie Neal, as was suggested in Garrison's obituary, was a marriage that produced three children (twins and Afton). However, this appears to be his fourth marriage as it was reported in 1893 of this union and of the birth of twins in 1894! The 1886 to 1892 marriage was instead his third. Researching this period, one can't escape the deep impression left by how death struck all ages, resulting in an average life expectancy significantly less than today. Those with heredity advantages and able to escape serious diseases and illnesses lived to an age that doesn't appear to be substantially different than today.

The year 1886 is approached with cautious anticipation that several developments were just waiting in the wings before appearing. The boomtown atmosphere was continuing in Centralia as the population was reported on June 19 as having grown to 1,537. It gets more interesting as the one performing the census was C. E. Chance. He also reported that the Centralia Schools had 563 students. Local news also reported that he was reelected to the schools' Board of Directors and gave a financial report of the Centralia School. C. E. was also a delegate to the Grand Lodge of A.O.U.W. (Ancient Order of United Workmen) that met in St. Louis in February. From a business perspective in the September 16 issue of the *Guard* included an editorial that listed the names of the various businesses in Centralia. Chance Brothers was not listed. There was an Assignee's Notice in February that related to four parties, including Chance Brothers and Chance and Lanier (Lenier?). This may have been relating to the original Chance Brothers business and the C. E. Chance and A. E. Lenier firm that developed from the Chance Brothers dissolution. So C. E. had become more of a public figure than a retailer, although it isn't clear what had occurred, if anything, to the retail business. Then he contacted typhoid malarial fever and became very ill in late October but appeared to have recovered by December. The only other Chance family news noticed was a report that G. H. Chance had left a large pie plant at the *Guard* office in mid-May.

In the spring of 1887 there was a rising political star in the Chance family, the newly elected Centralia mayor, C. E. Chance. Gaining an overwhelming margin over the other candidate, C. E. replaced T. S. Sneed in that position. But John Chance and family weren't to be outdone. At some point, possibly at the time John left the Chance Brothers business, he purchased or managed a hotel. The local paper reported on February 26, 1887, that, "J. A. Chance contemplates retiring from the hotel business and engaging in farming. Mr. Chance has proven himself to be a No. 1 hotel man, and we shall regret to lose him." In March, John sold a Centralia lot and then like a long-awaited premiere, John's only son enters the spotlight. Both instances referred to involving Bishop Chance are predictable, when his later life is taken into account. Although his mother and aunt may have determined where they would spend this February Sunday, when Bishop Chance traveled it was jokingly said he usually headed west as they did on this occasion to Moberly. No other children apparently accompanied them, only Mrs. J. A. Chance's fourteen-year-old son. There are many challenges facing a teenager, especially when their school performance is being publicly monitored. Room V of the Centralia Public School produced for the *Guard* of March 26, 1887, a ranking of students. Poring over the list, a young lady of the Chance family, Dottie, first appeared and going further, at number fifty out of the sixty-five students listed, was Bishop Chance! The rankings were based on scores, recitals, attendance, etc. and somehow Bishop's position isn't surprising. Socializing with classmates and trying to figure out a way to make an adjustable school desk may have put him at the top of the class, but of course these attributes weren't part of the ranking formula.

The year 1887 proved to be a smorgasbord of Chance family activity especially during the later half of that year. Sallie Chance, sister of Bishop, along with a group of local young ladies, agreed to ride on the floats in Centralia's big Fourth of July festivities. And C. E. again entered the news by performing another census in May. He reported lower figures of 491 school students off from over 500 the previous year, and a corresponding drop in the Centralia population from over 1,600 to 1,404. In comparison Columbia reported a population of 3,621. On August 4 it was reported that C. E. had performed a wedding! After the fireworks of July 4 and all the hustle and bustle in this frontier town, two local residents decided to get away to a more relaxing place. On November 12 it was reported that Jonas Chance would soon leave for Texas where he would spend the winter. J. A. Chance desiring to get some benefits from warm spring water, headed to Hot Springs, Arkansas, for a short visit. A pattern for travel and winter homes in Arizona/Florida for later Chances may have begun here in 1887.

Certain practices in 1888 remained similar to previous years, to the credit of C. E. Chance. His annual census revealed a further decline in Centralia's population to 1,136. But J. A. Chance was making waves, or at least this was what Centralia editor A. Rodemyre thought when John came into town promoting an out-of-town competing printing house. The February 25 *Guard* had this to say: "Columbian J. A. Chance, formerly of the Commercial Hotel Centralia, but now representing a Kirksville printing house, was here Monday. Mr. Chance is an excellent gentleman but we are sorry to say he is representing a very poor concern." J. A. Chance among other things was a salesman, and these talents and experiences must have rubbed off on his son. The hotel with which John was earlier connected is also identified. There were also other Chances in Centralia news during 1888. Butler Chance lost his home south of town to fire on October 25. G. H. Chance "has removed to town." This likely meant the aging father and grandfather figure of the Chance family had in November moved into town from his farm. Two younger Chances, Jonas (or Jona) and Charley left for "the land of gold" in Red Bluff, California. Jonas appeared to be sort of a free spirit as he reportedly had spent the previous winter in Texas and Charles seems to have the same traits as in October 1892 he returned to Centralia from a six-month stay in Wyoming. These were likely work-related adventures.

The desire of residents for Centralia to become more sophisticated took many forms, such as improving the appearance of the town, not chewing tobacco in church, socializing and supporting social events, and rail travel to other area towns and distant cities. A key driving force in what Centralia was and could become, frequently disdained and applauded by other editors, was the forty-seven-year-old charismatic *Guard* owner/editor Mr. Adam Rodemyre. His first marriage to a local Hallsville lady (Miss Tribble) had produced a daughter and two sons and influenced him to live in the Centralia area. Recently widowed, he married a twenty-seven-year-old Hallsville area resident, Miss Mattie Tucker, daughter of Mr. and Mrs. J. E. Tucker, in May of 1890. She became an important contributor to the weekly *Guard* for a time and like her husband, generated some waves concerning women's perceived position and spirituality in the late 1800s. One of the most courageous groups during those early times were young women who consented to marriage, thus relegating themselves to the possibilities of multiple childbirths with the potential of dying as a result. But before the notion becomes paramount that men were exempt from death and injury, a suggestion by the August 16, 1895 paper that there were fifty widows in Centralia, indicates otherwise! There was a harshness and hardness during those times, in which disease and illness were prevalent without vaccinations, anti-infection drugs and hygiene brought by modern sewer and water systems. Frequent deaths struck all ages, but it wasn't allowed to prevail as large families and a multitude of recorded marriages indicate. In late February 1889, sixty-six-year-old G. H. Chance had become ill but he apparently recovered as on June 7, 1890, he departed on a trip. "Mr. G. H. Chance, Miss Maudie Chance and Master Grover Chance left on Tuesday for an extended visit to Bluffs, Ill." Like disease, accidents were also a problem throughout this period. Runaway horses, horse kicks, sawmill accidents, fires, bridge collapses, and frequent railroad train derailments and head-on collisions were the results of predominately manual-labor methods, limited technology, lack of regulatory structure, lax safety rules, and a reflection of the rawness and speed of the revolutionary growth that fostered America's settlement westward. The August 3, 1889 *Guard* told of one such incident. "John Chance, who was in the railroad accident last week, has returned home. He was quite seriously hurt, and is confined to bed." On the tenth he was said to be "able to be about." To counter structure fires, brick construction had become the rage. The November 2, 1889 paper told of the new brick Christian Church that was scheduled for completion by year's end.

On February 22, 1890, the *Guard* reported that: "G. H. Chance has assumed charge of the Columbia Hotel, as R. G. Harris having failed to make both ends meet." This situation does raise several issues, one of which may indicate that Garrison Chance and not his son John was the financier of the Chance hotel. Was this the same business John had left three years earlier in 1887? The Columbia Hotel appears to have been an error and should have read Commercial Hotel. However, Centralia had more than one hotel through the early years of its existence, a fact that tends to generate some confusion, especially when the names of existing hotels changed. The paper suggested on September 25, 1891, in covering the earlier history of Centralia, that in the 1859 period the Eldorado Hotel of J. J. Collier was built between the tracks. T. S. Sneed also had a hotel that was burned out during the Civil War, but he rebuilt the Phoenix

H. C. Threlkeld, longtime Globe Hotel proprietor. (Centralia Historical Society)

Hotel in 1865 on the site of the present Globe Hotel. This structure may have also burned in the 1868 Centralia fire. Advertisements in the newspaper in the 1870s described the plush accommodations that could be had at the Centralia Merchant's Hotel. One of the earliest Centralia residents is mentioned again in the February 10, 1893 *Guard* which appears to identify him as the longtime Globe Hotel proprietor. "H. C. Threlkeld, the genial host of the Globe Hotel, has burned 900 bushels of coal since cool weather began." In August 1892 the editor and his wife had supper at the Globe with host Mr. Threlkeld, further confirming his relationship with the Globe. The December 16, 1892 paper further confirmed that the Globe and Commercial Hotels were separate entities. "The Commercial Hotel has been covered with an iron roof and otherwise improved for trade." Also suggesting there may have been a third hotel, the Merchants, the December 30, 1892 *Guard* stated, "Luther Brown is clerking at Merchants Hotel." On May 26, 1893, the Globe was being painted.

Looking back from mid-1890 until early 1891 reveals the grounding of the John Chance family to the Centralia community. As soon as Garrison Chance returned in late July from Bluffs, Illinois, the July 26, 1890 *Guard* carried this notice:

> *HOTEL FOR SALE The Commercial Hotel, Centralia, opposite the Wabash Depot, for sale. Good location, well furnished, everything first class. Terms reasonable. Address G. H. Chance, Centralia, Mo.*

This would suggest that the Commercial Hotel was likely purchased by Garrison Chance prior to his son John departing from its management in 1887. Garrison then apparently sold it but got it back when the new owner could not make a go of it. So without a capable manager he decided to sell it again. The "for sale" ad ran until early October, with apparently no takers. But in September there was a new development as an additional notice in the September 20, 1890 paper revealed:

> *THE COMMERCIAL HOTEL is being repaired throughout, and in the future will be run under a new management. Those desiring good board will do well to stop with us, for we have refurnished it new and will make it a first-class house. Ballew S. Smith.*

This appears to indicate that Smith purchased it but he instead must have been a contractor as the ads to sell the hotel continued as earlier stated. There appears to be some credibility with the hotel's stated condition in the for sale ad. But maybe this refurnishing was to change its image or even its name. On January 31, 1891, the *Guard* reported this short but noteworthy bit of news: "John A. Chance will return to Centralia and resume charge of the Commercial Hotel." This seems to indicate Garrison still owned the hotel, but had persuaded his son to return and manage it for him. On February 7 this move is confirmed but with two important revelations according to the *Guard*. "J. A. Chance, who has been running a hotel at Higginsville the past year, has returned to Centralia with his family and resumed charge of the Merchants Hotel." Apparently the refurnished Commercial Hotel had become the Merchants Hotel but was that in error? To add to the confusion, a news item in the September 29, 1893 paper stated, "G. H. Chance takes charge of the Merchants Hotel the first of next month." On November 3 this was confirmed: "Our old friend G. H. Chance is now duly installed as host of the Merchant's Hotel." The reference to the John Chance family living in Higginsville was new information. This apparently included his seventeen-year-old son Bishop. Higginsville is near Kansas City and on the C&A (Alton) Railroad. This information is revealing as it shows the seemingly constant turmoil (movings/changes in employment) in the John Chance family lives. This is clearly unexpected as a much more stable and predictable environment was visualized for a young A. Bishop Chance childhood.

C. E. Chance, who had become express messenger for the Columbia Branch Road (Railroad) by May of 1891 and a coal agent in November, also had connections with Bluffs, Illinois. The July 17, 1891 *Guard* reported that his wife and daughter traveled there. "Mrs. C. E. Chance and daughter are visiting relatives in Bluffs, Illinois." On April 21, 1893, another visit to Bluffs is covered, but it revealed that Mrs. Chance was visiting her mother and per C. E.'s obituary we learn that his wife was from that area. This is the area in which Garrison and Albert grew up after their parents moved to Oxville from Ohio. They then lived with their grandparents in nearby Pittsfield before coming to Boone County, Missouri. In May of 1894 C. E. and his daughter, Miss Elsie, visited Bluffs in Scott County, Illinois. The Wabash Railroad's Hannibal and Decatur main line ran through Bluffs but the Moberly/Hannibal line wasn't yet completed by the Missouri-Kansas-Texas Railroad (M-K-T, or Katy) which later was acquired by the Wabash and taken across the Mississippi to connect up with this original Illinois line. Bluffs also had the distinction according to George Law, a retired veteran Wabash employee, of having a Wabash branch line that ran to Keokuk, Iowa. When the Illinois River Bridge failed, the line was discontinued. The mid-year period of 1891 would also see the construction of the tallest and at that time the highest quality building in Centralia, utilizing a very smooth/shiny pressed brick. It was the new Farmers and Merchants Bank building that would prominently outline the northwest corner of the city square. The building looks as substantial one hundred years later as it did in 1891. The upstairs area would become a very important area in Chance history as the Chance telephone operations were housed there for a time.

The numerous employment changes of the Chances reflect difficult economic times. The local paper reported on January 8, 1892, that 1891 was one of the most difficult years financially since 1857. Business failures were common. A case in point was the *Guard* reporting on February 19, 1892, of two more Chances leaving the area. "C. H. Chance and his brother left last week for Seattle, Washington, where they will relocate." And following that in mid-1892, Centralia's most vocal and progressive spokesman, Adam Rodemyre, the *Guard*'s owner and editor, assumed the position of Centralia mayor. One of his latest notions on November 20, 1891, was, "Electric lights are being adopted by towns around us and it is about time for Centralia to be illuminated by something cheaper and better than kerosene." This was undoubtedly influencing a young Bishop Chance who would carry on the Rodemyre tradition of advancing technology in and around Centralia. But both would face economic hurdles and public resistance in their support of advanced technological developments to insure modern conveniences for their community.

The 1892–93 period was becoming richer in Chance family coverage. The October 14, 1892 paper told of Bingham Chance, son of Garrison, having broken his shoulder in a work incident when he fell off a stampeded horse. Tolerating the crankiness, unpredictability, and danger associated with animal power is reflective of the harshness of those times. Garrison had obtained a newfound title according to the January 27, 1893 *Guard*: "Uncle Garrison Chance is said to be the latest aspirant for postmaster." But the rising star of this time was Bishop Chance who with of a head full of new ideas would become involved in several different occupations. But they were only temporary interludes, unable to fully capture his youthful and lively spirit. After high school he secured a job in a drugstore, which may have briefly inspired him to become a pharmacist, according to his recollections, followed by a two-year stint in a clothing store. The October 28, 1892 *Guard* reported that, "Bishop Chance and Ned Stewart are salesmen at the Boston Store." Ads in the paper confirm this was a Centralia clothing store. This job was of short duration as a March 17, 1893 news item told of the closing of this store. Picking up bits and pieces, including pictures, A. Bishop was a well-attired man. The parallel that can be drawn from comparing these experiences and results to Harry S. Truman, for example, is revealing.

Per Centralia's weekly, Bishop Chance was becoming a busy, around-the-town sort of guy.

The clothing store where A. B. (second from right) clerked as a young man. The owner said he was doing A. B. a favor when he helped him get a job as hotel night clerk. A. B. agreed. (*Chanco News* Special Edition, 1949)

In November 1892, Bishop Chance and three friends spent a Sunday in Mexico (Missouri). In the January 6, 1893 *Guard* the report of another job, or maybe some early moonlighting was reported. "Bishop Chance is now agent for the Conger Bros. Steam Laundry. Take your laundry to him." But any savings apparently weren't destined to that of a matrimonial nature as suggested a week later on January 13, 1893, "Bishop Chance is likely to remain a bachelor." After a young person has been out in the world and has experienced the hardship and shattered some dreams and youthful illusions, the security of associating once again with one's parents doesn't look all that bad. Although Bishop appears to be a man about town, he may have continued to live at home.

Hotels catered to salesmen, travelers and visitors who were frequently embarking or had disembarked from one of Centralia's two railroads. E. G. (Granny) Ryman knew that A. Bishop:

> *created a signaling system or call system for the salesmen. You know the Globe Hotel [major Centralia hotel longtime operated by H. C. Threlkeld] was known all over this part of the country, not only in Centralia, but several counties because all of the traveling men would like to come in to the Globe Hotel for a place to spend the week-end. It was fine because they had excellent food and a nice room and it was a pretty fair size building for this part of Missouri in those days. Aunt Lucy was the cook. . . . Everybody called her Aunt Lucy. Then there was a place known as the Hulen Butcher Shop and Mr. Claude Hulen had a son by the name of Mike. He was a buddy of mine so we would go out to the slaughter house and they would kill hogs. We would be out there rendering the lard in a great big case iron vat. We would hook the pork tenderloin, tie a string around it, and dip it down in the vat. Then we would take it over to the hotel and take the cracklins and*

Wabash Street—Globe Hotel: This is a winter scene of the new Globe Hotel on Railroad (then Wabash) and Allen Streets. A. Bishop was employed there for a short time in 1895–96 where he apparently developed and perfected his hotel room calling mechanism. Another building in the extreme left background is the Anthony Building. (Centralia Historical Society)

Aunt Lucy would make cracklin corn bread.
Mr. Claude would get after us. (1979)

Cracklins were cubed pieces of fat cut up to make lard. The mention of the tenderloin cooked on a string was apparently for personal consumption. Those who recall the butchering day on the farm may have cooked a piece of tenderloin in the hot lard kettle for their own consumption. That is why Mr. Hulen may have been concerned as there was a loss in the quantity of tenderloin between the point of butchering and what arrived at the butcher shop, possibly destined for the Globe Hotel. All that now (2000) remains of the Globe at the corner of Allen and Railroad Streets is its foundation, which was retained for a parking lot.

The Hulens were an early Centralia family that a later relative Virgil Junior Hulen revealed that the early Hulen settlers had a home east of the F. Gano Chance art-deco (1939) home. One of the first local press notes about the Hulen Meat Market was the purchase from Yarrington and Son in the June 25, 1909 issue of the *Guard.* There were at least two brothers involved later and in this instance the purchasers were A. W. and Son. They then combined with a W. T. Hodge and it became Hulen and Hodge, according to the paper of July 9, 1909. On August 1, 1913, Claude Hulen was managing the Centralia butcher shop when his brother Willis bought out another butcher shop in Mexico (Missouri).

One reason why Bishop may have gotten interested in electricity was editor A. Rodemyre's insistence that Centralia keeps advancing in such areas as electrical lights and service in the downtown area. This began in November 1891 and by April 1893 Rodemyre had been elected to a second term as mayor. So along with his promotional comments in his paper about the need for electricity, he was in an influential position to get it done. The May 12, 1893 *Guard* told of the approval of the electrical light proposition by Centralia residents, but after visits to Vandalia, Mexico, Moberly, Fayette, and St. Louis to see their systems, the question arose of how to pay for a $10–12,000 generating plant and distribution facility. Some townsfolk were suggesting that a $800 fire engine and $2,500–4,000 dollar downtown water system was needed more than electrical lighting. The mayor promised this was second on the city's list of needs and priorities but its cost was even more of a burden (as a future mayor almost two decades removed would discover).

The year 1893 was a pivotal one in the life of Bishop Chance. Credibility can be tied to accomplishment and like Henry Ford who gained notoriety in racing to help launch his automobile manufacturing operation, Bishop Chance through racing also gained recognition. This sport also revealed the concentrated enthusiasm and endurance Bishop would throw against the many challenges confronted during his adult life. The *Guard* in its August 25, 1893 issue suggested that Bishop had now become "A. B. Chance," at the Mexico Cycle Club's meet the previous week. "A. B. Chance was entered in the mile and a half handicap, but owning to illness dropped out." Although his failure to race would appear to reflect negatively, he had already proven himself as a very capable "wheelman."By the time a subject such as bicycle riding and racing becomes popular enough to become newsworthy, it may have had years of preparatory activities. Regardless, wheeling as it was called, hit Centralia in 1893 as abruptly and unexpectedly as a winter blizzard sweeping across the Centralia prairie. On May 26, 1893, the *Guard*, as was Mr. Rodemeyre's pattern and style, issued a policy statement concerning the ins and outs of cycling.

> *The season of bicycle riding, when the roads are supposed to be in their best condition for wheeling may now be said to have opened even if they were not as acceptable as wheelmen desire to have them. More or less wheeling is done in the winter, but during summer and autumn are the periods when these outings are much indulged in and when they are delightful as contributions to physical development and good health.*

Maybe the editor even got into wheeling and canceled his almost weekly solo, weekend, business trips to St. Louis. Maybe they were inspirational voyages to taste the progressiveness of the big city and to obtain out-of-town advertisers. His July 21, 1893 paper began covering race events: "Bishop Chance went to Moberly last week with his bicycle and carried off the prize of five dollars in contest, winning by 1/8 of a mile ahead of all competitors. . . . Bishop Chance is the Nancy Hanks of Centralia bicyclists, and he makes the machine fairly hum." Hanks must have been the Barney Oldfield (early famous race car driver) of wheeling. A week later the *Guard* covered the organizing of a Centralia Cycle Club.

> *The Bicyclists of Centralia met in the City Hall Wednesday night and organized a Bicycle Club, with J. M. Phelps, Pres., Chas. Pickert, Sec'y., and Bishop Chance, Captain and Treasurer. A committee was appointed to draft a constitution and by-laws. The club adjourned to meet again Wed. night next. There are about forty wheelmen in Centralia and there is no reason why we should not have an organization of this kind here. The club proposes to grade the race track north of town and have a series of races this fall.*

The current Centralia Fairgrounds are north of town but it is unclear if this is the same area as the 1893 track. The *Guard* continued its coverage on August 4, 1893. "J. M. Phelps and Bishop Chance made the trip to Sturgeon, by the north road, 10 1/2 miles on their bicycles in 57 minutes." The north road may have been the old No. 22 (No. 151) by his future farm

(Sunnydale) on into Sturgeon. Apparently the local news hounds had picked up on a extra-curricular activity that Bishop and a friend had put their bikes to use for—a little sparking! In the August 11, 1893 Hallsville correspondence in the *Guard* was the acknowledgement of two visitors to their community, Bishop Chance and Orpheus Mayes. "Centralia champion bicycle riders coming down Saturday evening." In another section they continued to be tracked as they returned on Sunday for more of what only Hallsville offered which apparently was pleasing to their taste.

After the meteoric rise of Bishop Chance's competitive bicycle racing activities, it was expected that it would continue unabated for a period. But March 1894 began on a negative note as Bishop Chance was thrown from his bicycle and had his right knee badly bruised. This injury may have plagued his bike racing hopes, but it didn't stop his "wheeling." In late April he, J. M. Phelps, and Orpheus Mayes went to Moberly on their bicycles and returned the same day. No further developments were found that summer, except Centralia had decided to again have an elaborate Fourth of July celebration. One of the activities for the kids, reflecting the popularity of bicycle racing, were two contests for younger kids (under twelve) and for the twelve–sixteen age group. Bishop Chance was put in charge of these two events. But no further bicycle racing activities for Bishop Chance were found in the local paper. Maybe his competitive racing career was behind him as an October 1895 race might suggest. It was the "Boone County Championship" race won by Centralian W. T. Summers, against other county wheelmen, but not Bishop Chance.

Another competition that was heating up was the struggle between a multitude of bicycle manufacturers. The popularity of the bicycle in mid-1890 inspired a Hallsville resident, J. J. Armstrong, to become an agent for Safety Bicycles/Tricycles and began advertising in the *Guard*. Bicycle ads continued in the paper on an occasional basis, but in early 1894 it exploded. In March, two brands, the Weaver and the Elkhart Bicycle, both of Elkhart, Indiana, were advertised. Apparently a bike price war was occurring as the Weaver was advertised as a "$125 wheel for $59"! The Elkhart Carriage and Harness Manufacturing Company's "Elkhart" was lower at $55 with such features as twenty-eight-inch wheels, pneumatic tires, weldless steel tubing, and drop forgings. On April 6 two more ads appeared: the $85 Waverley of Indianapolis, Indiana, and the $75 Knight of the Knight Cycle Company of St. Louis.

In the February 15, 1895 *Guard* was an article of a Massachusetts manufacturer who was advertising corncob handles for bicycles. "He says they are 'durable, absorbent, clean and beautiful'. This makes another outlet for one of the by products of the farm." This was apparently before Washington, Missouri, became the corncob pipe center of the world. Centralia continued to be a hotbed of bicycle activity as in May it was suggested that there were more "wheelmen" than in the previous year. One of the most active industries in America at that time was bicycle manufacturing and with increased numbers the local paper suggested there would be an increased demand for better roads. At the same time there were growing pains with improper use of bicycles on city sidewalks reaching epidemic proportions and according to Mr. Rodemyre a bicycle has no more right on sidewalks than a wagon! But greater numbers of bicycles were predicted as prices were halved. Even the railroads were impacted as in December 1895 they began to charge twenty-five cents for hauling bicycles as they were said to have become a nuisance in the baggage cars.

Bishop found that the out-of-crate type bicycle needed a modification that would make it more user-friendly and suitable for racing and fast riding. The position of the handlebars was at a compromising angle that suited normal use but not racing. This solution would help give

A. Bishop Chance invented the reversible handlebars for bicycle racing. This picture was taken in Mexico, MO 1893. (*Chanco News* Special Edition 1949)

put another on, using bolts. He conceived the idea of fixing a clamp, coming off of the shaft (front wheel). You could loosen that and raise and lower the handlebars. Before he went to get a patent on it, he showed it to a bicycle company, and the next thing he knew the bicycle company was using it. [Paul Sturgeon suggested this was the Meade Cycle Company] *He never did get any compensation out of it. One of his first ideas. (1978)*

Bishop had just graduated from the college of "hard-knocks" and any naiveté before would now be followed by a more cautious level of professionalism in the legal handling of his future inventions. According to Bishop, he "designed the reversible handlebar" in 1893. His description would suggest that these handlebars could be installed in a low or high position instead of being the adjustable type. *Chance Tips* magazine of April 1947, the fortieth anniversary issue, stated A. Bishop Chance had invented the reversible handlebars for bicycle racing in 1895. His idea was later exploited by a Chicago manufacturer.

The census report, as carried in the August 11, 1893 *Guard*, revealed a steady Centralia growth that had increased to 1,617 residents (plus 94 dogs!) Apparently in 1893 things were booming again in Centralia. The paper reported that Centralia was a bustling metropolis of

Bishop an edge in bicycle racing competition. So the stage was set for a Centralian to consider a better way, a solution, to deal with a fixed, nonadjustable handle bar. This would result in what is considered to be the first prominent and practical invention that could be credited to Albert Bishop Chance. His son, F. Gano described what he had heard about it:

> *All the bicycles first had the big wheel. Then they went over to the present type, two wheels the same size. For racing you had to take the handle bars clear off and*

factories and businesses such as a cigar factory, creamery, buggy and carriage factory, chicken plants, picket fence factory, two broom factories, a mill, plus other assorted establishments. In October it was further revealed that there were four broom factories in and around Centralia. Speaking of "sweeping" changes, the editor suggested in November that Centralia was growing into a manufacturing community. For those who thought A. B. Chance Company brought that distinction to the Prairie Queen, let us reconsider that premise. Because of the loss of a clothing store job in March, Bishop Chance turned to an area where the action was, the building trade. The May 26, 1893 *Guard* stated, "Bishop Chance and Jim Younger have embarked in the carpenter business." The September 8, 1893 *Guard* told of Bishop taking some time off to visit friends in Gallatin, Missouri. Gallatin was on the Wabash Council Bluffs line completed in 1879.

In addition to the Centralia Chances, the Chances in the Union area were becoming established. C. H. Chance and Miss Lillie, apparently a daughter, were noted in the paper. In the Union news of September 29, 1893, it was noted that "E. B. and C. H. Chance bought a mill from O. A. Denham, and grind for one-sixth." In Centralia the former school activities of C. E. appeared to have now been continued by J. A. (John) Chance as he did an 1894 school census, in which students now totaled 558. However, the patriarch of the Boone County, Missouri Chance family, seventy-two-year-old G. H. Chance was still able to top any other family member in respect to longevity and accomplishments. One of the most noteworthy, besides a business trip to Nevada City in January, was the April 13, 1894 *Guard*'s acknowledgment that, "Uncle Garrison Chance has broken the record in his old age. On last Wednesday night his wife presented him with a pair of twins, a boy and girl." The year 1893 would bring a concession by Adam Rodemyre, that financial commitments didn't appear to be forthcoming for the Centralia electric plant as apparently electric lights for downtown Centralia had been overcome by drought and tight economic conditions. In November he noted the importance of the three public wells during that year's drought and suggested that a water system may now be needed. To produce more insult to injury, in June of 1894 the paper reported that Moberly and Mexico had shut off their electric lights "on account of hard times." Mr. Rodemyre pulled no punches but reported the news accurately even when it didn't necessarily shine a favorable light on himself.

The 1890s period would bring a new way for Centralia residents to communicate with the outside world, the telephone. The new technology and the mysteries of how it worked would intrigue an inquisitive young man of this community. It would not only involve the Chance family in the telephone business for some three decades but would also be the spark that eventually ignited the formation of Chance Company. Knowing editor A. Rodemyre's support of all types of advancements to date, there was a high level of confidence he would recognize the telephone as another milestone of modern conveniences that had finally reached Centralia. True to form, from his inside perspective as mayor, the November 3, 1893 *Guard* went to great detail in describing what was about to occur.

> *Centralia will soon be connected by telephone with Mexico, Sturgeon, Moberly and Kansas City. The company putting up the line was in Centralia last Wednesday consulting with the city officials, regarding to putting up poles, etc. A telephone office will be opened up in the Globe Hotel, and the charges will be 25¢ for five minutes conversation. It is the intention to continue the lines on to St. Louis. This will be a big thing for Centralia. The poles will be put upon the alley running east and west behind the Globe Hotel and county line road from Sturgeon Street on the south.*

Two surprises were a public telephone located in the Globe and a trunk network instead of a local only system with multiple phone locations.

A week later on November 10, 1893, the local paper covered details of the sprawling telephone system that was contemplated, built, or being built.

> *Contemplated line will extend from Macon to Moberly, thence to Higbee, Tipton where it will hookup with line from St. Louis. The Macon line connects up with a line running to Brookfield, Chillicothe, Trenton, and other points. It also continues including Glasgow. From Moberly the line will pass through Renick, Sturgeon, Centralia and Mexico, thence to St. Louis. The telephone line will follow the State roads. . . . The licensed Bell telephone will be used.*

According to early telephone development from the Independents perspective, Bell was frantically trying to tie up service before their patent expired in 1893–4. As a reflection of their predominately built systems in cities or high population centers, they had limited interests in serving smaller rural communities, especially the outlying rural areas. This would of course bring local independent telephone exchanges like what would develop in Centralia and Macon, eventually involving the Chances and Garys. By early December the poles and wires were up and on December 18, 1893, the *Guard* revealed further details. "The office of the telephone company will be in T. H. Woods's cigar store, a very convenient place."

The 1890s were a combination of the best and worst of times. The industrial revolution was accelerating with the dawning of each day, welcoming new advancements. It was inspiring and reassuring to see a fellow Monroe Countian, Mark Twain, promoting his latest publication, *The Tragedy of Pudd'nhead Wilson*. It was the Victorian period, an era that some associate with stuffiness while others see it as a time when courageous individuals attempted to pull their frontier society up out of the mud of mediocrity. Bishop Chance and celebrated editor A. Rodemeyre were such people. This can be seen in the editor's activities such as the electrical lighting project that had backfired on him. Back in the 1880s he was leading another movement, "cow control" within Centralia. Some owners installed bells on their yard mowers which unlike our modern mechanical versions would stroll around town all hours of the night. It got to the point that people's sleep was being affected so the suggestion went out to remove the bells. Confirming that on occasions our good intentions are transformed from positives to negatives, with the popularity of the bicycle in the mid-1890s came the problem of right-of-way on sidewalks and highways. An editorial concerning this issue decided that sidewalks were the pedestrian's domain but the streets were another matter. The pneumatic tires on bicycles were so quiet that they needed a bell to alert pedestrians and road users of their approach!

When a person makes a stand there will be obstacles and consequences. There is a constant struggle against good and evil, of right and wrong. Most people are not courageous enough to stand alone. Some are more concerned about self-preservation, popularity, or community status, getting ahead in their jobs, and not brave enough to stand against injustice without weighing the cost. Instead the majority of people find solace, comfort, even justification in allowing someone else to carry on a struggle or work. But there is a reward in serving others; it is a Christian principle and one could suggest that it is better to die courageously than to live as a coward. Mr. Rodemyre was a courageous individual who constantly confronted the wrongs he saw, and Centralia and area residents benefited. However, there can be a cost as his defeat for a third term as mayor suggests that not all the local people agreed with his stricter rules of enforcement and behavior, or his beliefs in progress. In the May 17, 1895 *Guard* he spoke to what could be interpreted as the lack of majority approval, not so much as desiring the mayoral position, but to be accepted by his public. Easily lost within the multitude of his paper's weekly material was a statement about people putting their hand on his shoulder and suggesting he fight this or that while they silently and confidentially huddled on the sidelines. Society tends to allow those in the forefront of this wedge to operate alone in confronting the wrongs in the world, and become the sacrificial lambs. This isn't an endorsement of protests in the streets or group conspiracies either in support of legal or illegal activities, but instead citizens standing up individually for law and order in our everyday lives. This can be as simple as voting at election time or more costly consequences of not caving in to pressures from evil sources.

It is unclear what Bishop's musical aptitude was but he and his wife's later support of his two sons traveling to Mexico for music lessons suggests an interest. In early 1895 the local newspaper arranged for a contest for area participants to win a music education. It was a raffle for those wishing to support one of a list of persons that had grown to over twenty. They could do so at the *Guard* office by buying voting tickets. Bishop Chance appeared on this list, but at the bottom. It is assumed that his involvement was on the up and up versus a dare by some of the young men in the community who Mr. Rodemyre criticized at times for some destruction or mischievous act. By March 8 it was clear to avoid further embarrassment several names with under one hundred votes, compared to the leader with two thousand, were pulled from the lists. During this time Bishop Chance was trying to define himself, but there were setbacks. This situation reveals that Bishop may not have been a very popular person at that time and this makes him sort of an underdog, who if success was to come, he would have to make it for himself. On March 22, 1895, the *Guard* recorded that he was continually to do just that. "Bishop Chance is getting to be quite an artistic sign painter. He painted the signs at the Blue Front." This was a local clothing store but there is no confirmation that Bishop was clerking there, but his multiple talented abilities are becoming clearer.

By 1895 new hotel developments had polarized in Centralia to the point that several conclusions could be reached. The most obvious was no ownership or managerial connections were found between the new Globe Hotel and the Chance family. However, apparently the "Old" Globe became the Merchants Hotel in which John Chance was involved. The Globe's manager, H. C. Threlkeld, was described by the September 23, 1904 *Guard* as the longest participant in the Centralia hotel business. Other pioneers were mentioned, but not the Chances. Henry Clay Threlkeld (1844–1915) was one of Centralia's early premier businessmen and someone who influenced A. Bishop Chance and even for a short time was his employer. Former Centralia resident Harvey Hulen (frequent *Guard* contributor) said he recalled attending Turner School with Clay Threlkeld and first knew him in 1855. Upon his death on June 9, 1915, the paper published details of his life. A native Boone Countian, he had worked on the railroad, driven a team, was in the grocery business, worked nine years in general merchandise and then entered the hotel business. He started the Globe Hotel on the site of A. R. Toalson's block yard and in 1895 leased the new Globe Hotel. He became a very successful hotel operator and eventually was involved in at least one other hotel in Carrollton. He retired around 1908 and moved to the San Diego, California area to be close to his daughter and son-in-law.

As previously discussed, there were several hotels that existed over the early years in Centralia. At this time in 1895 there were at least three and maybe four hotels including the Globe and Garrison Chance's Commercial Hotel, which apparently his son John was managing. A reference by the *Guard* on March

22, 1895, reported the Merchants Hotel was being renovated throughout. The Merchants surfaced again when a marriage was performed there in early July 1899. After its destruction in the fire that struck Centralia in August 1899, the Merchants Hotel was then operated by H. W. Johnson and owned by the Walker Brothers. The Winsor Hotel on Railroad Street was purchased midyear 1896 by the Hodges and was refurbished and its name changed to the Boone Hotel but it would be sold again in June of 1899. An 1897 article on Centralia businesses only listed the Globe, Boone, and Park Hotels. The Park Hotel originally built by T. S. Sneed, according to the *Guard* of April 14, 1899, was destroyed by fire the previous Sunday night. At the death of "Bud" Alexander (1861–1933), in early 1933, it was said he had run a hotel where Jones Motors (Chevrolet Dealer) was currently located on Railroad Street. He ended up in Washington State after the hotel had burned. The popularity of the old Globe was discussed in the January 11, 1895 paper revealing that during 1894 it averaged 850 guests a month or 10,000 during the year. Such popularity and economic justification brought a rumor on May 31, 1895, that a large three-story brick hotel [will be erected on the northeast corner lot now occupied by the Park Hotel, and owned by J. R. Bryson]. That corner lot opposite the Wabash Depot was said to be ideal but it was not on the site of the Park Hotel as it wasn't destroyed until 1899 in a fire. On July 5, 1895 confirmation and further details were forthcoming. "The foundation of new Wilson and Hall hotel building is laid." A separate *Guard* editorial said:

> *The new Wilson Building now being erected opposite the Wabash depot has been leased by H. C. Threlkeld, for five years will be used as a hotel. It will contain 34 rooms with all the modern conveniences and will be an ornament to our city. Centralia has long needed a first class hotel building.*

Revealing the latest developments and modern systems and conveniences to be incorporated into the new Globe the August 16, 1895 paper reported:

> *Wilson and Hall have let the contract for putting in steam heating apparatus in their new hotel building. They will also put in an electric annunciator, the wires connecting with every room from the office. The building will also be wired for electric lights.*

On October 25 it was reported that H. C. Threlkeld had re-leased the old Globe "which he will use as an annex to the new Midland Hotel." The new Globe was temporarily called the Midland. It was reported in the paper on October 28, 1904, that the old Globe or Merchant's Hotel burned. On December 6, 1895, it was stated that the new Globe would open on December 11. An opening celebration followed in late December.

On January 3, 1896, the *Guard* covered a special reception and banquet for the area's young ladies at the new Globe Hotel. A listing of the participants from as far away as Paris and Mexico, along with descriptions of their gowns and accessories were colorfully described. As the gala of young ladies arrived in their carriages, H. C. Threlkeld of the Globe had provided for a reception committee made up of several young men including Messrs. Chance. Later in the program Mr. Rodemyre identified which Chance this was: "Then our witty, original Bishop Chance paid "the visiting young ladies' a toast." The real A. Bishop Chance was emerging; his charming ways were developing and beginning to be recognized. It is not a certainty who enjoyed this elaborate and special evening more, the beautiful young ladies who participated, or one Bishop Chance and A. Rodemyre.

Bishop Chance's hotel calling system appears to have surfaced with the new Globe's electrical annunciator. This may have been the platform from which his elaborate desk mechanism, to call each room automatically at a prescribed time, could have developed. He said the development of it began in 1895 and he filed for a patent in 1896, which fits the new Globe's opening and Bishop's employment there. The buzzer call system would appear to have had certain similarities with a telephone switchboard that he was experimenting with and building during that time. The *Guard* of March 6, 1896, confirmed his Globe employment: "Bishop Chance is night clerk at the Globe Hotel."

The length of the job at the Globe is unknown, but it offered time enough for Bishop to work on some electrical ideas. There was the advantage of time to work on such activities without being chided for daydreaming and instructed to get back to work, as would be the case in his previous drug and clothing store clerking jobs. During the quiet of the night he did some studying according to Gano. "My father was the night clerk there and being at night, not too much activity, he was in a position of spending his time reading. . . . He took one of the first correspondence courses in electricity." (1981) The serenity of this rural community was broken not only by the nighttime wildlife sounds echoing off the surrounding prairie and an occasional distant rumblings of a train, but also by the visions of a young man destined to develop a new way or tool to do something better. All across America there was the emergence of a new kind of individual that had the strength to challenge the unknown and push the frontier of industrial progress with innovative ideas. Some jokingly called them tinkerers and dreamers, while others more accurately, usually after commercial success, called them inventors and a new breed of pioneer.

The Telephone Connection (1878–1899)

An important revelation in the early history of the Chance Company is that it was begun without the earth anchor and little or no support from the electrical utility industry. The founding of Chance Company was instead primarily associated with the telephone industry. This is significant and reflective of the fact that in many rural areas the telephone line preceded the electrical power line by some fifty years. Whether communication systems were more important or just simpler and less costly to perfect than low-cost electricity for the masses, is an intriguing question. Regardless, the telephone was quickly progressing in the late 1800s and early 1900s while electricity development was still in its infancy and progressing slowly, mainly in metropolitan areas.

The telegraph was a start but the telephone allowed rapid personal communication for the masses. This was not unlike a young man's dream in Dearborn, Michigan, to put the masses into automobiles. Although the so-called civilized world was quick to brag about their accomplishments, before either the telegraph or telephone, the smoke signals of the Indians or the drums of African natives were faster than Paul Revere's horse. Communication is highly prized today in America and the world even though it is now taken for granted like electricity. We owe much to those pioneers who brought telephone service and incidentally, because of the telephone's similar construction characteristics, it was natural that some of those same people would also help us replace the coal-oil lamp with the electric light bulb. Mr. A. Bishop Chance was one of these multitalented pioneers.

According to *The Show-Me State History*, telephone development in Missouri was some of the first in our nation. One of the true pioneers was George Durant who set up the St. Louis Telephone Exchange on April 19, 1878. This was just after the first one was established in New Haven, Connecticut, on January 28, 1878. The next year, the Kansas City Exchange was begun by J. D. Cruise on May 1, 1879. But the person who brought portions of Northeast Central Missouri telephone service was a man that A. Bishop Chance mirrored in certain ways. Charles McDaniel's (1849–1938) childhood in Hannibal was similar to that of Tom Sawyer, and some feel Mark Twain may have gotten certain impressions from the turbulent childhood of McDaniel. However, his adult life was a different proposition. His financial supporters were Hannibalian Colonel Hayward of the Burlington Railroad and Hannibal and St. Joseph Railroad fame and W. H. Loomis (Note the later Loomis and Gary family connection). But before the involvement of those men in his more substantial ventures and after becoming a "Tinner," McDaniel relocated several times before ending up in Corning, Iowa. There he subscribed to the *Scientific American*, studied inventions of Thomas Edison and Alexander Graham Bell and other electrical equipment such as call bells and signaling apparatus, which culminated in the building of a telephone line in the fall of 1878. Like many other early telephone people he soon learned what did not work financially and how tough it was to make a profit in the telephone business. Inspired, however, he returned to Hannibal and built the first commercial telephone line there. With Colonel Hayward's backing, the Hannibal Telephone Company was organized in September 1879. After winning a legal battle against Western Union, McDaniel began the construction of the exchange and after rejecting a switchboard that was too small he built his own with the help of an Indianapolis firm. After the operational phase had smoothed out, McDaniel built the first long distance lines in Missouri, a twelve-mile line from Hannibal to Palmyra and a ten-mile line to New London. Soon after that McDaniel, working under Bell licensing and Frank N. Clark, established a new exchange in Moberly that came on line in 1881. There were other isolated developments such as a line strung from Ashland to Guthrie in 1879. Activity in Columbia followed, but it was mostly of a private nature, although a line from Columbia to Rocheport was put in for business purposes in that time frame. The *Guard* on February 10, 1899, described a Columbia telephone line to Hinton. The exchange in Macon came on line about 1885 and so all this activity surely caught the attention of nearby neighbors in Macon (Gary) and Centralia (Chance). W. H. Loomis then enters the picture, financially supporting McDaniel in establishing exchanges elsewhere in the state. The southwest part of the state was the only immediate area available and exchanges were established in Joplin, Webb City, and Carthage, Missouri, and Galena, Kansas. Notice that Theodore Gary later purchased some or all of these exchanges and a W. H. Loomis Jr. became Mr. Gary's son-in-law in 1908!

Telephone development, as inspired by one of the earliest telephone builders in America, had a profound affect, especially in Northeast Central Missouri. Bishop Chance was only six years old when Charles McDaniel put the Hannibal Telephone Company into operation and only eight when the Moberly system was established. What Bishop later did, involving hotel call system and switchboard work, was covering areas that the genius of Charles McDaniel had already touched. This isn't meant to diminish the importance of later works, but instead giving credit to a trail-blazing pioneer who drove the early development of telephone technology and service to a level that took decades for others to catch up with. Although it has been suggested that Centralia had one of the first phone systems, in reality it was only average in relationship to the development of local phone exchanges in most other Missouri communities. The fact that Hannibal and Moberly preceded the exchange in Centralia by fifteen years puts this in its proper perspective, although as seen, Charles McDaniel was way ahead of his time. The impact of noting additional giants born, raised, or living in the Northeast Central Missouri region is phenomenal when one considers their contributions were often first in significance or second only to a few other areas and people in our great nation.

Some of the early private or business communication lines were non-voice carrying systems that rang bells or made other electrical signals. They were more akin to a telegraph than the voice carrying telephone. But these early primitive systems were clearly inadequate when people realized the ability to have a two-way conversation with a distant party utilizing the telephone. The development of independent phone companies sprang up all across the country, especially after the Bell patents expired in 1893–94. These Independents were quickly established by those of vision but just as quickly they sold out when operating/management problems and costs for expansion raised their ugly heads. This presented an opportunity for seasoned businessmen like Theodore Gary and up-and-comers such as Bishop Chance. Besides business prowess, Bishop Chance had an ace up his sleeve as he had the technical ability and inventiveness to improve and solve the physical challenges that primitive telephone plant/equipment presented. This is very important as those talents brought the Garys and Chances together and also led to the development of products that resulted in the birth of Chance Manufacturing Company.

Theodore Gary. (Robbins, Henry P., *Theodore Gary*, 1932)

Gano in 1968 talked of his father's first "claim to fame" in the telephone business:

> *He shared with my grandfather, John Chance, the spirit of the entrepreneur. Together they started Centralia's first telephone system in 1896. It was called the Home Telephone Company, and they operated it for thirty years, before they sold it in 1926. (1968)*

As revealed later, Home Telephone developed after absorbing the Chance Centralia phone system. In 1926 the Home Telephone Company was sold for $10,000 to E. O. Turner. He would later sell it to Missouri Telephone Company with offices in Columbia. Throughout the early 1900s, according to Gano:

> *As a sideline, my father installed telephone switchboards in other towns in Missouri, Iowa, and Kansas. Among those new switchboards was one that he sold to*

Theodore Gary in the city of Macon, Missouri. That was Theodore Gary's first telephone exchange. He went on to success as a financier and operator of a large chain of independent telephone companies which were merged in 1955 into the General Telephone Company. (1968)

Later Gano suggested that one of Gary's enterprises was Automatic Electric Company, which was later absorbed into G.T.E. (General Telephone and Electronics Corporation). Gano, in speaking of the various facets of his father's shop activities, included the repairing of telephone equipment by the same employees who would also build Sky-Rockets®, as time allowed. "They weren't separated until after the fire. He built the new exchange over the Missouri Power and Light office." (1978) Later information suggests the 1917 fire occurred after the Chance Manufacturing Company had been in operation several years.

Manufacturing Vice President Paul Sturgeon recalled that he knew A. Bishop had a relationship with Theodore Gary of Macon, Missouri, and that extended another generation as apparently Gano was acquainted with a younger Gary, who surely was Hunter. Paul knew that the Gary Company was involved in the formation of G.T.E. and he further explained, "Mr. A. B. built that telephone system in Macon. Theodore Gary was in the real estate business up there. Mr. A. B. sold that telephone system to Theodore Gary. The Garys later moved to Kansas City." (1979) The exchange in Macon and Mr. Chance's role and friendship with Mr. Gary were briefly discussed by Mr. A. Bishop in recalling historical information about his earlier life in 1897: "Assumed the title of 'Telephone Engineer.' Sold the idea of future possibilities of telephony—and got my first job in telephone exchange construction at Macon, Missouri, and sold the plant to my old friend, Theodore Gary." This confirms the friendship between the Chances and Garys, but as later seen Mr. Gary had hundreds of friends, although they were considered lifetime and not temporary relationships.

Another overview of how Mr. A. Bishop began and the developments along the way that established Chance Company is recalled by Gano:

by building telephone lines between he and one of his friends, importing the parts from Sweden. Then he got a job designing the first telephone line between Columbia and Centralia. . . . About that time . . . he built the telephone exchange in Macon. . . . That is the reason why there is no big place where he did a lot of manufacturing, etc. He was working between here and Macon, Columbia, etc. and finally settled down in the little frame building that was once a coffin storage, to make these lightning arrestors. That is where he had this storm in 1912 and the telephone business here and that is what started him to design the Never-Creep® Anchor. (1978)

The relationship that existed between the Garys and Chances and how it influenced A. Bishop Chance and Chance Company was described in *The Show-Me State Story: A History of Telephony in Missouri*. On page 108 the story, which covered the history of local phone companies seemingly in every town and city in Missouri, switched from the Macon story to Centralia. Because of the historical data, the conclusions reached and the writer's knowledge of the telephone industry, the text is reproduced here in its entirety.

Centralia's first switchboard was installed by John and A. B. Chance in 1896 over A. G. Wilson's Drug Store. It was called the Home Telephone Company. Today Kinkeads Drug Store occupies that same building in Centralia. The ingenuity and mechanical skill of A. B. Chance brought improvements to the young telephone industry in central Missouri. He was the rebuilder of the Macon exchange in 1898, [Note: Other sources suggest 1896 or 1897] *which Theodore Gary had acquired the previous year. The first line between Centralia and Columbia was strung in 1900 by Mr. Chance. These two projects started a strong lasting relationship between Mr. Chance and Mr. Gary.*

All telephones and switchboards in the early stage of communications were magneto or local battery. About 1920, common battery service came to city subscribers. The Chance family continued the Home Telephone operation until it was sold to Theodore Gary in 1926. Dial service to all customers was placed on November 30, 1954. The Chance Company, based in Centralia, manufactures and distributes products used by utilities throughout the world. An ice storm in 1912 took down many poles in central Missouri, due partly to inadequate anchoring of poles. Common practice at dead end points was to use logs, block of cement, disc blades or other large objects buried in the ground, to which a guy wire was attached. Most gave way under stress, plus digging large holes for these objects required much time, particularly in rocky or frozen soil. The Centralia, Columbia and Macon exchanges lost many poles. This brought a new idea to the creative mind of A. B. Chance, from which the never creep expandable anchor [Note: The Never-Creep® wasn't an expandable anchor] *was invented and patented. A small hole was bore in the earth into which was placed a log, and later a metal plate. To this rod with guy wire was inserted, at an angle, into the log or plate. Much less digging, less material and the holding power was always against solid earth. From this beginning improvements were made in the anchor that has been used continuously from that time on. Mr. Theodore Gary immediately saw the advantage in this new method of anchoring, used it extensively in his companies, and sold many units through his Automatic Electric plant in Chicago.*

Today the Chance Company in Centralia still builds and distributes items used by many companies world wide, with the anchor patented by A. B. Chance to better serve his telephone plant, being the basic item among many others. In this case a sleet storm was a blessing, bringing an inspiration and new idea to this early entrepreneur from which many people have benefited.

There are several important things this information reveals that is beneficial for our purposes. Mr. Hope, later in this section, confirmed the location of the early Centralia telephone switchboard in the Wilson Drug Store or the Baker Building mid-block, east side of Allen Street's downtown business district. The Kinkead Drug Store has long since moved across the street from that location. The rebuilding of the former Willis exchange for Mr. Gary in 1898 confirms Chance interview material. It does fall short in covering the rebuilding of other exchanges in Iowa, Kansas, and Missouri besides just those in North Central Missouri. Mr. A. Bishop's additional activity was very likely connected with Mr. Gary's wholesale purchase of other exchanges in the Midwest area. It's apparent there was cooperation between the Chances and Mr. Gary in connecting the Centralia and Columbia exchanges, contrary to the kind of cooperation that was occurring between the Independents and Bell properties in the 1900 time frame. Paul Sturgeon's comment about the friendship between the Garys and Chances is reinforced by the statement that, "a strong lasting relationship" was begun as a result of their business dealings. As reported in the 1926 *Guard*, another party purchased Home Telephone in Centralia, although it was connected to the Gary companies, as confirmed by the above telephone source. Home Telephone, of course, didn't originate in name until several years after the Chance exchange was begun. This employee recalls A. B. Chance Company documents including customer sales to Automatic Electric and one or more customer codes being set up for them. It is reassuring to understand the positive impact the Never-Creep® Anchor had on telephone people, including Mr. Gary.

For the moment, this will become the Theodore Gary Story. In the early sections of a recent book about G.T.E. (General Telephone and Electronics Corporation), Mr. Gary was discussed. His companies made up an important part of one of the three arms that consolidated and became G.T.E. It explained how important his contributions were to worldwide telephone service. Theodore Gary and Company controlled 80 percent of the world's automatic telephone equipment at one time and in one year (prior to 1943) did in excess of $150 million in business. Even more importantly, he and other telephone pioneers brought about relationship changes between the Independents and Bell that averted government intervention and helped lead to the modern and efficient telephone system America and the world enjoy. This generated even more curiosity and questions concerning his relationship to A. Bishop Chance and how they may have positively influenced each other. It was also fascinating to learn that Mr. Gary lived much of his life in Macon, a portion of it north of town at his sprawling *Oakhill* estate where he died on November 4, 1952, at age ninety-six. According to information from Oakwood Cemetery in Macon where he, Hunter, and other family members are entombed, Theodore Gary was born in 1856. Two book sources (*Theodore Gary* and *GTE*) supports this while another book source (Missouri and Missourians) plus his initial obituary suggest, apparently incorrectly, that he was born in 1854 and/or was ninety-eight when he passed away.

An initial concern about the Chances and A. B. Chance Company was the apparent lack of recognition of their contributions. Mr. Gary was a very important figure in early telephone development and the empire he controlled at his height was an immense operation. Mr. Gary wasn't included in the *History of Missouri* publication, but was found in the collection of *Missouri and Missourians*. Local historian Ralph Gerhard immediately knew who Mr. Gary was and had even been in his home *Oakhill* that was being refurbished. Ralph said it was huge with many rooms and had electronic gadgets all over the house, which relates to his telephone and electronic involvements.

One of the issues that *The History of G.T.E.* brought forth was Mr. Gary's purchase of the Macon Telephone Exchange in 1897 from a John Wills. (Note: Another source suggests Willis.) From the information about A. Bishop Chance building the Macon telephone system for Theodore Gary, the question arises as to what was built. This could have taken place in 1896. It appears more primitive systems were being replaced with more advanced and larger capacity systems. One development that points in that direction is his partner James A. (Alec) Hudson (1853–1922) exercising an option on the Columbia exchange and by 1907 they or Mr. Gary owned other telephone companies in Joplin, Nevada, Carthage, Webb City, and Carterville, Missouri. A year later Atchison, Kansas, was added. Less than a year later the Independence, Missouri, exchange was purchased. Topeka, Kansas, and St. Joseph, Missouri (1906) associations were later developed. Mr. Gary stated, "All these plants had to be refinanced and rebuilt, new capital being interested." Notice that A. Bishop Chance was a possible subcontractor who built, rebuilt, or worked on exchanges in Missouri, Iowa, and Kansas! Was there a connection? One indication of this was a casual mention by his son Gano that his father A. Bishop Chance was involved in the 1920s with the pioneering efforts of building paved roads across the state of Missouri. Is it not interesting that Mr. Gary was the first chairman of the new Missouri Highway Commission from 1921 to November of 1926. In fact this activity sort of signaled his retirement as he was sixty-five when he accepted the chairmanship from Governor Hyde in 1921 and became known as the "Father of Missouri's hard road system." In the selection process of the Board and other prominent highway positions Mr. Gary was involved in, there is no mention of the Chances. There were county support groups that Mr. Gary organized in support of the state Centennial road system that may have included A. Bishop Chance, as the names of those individuals weren't mentioned in Mr. Gary's biography. So this is disappointing as there was hoped to have been a closer recognized relationship. Mr. Gary was forty or forty-one when he bought the exchange in Macon and A. Bishop Chance was only twenty-four in 1897. This is an important period for A. Bishop as this was his first real and independent venture out into the business world. The question arises if he hooked his wagon to one of a special group of America's prolific entrepreneurs of the first quarter of the twentieth century.

Theodore Gary was an organizer who had an uncanny ability to pull people together, even those with seemingly opposite views. His impact upon the telephone industry is legendary. It was no accident that those at the Missouri state capitol called upon him to negotiate and organize one of the most difficult problems facing the state, the highway system. He did it not in an overbearing way but fairly, honestly, and respectfully so as not to divide, but bring together even warring factions. In 1911 the location of the St. Louis/Kansas City highway was hotly debated between the proponents for the south, central, and north routes. Mr. Gary gave a presentation on the merits of the three proposed routes. Even though the south route went through Jefferson City (state capitol), the committee wasn't politically influenced and chose the central route from St. Charles to Columbia and on through Glasgow, Marshall, Waverly, Lexington, and Independence. His Missouri Highway Commission appointment in 1921 was followed by his appointment in1929 to the State Survey Commission that analyzed the inadequacies of the various branches and institutions of Missouri government along with the imbalance in the state tax structure that supported same. In 1928 the University of Missouri honored Mr. Gary with "the degree of Doctor of Laws" (Robbins) for his public service to his local community, state, and nation. Although not having the early opportunity to be an educated man, his continued personal growth was phenomenal and worthy of recognition. One of his personal crowning achievements was his selection and training of key personnel to manage and operate his huge business complex. A January 25, 1929 letter to nineteen top Gary managers suggested they train their sons for some branch of the business but two qualifications were essential: dependability and honesty. He positively influenced many people who he had contact and association with. Recalling that Harry S. Truman was involved in the improvements of Jackson County roads, it raises the question if he may have developed an acquaintance with Mr. Gary, a frequent resident in the Kansas City area at that time.

Our prime focus, however, has to be how and if he influenced A. Bishop Chance. Further information suggests a certain pattern that is somewhat similar between these two men. They both were family men, Mr. Gary losing his first wife in 1881 after coming to Macon in 1878 from Cole County, Missouri. His family had moved in 1866 from Wayne County, Ohio. Both his parents were French immigrants, arriving in the United States as children. In 1883 he married Miss Helen Fairbanks Larrabee of Macon and like Mr. and Mrs. A. Bishop Chance they had two children: a son Hunter Larrabee Gary (1884–1946) and a daughter Mary B. Gary who became Mrs. Wesley H. Loomis Jr. in 1908 and had at least two sons: Theodore Gary Loomis and Wesley H. Loomis III. Hunter also had two children, a son Theodore Sarvinet Gary and a daughter Mary (Mrs. Edward T. Harrison). Mrs. Helen F. (Larrabee) Gary passed away at *Oakhill* on

May 9, 1936, with entombment in Oakwood Cemetery in Macon. The son of Helen and Theodore, Hunter Larrabee Gary, was Board chairman of the Theodore Gary Company of Kansas City and Associated Telephone and Telegraph when he passed away at Reno, Nevada, on November 30, 1946. He and his family had made Reno their home for the last decade.

A "precious possession" according to Mr. Gary in *Missouri and Missourians* was the loyalty and close relationships maintained between him and the many men and women associated with him, some for more than forty years. This may help explain A. Bishop's fascination with Kansas City where Mr. Gary had his offices and an apartment or home called *The Walnuts*. However, there has not been found any official reference to any visits between the two families and no mention of the Chances in any Gary books or telephone material. An earlier reorganization in 1919 resulted in the switching of his headquarters from Macon to Kansas City. It is significant that A. Bishop was involved with Theodore Gary early when he first became involved in the telephone industry. In a few short years Mr. Gary eclipsed most others in the telephone business, making him one of the telephone industry's most powerful forces, who in the end even impacted the huge and all-powerful Bell System. Although A. Bishop's technical skills were important early on to get Mr. Gary up and running, it quickly became secondary to service, acquisitions, accounting principles, selection of key personnel, and other facets of operation. The acquisition and expansion of Automatic Electric's automated equipment was a huge endeavor, and so as time passed and progress was made, the involvement between these two men undoubtedly distanced from that special early time although there may have been occasional contact.

In reference to the Macon telephone exchange Gary's biography suggested that there were other facets of its operation that were important besides the technical aspects:

> *The Macon plant was rebuilt and expanded and the number of subscribers in town and country grew rapidly. The service, based on the rule of courtesy that later became famous, was evidently satisfactory. In fact, Macon's telephone service became so renowned that business men in other places began to clamor for "the kind of telephone service Gary gives Macon." (Robbins)*

The idea of superior service to customers was one Chance Company always embraced. Another commonality was the modern accounting principles that Mr. Gary applied to his business, not unlike the material that accompanied a young F. Gano Chance to Columbia to secure a Depression loan to sustain Chance Company.

> *Mr. Gary was not a promoter selling stock. He was concerned in developing service. He installed the most accurate books. He made, at the very first, an adequate charge for repairs and depreciation. His company was on an absolutely sound basis. He had no difficulty borrowing money, for he had collateral that bankers could understand. During this period of his telephone business his credit was good at thirty banks, and he freely used it. (Robbins)*

Theodore Gary was a man brave enough to get involved in new challenges. Most people are content to hold onto what they become comfortable and successful with, but in forty years Mr. Gary was said to have had forty different undertakings. All of this energy was bound to catch notice. His aggressive acquisition program of buying out local telephone exchanges and possibly his most significant, Automatic Electric Company in Chicago, didn't go unnoticed by the Chances. How did the business prowess of Mr. Gary and the technical skills of John's son interrelate? Was the acquisition program of Chance Company influenced by Mr. Gary's examples?

In spite of the seemingly unlimited wheeling and dealing that Mr. Gary was involved in his real estate, banking, hotel, telephone, manufacturing, and other pursuits, there was a solid structure of beliefs and characteristics that not only drove him but regulated how and what he did. He displayed such a fairness and consideration of even competitors' points of view and interests that he obtained the trust of prominent business people around the world. It is interesting to see such similarities displayed by A. Bishop Chance. In advising his grandsons he suggested they think constructively, always be truthful, read and think about what you read, keep an open mind, and take care of your health. He suggested that one should hold your temper and tongue and be a positive thinker and person. It's interesting that Mr. A. Bishop's longest associate of over thirty years, Miss Ida Lou Traughber, said she had not heard him ever raise his voice in anger! Like the Chances in Centralia, Mr. Gary had a profound affect upon his longtime home of Macon. He was involved in building a lake for a clean water source for the city and railroads, golf course, tennis courts, city gas, library, Oakwood Cemetery, Samaritan Hospital, and other civic and charitable contributions to his local community. He even published at least two books, *The Macon County Hand Book* of 1888, *The Independent Telephony* in 1907, and prepared an article for the March 6, 1926 issue of *The Saturday Evening Post* about the independent telephone industry. A biography, *Theodore Gary*, was authored by Henry P. Robbins in 1932 and copyrighted by his son Hunter L. Gary.

The role of the family and how later generations relate to former prominent family members is intriguing and also puzzling at times. The Gary and Chance families seem to cover the broad spectrum of how this plays out. A recent visit to Macon (2003) was both gratifying and sad. Seeing homes built around the Gary's former *Oakhill* estate suggested change. A trip to Oakwood Cemetery was reflective of a family that had deserted their roots. Lack of maintenance on the mausoleum dedicated in 1926, where Gary family members are entombed, was troubling as in Mr. Gary's biography concerning Oakwood it was stated, "The endowment requires an annual expenditure for maintenance so specifically as to make neglect impossible." (Robbins) A similar arrangement with the Macon Public Library was eliminated after the death of Mr. Gary, according to a prominent local Macon historian. The Chance family situation after the deaths of A. Bishop or F. Gano Chance could have reflected the same scenario, but it hasn't. The leadership and concern of Jack Chance, the grandson of the founder, has been instrumental in maintaining a Chance family presence in Centralia, although there were ample reasons for such a posture not to exist after the sale of A. B. Chance Company. An example of Jack's commitment and strong family ties is his support of this book, which suggests a gathering of family instead of a separation. It was my understanding that when A. B. Chance III passed away, there were some questions of where he was to be buried, off the U.S. mainland or in Centralia within the Chance sepulcher. The Chance family leader F. Gano was said to have decided that question in short order. Another example by Mr. A. Bishop taking care of A. B. Chance III's financial interests after his father's death is reflective of the Chance family tradition displaying responsibility in close family matters. Mr. and Mrs. Theodore and Hunter Gary likewise performed many generous and thoughtful acts in support of family and substantial financial donations to the Macon community.

In the midst of all the dynamics that can exist within the family structure, there were two truths that emerged. The first is the loss of control after one has passed on. Legal agreements are often changeable and what one plans for and foresees in the future, after death, can be an effort built on quicksand. Two verses of Scripture come to mind: "And the rain descended, and the floods came, and the winds blew and beat upon that house; and it fell not: for it was founded upon a rock." (St. Matthew 7:25, KJV) But materialists have to accept the eventual loss of control: "Lay not up for yourselves treasures upon earth, where moth and rust doth corrupt; and where thieves break through and steal." (St. Matthew 6:19, KJV) These are reasons that the written word and books are so important as they have a better chance of surviving beyond our lives and material possessions and gives insight into previous times and people. The second truth is the disparity in the lasting recognition of a businessperson over a political leader. Both are important positions, especially at the level of Theodore Gary. The wisdom of the Chances comes into play as they realized the importance of recognizing business employees first came internally within Chance Company, so it would have an opportunity to be shared with the media. The same is true of statues and historical accounts for businesspeople. Don't expect local, state, or national governmental bodies to release funds to build a statue or to name a street or building after a businessperson. Great military and political leaders helped win World War II, but in truth business leaders, engineers, factory workers, railroaders, farmers, teachers, scientists, and inventors such as the Garys and the Chances played a huge role in helping to sustain our freedom, then and today.

The association of the Chances with the telephone seems to have gotten started when a young Bishop Chance built his own system. In F. Gano's words, "I understand that he ran a few lines to his friends here in Centralia . . . I'm sure that my father probably had the first telephone in Centralia, which he built and ran it to somebody else's house." (1978) The *History of Missouri* suggests that in 1895 A. Bishop "and another youth rigged the town's first telephone lines between their homes." The December 22, 1893 *Guard* revealed an earlier proposal. "John Wills and Bishop Chance propose to put up a private telegraph line between their respective residences." Is this the person from whom Mr. Gary purchased the Macon telephone exchange? John in the late 1890s was living in Macon. It is assumed the railroad already had telegraph lines for their operational purposes so the notation in the January 19, 1894 *Guard* that a telegraph line was put in must suggest a Western Union type service for the public. Western Union had tried to take over the telephone industry after seeing its potential, but that was halted with the patent-infringement suit of November 10, 1879, won by "Ma" Bell. From Bishop's early interest and study of electricity evolved an interest in the telephone, telegraph, and the development of hotel wake-up bell system. The similarities in these systems are obvious in requiring lines in which electrical impulses could be transmitted. The hotel bell system (annunciator) and the telegraph and telephone were operational and so actually performed meaningful tasks. This shows that Bishop was a practical young man who was selling his telegraph and telephone services first to his young friends. Gano suggested that, "the first thing you know he began setting up a telephone exchange. And this telephone exchange here started growing almost from a hobby. . . . My father was ten years in general telephone exchange construction in Iowa, Kansas, and Missouri. That was from 1898 to 1909." (1978) According to the *History of Missouri*: "By buying parts, experimenting with them, he made telephones and installed the first switchboard in Centralia."

F. Gano believed the Centralia telephone system developed as follows:

> *I presume my father had a few telephone lines and there was another party that came in here and started building telephone lines and we had two telephone exchanges. I think they called the other Mutual or something and so my father, through my grandfather, together, they bought out this other company. . . . My father built some of these telephone lines and built an exchange . . . . I remember my father telling about he had these two telephone companies in the same town. Well, that just doesn't work. (1978)*

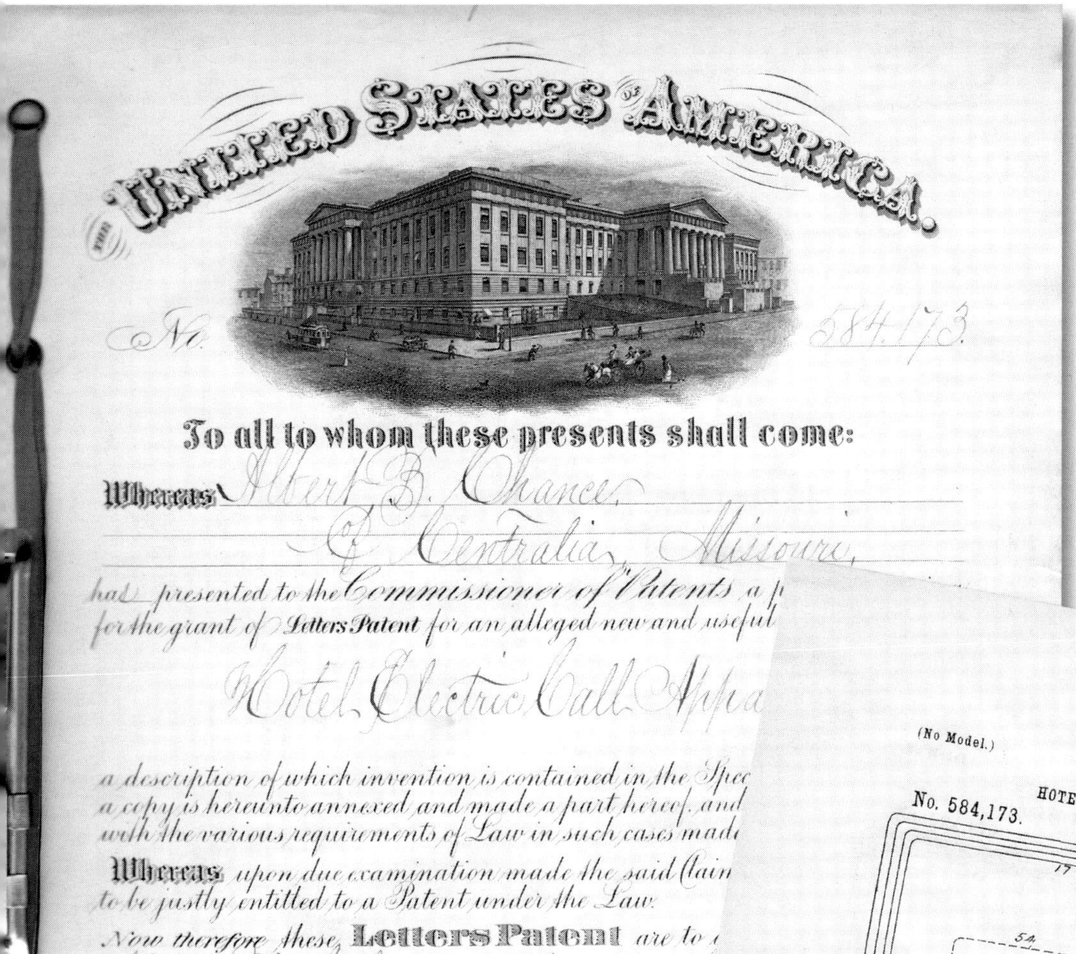
United States of America.

No. 584,173.

To all to whom these presents shall come:

Whereas Albert B. Chance,
of Centralia, Missouri,
has presented to the Commissioner of Patents a p
for the grant of Letters Patent for an alleged new and usefu
Hotel Electric Call Appa
a description of which invention is contained in the Spec
a copy is hereunto annexed and made a part hereof, and
with the various requirements of Law in such cases mad

Whereas upon due examination made the said Clai
to be justly entitled to a Patent under the Law.
Now therefore these Letters Patent are to g
Albert B. Chance, his
for the term of Seventeen years from the eigh
June one thousand eight hundred and
the exclusive right to make, use and vend the said in
United States and the Territories thereof.

In testimony whereof
hand and caused the s
to be affixed at the
this eighth
year of our Lo
and ninety Sev
the Independence
of America the o

Countersigned: Benj Butterworth
Commissione

1897
Feb 25 1911

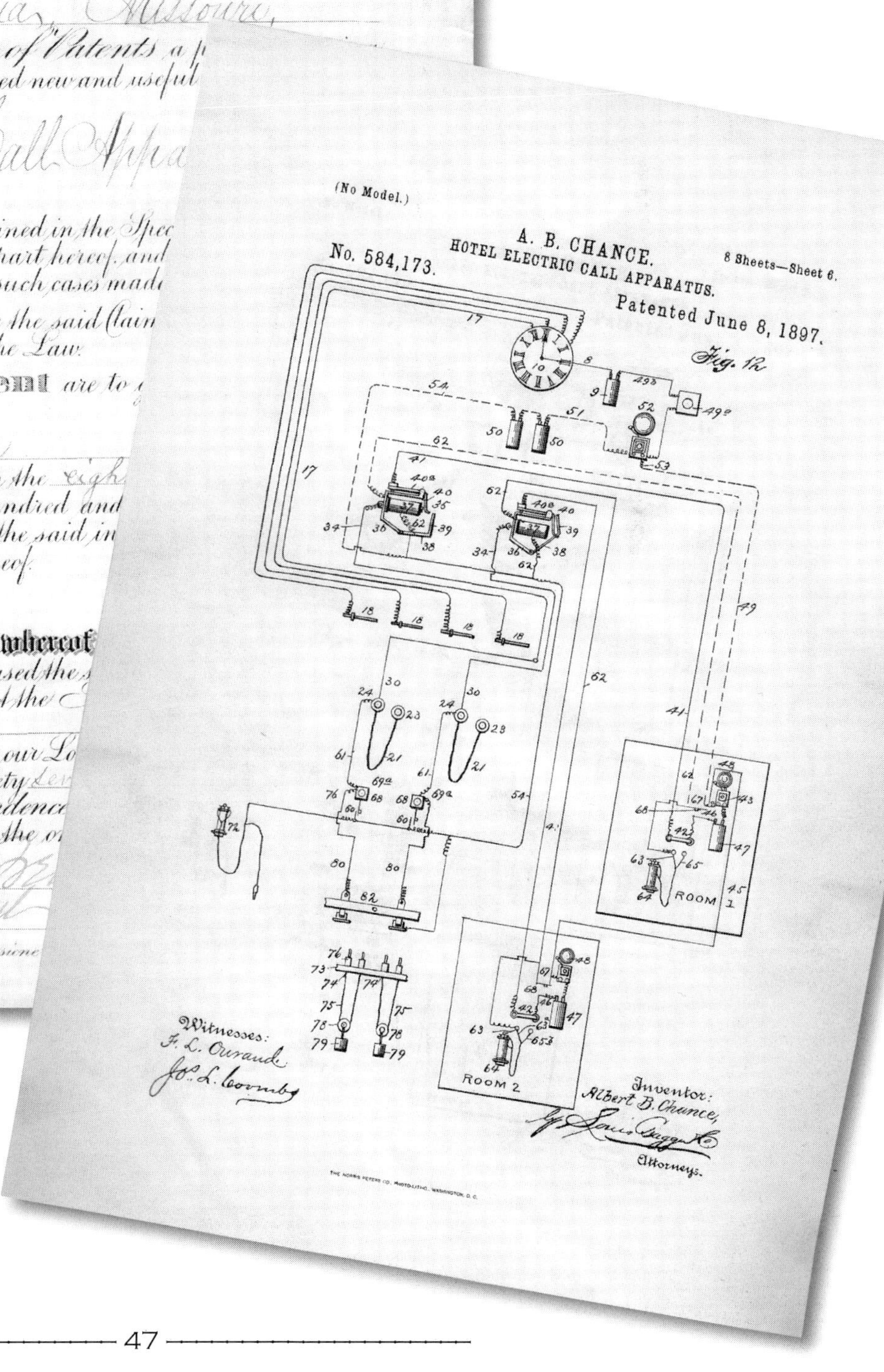

Patent No. 584,173 of June 8, 1897, for "Hotel Electric Call Apparatus" with the application filed on November 23, 1896, is one of the most highly publicized inventions of Albert B. Chance. This was due to the complexity of this electrical device and the earliness of the origination. The cover and sheet six of eight displaying the diagram of the circuitry is reproduced here from the original document developed by "The Morris Peters Co.: Photo-Litho, Washington, D.C." (Jack Chance Collection)

The early telephone systems were very confusing with Bell competing against the Independents. Covered earlier was a decision in 1893 by a Bell licensee, to bring long distance service to Centralia, but they did not have the inclination to provide local service in sparsely populated small communities because telephone sources suggest it didn't generate adequate revenue in relation to their investment. The Independents, such as Chance, usually provided such local service. As early as 1899, the LaCrosse telephone exchange (later Buffum) was operated alongside the Chance exchange. In 1904 there was word that the Mutual group also wanted to enter Centralia, which they did the following year. Neither of these two companies appeared to have been purchased by Chance, although during World War I the Mutual exchange was forced by the government war board to join Home Telephone. There was no record found of any transaction then or later between Home and Mutual. So at one time there may have been as many as four telephone systems in Centralia!

Jack Chance, during this interview with his father Gano, was referring to an unknown source that stated:

> *Centralia was one of the first small towns in Missouri to have an electric telephone system in 1887 or 1888. A party by the name of Lamber constructed a pole line connecting Centralia and some other towns in the area. This caused A. B. Chance to take up the study of telephones and small apparatus equipment. (1978)*

As earlier confirmed by the *Guard*, the long distance Centralia telephone lines and a phone was installed in 1893 or early 1894. Roy Hansen also referred to some material about Bishop's father, John Chance.

> *He operated the Merchants Hotel but disposed of his interests to become a salesman of farm machinery. Eventually, he turned to the development of the Centralia Telephone System, which was subsequently incorporated as the Home Telephone Company. In this enterprise he was associated with his son Albert. Mr. Chance (John) was actively associated for twenty-eight years with Home Telephone Company and its predecessor serving for several years as Manager, until 1926. (1978)*

Gano felt his father built the telephone exchange in Centralia in the 1896–97 time frame. "He started buying parts and started working on building the telephone equipment in '96 and '97." (1978) These parts were apparently destined for telephone exchange construction. Gano believed that his father's building of exchanges in other towns would have resulted in him first building one in Centralia. "Why would he be building telephone exchanges in other cities and not build it in Centralia?" (1978) In 1896, according to *The Show-Me State Story*: *A History of Telephony in Missouri*, John and A. B. Chance entered the telephone business in Centralia. The same source said A. Bishop rebuilt the Macon Exchange for Mr. Gary in 1898. According to grandson, F. Gano Chance, "He [John] more or less financed my father in the telephone business as a partner." (1978) Their firm was absorbed in 1905 by the Home Telephone Company, a commonly designated name for the multitude of small independent telephone exchanges popping up across the country. According to information that Jack Chance was sourcing, A. Bishop and his father constructed a two-hundred-line board in the Centralia telephone system. "Bishop was running the phone company and his sister Sallie (Chance) Fountain was the first telephone operator." (1978) This system was eventually replaced by an eight-hundred-line board. It was anticipated that the *Guard* would support this start-up information and the May 8, 1896 issue did: "Many of our business houses are putting in telephone." It was so new that the editor was a bit awkward in describing it as we today would put an "a" before telephone. Bishop Chance again made the paper on June 5 with his first recorded visit to Kansas City. Bishop Chance's world was exploding as on July 3, 1896, the *Guard* reported, "Bishop Chance has opened a general electric supply depot in the Bush Building." This suggests that the location described is in the old Christian Church building where he would later locate his manufacturing operation. But more surprisingly was his residence was now in Macon as the *Guard* of September 25 revealed: "Bishop Chance came down from Macon City last week to look after his telephone business."

The bicycle in 1896 isn't the center of Bishop's universe as it earlier seemed to be. An 1897 bike race in Mexico and at the Centralia Fair didn't include him, or he wasn't mentioned, as cyclists from as far away as Kansas City and St. Louis were now stealing the spotlight. A practice that Bishop Chance seems to have mastered was concentrating his efforts toward goals and when achieved or it was time to move on, he would quickly redirect himself to a new area.

The opening of an electrical supply house in Centralia in July 1896 may seem abrupt but in reality only a necessary step in handling such supplies in a more orderly manner, as his telephone and electrical work expanded. According to interviews, Bishop had been procuring electrical parts for some time prior to 1896. This "depot" appears to be a supply house for the start-up of Bishop's out of town telephone construction activities. There is no indication that this was a retail operation, as the report of Bishop coming from Macon in September 1896 indicates he was out of town for a time. This raises the issue if Bishop worked on the Macon telephone exchange for Mr. Wills (Willis) before Mr. Gary purchased it the following year in 1897. Further confirmation of his telephone activity was reported in the November 20, 1896 *Guard* and also revealed that Bishop was now living in Macon City (now known as Macon): "Bishop Chance, of Macon City is here this week looking after his telephone interests." All of this telephone activity by the fall of 1896 suggests that Bishop had quit his night clerking job at the Globe Hotel and was now a full-time telephone entrepreneur.

In Mr. Gary's *Independent Telephony*, there was a reference that his telephone activity began in Macon, "More than ten years ago." This 1896 reference is further confirmed by his book scanning the equipment progress in his Macon Exchange. It stated that equipment installed in the seventh and the eighth years of operation: "which continued in service until the end of 11th year. . . ." Since this book was copyrighted and dated in 1907 and shows the financial records through 1907, this suggest the first year would have begun in 1896. Because the financial records of the Macon exchange are charted from 1898, an incorrect assumption could be made that this indicates the purchase year. Records may have not been available or established to Mr. Gary's liking until 1898. From an investment performance standpoint, Mr. Gary may have chosen to ignore parts or all of the first two years. This represented his entrance into the telephone business during which time he was establishing his operating procedures and accumulating experience and expertise. His 1907 book reveals he was a very fast learner. His book shows at least seven central board/switch equipment changes or additions in the first eleven years. This would have begun in 1896 and continued in 1898, 1899, 1900, 1901, 1902, 1903, and 1907. Two pieces had multiple years listed. The reason 1907 is referenced is the 1902–03 equipment: "continued in service until the end of 11th year, when it was replaced. . . ."

The possible 1896 purchase by Mr. Gary of the Macon Exchange relates very well with the Macon activities of A. Bishop Chance in 1896. This is two years prior to when A. Bishop was said to have installed the switchboard in the Macon Exchange, but this is likely due to the same source, who used the apparent incorrect 1898

Gary purchase date. Just when it seems logical, a 1897 remark by A. Bishop in a highlights of his life (Chapter 6) states: "and got my first job in telephone exchange construction at Macon, Mo. and sold the plant to my old friend, Theo. Gary." This generates a question of whether he had temporary ownership of this exchange, or if he was referring to the equipment he had installed. This could suggest that he worked on that plant in 1896 and didn't complete it until 1897 when Gary may have taken over the actual operation of the refurbished exchange. A bigger question is how that experience influenced the Gary and Chance relationship. Did A. Bishop become Mr. Gary's future equipment installer and upgrader in the Macon and other Gary exchanges in Missouri and Kansas? This would help explain A. Bishop's some ten years of telephone installations in the tri-state area (Iowa/Kansas/Missouri), which is for the most part found to be specifically undocumented. The Macon experience is valuable because it may be the only such activity that is documented. This also is logical as the Garys were still located in Macon before relocating their operations to Kansas City in 1919. Other evidence of A. Bishop's many 1897 local activities suggest his work in Macon was already completed, or at least for the moment. It is only logical that he temporarily lived in Macon while he was performing extensive rebuilding of the exchange there.

The first telephone in Centralia was apparently connected to the Bell System. However, expansion of this to include several businesses, as reported in 1896, may have been the new Chance Exchange. There are several developments in 1897 that tend to clarify what was occurring as the *Guard* reported on August 27, 1897:

> *The Centralia telephone system now has thirty-five telephones with a central office. People now order groceries, etc., through the telephone and do not have to come up town shopping. On September 3, 1897: "A telephone company has been organized in Centralia."*

This could be the Chance's telephone company, but again no hard evidence was found to support this. Its official organization may have followed its actual beginning. On September 10 there were a couple of articles, the first sort of a braggadocio comment directed toward Slater (boyhood home of Steve McQueen) who had captured the Chicago and Alton (C&A) Railroad shops which Centralia and other towns tried to earlier confiscate. "Centralia has two thousand people and a telephone system. Slater has more than three thousand people and no phones." Notice the progress in this editor's descriptive terminology in describing the telephone as "phone," which we still commonly use a hundred years later. The second mention is a description of the benefits of the just-completed Centralia Fair where a survey revealed the dollar value of the Fair to Centralia businesses. "B. C. Chance, telephohe [telephone] . . . $20." There seems to be a couple of errors, but A. B. (Bishop) Chance appears to be Mr. Telephone in Centralia and by September 17 his subscribers had increased by fifteen in less than a month. "Bishop Chance now has 50 subscribers to his telephone." The Chance connection to the Centralia exchange is further confirmed by the Fair committee payments of expenses incurred by the fair. "Telephone, Chance . . . $5 00."(space, no decimal point is used) If there is any question remaining, a September 24, 1897 local piece in the *Guard* seals it. "Centralia now has a complete telephone system, connecting with all the surrounding towns, with A. B. Chance as manager." There are at least two revelations, first that the local Chance system was apparently connected to the long distance Bell lines, however, it would appear to be overstated as the Columbia line had not been built by Bishop Chance yet. A telephone historical source suggests this was accomplished in 1900 after Mr. Gary had become involved in the Columbia Exchange.

The Columbia telephone line was discussed in the October 29, 1897 *Guard*, "The construction of the telephone line between Centralia and Columbia will be commenced at once. Centralia always starts every progressive move in Boone County." This ending statement was sure to start another round of wars between Columbians and Centralia's most prolific spokesman and unflagging promoter of the Queen of the Prairie, Centralia's incomparable editor, A. Rodemyre. He clearly played a role in revealing and adding fire to the divisions between Centralia and Columbia residents over the years. However, the editor now had an ally in Bishop Chance who would continue to make Centralia a leader and later would takeover the role from the *Guard*'s celebrated editor in support of Centralia. No sooner had the Columbia line's construction been discussed than another long distance line was constructed according to the November 26, 1897 *Guard*. "The telephone line between Madison and Centralia has been completed." The lines were of little worth without being connected to a phone, but there was no need to call in another firm as Bishop Chance had the whole process covered, as this December 3, 1897 *Guard* news column from the Saling area reported: "Bishop Chance passed through here this week en route to Madison to put in the phones." The *Guard* promptly followed up on December 10 boasting that "Centralia has the most complete telephone system of any town of its size in the state." What is clear is Bishop Chance seems to be in charge and some questions arise as to his father's (John) involvement, if any, at this point. This is further confirmed on April 6, 1900, by the *Guard*: "The telephone company with Bishop Chance as manager has been incorporated under the State laws." Did his father contribute financially or is this another example of A. Bishop taking family/employees under his wing?

The following essay, as aired in the January 14, 1898 *Guard*, reveals the workings of the Centralia to Madison Exchange, and even how it was financed.

> *The telephone line between Centralia and Madison is owned and controlled by the farmers along the route and telephones have been placed in many residences. It is such a novelty to many that there seems to be an irresistible impulse to exchange greetings with neighbors. All manner of subjects are discussed from the health of the family to politics. The chief difficulty lies in the fact that all along the line people want to talk at the same time and as a result there is regular Babel of noises that keeps the wire hot. People are rung up at Centralia, Sailing (Saling), Tulip and Madison and all kinds of questions and answers fired back and forth. Being stockholders it costs nothing and they seem determined to get all the fun out of it possible. The poor devil at the Central Office is kept on the jump about half the night, but the farmers enjoy the fun and keep the bells ringing.*

A. B. Chance did have other interests, which may have been ignited, in one case, by Mr. Rodemyre bragging about the beautiful women in Hallsville. No sooner were those comments recorded in the *Guard* than the September 24, 1897 issue reported in the Hallsville section that three Centralia lads, A. B. Chance, W. T. Summers Jr., and R. B. Hulen came down from Centralia Sunday afternoon. It is doubtful if they were surveying a route for a Centralia to Hallsville telephone line! The *Guard* also reported the visits of young ladies to area families and friends such as this one in the September 24, 1897 issue. "Miss Mamie Gano, of Missouri City, is visiting Mrs. P. R. Diggs." Keep this visit in mind when 1898 is more closely examined. P. R. Diggs is identified in June 1904 as the Centralia Wabash Railroad ticket agent.

Although the telephone is considered the business that provided the platform, which eventually led to Chance Company's involvement in the electrical utility industry, some surprises have surfaced. As covered earlier,

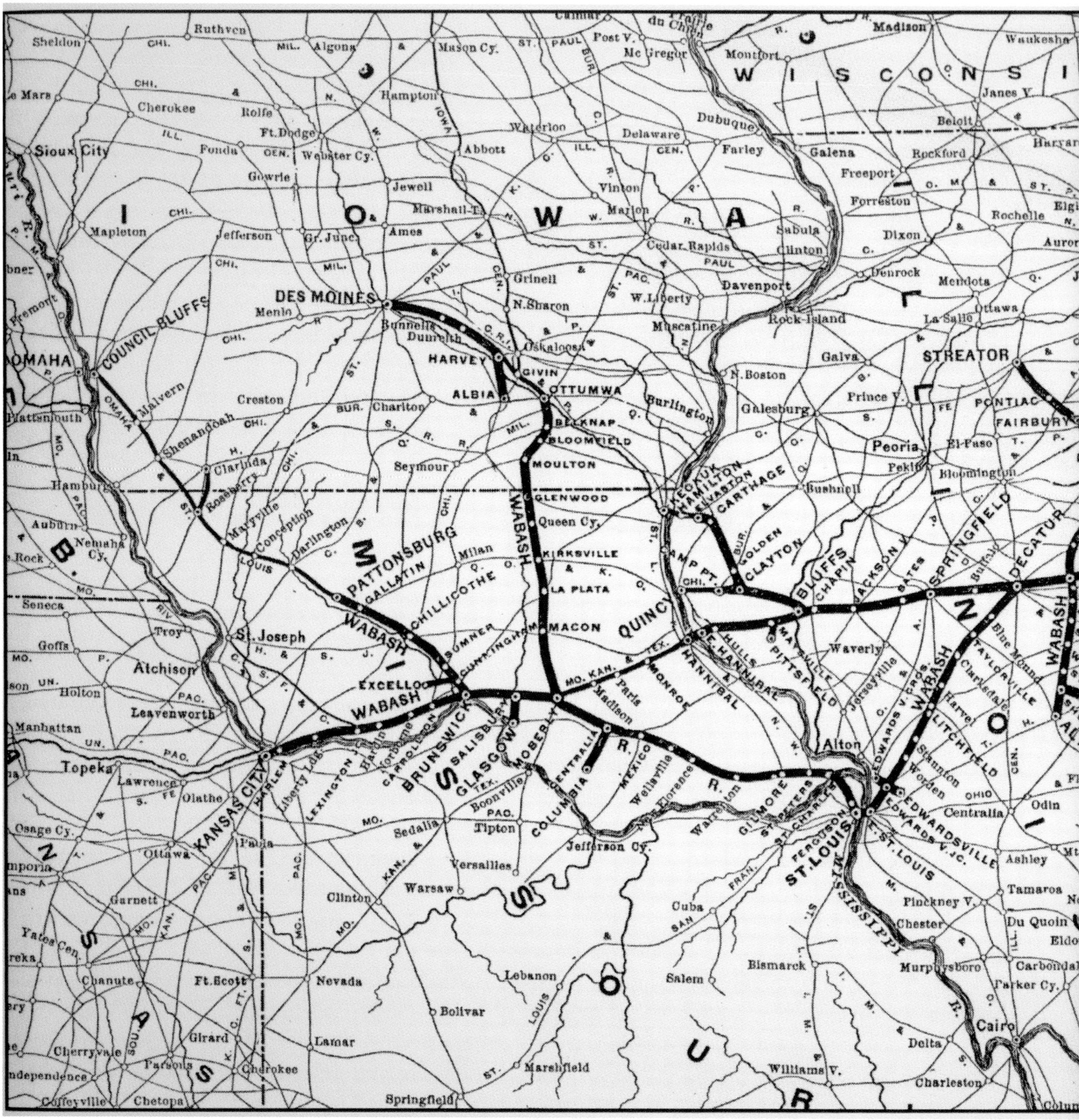

Circa 1890 Wabash Railroad map. (*Follow the Flag* by H.Roger Grant)

Mr. Rodemyre, Centralia's active, progress-minded citizen at that time, was pushing for electric lights and service. Confirmation that a decision had been made to proceed was covered in the November 6, 1896 *Guard*: "The vote on the electric light proposition carried by 246 in favor of lights and 21 against it." The electric light plant (power plant to operate generator) was to be located between the railroads, opposite the water tank as confirmed on November 20. This is the Wabash Railroad's water tank discussed earlier. Reports in December followed the progress of construction of the new plant, which was scheduled to be operational by early 1897. The December 25 *Guard* reported that the light poles were up and the businesses were wired for lights. On March 12, 1897, the *Guard* confirmed that the Globe Hotel "has

put in electric lights." But the telephone wasn't the only activity Bishop Chance was involved in as suggested by the December 18, 1896 *Guard*. "Bishop Chance is putting in electric light fixture [not plural] in the store buildings." To the many firsts Bishop Chance assailed to, add Centralia's first electrician! Electricity would also come to local residences as it had to businesses as the *Guard* reported one of the first such applications on March 4, 1898. "Jonas Berg has put electric lights in his residence."

The year 1896 had other developments as a census report by J. A. (John) Chance in May 1896 indicated a Centralia population of 1,700, with 601 school students. This is reminiscent of C. E. Chance who sold his Centralia home and moved his family to St. Louis in October. The census apparently came under scrutiny as to its coverage and redone by city authorities as the December 11 *Guard* reported the new census reported 2,005 residents. A familiar family name of an early Chance Company employee, Traughber, was placed in a couple of important positions. F. M. Traughber became the president of the Centralia School Board and was also elected to the presidency of the Building and Loan Association.

There had been several successful Fourth of July celebrations and the local fair also reflected Centralia's ability to attract huge crowds at the fairgrounds (see Volume II "Small Town U.S.A."). A February 1897 note in the *Guard* helps identify where the Centralia Fairgrounds were located, as it said the city was extending Hickman Street out to the fairgrounds. The previous fall something very special occurred which may have tended to stir the pot of rivalry between Columbia and Centralia. On November 13, 1896, the *Guard* said the University of Missouri Tiger and Iowa team played a football game on Saturday at the Centralia Fairgrounds! But astounding events were occurring around Centralia as a rural resident several miles out of town reportedly went rabbit hunting at night from the reflection off of Centralia's new downtown electrical streetlights. It wasn't reported if there was a full moon. However, area residents could see the lights at least four miles out of town according to the March 5, 1897 *Guard*. This may have been visible because of the highly elevated position of Centralia. For more sophisticated big-city-type entertainment, there were various plays and entertainments taking place in Centralia's Opera House. There seems to be no greater supporter and appreciator of such events than Mr. and Mrs. Adam Rodemyre.

As the Gay Nineties rushed toward a sure demise in ushering in the twentieth century, was there a chance that 1898 could upstage 1897? Could the seemingly maddening pace of one A. Bishop Chance be intensified? The world scene was also a powder keg, erupting in sensational new advancements but also locked into age-old problems of war. The Spanish-American War came quickly when the Battleship *Maine* sank or was sunk. Locals and the whole country were up in arms, but maybe this was a sign of healing, of reconciliation between the Civil War factions as now they could serve together and fight side by side against a common un-American enemy. Even St. Louisian Frank James' son was accepted in donning the uniform his family had fought against some three decades earlier. There was also the lure of gold and the cry: "North to Alaska to the Klondyke country." The younger set with little roots to keep them anchored were easily persuaded to head north and search for big nuggets they wouldn't find. But don't believe that some of the oldsters didn't have fire within their minds and hearts. Bishop Chance had definitely been stealing the spotlight and maybe his granddad thought enough is enough. If there is any wonder where Bishop may have inherited a portion of his strong spirit, a gold fever story in the November 19, 1897 *Guard* reveals all. "G. H. [Garrison] Chance, who is nearly 80 years old, talks of going to the Klondyke in the Spring." Maybe it was just talk—but maybe not. A couple of years later Garrison was still traveling, as the *Guard* in January 1900 recorded his return from a trip to Oklahoma!

Everybody seemed to be going somewhere, to war or north in search of gold. Even young ladies traveled alone to the Omaha Exposition. Today many people head to the South or Southwest as winter cold approaches but the *Guard* suggested in December 1897 that only the birds were hibernating in large numbers, indicating a harsh winter was fast approaching. Although smaller town residents, especially those along rail lines, flocked to the cities such as Kansas City and St. Louis on shopping sprees or to see the streetcars, there was also an exodus from the city to the surrounding areas. A visit by a Gano to the Centralia area was reported some ten months later when another Gano came to see the same party, a Mrs. P. R. Diggs. (Another visit reported in the May 10, 1901 *Guard* by Mrs. Diggs' sister, Miss Leora Gorin of Missouri City, confirms her maiden name.) The second recorded Gano visit was to have a profound affect on the Chance family, or that branch of the family led by A. Bishop Chance. Maybe we should accept the fact that Miss Gano's return is recorded, as an omen. Her arrival may have been inconspicuous as other stories suggest this was the way in which she and a local young man met for the first time. What we know is Miss Gano's return wasn't unnoticed and so it became public by those who saw and reported all, the *Centralia Guard* of July 15, 1898: "Miss Frances Gano of Kansas City, who has been visiting her cousin, Mrs. P. R. Diggs, has returned home." Everybody seemed to be going somewhere and A. Bishop Chance must have become captivated with the Kansas City and Missouri City areas, awestruck about the golden prospects west on the Wabash, whose motto became "Follow the Flag."

Still known simply as Bishop Chance, on April 22, 1898, the *Guard* confirmed further growth in his Centralia enterprise. "Bishop Chance has 80 subscribers to his telephone system." On August 26 the paper, speaking of what Centralia had to offer, listed several "haves" including: "We have a telephone system east, west, north and south, besides a local system." But the chance for change in Bishop's life was now a certainty as July slipped by and the fall of 1898 blossomed forth in its parade of colors. Although no record in the *Guard* is found of trips by Bishop Chance to the Kansas City area during the fall of 1898, one could sense that a lot was occurring. For the first time why weren't the prying eyes of the Rodemyre family privy to the coming and goings of its constituency, suggesting Bishop may have been departing on

ev. John Gano baptising George Washington (A. B. "Buff" Chance IV Collection) the midnight train. Then on December 23, 1898, the *Guard* conveyed some explosive news to its readers: "Married in Kansas City, December 17, Mr. A. B. Chance, of Centralia and Miss Frances Gano, of Kansas City." A gracious tribute in the Rodemyre tradition followed. "Mr. Chance is one of our most worthy industrious and popular young men. The bride is a beautiful accomplished young lady, highly esteemed. We wish the happy couple joy and happiness." No longer was there a young and struggling Bishop Chance but instead a struggling Mr. A. Bishop Chance with the support of his new bride.

On December 17, 1898, at age twenty-five, A. Bishop Chance and Frances (Fancy) Inglehart (Chanslor) Gano (1875–1970) were married. According to *Chanco News* of June 1957, the contact between the two occurred when Frances "had come to Centralia to visit a cousin, who arranged a 'date' with Bishop Chance." The same Boone County historical source listed Kansas City as her home but according to her son F. Gano, his dad met her in Missouri City where she was living although she may have left home in the meantime and moved to Kansas City. Miss Gano was born on February 14, 1875. in Missouri City, daughter of Aaron Price and Gertrude (Chanslor) Gano. The Ganos had come from Kentucky and her father was associated with the Baptist College in nearby Liberty. Missouri City is located on a bluff overlooking the Missouri River and is near Kansas City, a few miles southeast of Liberty. Like Centralia, Missouri City catered to two railroads, the former Burlington or CB&Q (Brookfield, Macon, and Quincy or Hannibal) and the North Missouri (Wabash) that also passed through Centralia. What a coincidence that Mr. A. Bishop may have frequently traveled west to Kansas City through Missouri City!

The Gano family had a very important member during colonial times according to F. Gano Chance, who gained the title "Fighting Parson of the Revolutionary War." Reverend John Gano was F. Gano's great-grandfather. Gano was the second son of Mr. A. Bishop and Mrs. Frances I. (Gano) Chance. He of course received the name Gano from his mother's side of the family and his first name is also similar to his mother's except with an "i" inserted for the "e." Jack, Gano's son, said in December 2005 that his father wouldn't use that name because he thought it was too feminine. According to his great-grandson Gano, Reverend John Gano was a very close friend of General George Washington and he went with him on many of his campaigns. He also became pastor of the First Baptist Church in New York City. Gano continued:

> *I have in my files at home a very interesting letter from George Washington to Rev. John Gano in his hand writing in which he said, if my men are as brave as my Chaplain, we will win this war! (1978).*

Reverend Gano reportedly baptized General Washington in the Delaware River. There is a chapel at William Jewell College at Liberty, Missouri, that is dedicated to Reverend John Gano and where F. Gano Chance spoke to students about his ancestors.

During the presentation of a program about A. Bishop Chance many years ago, Paul Sturgeon mentioned the attraction that Mr. Chance had for Kansas City. Paul suggested when he traveled; all roads seemed to lead to Kansas City. Since Centralia was closer to St. Louis, residents tended to head east instead of west to shop or for business purposes. This stands in stark contrast to A. Bishop's attraction to Kansas City as he was in certain ways going against the flow. But in examining his life, this wasn't an uncommon trait. If either of the two Centralia rail lines were used to travel west, the eventual destination of both was Kansas City. The early telephone work that A. Bishop Chance did was usually to the west of Centralia and the early friendship with the Garys in Macon was fostered by the opportunities of each being able to reach the other quickly, on the same rail

Reverend John Gano Memorial Chapel on the grounds of William Jewell College in Liberty, Missouri. (Jack Chance Collection)

line. The short-lived relocation or sales branch of Chance Company was also to the west in Kansas City. So along with the previous home of his wife and a business relationship developed with Mr. Theodore Gary, who later relocated his business to Kansas City, it is no wonder he was infatuated with the K. C. area.

Mr. A. Bishop had a sense of pride in being an American and a Missouri native. Although most Centralians preferred the diversity of St. Louis, Kansas City often described as a large "cow town" with its strong, Midwestern agricultural connections was the choice of Mr. Chance. Kansas City has its own dignified class and style and to his liking he also found the people to be friendly. The quick and easy way to get to Kansas City from Centralia was by train during the first half of the twentieth century due to the superior development in rail passenger service. However, with improved highways and automobiles after World War I, it became practical and more common for longer trips by auto. Mr. A. Bishop's preferred route to Kansas City was by U.S. Highway 40 and possibly 24. There were small towns with lowered speed limits to transverse and there was one incident involving Mr. A. Bishop on his way to Kansas City that was told by his grandson Jack Chance. After being stopped for excessive speed in this unknown town, the policeman wrote out a ticket, with a resultant stiff for the times, $5.00 fine. Unlike today when people have an opportunity to get mad before paying such a fine by mail, it was due on the spot or to the pokey you went. Mr. A. Bishop reached in his billfold and pulled out a ten-dollar bill, which he handed to the policeman. The officer said, "Wait, I'll get you your change," to which Mr. A. Bishop responded, "keep it for my return trip because when I come back through I'm going to be traveling at the same speed!" (May 30, 2000)

As the last year of the nineteenth century unfolds, it helps us define who A. Bishop Chance was and how he established himself. The growth of the telephone was the avenue on which he was traveling and clearly he was taking over where earlier pioneers had left off by promoting, operating, and building telephone systems and by the installation of phones in businesses and homes. This activity wasn't confined to just areas west of Centralia as the *Guard* suggested the editor and Bishop Chance had gone to St. Louis on business in early February 1899. This was a common occurrence over many years for Mr. Rodemyre and this was a way in which he could feel the pulse of progress and could keep abreast of new developments like the World's Fair coming to St. Louis while visiting his daughter. This news note: "Bishop Chance and A. Rodemyre were in St. Louis Saturday on business,"could suggest they were traveling together. Mr. Rodemyre may have been showing Bishop around town and this is significant in providing a direct example of the editor's influence on Bishop Chance. It had become apparent that they were alike in certain progressive ways and Mr. Rodemyre had become a positive and influential person in Central Missouri who touched many people, obviously including Bishop Chance. Mr. Rodemyre was also undoubtedly intrigued and drawn to this whirlwind of a man who was a driving force in bringing reality to ideas discussed in scientific publications or advancements found in major cities like St. Louis.

Bishop's movements east were also confirmed by a news report of March 10, 1899. "Claude Bass and Bishop Chance are in Montgomery City putting in telephones." It's not clear if this means a complete exchange (switchboard and lines) or if it can be taken literally to suggest simply telephone installations. New associates like Claude Bass appear in the newspaper, but little more is known of them. On January 27, 1899, another individual is discussed: "J. A. Johnson has sold his interest in the Columbia telephone line to Bishop Chance." There are suggestions that Bishop struggled financially during these times. He revealed in a personal history record that in 1894 he "got a job with E. Kahn at $20.00" and in 1896 "pushed my work as rapidly as my savings from $20.00 per month would permit." Regardless, much more is known about Mr. Johnson and his association with Bishop Chance as it apparently ended in controversy according to the January 6, 1899 *Guard*.

> *A Card from J. A. Johnson. I am not seeking a newspaper controversy, neither do I desire to cast reflections upon any one, but in justice to myself I deem it necessary to correct an erronous impression conveyed through Mr. Chance's card, which is calculated to injure a business enterprise. There is no depreciation in the value of the Columbia telephone line, but as a natural consequence the longer the line is operated the greater must be its value as a business enterprise, and the more coupon books that are redeemed the sooner the line will be on a cash basis. Myself and A. B. Chance own the Columbia telephone line and operated it until October 1st, 1898, when Mr. Chance leased the line. I sought no advantage, neither do wish to wrong any man intentionally, I understood the contract and it is not my fault that Mr.*

Chance construed it different owing to an error in copying by the typewriter from the original Mss.

During the nine months that the company operated the line we redeemed $162.50 in coupons and realized $72.93 net cash, making a total of $235.23 or $26.13 per month, being about 35 per cent on an investment of $900. This does not look like depreciation. The matter in controversy between myself and Mr. Chance was decided by an intelligent jury who weighed the evidence and decided on the merits of the case. I see no cause for criticism or necessity for re-trying the case through the press.

J. A. Johnson.

Bishop Chance was definitely getting his feet wet in business dealings as this Johnson correspondence indicates. It is unclear if a right or wrong distinction could be drawn in this case outside a jury ruling, as Bishop apparently was attempting to dispel in an earlier piece in the paper. It does reveal some valuable insight into the financial implications of the telephone business and Bishop's use of the newspaper to forward his agenda. Not long after on May 5, 1899, the *Guard* made another announcement concerning this line: "A. B. Chance has sold his telephone line between Centralia and Columbia to Frank Buffum of the LaCrosse Lumber Co., for $1,200 cash." It becomes obvious that Bishop Chance was wheeling and dealing and was not allowing any grass to grow under his feet. Any profits of this sale were very likely invested back into other current projects as his telephone construction activities reached full stride.

There was new information concerning the proposition that Centralia had two phone systems. This could earlier be explained by the long distant trunk line running through Centralia and Bishop Chance's local systems, but this idea became suspect when the *Guard* commented on this situation. The first mention on June 2, 1899, concerned LaCrosse interests who apparently controlled more than just the Columbia line. "The LaCrosse Lumber Company has the most complete telephone system in the state." This was followed on August 11 by: "with two telephone systems Centralia people ought to be enabled to make talk cheap." And then the announcement of even further telephone expansion on December 8, 1899. "The Bell Telephone Co., is building a long distance line between St. Louis and Kansas City and will have an office in Centralia." The location of the exchange office was later on January 1, 1900, identified as to occupy the second story of the Wood building. Again in the same issue, the extensive Centralia telephone system was summarized.

Centralia has the most complete telephone system of any town of its size in the State. We have a thorough local system and a line north to Audrain and Monroe counties with central office in charge of B. C. Chance. The LaCrosse line east, west, north and south, central office in charge of J. D. Orcutt at lumber yard; and the long line east and west, with office in corner restaurant.

Note that the reporter again suggested that A. B. Chance somehow became B. C. Chance, likely because A. B. was commonly addressed as "Bishop." It is also possible that for a time some used the initials B.C. as a nickname.

Area developments and talk of such local improvements as the need for better sidewalks, streets and roads continued to be covered and promoted by the *Guard*. They also were a driving force in city beautification, drainage and sanitary issues while making a stand against bad behavior such as the destruction of watermelon patches and "crap-shooting." Residents who were maintaining nice yards, however, were facing something that was fouling up the whole works, free running chickens! To provide the owners of these fowls an opportunity to corral their egg producers, the *Guard* suggested they could be "considered as wild game" unless the owners "put them up—and kept them up." In April 1899, the Rodemyres were advocating the need for a city waterworks and suggesting Centralia was the ideal place for the home of the Missouri State Fairgrounds. Columbia didn't need it as they had all they could handle in operating the University! This was sure to fuel renewed hostilities between these two towns. It seemed like every year was bringing out some form of competition among area towns to become the site of a variety of business and institutional locations, things that today are taken for granted. But those were the days when modern civilization was taking root. There was bitterness expressed by some of the towns that lost out to Sedalia in being awarded the Fairgrounds. To save face and to be realistic about something that was a more prestigious prize, the *Guard* in a prophetic manner suggested on June 9, 1899: "Good factories giving employment to many men would be of more benefit advantage to Centralia than the State Fair."

In January 1899 a report that Columbia had a new railroad brought thoughts of the long anticipated Katy, which would impact the Centralia/Columbia branch line and give Columbia two railroads like Centralia, although they would be secondary branches off the main lines. However, the short line into Columbia off of Katy's main line at McBaine, the main running along the north edge of the Missouri River toward St. Louis, was not completed until 1901. Later on May 12, 1899, a news item identified this new Columbia line: "The Missouri Midland Railroad is the name of the eight mile railroad between Columbia and Kennard." A later proposal suggested its potential completion on into Fulton. Continual promotional schemes and feelers were frequently presented to progressive communities; some appeared legitimate while others were of a more questionable nature. They usually hinged on the potential for the influx of local capital. The same things occur today although the quest is for state and federal funds and local tax and utility concessions for new additions.

What better way to close the 1800s than to focus on a subject that represents modern progress and involves Bishop Chance, the telephone. The *Guard* in a June 30, 1899 advertisement listed a phone number (61) for a local grocery store. What would come next?

Chapter 2
The Birth of Chance Company
(1900–1910)

A Tradition of Winning

There was a competitive, challenging spirit that existed within A. Bishop Chance, which he apparently inherited naturally, but it was also something he nourished and grew. This rubbed off onto his two sons and other people around him. These characteristics included inquisitiveness in knowing how different things worked, a continual interest in new advancements of the day, and becoming involved in numerous activities including competitive events.

Bicycles were popular machines in the late 1800s and early 1900s, prior to the automobile revolution. They were functional and represented a distinct mode of transportation although there were weather/road limitations that also plagued the early automobile. An influence of early "wheelmen" was their support of better roads benefited by the organizations they spawned such as a "Missouri Division of the League of American Wheelmen," as described in the November 24, 1893 *Guard.* Theodore Gary also recognized the American Wheelmen's contributions for improved roads in his highway papers. It could be heard that the bicycle would never replace the horse or the iron horse, but it was a transitional machine that was a sign of the machine age in its infancy. It led to motorized cycles and so spawned a whole two-wheel industry that thrives a hundred years later. It also had other influences, such as Henry Ford's first automobile utilizing bicycle-like components and their use by two brothers in the bicycle business in Ohio that resulted in a short but historical flight at Kitty Hawk. But the bicycle had its own applications, such as racing and it could have been the first propelled machines used in that sport. The August 22, 1885 *Guard* told of arrangements that had been made to have a bicycle race on the first and second days of the Columbia Fair. On the second day, one of the races would be against a fast horse, and promised to be very exciting. Caught up in all of this trendsetting activity, out where New Yorkers would call the edge of the frontier was a young twelve-year-old, energetic Centralian by the name of Albert Bishop Chance. This is not a surprise as he can be pictured as one who had a rich thirst for life. Racing can be a social thing that draws people together and fits the competitive nature of young men. There can also be generosity involved because it can serve others by entertaining them. It is revealing that F. Gano Chance became a magician, supporting the Chance trait of desiring to socialize with, give

This A. Bishop Chance portrait is similar with those in the 1909 *Guard* when he ran for the Centralia mayor position. (Jack Chance Collection)

to, and entertain people. In a small northwestern Missouri town, two Mennonite boys were seen racing their parent's horse and buggy up the street. It was entertainment generated from a little boredom and mischievousness mixed with youthful energy, in spite of the stringent guidelines of their religion. The July 28, 1893 *Guard* suggested that horse racing on public roads around Centralia had become a problem and a nuisance for people traveling those roads. It had to be stopped!

To those who have never won anything, the success of the Chances in winning prizes and contests can more readily be appreciated. A. Bishop and F. Gano were quite successful as they both won automobiles, a new, unproven, and sometimes unwelcome "contraption" in the early 1900s. During A. Bishop's lifetime, travel was common by train, especially for any significant distance on the North American continent, because of poor roads and the late development of the automobile. Dirt roads were the norm and were even considered highways in some instances. If a road was graveled, it surely would be labeled a highway. The *Guard* on March 16, 1906, addressed the poor condition of Centralia streets: "The roads are worse on Main Street than they are in the country." They said this didn't speak well—they needed to do something about them. A petition on April 27, 1906, to pave Allen Street was circulated, but no Chance names appeared on it. To put this in perspective, the founding of Chance Company in 1907 was during a time when the auto was predominantly an experimental, rich man's toy, existing in small numbers in large cities. As a consequence, the impact that the two railroads passing through Centralia had on local residents was significant. It not only brought the world to Centralia's doorstep, but it opened the world to local residents as well. The railroads continued to improve like Burlington's new St. Louis short branch, completed in January 1905 via Mexico to Old Monroe. The Chicago and Alton (C&A) and Burlington shared their track from Kansas City to St. Louis for passenger service. In early 1908, by combining their tracks west out of Carrollton, the Wabash and Santa Fe enjoyed the advantages of a double track railroad, but the Alton and Wabash failed to come to a like agreement from Mexico to Clark in 1907–08. But double track was not uncommon, as the June 26, 1914 *Guard* said that the Wabash had two main line tracks from Clark to Huntsville and from Salisbury to Brunswick in addition to the Santa Fe Carrollton to Kansas City arrangement. On May 7, 1909, the *Guard* revealed that the Wabash had bought land west of town to put in a wye so the branch trains could be turned around for their return trip to Columbia. But this was a premature report that did not occur until a decade later. In short, there appeared to be much greater growth in railroad service than in progress on roadways for automobiles, in the first two decades of the twentieth century.

There is a notion that there are many overnight sensations, but most success is achieved by laboring in an apprentice-type role for a time. During Gano's childhood he read his dad's magazines and said: "There were ads in *Popular Mechanics* of a magic lantern, if you sold so many bars of Cuticera Soap, and I remember doing that." This kind of experience was valuable and undoubtedly benefited and influenced Gano later when he became a successful business leader. This writer's recollection of selling garden seed brought back memories of poring over the prize catalog which showed progressively more tantalizing rewards for increased sales effort. Gano further recalled another one of his early business kind of interests:

> *I got into another sales contest, which a local department store sponsored, in which every time you sold something, you got votes. It ran for a month or so and the one that got the most votes got a pedal type automobile. I did win it, a beautiful auto that I operated for a year or two.*

Gano was around eight years old at the time and

soon his interests would turn to the electronic mysteries of more sophisticated toys such as the telegraph, telephone, and x-ray machine. Undoubtedly, young Gano's interest in the pedal car had been instigated by his family's motorcar. The idea of selling kids on miniature versions of the real thing obviously wasn't hatched recently. As Gano recalled, "The first car in Centralia was one that my father won in a contest." A small picture of the car was shown in the newspaper and the plates for printing were fairly coarse and made up of many little dots. The contestants were to guess the number of dots within the picture. His father took the picture to a photographer friend and had it blown up. With a pen he numbered each dot and won the car.

Mrs. A. B. Chance was reported by the *Guard* on June 24, 1904, as the winner of the dot contest, which contradicts her husband being the prize winner:

> *Mrs. A. B. Chance has been awarded the first prize on the dot contest in* American Home Magazine, *published at Des Moines, Iowa. The prize is $50.00 in cash and the choice of an automobile, 80-acre farm or horse and buggy.*

Since this appears to be a female-oriented publication, it may have been considered expedient for Mrs. Chance to be the designated contestant and winner. She may have also been the subscriber. As Gano recalled, the method that his dad used and his involvement in the contest was ingenious. Besides the automatic prize of $50 was a choice of a car, eighty-acre farm, or a horse and buggy. This choice was made easier by disparity in the value of a horse and buggy versus the other two prizes and Bishop's problem with a horse dying from heat exhaustion in 1901, gives impetus to the "horseless" carriage choice. His mechanical and progress-oriented aptitude and the wide variation in land values further influenced the selection of the automobile.

Mr. and Mrs. A. Bishop Chance are pictured in the auto of this 1905 photo. This may be the new Reliance delivered in 1905. Their home in the background was on Snow Avenue, now South Rollins. It was destroyed by fire in 1915. (Jack Chance Collection)

But even before the A. Bishop Chance family had received their automobile, there was trouble brewing between horses and mechanical horsepower. The introduction of the automobile to Centralia stirred up more than dust and as expected the *Guard* got into the thick of the controversy, as this September 9, 1904 editorial testifies:

> *As Geo. Mitchell, southeast of town was going out home Thursday evening of last week he saw an automobile coming towards him and as he knew his team would not stand for their new competitor in traffic, he warned the chaffeur* [chauffeur] *to stop. The auto came steadily and swiftly on and Mr. Mitchell's team turned to run, breaking the surrey tongue, dashboard and otherwise damaging the vehicle. After the damage was done the chaffeur stopped the auto, which was the big Winton tonneau damaged here last week on a culvert, and was laid up for repairs. An automobile has a right to the public highways, but the owners and runners of these machines should not attempt to run upon a team that is frightened at the machine.*

There was one incident covered in the newspaper on September 24, 1909, when an unknown party on a county road near Centralia, shot and damaged a passing car with a shotgun. Frightened and concerned for their lives, the auto party was said to have been out touring and not intruding on someone's watermelon patch.

However, there were also those who were looking for a little excitement and found that the horse and buggy provided an avenue in which they could escape the prying eye of townspeople. Young men and boys were frequently criticized for generating disturbances at church or on downtown streets, recklessly operating their bicycles on city sidewalks, horse racing, and a multitude of other sins. But certain local young ladies also had an appetite for some of society's unaccepted practices such as smoking, aggressively flirting and drinking. The *Guard* in attempting to stamp out such behavior would somehow find out about these activities and make the populace aware. This eventually would lead to the exposure of the participants if they were continued. On September 6, 1901, the community was alerted to the reckless behavior concerning some of their young adults. "We learn that certain young ladies who go out buggy riding are in the habit of guzzling beer and coming back boozy." Can you imagine the questionable looks that local parents gave their daughters after reading their hometown newspaper?

A day the A. Bishop Chance family had been anticipating for almost a year finally arrived according to the April 7, 1905 *Guard*:

> *A. B. Chance received his automobile Thursday morning, and it will be well for the natives to trim their ears for the goose-hawk of the auto horn. The new "buzz-Wagon" is a beauty, and will make all the near-by world Mr. Chance's neighbors by a few turns of the wheels.*

This car turned out to be as much an oddity and unwelcome object as a horse and carriage are today on our streets and highways. So much did the Chance's frequent evening drives irritate locals that according to Gano, they would be asked which direction they were traveling that evening so townspeople could surrey in the opposite direction. It appeared rather common for people to travel to see friends on summer evenings or to simply enjoy their fancy carriages on a trip out through the countryside. The problem with a car was its appearance, noise, clatter, and smoke and smell of oil and gasoline highly irritated horses. Horses would sometimes buck, rear up, and generate quite a disturbance. This kind of excitement didn't fit in very well with the leisurely sort of lifestyle of the period, but more importantly, runaway horses were a life-threatening situation. In

April of 1906 the newspaper told of Edison's latest invention, the storage battery. It promised to revolutionize the automobile industry by resulting in a "noiseless" car, welcome news to the majority still depending on the horse for their local transportation.

Gano recalled one special trip to Jefferson City, about a sixty-mile journey that took all day. As they were returning, a surrey came around a corner and two horses became spooked by the car. It was all the teamster could do to hold the two bucking horses. "This fellow was awfully mad," and Gano was worried that his dad was going to get into a fight, but he added that his dad didn't have a quick temper and chuckled as he got out of the car. Apparently, the threat by this person that "he was going to lick Mr. Chance" was dampened considerably by Mr. Chance's calm demeanor. The upset person continued, "You have no business being out in that crazy machine." This episode made quite an impression on a young F. Gano Chance. Trying to control horses was a very challenging proposition and serious injuries and even deaths weren't uncommon with runaways. The *Guard* editor, in the 1890s, even denounced football games in city streets, as it had almost caused several such incidents. One recorded runaway was caused by the noise from the unloading of steel railroad track materials near the Centralia depot. The two ladies involved threw off their two young children before the wagon was overturned and destroyed against a tree, leaving the ladies with serious injuries.

Instead of the area's surreying crowd on Sunday contending with just the Chance auto, there was an invasion according to the *Guard* on April 28, 1905: "The skittish horse had a great time Sunday. There were three automobiles puffing up and down the streets in the afternoon. One machine made the trip from St. Louis to this city in twelve hours." But as the paper wisely cautioned, the auto like other advancements of the day was here to stay and local folks would need to adjust and tolerate their increasing presence. This was the advice the Rodemyres gave on May 12, 1905:

> *The automobile has come to stay, and it is well enough to get used to it. Some years ago a bicycle was the terror to people who drove skittish horses, but now it is a fool horse indeed that would shy at a wheel. In less than a year the auto will have become a very common sight upon the highways in even the remote backwoods sections. Indeed even today the chauffeurs rather seek the deserted roads and leafy country lanes. The automobile clubs of larger cities have planned a series of runs through the state and Centralia is on the route of possibly three of these clubs and we may expect to see the latest in autos during this season.*

A week earlier in the May 5 edition, the city fathers had acknowledged the horseless carriages presence and the need for control of what some would suggest was the latest nuisance. "Centralia has passed an automobile ordinance. We are fast assuming metropolitan airs." The idea that Centralia was quickly following the lead of advancements common to a St. Louis or Kansas City was a source of pride for the Rodemyres and A. Bishop Chance.

A. Bishop wasn't the only Chance making noise as young Grover (Garrison's son by his second marriage) may have preceded the mold of Buford (the A. B. Chance's eldest son) by getting his own personal set of wheels. Later in the summer the *Guard* recorded on August 4, 1905, that even with only two wheels, Grover's newfound mode of transportation had certain unwelcome capabilities equal to the early automobiles: "Grover Chance has his motor cycle in running order, and can make about as much noise, speed and smell as any full grown auto." A. Bishop must have felt public pressure against the automobile, especially since he was a prominent Christian Church member, and his horseless carriage touring was not entirely socially acceptable in that early time. To counter some of the objections, one of which was noise, the September 29, 1905 paper had an interesting news item: "A. B. Chance has ordered a new automobile which will be in the first of next week. The new machine is a Reliance and is almost noiseless." However, had the Reliance also conquered the smell problem? To show the resistance toward the internal combustion engine even for newspapers, the *Madison Times* in the December 1904 period had to be printed in the Timbrook Blacksmith Shop because their landlord didn't want the *Times* new press engine in their building.

Like other discoveries, this second auto is new information. The idea of Bishop Chance having the first Centralia auto, however, has become suspect according to a front-page editorial on the May 12, 1905 paper entitled "Herbert Hutchinson Doing Paris." This is not Paris, France, but a neighboring Missouri town and neither was it a *Guard* exclusive, but instead gleaned from the *Appeal* and *Mercury*:

> *Herbert Hutchi(n)son of Centralia is still here with his automobile and the boys and girls who fancy something a little faster than old Dobbin and the family surrey are trying to persuade him to remain. He threatens to do so, being fairly well pleased himself; having had the pleasure of teaching a score of pretty Paris girls to become expert chauffeurs. And who wouldn't be pleased? "Hutch," as his friends call him, is certainly a most affable, accommodating and gentlemanly young man. (Paris Mercury)*

The disposition of the Chance automobile had kind of an unproductive and unexciting ending when traded for some 160 acres of land in the Texas Panhandle. It did not appear to be swampland, but neither did it produce any gushers. Gano recalled that his father acquired the tract of real estate in exchange for the car:

> *After how many years we had it, someone in Texas traded him the Texas land for the car. Of course we were told there were great possibilities of oil and we got it and paid taxes on it many years. Nothing ever happened. (1978)*

Oil fever may have first been encountered by area residents right in Centralia. Granny Ryman told of oil that occasionally oozed out of a hole on 8 Mile Lane (Highway Z), in south Centralia near the high school. This intrigued the local kids enough to go out and throw rocks in the hole trying to detect how far down the oil and or water level was located. On September 21, 1906, the *Guard* ran a report that may connect all of this together. "A. B. Chance, F. S. Sappington, Chas. Meffert and Mr. LeGrasse left Tuesday for Pecos, Texas on a prospecting trip." Although it may have taken a couple of years on August 21, 1908, the *Guard* confirmed what Gano said: "A. B. Chance bought a farm in the Panhandle, this week, giving his automobile in exchange." It's not clear if A. Bishop traded in his first car on the Reliance or kept both.

One of the biggest prizes for townspeople would have been the discovery of oil and gas deposits. As the oil/gas fever hit Centralia in July 1909 there was talk of the possibilities of the doubling of the population overnight and increases in land values. Noting the previous involvement and apparent fascination of A. Bishop Chance and others in Centralia with discovering oil deposits, this was a natural consequence. Like the telephone and electricity, the petroleum energy explosion was an overnight sensation in America, as the popularity of the automobile and other internal combustion sources grew. A group, "The Centralia Development Corporation," was organized to pursue local oil exploration and a well-digging outfit was moved in during 1909 and pictured in late 1909 and on the front page of the March 1910 *Guard*. After the "sinking" of the shaft, the progress of its downward movement was monitored with little news to generate hope for a "gusher." When you have a lemon, the most promising alternative is to produce some lemonade. As the waterworks progressed, there was a need for a deep well and what a golden opportunity to utilize the oil-drilling rig in an attempt to hit a source of water. This was the conclusion of city administrators in March of 1910. The big concern—or maybe an outside hope—was the chance of hitting a gas or oil deposit. But instead they hit water and now Centralia had a new source of a most valuable commodity, water!

The Centralia waterworks was not without problems. Ironically, they did strike oil at a most inopportune time, when they instead expected more water. The local paper in April of 1915 said an obtrusion in the water pipe forced the use of air pressure to correct the problem. Oil was found to be the culprit, but it wasn't expected to be a paying proposition, just an annoyance. The year before, in May 1914, the city was losing water from a big leak or meter problems and water receipts were not covering operational/overhead costs. In 1916, the pump was found inadequate and a new one had to be obtained. It was difficult getting the necessary funding approved and administrators were reluctant to be responsible for the purchase of a new pump that might not perform the job adequately. It did prove more than up to its task in providing the convenience of running water to businesses and its citizenry, plus offering fire fighting protection. Although the Centralia Development Corporation (C.D.C.) was dissolved in the 1911 period and appears to have been a failure, without it Centralia may not have quickly gotten a deep well. So even in what appears to be a situation of failure, prominent C.D.C. officials A. Bishop Chance and Dr. R. L. Hope contributed to their local community.

Waterworks in towns and cities generated employment and boosted America's standard of living, health benefits, and economy. Engineer Linwood Hope, son of the longtime operators of the Centralia Rexall Drugstore, helped put in a waterworks in Madisonville, Kentuky. He later moved to Arkansas to accept a Missouri Pacific Railroad engineering position. But it wasn't long that Centralia water demand had outgrown the pump equipment again. The February 9, 1917 *Guard* reported the volume of water being used, in slightly over a two-month period, was dominated by the usage of the C&A Railroad. Residents used about a third of that total, not including fire and other public uses. On the front page of the March 9, 1917 *Guard*, immediately below the article about the possible Chance plant, was news of the city purchasing a new compressor to backup current equipment. This was amid comments from the Wabash, a potential new customer, tiring from cleaning mud out of their engine boilers, especially the Columbia Branch engine utilizing water pumped from their lake east of town. By mid-September, the Wabash started using city water.

There was another Chance involvement in the oil business that coincides with the 1937 Tips Tool acquisition. When Gano thought of the Taylorville, Illinois area, he remembered their Illinois oil investments and cousin Charlie Kanatzer:

> *I got interested in the oil business over there because my father's cousin, whose name was Kanatzer, who used to live here, would write letters to my father saying, "come on over here, there is oil here." My father finally told me, "Gano go over and see what Cousin Charlie wants." Well I went over there and Charlie just knew there was oil there, just had to be. Charlie had been in the oil business in Texas and lost his shirt. He lost his shirt again in the bank in 1932 and this was about 1937, but Charlie was sure there was oil there. Well really in the interest of Charlie's enthusiasm, I didn't believe there was anything to it. I said "OK Charlie, what is your deal" and he said, "well I will do all the leg work to find out where all the major oil companies are leasing because I am buying some for them*

> *and you buy a chunk." I talked to them and they said it is all right. Well I said, "all right Charlie, here is $500.00 to start with, see what you can do." He started buying leases and royalties. It finally grew of course, we had about $3,000 or $3,500 invested in it, and beyond that we being the principle independent investor in that area. Well, it was under the name of my father, A. B. Chance of Tulsa, Oklahoma, and the oil companies then began to try to find out who is this guy A. B. Chance of Tulsa, Oklahoma. They could not find him. They finally found A. B. Chance of Centralia. So my father sent me over to the county seat of Illinois to see the Carter Oil Company that is the operating division of Standard. So I took a lawyer and went over and they wanted to buy a Texas strip of land, 110 acres which we had a lease on the right to drill. We did not intend to drill but we thought we might sell it to somebody that would. The royalty is part of the original owner's 1/8. When you own land you cannot sell more than 7/8 to the driller. 1/8 is yours, and then you can take and sell that if you want to but the driller cannot buy more than 7/8 of your land. So the 1/8 is called royalty. Then when the Carter Oil Company found A. B. Chance Company of Missouri I went over there and worked out a deal with Carter Oil in which they put the land in the contract to drill within a mile and a half of its borders, within 90 days and go down 3,000 feet. (1978)*

As it turned out, Gano recalled that they, "sold the right to drill for $6,750 which covered all we put in it plus we were able to buy some royalties." (1978) So it would appear that the Chance family came out winners in this oil venture owing to the wheeling and dealings of their cousin, the insertion of Chance capital, and the persistence and daring that permeated the Chance spirit in seeing and supporting things through to a conclusion.

On March 8, 1907, the *Guard* recorded yet another prize won by the A. Bishop Chance family:

> *Mr. A. B. Chance of this city received notice this week that he had been awarded a piano in the recent guessing contest of Hearst's Home & Farm. The contest was a dot guessing scheme similar to the one in which Mr. Chance won an automobile in 1904.*

One difference between this announcement and the earlier automobile contest was A. Bishop became the designated winner. It also stated that he was the winner of the earlier auto, which confirms what Gano earlier suggested. The positive approach that surrounded the spirit of A. Bishop would serve him well as he entered the difficult task of start-up and growing the Chance Manufacturing Company.

One of the most coveted prizes for those who place their name before the public, is the election to a political office. The February 19, 1909 *Guard* announced the candidacy of Mr. A. B. Chance for mayor of Centralia. In an unusual move, they were throwing their support to a city office candidate. They cited ties to Mr. Chance through a "warm and longstanding friendship; when he happens to be what we consider a model man for the place, and last but not least a consistent Christian gentleman." Solving problems in the implementation of the waterworks proposal and bringing together various factions within Centralia were cited as benefits that Mr. Chance would bring. The local business people were said to be pro-Chance. He was described as follows: "The name of A. B. Chance stands in Centralia as a synonym for progressiveness." The successes of A. B. Chance had led to his acceptance by the community. He was also much more traveled and experienced and certainly any naiveness that was commonly associated with his early years tended to evaporate as his life evolved. Partially because of the *Guard*'s support of A. Bishop's candidacy,

Stumped: In a 1909 paper, a comment was made about A. B. Chance running like a scared rabbit in the election for mayor. This photo of Chance in the hollow of a tree stump paints a picture for us to absorb. (*Chanco News* Special Edition 1949)

they expounded on his virtues, especially his business inventiveness and leadership prowess. They suggested that with "his own indomitable spirit, native ability and grit..." he had developed Chance Manufacturing Company. A February 26 newspaper piece continued the enthusiasm for his candidacy:

> *as there is everywhere an expression of satisfaction and rejoicing over the announcement of his candidacy. Mr. Chance is making a regular canvass for the office and he is going after it in the same forceful manner that has always characterized him in his own business. There is no lukewarmness or half-heartedness about Mr. Chance and he is in this race to win.*

On April 9 the paper made a comment about A. B. Chance running like a scared rabbit in the election for mayor. A valuable picture of the kind of man he was is painted for us to absorb just as the electric theater showing films later in October 1909 burst upon the scene. A year later in May of 1910, his return is covered after spending three weeks in Kansas City, likely at his "stock sales department" he had established there in late 1909 or early 1910. It may have been beneficial for his business interests to have absorbed much of his time. Like the famed *Guard* editor and mayor before him, Mayor Chance said such things on July 23, 1909, as "boys and girls shouldn't be raised on the city streets." When he decided in August of 1909 to enforce city ordinances to the letter, the repercussions ended in the resignation of two marshals in rapid-fire succession. In October 1909 the local paper covered a meeting of the state and surrounding states mayors in St. Louis, attended by Mayor Chance. He ran into the usual opposition in trying to clean up Centralia and even took one day off to personally pick and fix up the alley area behind the Globe. Adam Rodemyre would have loved this as such activities were ones he had also earlier campaigned for. Confronting the trash problem brought A. Bishop's innovative mind to action and four trash containers were made up and placed on the Centralia streets. One Chance Company rule was when problems were presented, it was also expected that one or more solutions accompany the presentation. As progress on the waterworks development rushed forward in 1910 (digging of sewer/water lines), a locked gate into a cornfield was not going to halt the forward momentum. Mayor Chance had the lock broken and the steam digger entered to carry out its assignment. Later that year in December, Dr. Hampton, the owner of the land, brought suit for damages. A. Bishop's straight-line approach sometimes brought challenges.

Roads and automobiles were making progress in 1910. Automobiles were becoming so popular in Centralia that A. Bishop and other members of the local motoring public began their own auto club in 1911. Auto trips by local residents became adventures that gained recognition in news items of the *Fireside Guard*. On September 13, 1912, when Gano was seven, the paper covered one fall trip, which may have been instigated to see the color of the fall foliage south of Columbia along the Missouri River: "Mr. and Mrs. A. B. Chance, Mr. and Mrs. R. Pool and Mrs. T. L. McBride, motored over to Jefferson City Sunday." The Pools were next-door neighbors to the Chances three years later when one of the Pool's outbuildings caught fire, destroying both their properties. A November 22, 1918 article told of T. L. McBride selling his ice company and in December they moved to Kansas City. The earlier trip that Gano remembered to Jefferson City appeared to have been a solo effort. The October 4, 1912 paper carried a trip by Mrs. A. B. Chance to Kansas City to visit her mother, Mrs. Gano. But this trip was apparently by rail, which was still king in speed and convenience of traveling long distances. This also confirms that Mrs. Gano had not yet come to live in Centralia. During those times such things as highway route signs were nonexistent and the North State Highway through the Centralia area was labeled "The Three Ring Route!" The telephone came into play by allowing the third pole on the road side line to be continually marked with three white rings, which allowed motorists the convenience of knowing they hadn't made a wrong turn.

Advancements can bring the need for regulations when they endanger the welfare of the populace. The paper on February 25, 1913, brought news of a new Centralia ordinance, which would address auto speed and noise issues along with an opportunity to generate a little revenue through licensing. This would help pay for the new streetlights that were said to be more efficient than the old ones. The ordinance would hopefully help local authorities deal with the phantom auto operator who sped through town on occasion nearly running over unsuspecting pedestrians and spooking horses. This surely wasn't a former mayor speeding to reattach a downed line or to fix a broken plug in one of the switchboards of a telephone exchange. This sounds more like one of the younger crowd, such as Uncle Grover mimicking the exploits of Barney Oldfield, who died in an auto race accident in late 1913. There were a lot of car accidents with serious injuries resulting even to the regular folks. Turnovers were common when corners were missed and cars ended up in a ditch. Operator inexperience, bad roads, and poor handling vehicles all added to the problem. Roads were a big issue everywhere with talk of the Lincoln Highway taking center stage nationally, and even the dragging and blading to smooth up area roads was newsworthy to the local paper.

In Centralia, Mayor Bruton and the Commercial Club President A. Bishop Chance became allies, if only temporarily, in advocating that the North State Highway not go through Sturgeon, unless those folks improved the road above that of the north or "Lawson" route. The west (Sturgeon route) appears to be the highway running west of Centralia along the Wabash into Sturgeon. The North Route appears to have been the old state road that ran north and then west by the Chance Farm or Sunnydale. The route from Mexico through Thompson into Centralia, or the North Route, apparently did not cross the railroad tracks as the railroads desired, because of the numerous crossing accidents occurring. But a controversy arose when Mexico and Columbia interests changed the north route, bypassing Centralia to the south, from Mexico to Columbia. This party's painting of the three rings on telephone poles south out of Mexico led to an entourage of Centralians, in two cars, who set out to paint or repaint the poles, leading motorists instead through the Prairie Queen. The fights over which town was going to get a rail line had now turned to struggles over who was going to get the state highways. Amongst all the turmoil, the importance that Mr. Gary's leadership in his bipartisan efforts to bring sanity and logic to Missouri's highway system a few years later, looms very important indeed.

"On-Line" Touching America and the World

(CHANCE MANUFACTURING COMPANY)

The great adventure and grand voyage of Chance Company's trip through the twentieth century would begin in what is an unlikely setting. Located in a sometimes wet and swampy area, not unlike Chicago, surrounded by a yet-to-be-tamed prairie area, the location of Centralia was marred with difficulties and its survival threatened time and time again. The birth of Chance Company, born in a spirit of boldness and optimism in the 1905 time frame, would also be tested. At first insecurely confined by the City Square, then amidst a call for relocation, it eventually would become an island unto itself, located between and restricted by the steel rails to its north and south. It would prosper amid the many challenges of an uncertain future and like the railroads early influences; Chance Company would open the doors of the Centralia area to the world. Distancing itself further and further from its early struggle for recognition, stability, and credibility, the world would eventually beat a path to Centralia, Missouri, U.S.A., and everyone would benefit as a result.

Few industries, institutions, or firms can boast of touching all of the population occasionally or even over a substantial period of time. In contrast, hardly a day passes that electricity doesn't touch all Americans during a twenty-four-hour period. Our heating and cooling systems, refrigerators, and our not-so-admired alarm clocks, all work even while we sleep. Along that electrical distribution line, supplying power to wherever a person might be, odds favor the possibility that many of the vital components on that line carry a Chance® trademark. This is awesome to touch everyone, twenty-four hours a day, everyday, even if indirectly, except for say a wilderness camper.

What is equally remarkable is that Chance Company is not a household word and never has been. Even local people scratch their head in complete puzzlement about "what do you all make down there?" So although Chance employees help put the glow in your light bulb, the heat in your curling iron, the rinse cycle in your washing machine, and a colored picture on your computer and TV screen, it is likely that the Chance firm will remain anonymous to the populace. Like the silent majority or even the mysterious nature of electricity itself, it can't be seen, but like Kilroy, we were and are there!

On-line in the early 1900s had a different meaning than it does a century later. In 1904 it could refer to the telephone and the lines that brought a direct communication link with local and distant family, friends, and business associates. In fact, it had become a vital tool in doing business, as a May 20, 1904 piece from the *Guard* suggested:

> *The value of a telephone in a country store is put at $10,000 by Porter & Porter, feed merchants at Carrollton, who last Friday filed suit in the circuit court of Livingston county, on a change of venue, against the Missouri Valley Telephone company, for that amount on the claim that the defendant removed the telephone from their store in Carrollton last September and refused to restore it when they had been patrons of the system and obeyed the reasonable rules of the company.*

As 1900 dawns, the telephone was still where A. Bishop Chance was concentrating his efforts and attention. But the event that captured the romance and spirit of that early time in a most vivid manner, especially with rail connections directly to St. Louis and its close proximity to Centralia, was the St. Louis World's Fair. Beginning in 1901, there was talk about the developmental work necessary to bring the Louisiana Purchase Exposition to the Gateway City. After several delays it would become a reality in 1904, as spring and early summer *Guard* issues displayed pictures of the magnificent buildings being completed to house the exhibits. The railroads were busy advertising special rates while the Wabash upgraded their track, which was raised four inches through Centralia. Extra trains were put on while the Alton installed block signals and added four new passenger trains between Kansas City and St. Louis. The *Guard* of course covered many visits by locals to the Fair including this one of July 22, 1904. "A. B. Chance and wife spent several days at the World's Fair in St. Louis." So it was "Meet You in St. Louis," a time when all eyes were on St. Louis until the Exposition closed in late November. Then the seemingly wasteful act of destroying most of what had amazed throngs of visitors, with some of the debris finding itself embedded in fills along the Wabash Railroad's right-of-way. One of the last structures razed was the giant 265-foot-high (twenty-six stories) Ferris wheel that had so captured the grandiosity of this progressive event. Like the Columbian Exposition in Chicago of 1893, the Louisiana Purchase Exposition gave the world a glimpse of the growth of the United States into a super power and the exposure to technological developments of the day that are still vital to our society, a century later.

Copies of the *Guard* from 1902 are missing from local archives, making it difficult to fully discern what was occurring in Centralia during that time. But coverage beginning again on March 27, 1903, reveals Mrs. A. P. Gano of Kansas City visiting her daughter in Centralia, Mrs. A. B. Chance. The favor was returned as on September 30 Mrs. Chance returns home after a month's visit with her mother in Kansas City, who had been widowed since 1902. A March 25, 1904 newspaper piece told of A. B. Chance, Dr. O. B. Mayes, and the Berg brothers leaving for Hot Springs, Arkansas, to remove any lingering winter chills in the warm spring waters of that resort town. Another longtime friend of Bishop Chance, prominent Centralia druggist R. L. Hope returned from Hot Springs in early April, indicating their presence had overlapped for a week or so. The heating systems of the day did not maintain even temperatures, especially at night, and travel other than by train was by means that exposed people to the elements, resulting in many respiratory problems that brought death to many during the cold, wet, and damp weather of the Midwestern winters. To counteract consumption and other dreaded life-threatening conditions, many remedies were established. Patent medicines, various home remedies, and the healing affects of hot spring waters were promoted and developed. There were prominent resorts like Hot Springs and Excelsior Springs, Missouri, but local historian Ralph Gerhard reported there were at least two such springs near Moberly that catered to those without the resources to travel to distant locations. These springs sprang up all across the country, their healing and curing of all types of ailments either real or imagined.

In 1913 there were ads run in the *Guard* by the Wabash Railroad, promoting the resort town of Excelsior Springs. To enhance ridership on their trains, it was common for railroads to promote various points of interest along their routes. For example, the Santa Fe brazenly suggested on their freight cars that they were the "Grand Canyon Line" and they built spur trackage to reach the Grand Canyon with their passenger trains. The Wabash likewise built a rail line near Missouri City, north up to Excelsior Springs. According to a news item in the November 24, 1911 *Guard*, the cost of putting in the Excelsior branch was richly rewarded according to an article captured from the *Excelsior Springs Call*. They reported that the Wabash had sold 280,000 tickets between Kansas City and the Springs that season. The *Guard* added that "this branch and the Centralia-Columbia Branch are two of the best paying pieces of track on the Wabash system." The health resort at Excelsior seemed to grow in popularity as it was mentioned frequently in the paper during the 1928 period. Although the Springs line was taken out many years ago, along Highway 10 and the old Wabash main line the placard still exists (2001) indicating the Excelsior Springs Junction. The old depot in Excelsior was reborn in the new millennium as a barbecue restaurant. Although A. Bishop went to Hot Springs early on, it was just a matter of time before he ended up at Excelsior, near Kansas City and his in-laws' former home. President Harry S. Truman, from neighboring Independence, was said to have stayed overnight in Excelsior at the Elms the night the *Chicago Tribune* was busy printing a paper with the headlines depicting Dewey as the winner of the 1948 presidential election!

The local newspaper led the charge for all sorts of crusades. It had been sniping at the tavern crowd for a long time and on March 31, 1905, they called for the closing of taverns on Sundays. There was a movement afoot in the state to do just that. Earlier on January 20 the *Guard* addressed the alcohol problem more directly:

> *The young men who get drunk every Saturday night and throw their hard earned money into the till of the rumseller are sowing naught but ruin. Don't do it boys, it is not manly and it will result only in evil.*

In the same issue they were pushing the need for waterworks to fight fires like the April 26, 1903 Allen Street fire which destroyed much of the business district. They also suggested that Centralia "can't get factories without waterworks." There was also a message to the young ladies of Centralia on May 29, 1903: "Complaint is being made that too many half grown girls are out on the streets, after dark, buggy riding and promenading with young men." Another of the early signs of the march forward by civilized society was the story of Johnny Appleseed spreading apple saplings far and wide, maybe even in Centralia. On June 30, 1905, the *Guard* described the difficulty residents in Centralia were having reaping benefits from growing their own food. "It is almost impossible to try to raise fruit in town. Boys rob and pilfer it. They are worse than the birds."

Adam Rodemyre had reduced responsibilty, as by 1904 his two sons E. T. and H. B. appear to be in charge of the *Fireside Guard* as they were the designated coeditors. E. T.'s obituary said the two sons bought out their father's newspaper interests in 1903. They apparently had problems with their father's remarriage, according to a friend, Clarence L. Northcutt. A news item of August 1904 commented about Mrs. (Tucker) Rodemyre traveling with a tent show in Kirksville. Because of her past strict religious views as expressed in local editorials, it was most likely a traveling evangelistic type group. According to the March 18, 1910 *Guard*, Mrs. Mattie W. (Tucker) Rodemyre passed away, apparently from peritonitis, at age forty-six. Because of frequent bouts with illness, it is suspected she didn't enjoy the best of health. Her parents, three brothers, and two sisters survived her.

Returning to a lighter moment during the times when Chance Company was about to become a reality, the calf-and-dog story evolves in the July 15, 1904 *Guard*. A calf was following a cow led by a boy along a downtown street. An unleashed dog bites the calf's leg, then the calf bolted through the plate-glass window of the Post Office. As the reporter concluded, "The question now is who pays for the glass." It's unclear if an all-encompassing leash law was then passed by the city fathers to cover such situations. This story has the earmarks of Mr. Adam Rodemyre, sixty-four, who passed away on July 27, 1906, at his home in west Centralia, after a lengthy illness that had prevented his presence at his beloved newspaper the last several months. He was a Christian by faith and was buried in the Centralia City Cemetery. His family suggested that a trust broken by a person he had trusted fifteen years earlier and other personal problems left a man who never fully recovered and who wasn't the same person they had earlier known. Adam Rodemyre was survived by his two sons E. T. and H. B., and his and Marie's oldest child and daughter Mrs. Lola V. Hays (1866–1919). Lola had ten children, seven girls and three boys, and was said to have become an author. Her death was reported in the January 3, 1919 paper and she was buried in the Centralia Cemetery. It seems appropriate to close the A. Rodemyre chapter of early Centralia history with the dog and calf story; although it may have been written by E. T., it depicts Adam's style of entertaining with the printed word while at the same time purposely picturing the rawness of a frontier town with a motive of bringing a kinder, more civilized and progressive local environment. He was a thoughtful, fearless, and principled man in his prime who loved and positively influenced the Centralia community. As the twentieth century begins to unfold, this was the basis from which Centralia and its people had and would continue to develop in such a special way.

Adam Rodemyre, editor of the *Centralia Fireside Guard,* passed away on July 27, 1906. (Centralia Historical Society)

The local paper was also capturing news about the Chance family. On August 28, 1903, A. B. Chance was found traveling to Gant on a Saturday for business purposes. A short time after winning the new automobile, this ad appeared on August 12, 1904: "Lost—An eagle watch charm in Centralia, Wednesday. Finder return to A. B. Chance and receive reward." In life you win and lose some. The long absence of news concerning John Chance was broken on October 14, 1904, when it was revealed his election as vice president of the Centralia Lodge of the Boone County Mutual Aid Society. A report on the Centralia Lodge of January 6, 1905, included both J. A. and A. B. Chance. Later that month confirmation was found of A. B. Chance's affiliation with the Christian Church: "A. B. Chance is the new superintendent of the Christian Sunday School." There was an emerging A. B. Chance who was becoming involved with the community in an official and service-minded manner. Grover Chance had been mentioned earlier in a trip to Bluffs, Illinois, in 1890 with his father and sister. Grover's mother was Miss Lurinda (Winscot) Chance (1837–1883), the second wife of Garrison. Although younger than his nephew, A. Bishop, he wasn't going to be outdone by his nephew's new automobile as he had purchased a motorcycle in the summer of 1905. Shortly after that he was injured per the August 29 *Guard*: "Grover Chance fell from a telephone [pole] north of town Monday and was badly hurt. He is able to be about again but his arm is in a sling." It is not clear if he was working for the newly formed Home Telephone with his half brother, manager J. A. (John), or one of the other exchanges. Mrs. A. B. Chance was visiting in Kansas City in early October after reports of being ill in May. In another legal case, this one concerning a suit brought by a Dr. Edmondson against J. A. Chance, the November 3, 1905 *Guard* carried a half-column story of J. A.'s victory against medical services transferred to another doctor and not rendered by Dr. Edmondson. It involved a cancer medicine treatment that apparently cured him. On February 2, 1906, the local paper found "A Bishop Chance was in Higbee and Clark last Friday on business." But neither was the patriarch of the Chance family inactive, according to a trip the *Guard* captured in that same issue: "Garrison H. Chance hale and hearty though 83 years of age, is visiting his elder brother and sister at Bluffs, Illinois." This further suggests that the Bluffs area was where a contingent of the Chance family still resided. Two weeks later on a Monday, A. B. was in Columbia on business.

Verification of A. B. Chance's out-of-state switchboard work has been weak, but local newspaper coverage of Centralia telephone activities has been rich in content and diversity. Advantageously, this became vital in tracking the start-up of Chance Manufacturing Company. From a somewhat state of normalcy to a seemingly chaotic sale and reorganization of area telephone companies in the 1904–06 time frame, was gleaned information about the direction A. Bishop could pursue. This important all encompassing piece in the February 10, 1905 *Guard* tells all:

> *The Centralia Telephone Co., has sold out and merged into "The Home Telephone Co," a company composed of some of the leading business men and farmers of Centralia. The new company will take charge of the system February 15th. The new company has been incorporated under the State Laws of Missouri with $60,000 capital. The work of reconstruction will be continued and all metal circuit put in. Mr. Chance, we understand will turn the plumbing shop into a manufactory of telephones and supplies. He has the Western agency of the United States for the Ericsson Telephone attachments, made in Stockholm, Sweden, and will handle these goods at wholesale.*

The most important revelation in this material is the decision by A. Bishop Chance to begin the Chance Manufacturing Company in 1905. Of course verification that this actually occurred immediately would be necessary in stating that A. B. Chance Company was two years older than the 1907 date that was chosen. It is a certainty that A. Bishop was building telephone products in a shop or at his residence even prior to that time. There is some question according to Gano, about when the Chance Company came into being but 1907 was chosen:

> *My father started so gradual that it is very difficult to find an actual date. . . . It is difficult to separate the telephone company from the start of the Chance Company because they were all in the same building with practically the same employees. (1978)*

Gano added that the telephone business required repairmen and a shop but at the same time,

> *he was making, assembling, and selling lightening arresters so some of the same fellows were working in both areas. . . . So we finally decided, it was about 1907 when he got enough business to say he was in the manufacturing business as well as the telephone business. (1978)*

Another item in the early Chance business operations was the plumbing activities. A special November 25, 1904 *Guard* issue spotlighted Centralia businesses including the Centralia Telephone Company. Describing that operation, it closed by noting their involvement in a plumbing business where all kinds of steam/water heating work were performed. This could have anticipated business opportunities when Centralia got its waterworks, but its approval by residents wasn't until 1907. The plumbing operation surely had employees and shop space that was shared with the telephone manufacturing and maintenance operations. The Chance plumbing business could be construed as the true forerunner of the Chance Manufacturing Company! Another revelation is verification of interview material about the Chance use and sale of Swedish telephone parts and the specific identity of whom A. Bishop represented. The telephone information is also important as it indicates that the Chance's sole ownership of the local exchange ceased and a group of local investors, including the Chances, became the initial owners of Home Telephone. It also confirms the late date of Home Telephone's organization, an offshoot of the Centralia Telephone Exchange, begun by A. Bishop Chance, with or without J. A. Chance's involvement.

The economic impact of the two Centralia railroads, such as the frequent carloads of livestock being shipped out, reflects a strong local agricultural base, but the push for factories to locate in Centralia had not ceased at the *Guard*. An August 11, 1905 piece praised the laurels of several small plants as beneficial because obviously the larger plants were more difficult for small towns to obtain. They also suggested that local business people should ban together and develop an improvement club for the advancement of the community and in securing factories. J. A. Chance described such a group as a Commercial Club who sees to it that their town is not left behind. An October 27, 1905 editorial told of John's recent trip to Vinita, I.T. (Indian Territory, later Oklahoma) where they had just completed a waterworks/sewage system, benefited by their Commercial Club. A key to where Vinita was located was found later, as Garrison had membership in a Vinita, Oklahoma Christian Church, although he did not move to Oklahoma until 1908. So John was visiting members of his family (brothers/sisters) living there as no visits by he or the A. B. Chance family are verified after Garrison moved to Oklahoma. The July 20, 1906 paper told of the organization of a Centralia Commercial Club, and although there were no Chances named as officials, it appears that the influence of John and A. Bishop was present in making it a reality. The group immediately began to push for waterworks and electric lights according to the November 16, 1906 paper. It was felt these developments were necessary to bring industry to Centralia.

It was more realistic for Centralia to focus on attracting small firms, as larger ones were concentrated in metropolitan areas because of their advantages over smaller towns. The benefits of the two railroads could only take Centralia so far. Small businesses can grow into large ones and during the twentieth century this would occur time and time again across America in some of the most unexpected locations. The owner couldn't always sell their vision for their enterprises to local residents and it is not clear how area residents responded to A. Bishop Chance's decision to enter into the manufacturing arena in 1905. It had more organization than production at that early stage, but things would begin popping in a few short years. But the Chance Manufacturing Company wasn't the only action in town as both a canning and broom factory were attracted to Centralia. First the canning factory machinery arrived in mid-June of 1907. It became operational by late August and a run of one hundred cans of tomatoes in four hours was a successful test of its intended purpose and capabilities. Locals were chosen for board positions including four directors, one of which was A. B. Chance. The canning factory was apparently a seasonal venture when the tomatoes were ripe. Only tomatoes are mentioned, and it was to start back up in August of 1908 with about fifty temporary employees. The Wabash was optimistic as they built a spur southwest to the plant and some people advocated the growing of tomatoes to locally supply the plant's needs. But it did not last as by April of 1910 there was a report of the firm having lost money every year of its operation and they had leftover, unsold tomatoes stored in town. The local paper covered its sale on January 6, 1911, by H. F. Winkelhake to Mr. Northcutt, a St. Louis dentist. It later caught fire. There was also talk of a broom factory in the October 1, 1907 paper that was sending feelers out to Mexico, Centralia, and Moberly. A July 1908 *Guard* piece said the new Centralia factory could turn out one hundred dozen brooms in a ten-hour day. By September, a broom factory in Moberly had apparently come upon hard times and A. L. Hulen bought their machinery to move to the Centralia plant. It was speculated this enlargement would increase employment over the twenty already working.

The question arises as to how much the *Guard* molded local public opinion. For example, many principles that the Chance family passed down were associated with ideas and opinions that were expressed in that newspaper, although this reflected what the community believed and stood for. There are regional beliefs then and today which can be very influential. In the January 18, 1907 *Guard*, an editorial that could be labeled "Out of Town Shopping," had some stinging remarks that were even prophetic. The watchword of local businessmen to spend your money locally, still applies today. The article posed the question if residents really wanted factories in Centralia. They asked if the small industries and businesses already in town were being faithfully patronized by locals:

> *Do we patronize and encourage those industries, however small, that contribute so much to our material wealth or do we despise the small home man and dump our home-made wealth into the lap of the big Eastern concerns, and wait for years for the money to come back to us through the constantly narrowing channels of agricultural and stock produce raised in and around Boone County.*

The economic power based in the East has had its impact on the Midwest and this was a concern some sixty years later in the sale of A. B. Chance Company.

As the article continued, another truth that emerged was the pressuring taking place in choosing to locate a firm in a particular location. Like lawsuits, it's surprising how "alive and well" these entities were at the turn of the century. One inducement that wouldn't cost the local community anything in attracting new industries was this idea of local prosperity. "The prosperity of the small factories and small industries will be the greatest factor in inducing good healthy factories to locate among us. Remember you have paid no bonus to the small industries here, and you owe them your encouragement." These comments relate to Chance Manufacturing Company and the involvement of the Chances in the Centralia Telephone Company. Accumulated information suggests how well versed A. B. Chance was becoming in various ownership arrangements, promotional schemes, and attempts of getting local economic support to maintain the Chance plant in Centralia. Any idea about J. A. and A. B. Chance being unaware or uninvolved in an unprofessional, "country boy" sort of way has long since evaporated. This isn't to brand them in a negative way but to demonstrate that it could be incorrect to label them as naive or inexperienced in business and legal dealings that would arise and challenge them in the future.

To support the idea that the Chances were gaining experience with the intricacies of business dealings, we turn first to their telephone business. It has been difficult to know what role J. A. Chance played as coverage of his telephone activities was minimal. A. Bishop has been noted as the manager and developer of the Centralia Telephone Company and his father's role is a clouded issue when he is finally mentioned in the summer of 1904. Later in November, J. A. is identified as president and A. Bishop as the secretary and manager. J. A. appeared to be a seasoned businessperson who wasn't as accessible as his son A. Bishop, so he received little newspaper coverage. Another reason could have been the so-called "Generation Gap" that Vice President Spiro Agnew made famous in 1970. The two Rodemyre sons were in A. Bishop's age group and there were comments of a youthful group of Centralians that were in all facets of community positions and activities, sealed with A. Bishop's election as mayor in 1909. By the turn of the century the Rodemyre sons were likely in charge of the daily operation and content of the *Guard*. They may have found a cordial A. Bishop open to contact with a desire to promote his telephone company and report progress to his local contingency of current and potential customers. Neither has confirmation been found of John's financial capability to support his son's telephone activities and growth, as other information tends to suggest. In fact, it appeared as if A. Bishop was on his own, but since J. A. was president in 1904 this wouldn't appear to be entirely true. Garrison was the only early Chance family member who suggested that he had significant financial resources.

Confirmation that J. A. Chance was involved with his son in the Centralia Telephone Company was revealed in a July 8, 1904 front-page editorial in the *Guard* concerning the conversion of it into a Farmers and Merchants Cooperation Exchange. It appeared that expansion demands requiring more capital was becoming a burden to the Chances, and so the cooperative plan was born. There would be $40,000 of capital stock and all subscribers would be eligible for purchase of stock. A week later on July 15, 1904, the paper carried some of the bylaws of the new cooperative and they again stated that it appeared all the stock would be sold within a few days and an organizational meeting would then be held. The following week of July 22, the *Guard* carried two short pieces praising the cooperative idea:

> *All the people of Centralia and vicinity should be interested in the cooperative plan exchange for there is no reason for it not being one of the best enterprises in Centralia according to the proposed plan of The Centralia Telephone Company. . . . The co-operative idea and plan of the Centralia Telephone Company is a good one and is bound to be a success both to the stock holders and the people in general as it will lead to the concentration of telephone interest.*

Unexpectedly, on February 10, 1905, the Centralia Telephone Company became the Home Telephone Company. A week later, the list of officers was revealed, which indicated that J. A. Chance's position of management continued to be above that of his son. The president was J. W. Bryson, the president of Farmers and Merchants Bank. A vice president, secretary, and treasurer were also named along with J. A. Chance as manager of the Centralia Exchange, and A. B. Chance as electrician. This didn't appear to be a very important position for Bishop, who appeared to be the driving force in the development of Centralia's first telephone exchange. But the position of electrician isn't technically surprising when considering A. Bishop's contributions to the equipment and plant aspects of the telephone industry. Although his switchboard work in distant places hasn't been discovered as was expected, there were a couple of local jobs that confirm such talents. The first concerns the announcement of the consolidation of the La Crosse switchboard with the Centralia Telephone Exchange on April 22, 1904. On September 23, 1904, the *Guard* confirmed the move had been completed to the former office of Dr. DeJarnett. Earlier on August 19, 1904, the paper had this to say about the new switchboard, that is very likely the result of the accumulated knowledge and skills of one A. Bishop Chance:

> *The new switchboard at the Centralia Telephone Company's Central, is a thing of beauty. It is finished in oak, has red and white lights which flash up at intervals when Central is called and indicate the line calling. The board has all the modern conveniences and will greatly simplify the work and increase the efficacy of the service. The LaCrosse Central will be moved to the Centralia Telephone Company's building and the two systems will be worked from the one Central, requiring the services of several "Hello girls."*

Utilizing Bishop's talents was thus exercised not only by his own firm, but also a competing exchange in the Centralia area. Fast-forwarding to January 15, 1907, the paper reported that A. B. Chance was in Middle Grove for two days testing out a switchboard for the Mutual Company of that place. This confirms he was performing switchboard work at other exchanges.

When examining the growth and business sophistication of the Chances, what is most remarkable about A. Bishop is his multiple talents. His technical, inventive side was displayed in growing the telephone networks in and around Centralia. His business prowess was reflected in aggressively pursued opportunities of the day, regardless if they were prizes in contests; get rich schemes, promotion of his inventions, or establishing his own manufacturing firm. The building and reorganization of the Chance telephone system was a prelude of what would occur in the Chance Manufacturing Company. This new venture appears to be solely A. Bishop's and wasn't shared with his father, as was the telephone exchange. The idea that Centralia would be its natural, automatic home is a commonly held, but apparently flawed view. Like other entrepreneurs of the day, A. Bishop attempted to get a new plant by promising to locate in the Centralia community. So the following news items may have convinced local folks to believe they might lose this opportunity if they didn't satisfy certain demands. To the advantage of Centralia, the Chance Manufacturing Company was already located and operating in town, although it may have been more difficult to sell the idea of expansion when there were a few or no employees solely dedicated to the manufacturing operation, without telephone or plumbing duties. But A. Bishop was capital short and to jump-start his company to bigger and better things, he needed help. So he began exploring the various ways of obtaining local or distant support. The *Guard* of June 12, 1908, told of one of his out-of-town expeditions: "A. B. Chance and Dr. O. B. Mayes arrived home from Fort Worth, Tex., Sunday, where they went to see about starting a manufacturing establishment. They were water-bound for several days." Dr. Mayes was a frequent companion and business ally to A. Bishop. Texas seems to have been a popular destination as it was for others during this time when oil exploration and overnight millionaires were being born. Business people tend to congregate where there is a healthy business climate and like California and Alaska, which had previously provided promises of gold, Texas and Oklahoma now held promises of black gold.

A. Bishop was busy making business trips to promote his company products and the company was gaining momentum and getting more aggressive, culminating in an announcement of a reorganization. The *Guard* carried the multiple-column announcements on November 12, 1909, entitled "Capitalized at $750,000.00." There would be $500,000 of stock offered to the public at a par value of $10 per share. The sale of these stocks had been placed in the hands of Dennis and Sweeney, stock financiers of St. Louis, who would put twenty men "in the field at once taking subscriptions." The balance of the article covered the products of the company. There wasn't a better way to make a big splash than this sign of determination displayed by A. Bishop Chance. Undoubtedly there were multiple reasons why the move was made including trying to sell the Centralia community on Chance Manufacturing Company's importance to the local area. A *Guard* editorial a week later on November 19, 1909, supported this:

> *A large factory with a big pay roll would be a valuable asset to the city. When you figure that the factory is bringing lots of money into the town and spending lots of money with her people, besides advertising the town to the world, you can see where the city would profit by keeping the factory of the Chance specialities here in Centralia.*

This subject must have been the talk of the town, but was Centralia large enough with the resources necessary to capture such a prize? Some may have concluded that this was a promotional scheme that's future lay down a road of uncertainty. But there were impressive examples of the far-reaching impact of the company, such as this pro-advertising piece carried by the *Guard* on October 22, 1909:

> *The Chance Manufacturing Company of this city, sent out a package of their Sky-Rocket® lightning arresters to Johannesburg, South Africa Thursday morning of this week. . . . Mr. Chance is an advertiser of the kind that believes in printer's ink and the orders that are coming in to him from Canada, British Columbia, South Africa, and other countries attest the merits of his advertising as well as the articles he manufactures.*

This reveals that Chance Company was involved in international sales, almost from its very beginning. The tension increased, however, with no firm commitments made by local interests in providing plant and ground for Chance Company. In addition another firm became involved, muddling the waters even further leading to an editorial comment soon after on December 3, 1909:

> *What will Centralia do about the two factory propositions, the Chance Manufacturing Co. and the Union Buggy Co.? We have no direct proposition as yet from either of these concerns, but we may as well get ready to make a light for them, as a great many towns will want these factories, if we do not.*

Centralia was still progressing as represented by the new and substantial First Baptist Church. It replaced one recently destroyed by fire and was dedicated on February 14, 1909. This was the location of this employee's first church membership.

As the year 1910 dawned there were some changes involving the direction of Chance Manufacturing Company and the so-called Kansas City connection. On January 21, 1910, a local news piece in the *Guard* carried this seemingly, unimportant item: "A. B. Chance was in Kansas City the first of the week attending to his stock sales department recently established there." This is significant information, with no apparent public announcement made. It is possible it could have been missed, but the way to get attention is sometimes to downplay it. On November 26, 1909, the *Guard* reported Mayor Chance had business in Kansas City the first of the week, which could have been associated with this new development. Because this wasn't necessarily a positive move for Centralians, it may not have been promoted. Does it show the resolve of the Chance Company founder to locate his firm out of town? It was a reality that all or part of the sales department of his new company was now located in Kansas City. The fact that it was recently established must point to a timetable that began at or before the big November announcement of company capitalization. The idea of a St. Louis stock firm and a Kansas City sales office doesn't seem logical on the surface, but it does reflect a broad base for the Chance firm. St. Louis was a greater financial center than was Kansas City, but the financial support of one or more parties in Kansas City likely led to a portion of the Chance Manufacturing Company being located there instead of St. Louis.

The Kansas City connection may have ended in 1914 when the major investor passed away. In Mr. A. Bishop's highlights document (See Chapter 7), there was a notation that raised intriguing questions: "1914: Take back of stocks sold in 1913." Within the handwritten notes that generated the highlights was additional information in brackets: "Chance Manufacturing Co., Phoenix, Ariz." In response to this, F. Gano Chance said:

> *That's where we all seem to have hardly any record. Apparently, in 1913, my father had already conceived the Never-Creep® and other items for the telephone business and this Kansas City financier saw the possibilities. They formed the company known as The Chance Manufacturing Company and they issued stock set up in the State of Arizona. Now, that checks out very well. Arizona became a state some time in 1912. So shortly after it became a State, evidently it was a good State to incorporate in, like New Jersey. So he set his company up in Kansas City and he hadn't been operating there for very many months when this fellow died. So he moves back to Centralia and he buys the stock of the little company that he had sold in Centralia. Dr. Mayes had some stock and four or five people had some shares of the stock. So he bought that and dissolved the corporation. (1978)*

Other local investors were said to be John A. Chance and a local banker. Notes from Mr. A. Bishop indicate an operation in Kansas City, either office or factory, or both. A catalog of approximately 3.5 by 5.5 inches was issued. What's even more remarkable about this information is the fact Chance Company was first incorporated in or around 1913, but because of the death of the Kansas City financier and the limited stock sales success, Mr. A. Bishop was back in control of basically a family operation by 1914!

The groundwork for a relationship between Mayes and Bishop Chance may have begun earlier in 1904 as reported by the March 25, 1904 *Guard*: "Ben Berg, A. B. Chance, and Dr. O. B. Mayes left Thursday for Hot Springs, Arkansas to spend several weeks." The next paper a week later on April 1 suggested that Mannie Berg also accompanied his brother Ben and the group was spending a month in this resort area. Jonas Berg (1848–1921) and Sons were Centralia businessmen who bought everything from scrap metal and hides to beeswax. On November 18, 1904, it was reported they had shipped twenty railroad cars of scrap metal to Pennsylvania mills. Jonas was a thirty-eight-year resident and started in business in 1883.

Share of stock. (Chance Foundation Files)

This information suggests that Mayes was a doctor and local advertisements stated he was a physician and surgeon who even fitted glasses. It also suggests that A. B. Chance began developing friendships early, which would prove helpful in later ventures. Per the August 30, 1918 newspaper, Dr. Mayes and his son Harold traveled to St. Louis. The death of the father of O. B., Dr. J. J. Mayes, eighty-four, was recorded in February 1920. He developed some of the early brick buildings west of the square and was associated with the Church.

Chance personal papers revealed a copy of a share of stock that was issued from the state of Arizona. It identified the corporation as Chance Manufacturing Company with capital stock set at $750,000. This is the same figure cited earlier in the 1909 corporate venture. This particular share was of $10 value each, which also corresponds to the November 1909 announcement. There was no assignee or owner, suggesting it was an example or an unsold share. There was only one signature, that of the secretary, O. B. Mayes. The president's line was unsigned and no specific date was assigned except "AD19" indicating it was in the 1900s. It was also No. 54, which could imply how few shares were outstanding. It is reproduced above.

The Chance Manufacturing Company was a mystery to those doing the company interviews, although this was also the name of the original Chance Company or A. B. Chance Company. The uncertainty, however, generated enough questions to have company attorney Jim Ausmus check into its background with the state of Arizona. Gano raised some of these issues:

> *Many corporations throughout the U.S. back in those days were incorporated in the state of New Jersey, I do not know. I am a little in conflict there when I looked up things in Arizona. If he incorporated in 1907-8-9 or 10, I would say it would be the State of Arizona. That is where my folks sent checks. There is not much known about the Chance Manufacturing Company and there ought to be, somewhere, minutes because of it. (1978)*

The name Chance Manufacturing Company appears to have originated in the 1905–07 period and continued to be the early name of A. Bishop Chance's manufacturing firm up into the early 1930s although the abbreviated "Chance Co." was frequently and more formally used following that time. Then Mr. A. Bishop's initials were added to make it the A. B. Chance Manufacturing Company until it officially became the A. B. Chance Company in August 1936.

The status of the sales of Chance Manufacturing Company stock remained unknown, as 1910 progressed. It could be assumed that it wasn't highly successful except for some local support and interests in Kansas City. A buy back and retrieving of stock in 1913 would suggest limited success and a desire to terminate and correct any outstanding remnants of that association. This opinion is supported by A. Bishop's renewed efforts by mid-year to get local support for a plant. A meeting of the shareholders of the Centralia Development Company was called for the purpose of discussing the well-drilling project, but it was also an opportunity for Mayor Chance to present the Chance Company proposition before closing the meeting. The *Guard* of June 24, 1910, covered the gist of his remarks:

> *They are about ready to build a factory and begin business on a scale commensurate with their importance in the electrical world. This factory will at first employ 35 men, but as business grows they will employ 200, possibly in less than two years.*

These were big numbers indicating an optimism that exceeded Chance Company performance until the 1920s. As it continued:

> *The factory will cost $25,000. The company will make a proposition or receive one. The company would like for a committee to be appointed to investigate the matter. Personally Mr. Chance prefers Centralia as the point of locating this factory, but it is a business proposition and the factory will go where the best offer is made.*

The last part of A. Bishop's reported remarks raises two thoughts. The first is the contrast with what actually occurred. It would be 1917 before a new plant was built and even then the Chance operation wasn't large enough to utilize all of its two-story space. One has to wonder how much this would have changed company history had his request been fulfilled in 1910. This is a reminder of the blessings that can come from unanswered prayers. Had A. Bishop got what he wanted and then failed, the A. B. Chance Company may not have become a reality. The other point is A. Bishop considered this a business proposition and had another town or city made an acceptable offer, he would have apparently located his plant there. There is an outside chance that he was bluffing, but maybe not. He had a powerful drive to creditably establish his firm with adequate facilities, but for some reason he was unable to get adequate or lasting support either locally or outside. This may have negatively impacted the growth of his operation. This is a sad situation in certain respects that reflects another of the many occasions when Mr. A. Bishop could have thrown up his arms in disgust and quit. Business people can face many challenges, but they only become successful by overcoming them.

In support of the increased activity at Chance Manufacturing Company, there were occasional

Big order (*Chanco News* Special Edition, 1949)

articles in the *Guard* about the firm obtaining significant customer orders. On July 15, 1910, an order from Canada was reported being received that week, which was "the largest single order for protection-apparatus ever placed by one electrical house with any factory in the world." The order was for five-thousand-plus Sky-Rockets®, which was the name of the first successful Chance product, a lightning arrestor. A reflection of the capacity of this small plant, although pressing, they reportedly would make up and ship this order in a week. The Canadians used these protective devices on government telephone and telegraph service lines and the reporter concluded that there was a connection in the earlier mentioned South Africa order, as both were under British rule at that time. The article continued by explaining the many items in the electrical protection area that Chance controlled and further explained its worldwide importance. All of this enhanced the position of the Chance firm in the eyes of Centralians. A news item three months later on October 21, 1910, told of the efforts of A. Bishop to secure local financial support for the expansion of his firm. A proposition was presented to the Commercial Club, which included a request to the city for a building and site, costing about $12,000. The city would deed this property to Chance after the payroll had reached $120,000 in a given length of time. Notice the building cost had been reduced to $12,000 from $25,000 in the June 1910 proposal. This indicates A. Bishop wasn't having much success in selling his company either in Centralia or elsewhere. It could have simply been the lack of capital, as the Chance operation seemed to be a growing and going concern, although at a level below earlier optimistic predictions. The *Guard* suggested the firm's local importance in an editorial that same issue:

> *The city has the proposition of the Chance Manufacturing Co., and from its liberal terms we are sure the city will be able to land this factory without seriously*

> *embarrassing anyone. Their present factory is located here and the proposition is not only to retain this industry, but to enlarge the business.*

By October 28, 1910, a Commercial Club Committee had come up with a plan to fulfill the proposition proposed by the Chance Manufacturing Company. The local paper carried the minutes of the meetings in which a plan was born to secure a tract of land adjoining Centralia, and divide it up into town lots for sale to the public. The key was for these lot sales to equal the $12,000 plant cost. The novelty of this plan was to allow those wanting to contribute to obtain financing for real property. The *Guard* explains it like this:

> *The lots could be sold for more than they are actually worth. The idea as suggested is a good one as many of those who could not afford to give a sum of money to the enterprise as a gift, could purchase one of these lots and in time at least get his money back if he did not make a profit. The land selected will be made known to the club shortly.*

This group had gone about as far as innovative thinking and planning could produce.

After plans to begin Chance Manufacturing Company were announced in 1905, apparently nothing occurred that was significant enough to appear in the *Guard,* although a telephone lightning arrester was advertised in the July 13, 1906 issue. It did not look like the three arresters pictured in Chance Patent material, and a H. B. Adams of Centralia was the agent for Boone County and W. White of Centralia was the agent for Audrain County, implying that it was a non-Chance item. Several months later, the February 22, 1907 paper carried a story entitled "A Centralia Invention" in which is shown a lightning arrester which would soon be given the brand name: Sky-Rocket®. The Chance Manufacturing Company would assemble the Sky-Rocket® that had been issued a patent and A. Bishop was said to be prepared to push and market his new product. He began doing that as a March 22 and 29, 1907 *Guard* news item indicated: "A. Bishop Chance left Saturday for Kansas City to attend a convention of telephone men. He is introducing his new invention, the Sky-Rocket® Lightning Arrestor and is meeting with much success." "A. B. Chance has returned from a trip to the convention at Sioux City, Iowa, and other points, where he has been advertising his new 'Sky-Rocket®' lightning arrestor." The key question that this material raises is what sellable products did the company have in 1905 when public acknowledgment was made of its establishment? One issue the 1907 introduction of the Sky-Rocket® adds credence to is the year chosen as the birth of Chance Manufacturing Company.

There are two acceptable spelling variations of *arrestor*: *arrestor* and *arrester*. Because some trademarks do have spelling variations of standard words to attract attention and for the public to more easily remember and identify them, this becomes a consideration in what is the correct spelling. *Arrestor* in this case seems to be a generic description of a number of these products on the market, as Sky-Rocket® was the trademark name given to the Chance version. *Arrester* will be used, however, as this is how it is described by Chance patent papers.

The lightning arrester was an important item and the most prominent product the first several years of the company's existence. People injured or killed by lightning is today associated with someone getting caught outside during an electrical storm. This seemed to be a more common occurrence in earlier times, possibly due to different weather conditions, the almost treeless early prairie area, or the fact that people walked more and spent more time outside. Today (2000), the grounding and protective devices used on telephone, cable TV and electrical lines dissipate lightning strikes so normally no human injury is sustained.

Lightning rods installed on buildings were an early attempt to protect structures, but in the late 1800s there were some unscrupulous salesmen that sometimes gave the product a black eye. But lightning was generating danger, especially after man began stringing metal wires and lines. On May 26, 1899, the *Guard* reported Paris (Missouri) had sixty-two telephones burned out during a storm! In August there was another report of a Huntsville man being killed by lightning while sitting on the edge of his bed. His employer had run a line from his house to his servant's quarters to operate a signaling bell. Even in the mid-century period, before some systems were modernized, stories were told of balls of fire coming off the old wall-mounted telephones. During a storm it was wise to stay off the phone and go to another room. So this should sufficiently whet our appetite as to what A. Bishop was marketing to help overcome the danger of telephone lines suddenly being energized by a bolt of lightning.

As the Centralia firm began to gather momentum in the 1908–09 period, it became more forthcoming about product information. A year after the Sky-Rocket® introduction, a January 24, 1908 article in the *Guard* covered a business trip to Chicago: "A. B. Chance was in Chicago this week demonstrating the utility and practicability of the Sky-Rocket® lightning arrestor in its several forms, his telephone test panel and other patents of the Chance Manufacturing Company." This suggests the possibility of more than one arrester and several other products. However, it would be some two years later, on November 12, 1909, before this was clarified and confirmed with some detail. The details of the expansion and promotional announcement were covered a few pages back but some of the product information revealed will prove helpful in exploring the Chance product picture:

> *The company will handle standard electrical and telephone specialties including the inventions of Mr. A. B. Chance and patents owned and controlled by him, comprising twenty or more of the most reliable of this line of specialties on the world's market. Mr. A. B. Chance is the creator of the pioneer or basic "Sky-Rocket®" lightning arrestor and its numerous family of electrical appliance protectors. He is the original inventor of the hotel annunciator system and still holds the patents to this contrivance.*

This scenario, of buying out competitors, although appearing a little far out for a hometown sort of entrepreneur, would become a familiar practice of A. Bishop. A July 15, 1910 *Guard* report on the large Canadian order confirms and gives further details about the Chance product line up:

> *This factory is engaged in turning out many lines of apparatus for lightning protection and electrical novelties and make shipments to various parts of the globe. They own the exclusive patents and rights to some of the latest and probably all of the best lightning protectors or arrestors on the world's market today and are also manufacturing many novelties and necessities in this field that stand at the head of the world's best today.*

This may represent the first acquisition by Chance that was expanded on the strength of acquiring products, whole product lines, and entire companies to broaden their offerings and to eliminate competition. This appears to be A. Bishop Chance's greatest hour to date, when the platform and business practices of Chance Company were being established and fine-tuned, besides his numerous displays of electrical wizardry. Note the multiple references to electrical specialties implying that the Tips acquisition was not the beginning of Chance Company's involvement in electrical products, although their market was more clearly defined as being the electrical utility industry in 1937.

Details confirming the early Chance product family are contained in three patent files.

The patented items are telephone related and were processed for Albert Bishop Chance of Centralia, Missouri, by patent attorney H. B. Wilson Company of Washington, D.C. Two patents were applied for on July 25, 1901, implying they both were likely handled on one trip to our national capitol. Patent No. 693,784 was for a "Transfer-Board for Telephone or Telegraph Circuits," issued on February 18, 1902. It would appear to be an item developed from Mr. Chance's switchboard work during that time. The second patent, No. 707,761 was for "Circuit-Closing Devices" and was issued on August 26, 1902. This was a "signaling apparatus for telephone systems." Another patent, No. 763,464, for the same-type device was filed on April 6, 1903, and the patent issued on June 28, 1904. All three of these patents were for a period of seventeen years and appear to be items developed during Mr. Chance's telephone and switchboard work. The filing text of the last patent application had two interesting witnesses—a Horace Chanslor Gano and Frances Chance (Mrs. A. B. Chance). Horace would appear to be a relative of Frances, possibly a brother, as her father had passed away in 1902.

The third patent related to arresters contained three separate patents, the last of which was filed by Albert Bishop Chance on September 12, 1906. This would appear to be the famous Sky-Rocket® Lightning Arrester given patent No. 894,723 on July 28, 1908. A patent document picturing the arrester is reproduced on page 90. It was said to be an "improved version" in a tubular body. It would have had to contain unique features as the two other arresters had older patents, originated by two other inventors! Possession of these non-Chance patents confirms the *Guard* article reference of Chance Manufacturing Company acquiring and marketing other arrester designs. The attorney for the Chance arrester was a Victor J. Evans, with no address given. A Frank Hough and N. M. Bagger witnessed the engineering drawing in the filing, and F. M. Burk and C. W. Settle witnessed the filing text on the patent application. These would appear to be local people, C. W. Settle being the secretary-treasurer of Home Telephone. The oldest lightning arrester of the three in the file reveals an application for a patent on March 23, 1900, and the seventeen-year patent No. 666,545 was issued on January 22, 1901. The inventor was William Norman McAnge of Suffolk, Virginia. An assignment by Mr. McAnge of his patent to Chance Manufacturing Company for $200 represents one, if not the earliest, Chance Company acquisition! It undoubtedly was also one of the least costly. The assignment wasn't made until October 4, 1909, which is later than expected suggesting at least two possibilities, an earlier agreement or the Sky-Rocket® was stealing the market. An examination of the three separate arresters in the Chance file indicates the McAnge product was a more complicated board arrangement with two arresters and assorted hardware. This would cost more to produce because of its many components and could be undercut in price by a simple version represented by the Sky-Rocket®. This is reflective of Mr. A. Bishop's business "savvy" of knowing that an equal or superior quality product that is simpler and less costly to manufacture, resulting in a lower selling price, would steal the market. Note this is during the same time that Henry Ford placed this principle into action with his Model T! It was a simple, but quality product priced so it was affordable by the masses. The second "Lightning Arrester" patent was applied for on May 9, 1905, and a seventeen-year patent No. 807,847 was issued on December 19, 1905. The inventor was Ernest Samuel Moorer of Anderson, South Carolina, from the same general areas as the first. All three were plugged into the same reasoning, a new and improved version of the lightning arrester. Incidentally, all three spelled arrester with the *e* instead of an *o*. No assignment was found in the file but the original copy of the patent, like Mr. McAnge's patent papers, were included in the Chance file suggesting Mr. A. Bishop had acquired or reached some binding

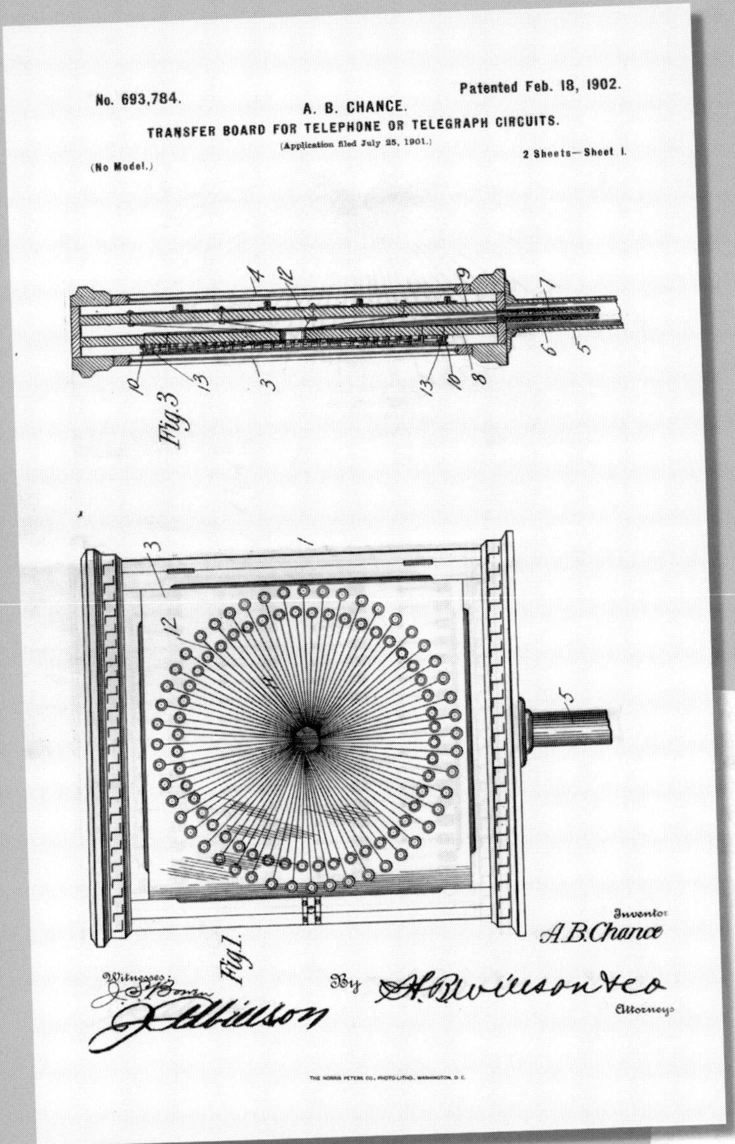

Tranfer board. (A. B. Chance Co.)

Circuit-closing device. (A. B. Chance Co.)

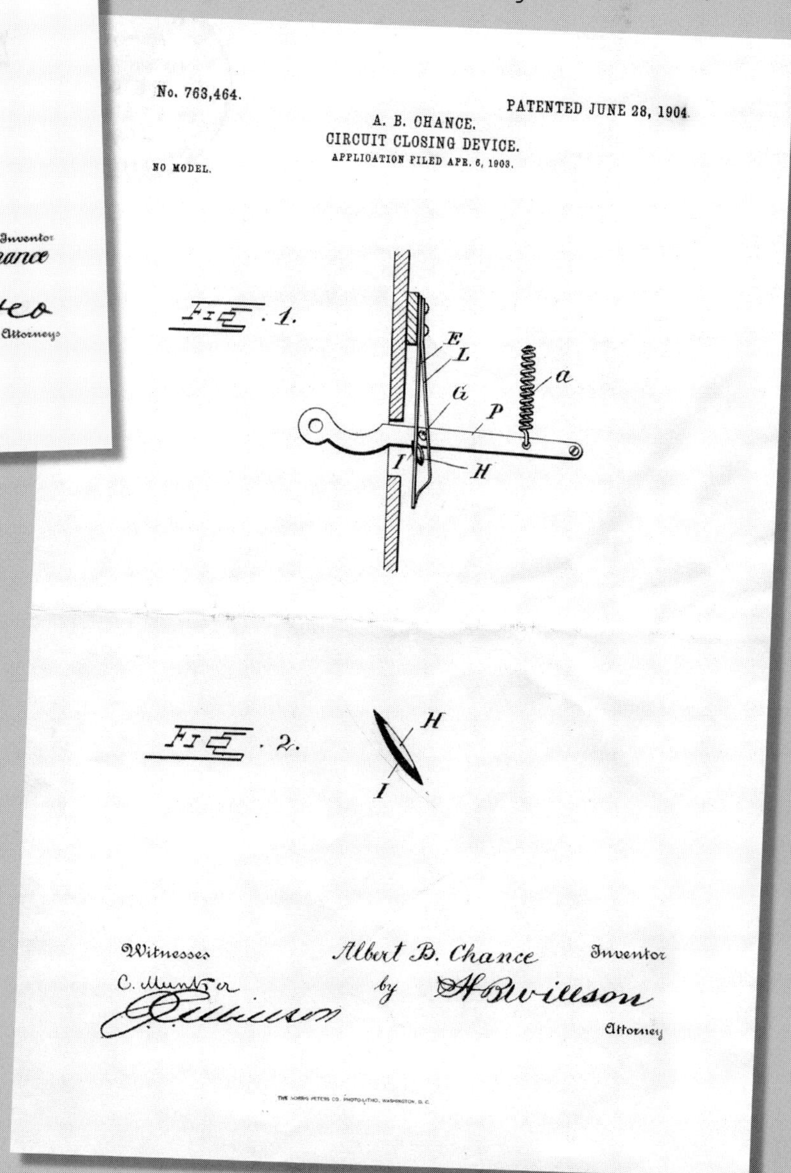

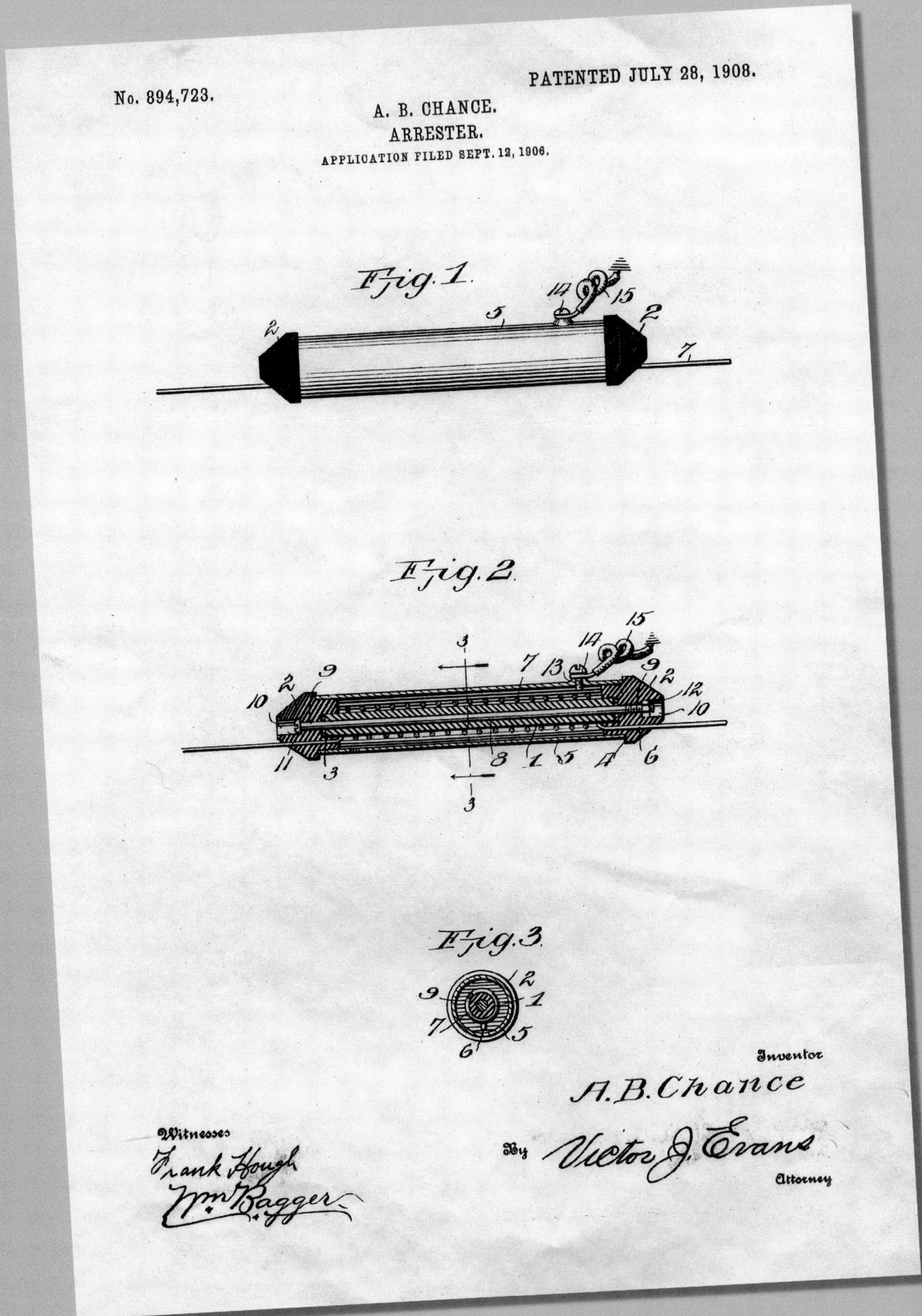

Patent No. 894,723. (A. B. Chance Co.)

agreement with the inventor. Although F. Gano Chance may be considered the father of the acquisition activity at Chance Company, it is clear that he was following in his father's footprints. Although the firms and products acquired generally grew in substance and cost during Gano's reign, he was taught well by the Chance family professor, The Centralia Wizard, Mr. Albert Bishop Chance!

Accumulated material continues to point to A. Bishop's genius in applying his skills in the development and sophistication of various items and also in facing and solving everyday problems. The July 27, 1906 *Guard* apologized to local telephone people, revealing the problems they routinely encountered with visitors who weren't satisfied with just looking at their many pieces of machinery and equipment:

> *The* Guard *editor feels that he owes the Home central girls an explanation. We have a new desk telephone and the ring-up crank is on the visitor's side of the desk. During the past two weeks it seems that every visitor who spied the crank was tempted to turn it, to the consternation of the editor. We are all right now as A. B. Chance, the wizard has hypnotized the crank into silence, and central will get an answer now when the drop falls. If the central girls in Centralia were not the most angelic tempered girls on earth the* GUARD*'s telephone would have been burned out.*

The use of "wizard" initiated the idea of bestowing on Mr. A. Bishop the title of "The Centralia Wizard!" His kind of talent was needed in San Francisco in April 1906 after the devastating earthquake.

Being service-minded was an A. Bishop Chance trait that was handed down to his company. Although some of the special programs later initiated and supported by A. Bishop would shock and awe you, they were examples of a pattern he established early. For example, the *Guard* announcement of a special program in their September 11, 1908 issue has the trademark of the Centralia Wizard: "A. B. Chance informs us that the Home Telephone Co. will shortly establish a bureau of information at their central. All patrons may get information regarding the weather, train time, markets, and all current events." And we have been led to believe the information age was established a century later! Several months after that, on April 23, 1909, the paper covered the planned installation of a new dynamo-generator by Home Telephone, which would pull power from storage cells instead of the older dry batteries. No doubt the progressiveness of their "Electrician" was responsible for updating their power source. The June 4, 1909 *Guard* confirmed this new system and the one responsible:

> *The Home Telephone Company have installed their new electric motor, dynamo and storage cells in their central office and will do away with the troublesome and unreliable dry cells. The new innovation will give the patrons of the Home Telephone Company better service as soon as the connections are made. In addition to the new arrangement, Mr. A. B. Chance, the electrician of the company, is planning many other improvements and a better switchboard service as the current is now strong enough to allow of the use of many features not hitherto used. The Home Company are not sparing trouble or expense to make the service good and if there is one thing Centralia has that she should be proud of, it is her most excellent telephone service.*

Gano suggested progress was being made in the small Chance operation, as the first significant separation of the employees from the telephone and manufacturing operations began: "When he built the telephone exchange up over the Missouri Power and Light Office." (1978) This is the building on the northwest corner of the

The first official home of Chance Manufacturing Company was in the old high-front former Christian Church building on the south side of the City Square. The Never-Creep® Anchor and Sky-Rocket® Lightning Arrester sign and fading trademark symbol, along with its empty appearance, suggest a post-1917 photo. (Jack Chance Collection)

City Square and the exchange was upstairs on the east end of that brick building. A fire in 1917 resulted in this move and the construction of this new exchange, but it appears that the separation of the telephone and manufacturing operations had already occurred, although as Gano suggests there was an overlap of employees and activities for several years. Gano said it was around 1912, when the Never-Creep® Anchor was invented, "his telephone exchange was up over the second floor, which is now Kinkead Drug Store." (1978) This is the Baker Building located mid-block on the east side of downtown Allen Street. This was the location of the exchange before the later north, City Square location, according to historical telephone sources and local material.

Amidst all the hopes and aspirations for a new factory, Chance Manufacturing Company's first real home base was instead a humble setting. The building that A. Bishop Chance occupied, as early as 1913, had seemingly outgrown its usefulness and was down on its luck. By 1905 Mr. Bush had relocated his business to his new brick corner building and used the old church for storage. Chance historical pictures show a tall-front wooden structure appearing like those in a frontier town depicted in a western movie. It would again gain notoriety but not for spiritual rebirths, but in becoming the pictorial birthplace of A. B. Chance Company. In speaking about the Chance business return to Centralia from Kansas City Gano remarked:

> *When he came back to Centralia, he rented the old Bush Building that M. S. Bush*

furniture store had and where he stored coffins. There were a lot of angles on that thing. We lived in a home down here which is now the Baptist Annex, which was a mortuary from 1914 to 1930. Why I laugh is that we moved out of our home and a mortuary undertaker took it over. I always said that Chances just pretty well stayed ahead of the undertakers. (1978)

The Bush Building was near the new Bush/McDonald furniture and undertaker business, later to become the *Guard* Office Building. According to a discussion by Gano and Jack Chance, one or both of these buildings were separated by a tall wooden fence. Gano recalled that between those two buildings was built a wooden fence about ten feet tall, and one of the older buildings became a nickelodeon. A comment in the July 19, 1895 *Guard* suggested that the old Christian Church building was occupied by M. S. Bush (1856–1926), which confirms Gano's remarks. Christian Church details, which appear to fit with accumulated information, were revealed in a special September 14, 1934 *Guard* feature of the history of the Centralia First Christian Church. The first church was built in 1875, which may not have been completed until early 1876. It was moved from the Post Office location in 1889 further west and used by Mr. Bush before becoming the first location for the Chance Manufacturing Company. The second church was built for $8,000 on the location of the first and dedicated in June 1890. A June 1925 *Guard* article stated it was torn down in 1910 after the new brick church was completed. Returning to the 1934 article, outgrowing the second building, the Christian membership began the construction of a much more substantial brick structure in 1910, costing $30,000. It was dedicated in May of 1911. Another interesting angle that may have involved the old church building was in 1896 A. Bishop occupied space in the "Bush Building" for his "general electrical supply depot." This would appear to be the same location where his manufacturing operation would be located over a decade later, or did he continually maintain space from 1896 to 1917? The April 28, 1905 *Guard* spoke of one of the most successful early Centralia businessmen, M. S. Bush, who established his furniture store in 1884. His "new store" (fifty by one hundred feet), a two-story brick building, appears to be the later *Guard* home on the southwest corner of the City Square. It is unclear where Mr. Bush's original store was from 1884 to 1889. A mid-year 1925 news article reported Mr. Bush's retirement and sale of his business to M. J. McDonald. Mr. Bush was the oldest Centralia businessman at that time.

Centralia businessman M. S. Bush owned the old Christian Church building that A. Bishop rented on at least two occasions. (Centralia Historical Society)

A home is an important part of a person's life as well as that of a business. This becomes an inviting subject in closing out the birth and early growth of the Chance Manufacturing Company. When LaCrosse Lumber decided to tear down and build a new, larger yard, a February 14, 1913 *Guard* article revealed the previous yard and the new one were located where the "first frame church building ever erected in Centralia" had been. The first church was of log construction, erected in 1854 and used by various denominations. Then the Methodists built a large frame church in 1868 on the 1913 LaCrosse lot that was again used by other denominations "until the Christians built a large frame building on the corner opposite from the City Hall on the southeast, or on Rollins and Sneed Streets. This building is still standing, but has been moved on the Bush and McCallister lots just west of the old site, and is occupied by the Chance Manufacturing Company as a factory building. The latter building was erected in 1871." Several previous *Guard* references suggest this church building was built in 1875–76, but otherwise this is valuable information in confirming that in February 1913 the Chance plant was located on the south side of the square in the old church building. Mr. A. Bishop Chance was at home each workday in a structure originally built for the gathering of Christian members, such as himself, but it now was the home of his fledgling firm.

This 1910-era photo pictures the Home Telephone staff then located in the Baker Building. John Chance is on the left and A. Bishop with Buford and Gano are on the right. Note the M. S. Bush poster, Emerson-style overhead fan and fancy heating stove in right forefront. (Jack Chance Collection)

Centralia Telephone Company and Home Telephone Company

Mr. Robert L. (Linwood) Hope, son of Robert Lee and Mary Downing Hope, revealed more information about the early Centralia telephone system:

> *The Home Telephone Company was over the drugstore which was operated for a time by a man named W. O. Baker on the ground floor. Upstairs was a little office which was occupied by Dr. Harley. Going on past his office, to the back was where two girls were working. There was a small bench where you made repairs or something like that. I remember going over there and using it a good deal to repair some toys. I was always making toys and it seemed to come natural to me. My father subscribed to the* Scientific American *magazine when I was about ten or twelve years old and it was just a plain magazine about toys and things, very elementary. (1979)*

Mr. McDaniel, the early Missouri telephone pioneer, also studied this magazine that covered early scientific development. The *Scientific American* ran advertisements in the *Guard* during 1897–98, for example, promoting the services they provided on securing patents, which Bishop Chance surely noted. Linwood further described the location of the Centralia exchange along side a ten-foot alley. Today these building features are still present with the W. O. Baker name high on the front of this mid-block, west-facing Allen Street brick building.

At the turn of the century A. Bishop Chance was struggling against an old foe that businessmen face: competition. The telephone business in and around Centralia was highly contested. The LaCrosse Lumber Company of Louisiana (Missouri) in June of 1900 had expanded further by buying Fayette Lumber, making eighteen total yards in the state. They were also said to be operating three hundred miles of telephone lines. By October the local paper reported that Manager Buffum of LaCrosse and the Louisiana Telephone Company now controlled over four hundred miles of lines. The innovative ways of Bishop Chance were put to the test and within the June 1, 1900 *Guard,* he didn't disappoint: "Bishop Chance, projected the election returns on a canvas screen in front of the telephone office, where the returns were received by wire. A large crowd availed themselves of the opportunity to get the early returns." But this was not enough to deter LaCrosse or even the Rodemyres whose support was also wavering. On June 15, 1900, they wrote: "Local telephone rates in Centralia are too high, being double those of Columbia and other towns." This may be the first acknowledgment that Columbia outdid Centralia on anything. To add further insult to injury the local paper on October 15, 1900, wrote another condescending piece:

> *The LaCrosse Telephone office has a neat and convenient desk phone which can be moved about or adjusted to any angle or degree to suit the user. Placed on a table or desk a person can sit down and carry on a conversation without any more inconvenience than if the person was in the room. The absence of whirring and singing sounds in the receiver makes the LaCrosse service a pleasant and satisfactory one to use.*

The truth can hurt and Bishop Chance was being christened early concerning pricing and the quality of his product, and was becoming alert to the fact that progress doesn't stand still.

Every year was bringing changes within the framework of the telephone business in and around Centralia. It was expected that there would eventually be a consolidation of what may have been at least three different phone companies in Centralia, or four if the new 1899 Bell St. Louis/Kansas City line is included. In the June 21, 1901 *Guard*, a piece from the *Sturgeon Leader* suggests the disposal and splitting up of what appears to be the late 1893 trunk line as described in Chapter 1.

> *The* Sturgeon Leader *says: The New Century Telephone Company in full partnership with the Centralia Telephone Company bought at trustee's sale in Kansas City last week the old Standard telephone line running from Mexico to Moberly. The Moberly, Fayette and Macon telephone companies also bought sections of the line running out of Moberly south and west. Arrangements were made to establish a co-operative toll line system over all the lines purchased by the parties above mentioned.*

The molding of the old Standard lines into the local systems generated further drama for Bishop Chance's involvement representing his Centralia Telephone Company. It is unclear who the New Century group were but it appears to be controlled by area Independents. A month later on August 9, 1901, after the Standard sale and purchase, the *Guard* reported: "Bishop Chance hired a buggy and a double team Wednesday to inspect the telephone line between Centralia and Renick. On his return one of the horses worth about $100 dropped dead." This may have been a business assignment for unknown parties or Bishop was inspecting the line for himself. The summer of 1901 was hot and dry with temperatures reported as high as 114 degrees. It was a drought year that later broke with late summer rains. Controversy was becoming a frequent companion of a hard-driving Bishop Chance. This incident is reminiscent of another event seven years earlier in April of 1894 when Bishop and two friends bicycled to Moberly and back on the same day. However, there is usually quite a difference in mid-Missouri heat and humidity levels between April and August. It is also easier to pace oneself than to know when an animal needs rest. This incident may have given impetus to Bishop becoming an early automobile owner and operator.

There was a sense that the *Guard*'s admiration of Bishop, dramatically fueled by his bicycle racing and early telephone developments, had subsided somewhat during 1899. The Johnson incident and the LaCrosse telephone progress revealed how intense the telephone competition in and around Centralia had become. For someone that was initially considered naive in legal matters, Bishop Chance appears to have quickly made up for lost time. The Johnson jury ruling in 1899 and the 1901 horse incident, both of which ended in court, support this premise. One surprise was that legal confrontations in this period were not uncommon. These local trials served as preparations for later contests against more seasoned foes that confronted Chance Company on a national scale. On November 22, 1901, the *Guard* revealed a lawsuit: "Dixon [Dickerson?], the livery man has sued Bishop Chance for killing a horse by fast driving. Lawyers Cupp and Dickey of Mexico are on the prosecution and H. S. Booth and T. S. Carter for the defense." A week later in the November 29 edition it was reported: "Josiah Hall, Justice of the Peace, this week awarded Mr. Dickerson $100 and costs against Bishop Chance for driving his horse to death. It seems that he drove the horse from Centralia to Moberly and back on the same day."

It was twenty-five miles between those two points or fifty miles for the round trip. The image of Bishop Chance at this juncture is not an innocent schoolboy showing affection toward his teacher by bringing an apple. Bishop had a firmer side that developed during a frontier childhood in which his father moved from place to place to obtain work, and a hard-driving pursuit of business success in young adulthood.

Like many aspects of life, the area telephone system was different in earlier times. It was a simple one-wire system that grouped households together into the celebrated "party line." A user never wanted to reveal any deep, dark secrets on the line as a distant but distinct click could mean there were, or had been uninvited guests also on the line. At the top of the pyramid of those who delved purposely or were exposed to the scuttlebutt impacting an area's social conscience, was the local operator. This was a time when the coffee shop group, barber, and hairdresser played second fiddle on the information hotline. Operators were professionals, however, who served their community in many important ways, from their privileged positions. One such function was paging. A future Chance employee, Virgil Blanton's mother, was one of the operators as Granny Ryman recalled. "You could call up on the phone and ask whoever the operator was when they were in the old Farmers and Merchants Bank, and they would look out the window." (1979) Nathan Toalson also suggested that as a youth your whereabouts and activities in the downtown area were likely being scrutinized by the prying eyes of a telephone operator: "I remember as far, as that it was back when they had the individual operators. If your mother or father wanted to find out anything about you they would call the operators and they would probably know where you were." (1979) From their vantage point there was some monitoring of the comings and goings of customer traffic down on the street. Granny said they would even call a store in attempting to locate someone for a caller. Can you imagine trying to find an operator today to call Hope's Drugstore to see if A. Bishop Chance was having his mid-morning coffee?

A Centralia native, Granny Ryman had a connection to the Hopes through his mother. Granny's father was a prominent farmer with a son and four daughters, and then he lost his wife. He remarried and Granny was the only child of his second marriage. Granny's mother was the Hopes' maid and lived with them, according to Granny, before she married Mr. Ryman. The Hopes' home is thought to be the A. B. Chance home, currently the site of the Centralia Museum. Mr. Hope and Mr. A. Bishop were close friends, especially in their later days in Florida. A news item in the April 16, 1915 *Guard* suggests that this marriage didn't last. "Mrs. R. L. Hope was in Columbia Monday as a witness in the Ryman divorce case. Mrs. Ryman was formerly employed in her home." Granny did not mention this in his interviews so it is unclear if he lived with his mother or was raised by his father. He would have been about eight years old at the time of the divorce. The *Guard* captured three Ryman news items in 1912, which suggest that F. G. Ryman was already living in Neosho Rapids, Kansas (southeast corner), as his brother R. F. Ryman from Jacksonville, Illinois, visited F. G. in Centralia in early June. This corresponds with Granny's comments about living a portion of his childhood in Kansas and suggests that F. G. was likely his father. Later in October, F. G. bought a farm near Clinton (Missouri) and "may move there from Neosho Rapids." This was some 150 miles northeast of Neosho County. In late December, F. G. was again visiting in Centralia so there were apparently strong Centralia connections.

In 1903 there were mixed signals continuing to surface about the quality of telephone service and its pricing to customers. This was a time when big business and unscrupulous power brokers were coming under attack. The August 1903 *Guard* suggested the telephone business paid big profits to the investors." In view of earlier comments about Columbia telephone rates being lower than Centralia, this may have negative connotations, but it also could be a positive, supportive kind of comment in growing a better telephone system. That is supported by another item in that same weekly paper: "The new telephone between Centralia and Tulip is considered one of the best in the state." At this point there would appear to be an end to development and change in the Centralia area's telephone systems, but nothing could be further from the truth. In March of 1904 the local paper revealed the Centralia directory was to be included in the St. Louis directory and Centralians would be connected to other large cities. That same *Guard* issue of March 11, 1904, confirmed yet another incoming system: "The Farmers Mutual Telephone Company from Hallsville to Centralia made application to come into Centralia. The Board met and decided to let them come in." This was followed a week later by almost a full column article on "The Telephone Situation Today." It was an article that supported the idea that telephone service had become not only a modern convenience but a necessary and very useful tool. However, the start-up structure of multiple exchanges in the same community had generated problems for subscribers in being able to reach parties on other exchanges. The government was now getting involved so that customers were not forced to subscribe to multiple exchanges with additional cost burdens or limitations in reaching other parties. In fact at that time, the state of Missouri had a law pending to further regulate and reduce local service to one exchange.

Telephone owners undoubtedly saw the handwriting on the wall and immediately began to take steps to be in compliance with potential new laws and to reduce or eliminate complaints on service issues. The LaCrosse group agreed to release the operation of their switchboard to the Centralia Telephone Exchange, as covered in an April 22, 1904 news item. An elaborate switchboard resulted, per an August 19, 1904 news article covered in the previous section. This was a preview of additional, significant realignments and even ownership changes. Driven by the threat of governmental intervention or capital shortages, the July 8, 1904 *Guard* covered an announcement by J. A. and A. B. Chance: "The Centralia Telephone Company have decided to convert the exchange into a co-operative system, to be owned and controlled by the subscribers. They will guarantee the earnings to be not less than 7 per cent." The following week a list of by-laws was printed in the paper. Some of the important points cited were the right for *all* customers to own stock, the control of the amount of stock any one person could own (25 percent), and the capital stock set at $40,000 with $10,000 of that being set aside for exchange work only. The paper a couple of weeks later on July 22, 1904, stated that "the Centralia Telephone Company is offering the small capitalists a fine opportunity for profitable investment." In the midst of all these developments, A. Bishop and wife spent several days at the World's Fair in St. Louis.

One of the noticeable results of the man-made advancements taking place, especially in Centralia where frequent remarks were made about the beauty of the town, was overhead congestion. The paper in August suggested that the diverging telephone lines in Centralia were taking up all vacant space. Part of this was due to the technology of the day that required more individual lines and the multiple exchanges existing at that time. But even then, increased phone activity required still more lines as the August 26, 1904 *Guard* confirmed: "The Centralia Telephone Company is running a new cable from Central to the west part of town." The special Thanksgiving issue on November 25, 1904, carried articles about each of the prominent businesses in Centralia. There are several interesting points made by the Centralia Telephone Company piece. Obviously its name hadn't been changed to the co-operative name, so questions arise if that conversion actually took place. Another revelation is it lists J. A. Chance as president and A. B. Chance as secretary and manager. It was established in1896 and currently had six hundred patrons, four day-operators, and one night-operator. In conjunction with the telephone operations they also operated a plumbing business that was oriented toward steam and water heating systems. Like John's 1905 support of a Commercial Club, there

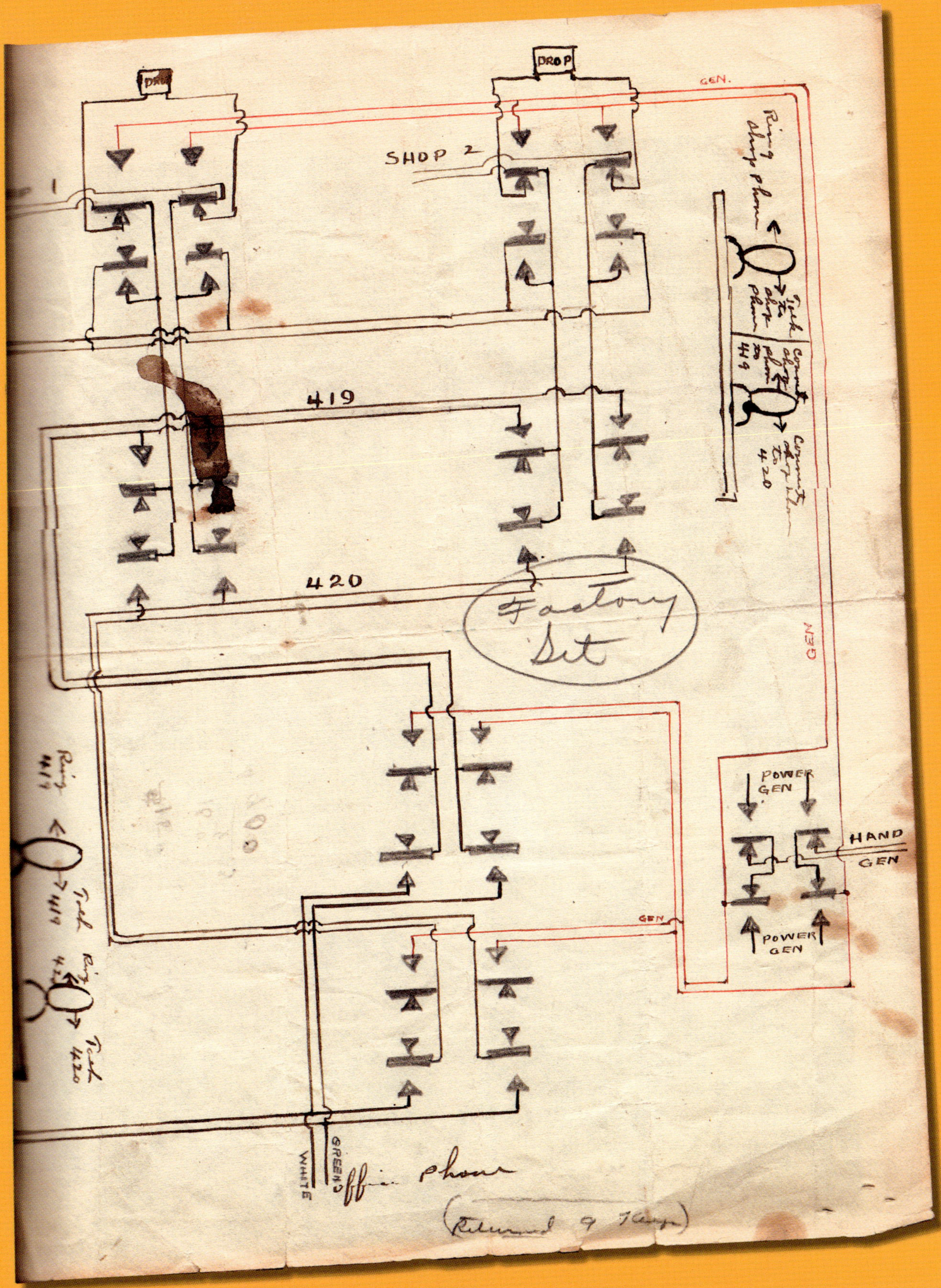

This A. Bishop Chance originating diagram represents the early office and factory telephone network likely at 210 North Allen. It appears to have consisted of four phones, 419 and 420 in the office, and one each for Shop 1 and Shop 2. (Jack Chance Collection)

were indications suggesting he was a good businessman with a steady and loyal dedication to those he was involved with. Although John A. Chance's role in the telephone business is referred to sparingly, there were other references that reveal more about A. Bishop's father. A January 1913 telephone operator piece in the *Guard* portrayed a positive picture of him: "Mr. J. A. Chance, the efficient manager, gave the girls each a present in cash as an evidence of the high esteem in which he holds the girls who face the switchboard and manipulate the plugs." Another positive glimpse of John A. Chance in the December 3, 1915 paper told of his record of service in Lodge work surpassed by few others. He performed duties in a position for twenty years that most other refused to fill. He was said to be, "one of the men always on duty when Monday night came around." Recorded social activity of Mr. and Mrs. John Chance included visits to family members and their son-in-law and daughter, Mr. and Mrs. Willis Fountain, who lived northeast of town. In the October 20, 1916 *Guard* was a continuing coverage of historical remembrances, one of which involved our subject: "J. A. Chance was a good-sized boy in 1870 and lived with his parents on the old Dunn place west of town, then the Chance farm." The story continued about the boys picking about a hundred baskets of apples during a heavy, wet snowfall, a prelude to a huge fall snowstorm on October 18, 1870.

The continuing expansion of the Centralia telephone system appeared to have no end as 1905 began with this January 20 news item: "The Mutual Telephone Company of Rowena held a meeting this week and discussed a proposition to build a line to Centralia." On February 3 it was noted that the Columbia telephone line had reduced rates. But a significant announcement a week later on February 10, 1905, by the *Guard* suggests the co-operative may not have become a reality as the merging company's name, The Centralia Telephone Company, remained the same when it was absorbed by The Home Telephone Company. It is significant that the Home Telephone Company did not exist for the first nine years of the Chance involvement in the local telephone system. In addition, it appears that J. A. and A. B. Chance were only partial owners of this reorganized telephone exchange. The three top officers selected to manage it appeared to be financial people with the operational end of the business positions filled by John and A. Bishop Chance. The February 17 paper identified the Home Telephone officers as J. W. Bryson, president (president of Farmers and Merchants Bank); vice president W. S. McBride; and secretary-treasurer C. W. Settle. This didn't deter the growth of the Mutual group as the *Guard* noted on March 3, 1905:

> *The Mutual Telephone Company is preparing to put in a system of town phones. The supplies have been ordered and with the opening up of the weather work will be commenced. The work of constructing the Rowena line will be started also and brought in to Centralia. The Mutual Telephone Centralia switchboard will probably be removed to a larger room.*

A month later the Centralia skyline became even more cluttered as the Mutual group installed new poles and phones in the city. There was a work force of twenty workmen. To add further insult to injury, Mother Nature raised its ugly side with a lightning strike in May on a Home Telephone line, putting some of their patrons out-of-service for several days.

The Mutual group must have seen opportunity as they proceeded to place themselves into the local Centralia telephone landscape. It's unclear if the use of Union in their name was new as the June 9, 1905 *Guard* article indicates: "The Union Mutual Telephone system has published their first directory and it contains over sixty phones city and country. They expect to put in their long distance lines on a separate circuit in order (to) get more satisfactory service." Amongst all the commotion was a growing sentiment for public ownership of railroads and telephone and telegraph services, according to the local paper. In quick order, possibly to counter the new Mutual directory, Home Telephone issued a new directory the first of July. There was also a reference to "get a new card, and save the wear and tear on the hello girl." Apparently Mutual was growing with the purchase of an office in Hallsville for a switchboard in May of 1913. On November 7, 1913 the *Guard* carried an extensive article about a new line, Centralia–Mexico trunk line that was being built by the Centralia Mutual group to connect with the Mexico Mutual group lines. It had the support of Mexico businessmen and was built or attached to the existing electrical poles of the Mexico Power Company, which came into Centralia from the east. A news piece in the January 1915 paper suggested that the Mutual group had grown considerably. Confirmation of at least three local exchanges in and around Centralia was indicated plus once again a rumor surfaced that the Buffum Telephone Company was selling out. This is what the February 23, 1906 paper had to report:

> *Frank W. Buffum of Louisiana, president of the Buffum Telephone Company, operating exchanges in Fulton and other Missouri towns, denies that he has sold out to the Bell people, or to any other telephone company. Mr. A. B. Chance of the Home Telephone Company informed the Guard man Friday that there was no foundation to the report that Buffum had sold out, as he had information from the inside.*

As March 1906 rolled around and springtime became more promising, an idea was hatched by Home Telephone to gain an advantage over competitors. Developing an edge over your competition is always a very desirable posture and this is the kind of approach that one could expect from A. Bishop Chance. The paper reported that Home Telephone was painting their poles brown at the base and white on top. This employee's recollections in later years of the public's insistence that insulators and other pole line hardware items blend into the sky on electrical power lines, suggest Centralians were apparently an early proactive group. Since Home Telephone appears to have been the largest exchange in town, it was surely impressive to see all their poles and note the fewer remaining. One sure purpose was to gain favor with local town folks, aesthetically speaking, in making all their telephone poles look nice. This was the jest of an April 27 article in the paper that called out to all other pole owners to follow suit. The paint scheme changed to a white with an olive green base. It is the old idea of taking a negative and producing a positive. There is nothing quite like the benefits reaped from a few gallons of paint. Of course maintaining that appearance is another matter, but often our focus is on the short-term.

Although A. Bishop may have been correct on the prematurity of Buffum's sale, the sale was consummated by April 1906. It appears that the Independents represented by the Chances, Gary, and others were also looking at the elaborate system that Buffum had developed. To indicate what was occurring, the pressures on such small exchanges as the Home Telephone Company and the Bell versus Independent battles, the announcement of the Buffums sale in the April 6, 1906 *Guard* is covered here:

> *The Buffum Telephone Company Sold to the Bell System. For some days past rumors have been current on the streets of Louisiana that there was some change about to take place in connection with the Buffum Telephone Co., and this morning a representative of the* Press-Journal *called on Mr. Frank W. Buffum, who informed him that the capital stock of the Buffum Telephone Co., had been transferred Friday to Kansas City Parties.*
>
> *Mr. Buffum stated that this sale had been made owing to the fact that during the past few months quite a number of exchanges*

> *that control entire counties had sold all of their properties to what is generally known as the Bell Telephone System and that these sales were mostly in the central, western and northern portions of the state, and other deals were now pending whereby more exchanges, now operated as Independent, would be liable very soon to be purchased by the same people, as these exchanges had already been offered to them, and naturally the result would be, if such was the case, that if these exchanges did sell to the Bell Telephone System that the subscribers of the Buffum Telephone Co., would for that reason be cut off from telephone service on many of the exchanges that are now used to a considerable extent, at the present time by the patrons of the company. He also stated that the construction in the territory occupied by the company, and much of which was generally looked upon by other Independent people as necessary for the Buffum Telephone Company to develop, would, if built, take a very large sum of money before properly developed and which the public were demanding to be done at once, and would take a much larger amount than could be reasonable expected for one Independent Telephone Company to expend; that the property of the Buffum Telephone Co., had been tendered to and examined by one of the leading Independent Telephone Companies in the state, who did not feel that they cared to purchase the same. (Louisiana Press-Journal)*

The remarks by Mr. Buffum reflected the challenges facing telephone companies including the huge amount of capital that was needed to expand telephone coverage that the public was now demanding. But this may have been a secondary consideration in view of what could be described as a "bail-out" type of sale. This was brought about by the business war that raged between the Independents and Bell interests. When the Bell system began buying out local area independents and gaining more clout, some of the independents became concerned of being left out and some like Buffum accepted Bell's offer and exited the business. From the remarks made, an independent group represented by Gary and those like Home Telephone had also unsuccessfully made an offer to purchase the Buffum telephone system.

As 1907 dawned what better way to terminate a case of cabin fever than to head to a favorite city where a convention was being held. On February 15 A. B. Chance was found in Kansas City attending the meeting of telephone operators. A week later the *Guard* would host a public announcement of A. B. Chance's latest invention, the Sky-Rocket® Lightning Arrester. Business trips to telephone groups would follow in promotion of his new product. After late March trips to Kansas City and Sioux City, Iowa, the local paper on June 7, 1907, carried the following story: "A. B. Chance has been in Chicago this week attending the International Telephone Association's convention being held in that city." He was also in Moberly in September, maybe to see their new reservoir, which would provide water for the area including a very important tenant, the Wabash Railroad with their fleet of water-hungry steam locomotives. Interestingly, later in February of 1911, the Wabash was forced to buy water from the city of Centralia when their local water pump failed. A. Bishop was instrumental in pushing the waterworks issue during his term as mayor. Earlier in May the *Guard* covered the merger of Home Telephone and "the country lines under the Gray system of Sturgeon." It is possible that the name *Gray* should have instead read *Gary*, but there was a Gray family, as a news article in the March 30, 1917 *Guard* mentioned a Sam Gray, originally from the Chicago area. Regardless, these were two local Independents closing ranks to cover a larger area giving access to better service for both of its patrons.

Expansion was still alive and well in 1908, as a new Home Telephone line was being strung

Home Telephone. (*Centralia Historical Society*)

from Centralia to two miles beyond Skinner. It would open up a new area, implying that territory was not being serviced. Within this kind of backdrop of trying to get all locals a phone and the automobile still about as common as four-leaf clovers, a startling announcement was made in the *Guard* of an eastern group investigating the possibilities of a dam on Salt River to generate electricity! This would occur over a half century later, the resulting lake in sight of Samuel Clemens' birthplace. This may have been a 1908 seed that germinated the Corps of Engineers eventual completion of such a dam. This precedes the construction of a similar electrical generating dam, twenty years later in 1928 that gave birth to the Lake of the Ozarks. In the Rodemyre tradition, Adam's sons positively summed up the 1908 telephone situation existing in their hometown on August 28, 1908. "Centralia has about fifty telephone lines coming into and through Centralia besides the two local companies, and one can talk almost anywhere in the State, and in adjoining States." The *Guard* was a booster of efficiently operated local businesses such as Home Telephone. Their printing operation inherited additional business as they began printing phone directories for the local telephone companies. On June 2, 1916, they confirmed the printing of a new directory for Home Telephone.

Not only was A. Bishop a telephone man, but he also had the responsibility of operating and promoting his new manufacturing company. An old adage suggests if you need something done, then find someone that's busy, is no truer than when applied to the Centralia Wizard. A. Bishop certainly fit this qualification and it is puzzling why he would entertain the thought of running for Centralia Mayor, but he did. The *Guard* on May 21, 1909, however, found him with his Chance Manufacturing hat on:

> *A. B. Chance is attending the convention of the Independent Telephone companies at Kansas City this week. Mr. Chance is there in the interest of Chance Manufacturing Company, of this place, and has samples of all the novelties and utilities made by this concern, consisting of the numerous and rapidly growing Sky-Rocket® family.*

It is so beneficial to this work that these early years are so excellently documented by the Rodemyre family, which allows the charting of the progress and development of the Chance Company. The energy and innovative nature of A. Bishop Chance is also being captured in various areas plus in growing and solving the problems of Chance Company. Up to the time of the family world cruise in the late 1920s, he would not again allow major distractions such as a public office to interfere with the early operations of his company. After the ground-shaking promotional scheme to sell stocks in his company in November, a neat advertisement (right) is carried by the December 17, 1909 *Guard* just in time for Christmas to close out the year in style.

In 1917 when F. Gano Chance was twelve years old, he climbed a telephone pole. The ascent was apparently as normal as an inexperienced lineman with climbing irons could muster, but his descent was a different manner. It may have looked like a logger contestant trying to beat the clock, but Gano was in more of a free fall, losing some hide on the way down. One advantage the telephone lineman had over the electrical lineman was the telephone line wouldn't zap you unless it was energized by a bolt of lightning. But the early telephone poles were untreated, so they decayed rather quickly and weren't guyed very well. Gano may have been told the dangers of climbing poles as there were at least two local reports of linemen being injured years earlier. The first was his great-uncle Grover Chance who hurt himself when he fell off a telephone pole in 1905. The *Guard* covered a more serious accident on November 13, 1908. Denton Palmer, one of Home Telephone's linemen, was up about thirty-five feet when the rotten pole he was on tottered, so he jumped instead of riding the

The December 17, 1909 *Guard* ad for their client Home Telephone Company was just in time for Christmas. It reveals the earliness of the commercialization by the telephone compny of promoting additional and decorative phone sets, etc. *(Guard)*

pole down. His anklebones were both crushed and he wasn't expected to gain the complete use of his feet or his ankles would be stiff. A 1913 report said that he was employed by Chance Manufacturing Company, which may be another example of Mr. A. Bishop Chance taking care of his own.

By August 1921, telephone problems were plaguing the Centralia area. Drooping wires were becoming a threatening situation, with several accidents being narrowly averted. One young boy was injured, but luckily telephone wires didn't carry the life-endangering electrical current of a power line. This may have allowed all parties to be more lax in replacing rotten poles and delaying other plant maintenance, which in the end endangered the public and reduced the quality of phone service. On October 7, 1921, the *Guard* carried two news items which relate to the Chances and Home Telephone. The first item concerned A. B. Chance who was representing Home Telephone in a rate increase meeting. The management of the phone company, represented by his father John Chance must have relied on A. Bishop's talents in such business dealings. John Chance was one of Centralia's senior residents, as suggested that same week when Mr. and Mrs. John A. Chance celebrated their fiftieth wedding anniversary! Early in February 1922, Home Telephone was sued by the Missouri Public Service Commission for overcharging on previously set rates. The March 10 paper said Home was paying the overcharge back to customers. The state's regulatory agency was displaying their teeth at a time when increased rates were needed to repair and upgrade the deteriorating telephone plant. These kind of problems were an early hint of why the local exchange would be sold and the Chances would eventually diverge themselves of involvement in a business they had been in for some three decades. But later in the year John, like his father, couldn't be counted out quite yet. In late September, he attended a telephone meeting in Mexico, Missouri. On November 17, 1922, the *Guard* carried this item:

> *J. A. Chance is attending a big telephone convention in Jefferson City this week. "Uncle John" is one of the pioneers of the telephone business in Missouri. He is a thorough telephone man and keeps abreast of the times in his line.*

On December 15 another article found Mr. Chance suffering illness. A reflection of this piece suggests Mr. J. A. Chance was held in high esteem and suggests his early telephone connections:

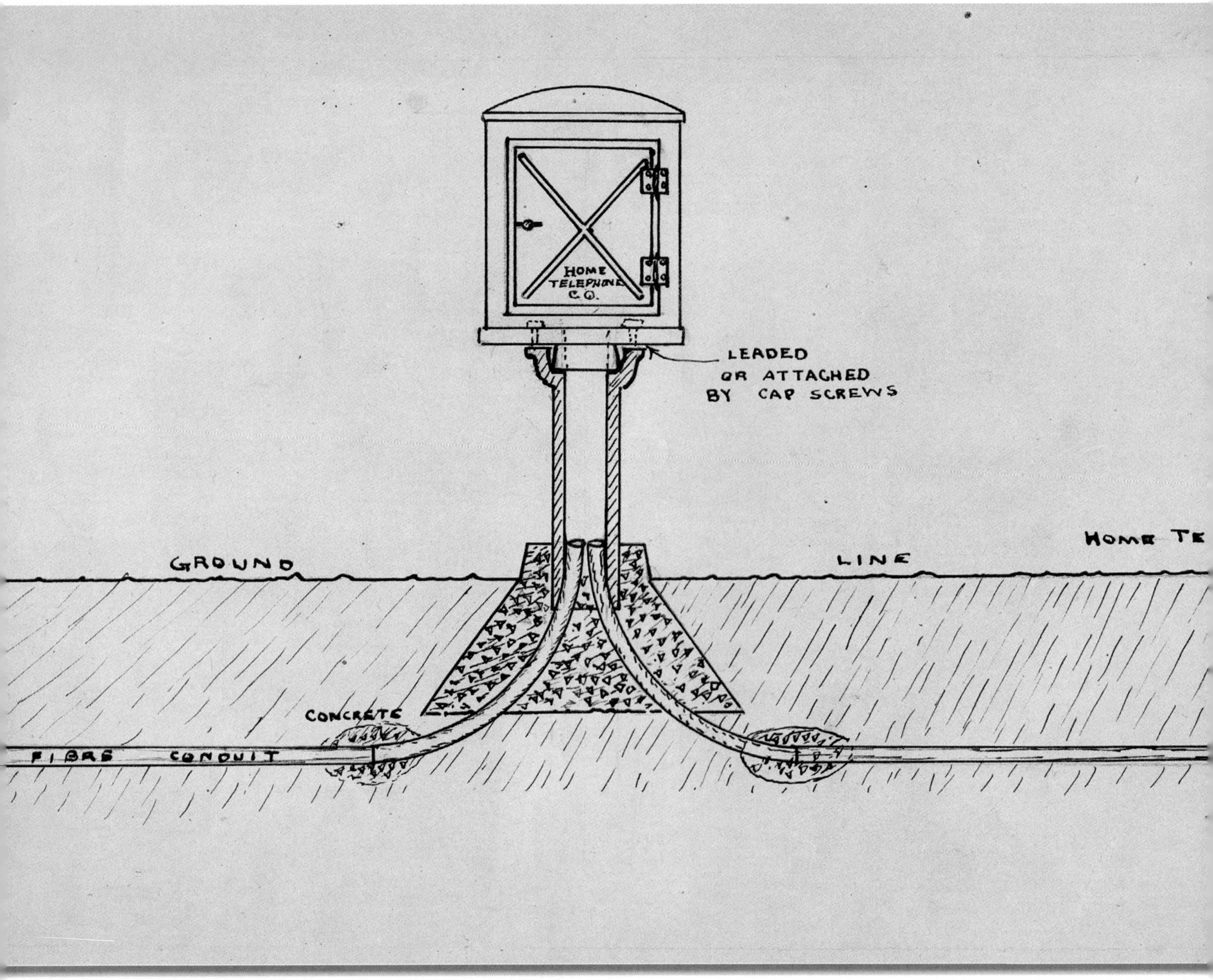

This is likely a Mr. A Bishop Chance originating sketch of the Home Telephone underground work in downtown Centralia, artistically enhanced by an artist. (Jack Chance Collection)

> *J. A. Chance, the popular manager of the Home Telephone Co., came up to his work as usual Monday morning, but was forced to go home on account of sickness. It takes more than a minor complaint to put "Uncle John" on the shelf, and his many friends are glad to see him up and around again after his sick spell.*

Besides the threat of decapitation, the appearance of the concentrated mass of telephone poles and hardware in the spaceless downtown area was an eyesore. However, relief was in sight as the July 7, 1922 *Guard* carried two articles about the plan for removal of the poles along downtown Allen Street. A new line across the center of the park as well as lines down the center of Lakeview Avenue wouldn't be allowed by the City Board. But the most eye-catching and surprising part of this project was the mention of the use of lead underground cable. A. Bishop discussed this further in an article a week later on July 14. He said the cable was enclosed in a vitrified conduit and would leave the central office from the alley west of the *Guard* office. It would then cross Singleton and the city park (City Square) to the west of City Hall across Sneed Street and be taken on poles at the alley west of the government lot. It was apparently approved because it would be underground, but even this was not the advent of underground usage as in Mr. Gary's 1907 *Independent Telephony* is pictured underground telephone cable within a manhole in Topeka, Kansas. This suggests that underground cable technology was driven by telephone usage, which later influenced electrical power use. In spite of seeing such advancements much earlier than expected, it's more normal to see a vision for the future like: "No wireless telephony can ever be made a commercial success." Mr. Gary was utilizing Mr. Thomas Edison's opinion in suggesting an investment in his present system, wasn't threatened. Mr. Gary was correct in that those systems weren't threatened during their normal life by new technology.

Along with the external problems of telephone lines and the underground work around the City Square to commence upon arrival of the materials, a July 14, 1922 *Guard* article also found internal problems. There was an economic business law discussed by Theodore Gary in his 1907 *Independent Telephony* publication stating that growth in the number of phones in a system proportionally increased the cost of construction and plant. That is, a cost of $1 per phone in a five-hundred-phone system might instead cost $1.50 per phone in a two-thousand-phone exchange. This is opposite a reverse law of volume reducing unit costs in other areas such as in manufacturing earth anchors or automobiles. This impacted many small independent phone exchanges that began with pocket change and as they grew the need for inserting larger amounts of capital convinced many small businessmen, without a bank friend, to sell their small plants to larger telephone groups. This was always a challenge to the early Chance telephone operation and even later to A. B. Chance Company, to have a source for larger amounts of capital to continue growth and expansion. That was especially important to F. Gano Chance, who believed a firm either grew or stood a good chance of dying. Back to 1922, A. Bishop Chance found that growth had generated challenges and in his opinion, "the telephone service is poor owing to the fact that the switchboard, designed for handling the business some five years ago, is now inadequate for the city's needs." The switchboard didn't have enough lines to handle the increased volume, especially at the normally busy, peak periods. We get an insight into another side of a usually cordial and gentle A. Bishop Chance, in his discussion with the writer of the July 14 article. On this hot July day, he became so aggravated when he couldn't get service that he left his plant and quickly moved across the Wabash tracks, up Allen through the business area, and upstairs to the phone exchange, facing the City Square. He "went up to flay the two or three operators . . ." and he "was mad as a wet hen!" However, instead of finding inactive and unconcerned operators, he found them doing their best with inadequate equipment and "his heart softened." He didn't even speak to the overly involved girls but instead turned, went out and traveled over to talk with his party without using the telephone. But he already had in mind a new type of three-panel switchboard, which was needed to increase the local telephone efficiency. This was important to a growing Chance Manufacturing Company with increased telephone volume and its remote location. An efficient communication system represented by the telephone had become a necessity in carrying on business. Ironically, A. Bishop could directly impact the level of service his manufacturing firm obtained because of his direct involvement with Home Telephone. In retrospect, this was a unique situation.

The August 6, 1926 *Guard* carried the news of the sale of Home Telephone. Not unlike other grassroots firms started up by an enterprising local individual, the ownership was being transferred to an out-of-town party. Home Telephone was a going concern, or it appeared that way, and was one of the largest Independents in the western part of the county according to the article, with some one thousand patrons. Again, it's confirmed that John A. and A. B. Chance were part owners along with C. W. Settle and W. S. McBride, although the Chances managed it. It was established thirty years before on the fifteenth of August. This would put its establishment back to 1896 although as previously pointed out the early years seem to have been a sole venture by A. Bishop with the support of his father and didn't become officially Home Telephone until 1905. The article confirmed that arrangement, except in name, and told of the first switchboard with the party line carrying seventeen telephones. It was first located on Allen Street in the Berg Building, which was on the north end of Allen, south of the tracks or near the new Globe Hotel. As it grew, a fire occurred (which would be in the Baker Building) and it was then moved above Farmers and Merchants Bank on the City Square. The corner lot on Rollins and Sneed just south of City Hall was purchased in 1925 and a new central with offices were built (Meador Funeral Home location). It was suggested the Home Telephone sale was a substantial financial transaction, especially for the area. The buyer and sale amount was being withheld at that time. The sale appears to have been to an E. O. Turner, a friend of Mr. Gary's, although it was later revealed there were connections back to the Gary group. Mr. Turner was instrumental in pointing out to his friend Mr. Gary in early November 1926 that Centralia was isolated from current or planned hard surface highways and Mr. Gary as head of Missouri Highway Commission became convinced that this would be corrected. The eventual connection of this eastern Illinois originating road (Highway 22) with Highway 40 (via 63 or 7) was an important consideration. Amid all the challenges still facing each generation, some are the same, just adorned a little differently.

A February 28, 1930 *Guard* news article cleared up several questions concerning Mr. E. O. Turner. He was leaving Centralia and resigning his position effective March 1 as vice president and manager of the Missouri Telephone Company with exchanges at Centralia, Sturgeon, Clark, and Columbia. He was going out on his own, having personally purchased the Frederickstown, Missouri exchange. He had come to Centralia in July 1926 having handled the negotiations for the purchase of Home Telephone Company. The article further stated that Gary interests owned the Centralia, Columbia, Sturgeon, Clark, Queen City, and Macon exchanges. So as earlier suspected, Mr. Turner had been directly connected to the Gary group who actually purchased Home Telephone in 1926.

What is amazing is the technology being developed and utilized at the turn of the century, or shortly thereafter. Although each generation may believe that they are the only people who have had to adjust to newfangled gadgets and systems, the truth is that every age has to do so. The fear of advancements likely got its start when the cavemen discovered fire. Until they understood its danger and made provisions to manage it properly, could they fully utilize and appreciate its advantages and benefits. There are critics of advancements today as it has changed the appearance of our nation from a natural, pristine face to one that has been marred by our modern civilization. Today, new construction usually omits power and telephone poles, reducing the consequences of modern civilization's march forward. Like poverty, maybe we don't want to see the full impact of our technological developments and feel guilty. So what takes place is hiding what has been developed. It is not important that we see water and sewer lines or telephone and electrical lines, as long as they work. Some function better underground anyway and in addition nature's beauty is richly enhanced.

It is unclear what the depth of relationship existed between Mr. Gary and A. Bishop Chance. Maybe that is unimportant in comparison to knowing what A. Bishop was exposed to in his involvement in the telephone business. Working for Mr. Gary and others in the 1907 time frame on the larger telephone exchanges, he was working with and seeing the latest developments. It was a relatively new area where constant growth opened the door continually to new methods to better perform certain operations. Tacking on new lines to an existing plant reached a saturation point where new technology such as multiple cables were enclosed in a single line or it went underground in heavily populated, cluttered metropolitan areas. Speaking on that subject are the words of one of the most forward thinking businessmen of that time, Theodore Gary on page 79 of his 1907 book, *Independent Telephony*:

> *In the first place the old open iron wire methods of construction have been abandoned. The new method being so much less conspicuous has never made much impression upon the mind, since in all the larger plants 80% of a plant is underground. But little of such plants is where it can be seen. Even the up-to-date aerial construction is hardly noticeable. The mind remembers only the tall poles, with eight or ten cross arms, with eighty or one hundred exposed wires, with the unsubstantial appearance. Now, a much shorter pole will carry cables with 800 wires, the material being copper and lead that have a very large per cent of indestructible value. Besides the danger from storm and sleet is removed.*

These are the sort of observations that could easily be mistaken for the latter half of the century, when the appearances of electrical power lines became a hot issue. These were concerns for F. Gano Chance and his engineers, as a manufacturer of electrical power line items that had to be designed to disappear in the sky or become more attractive, but A. Bishop and the Home Telephone people had already visited that problem with painted poles and underground lines, half a century earlier!

There is the suggestion that nothing is new in the universe and man simply reuses what is already here. Belief in this law leads to the rejection of the idea that man "creates" anything as only God can create from nothing while man only produces and develops from existing resources. Likewise, when someone develops a new, innovative idea, method, or product, chances are someone's earlier footprints will be partially or wholly retraced. However, during the long history of Chance Company, man has accomplished much that is considered new. When longtime employee Verna Palmer mentioned Granville Ryman's early duties of selling over the telephone in the 1930s, it reminded this employee of the *new* telephone marketing group that originated in the late 1980s, early 1990s, at Chance. Harold Ward, Ed Miller, Greg Davenport, and Tony Sneed were members of this special group. Although the approach may have been different, the telephone had been utilized earlier in the sales effort. The practice of telemarketing is altogether appropriate at Chance because company roots are traceable back to the telephone industry. In fact, a portion of Chance business has always been with the telephone industry. The telephone played an important role in connecting the world to 210 North Allen Street, even though it was remotely located in central Missouri away from major municipalities. Although the railroad and its partner the telegraph, moved communications forward, the phenomenal telephone truly opened the door to fast, inexpensive, mass personal and business communication capabilities. Another remarkable aspect of the telephone, besides it being an ideal type of communications developed and perfected in America, it is well over a hundred years old. Not only did it help drive progress over the last one hundred years, it is being utilized even more today. If you need proof, simply ask a parent of a teenager or notice cellular phone usage in every place imaginable at the turn of the millennium. It's another area that thanks should go out to the Chance family and their contributions, as well as other Missourians like Mr. Gary and Mr. McDaniel. They were early pioneers of Alexander Graham Bell's invention that impacted their local areas, the state, and eventually the nation and the world. We have been surrounded by giants, but they have become lost in time, and deserve a reintroduction.

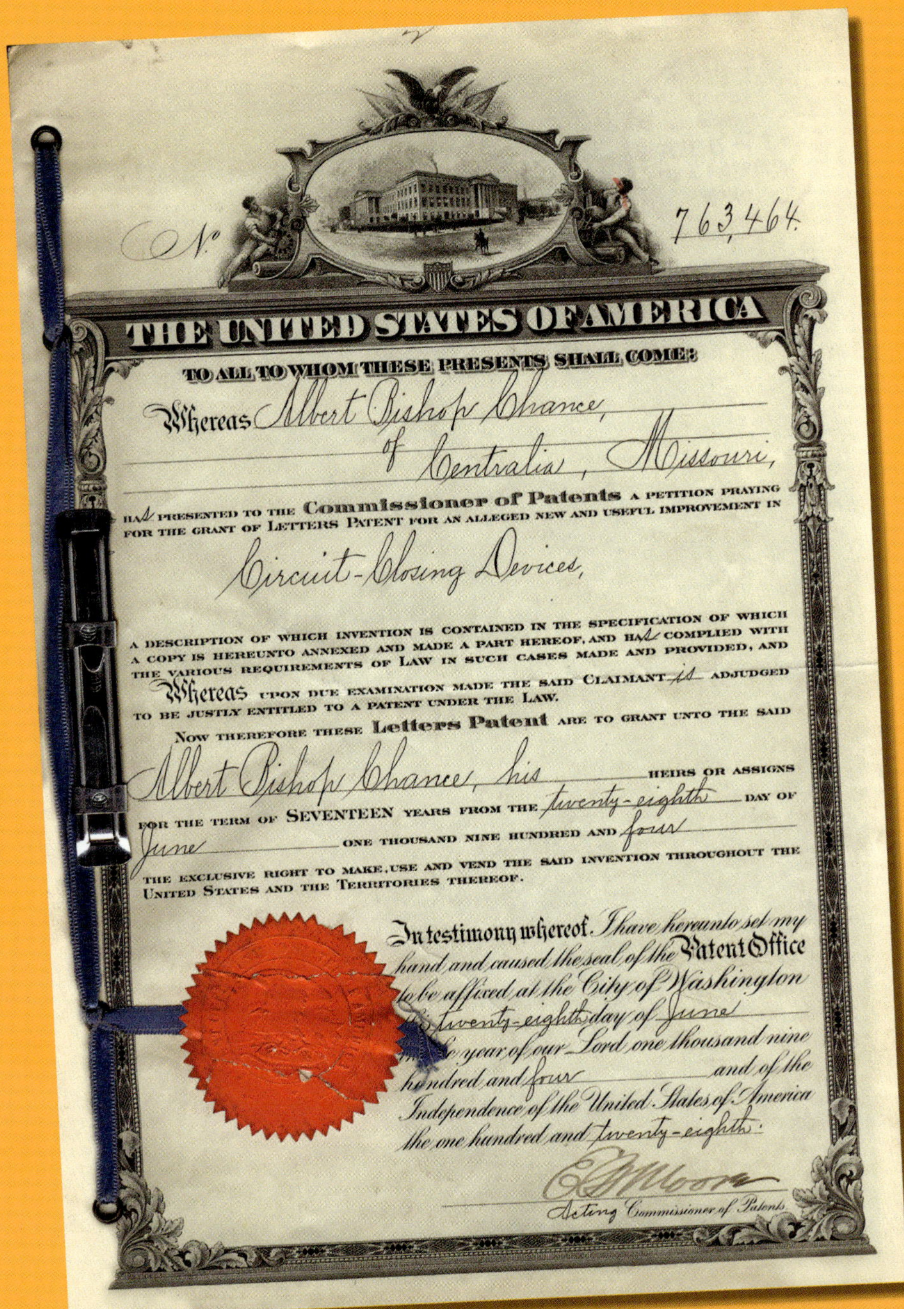

No. 763,464.

THE UNITED STATES OF AMERICA

TO ALL TO WHOM THESE PRESENTS SHALL COME:

Whereas Albert Bishop Chance, of Centralia, Missouri,

has presented to the Commissioner of Patents a petition praying for the grant of Letters Patent for an alleged new and useful improvement in

Circuit-Closing Devices,

a description of which invention is contained in the specification of which a copy is hereunto annexed and made a part hereof, and has complied with the various requirements of Law in such cases made and provided, and

Whereas upon due examination made the said Claimant is adjudged to be justly entitled to a patent under the Law.

Now therefore these Letters Patent are to grant unto the said Albert Bishop Chance, his heirs or assigns for the term of Seventeen years from the twenty-eighth day of June one thousand nine hundred and four the exclusive right to make, use and vend the said invention throughout the United States and the Territories thereof.

In testimony whereof I have hereunto set my hand and caused the seal of the Patent Office to be affixed at the City of Washington this twenty-eighth day of June in the year of our Lord one thousand nine hundred and four and of the Independence of the United States of America the one hundred and twenty-eighth.

E. B. Moore
Acting Commissioner of Patents.

The front page for Patent 763,464 for the circuit-closing device. (A. B. Chance Co.)

Chapter 3
THE FORMATIVE YEARS
(1911–1916)

The Never-Creep® Anchor

In 1911 the Guard *introduced the New Year by reporting on January 6 the travels of a prominent Centralian: "A. B. Chance has returned home from Washington D.C. and New York, where he has been on business."*

The headquarters of certain utility corporations were located in New York City and Mr. Chance was also a client of a Washington, D.C. based patent attorney. He apparently was still based in Centralia, in view of the then-existing Kansas City connection. A January 13, 1911 editorial explained that the Commercial Club was still working on the Chance Manufacturing Company proposal. A February 3 report said they were waiting on a survey by engineer Harding on the Chance Company propositions. It is unclear if this concerned the lots that were to be made available for sale to raise capital for the city-built factory or the site where the plant would be located.

The year 1911 would forever change A. Bishop's life. There were impending circumstances that would generate one of the most difficult periods in his life, but he did not buckle under. There was political turmoil brewing in Centralia and Mayor Chance seemed to be in the middle of it. Some people find working beneficial when facing difficulties instead of dwelling on those things we have little control over. A. Bishop was a person who got caught up in doing instead of wasting time worrying and he appears to have been comforted by such activities. A revealing piece in the *Guard* on February 10, 1911, began this period when various problems would come his way: "Mayor A. Bishop Chance had the misfortune to mash one of the fingers of his left hand recently while running a power saw in the factory and on Monday, February 6, it was found necessary to amputate the finger." He could frequently be found in the factory operating various machines in shaping and making various items. Quickly following on the heels of this was the burden of making a decision whether to run for mayor again. A *Guard* editorial of April 7, 1911, about A. Bishop's mayoral situation is revealing. Any brashness that he may have had coming into the mayor's position two years earlier had been replaced by a more experienced individual:

> *A great effort was necessary to induce A. Bishop Chance to run for re-election for the office of Mayor of Centralia. He said that he no longer thirsted for the empty political honors that turn to ashes upon the lips. It is known there is no*

cleaner man in Centralia than A. Bishop Chance, and while in human blindness he may have erred with the balance of his administration in overzealous reaching out for the promotion of the city's welfare, it can not be charged that he has in the least tried to further a selfish end. Just now he is nursing a large number of sore spots where adverse criticism and unjust abuse has taken off the hide in spots, and just now in the turmoil of a modern and expensive innovation the people are blinded to much of the good Mr. Chance has accomplished by his aggressiveness, but as time rolls on we are sure an awakened conscience and a judgement free from personal prejudice will accord to him the tardy credit for a splendid accomplishment.

Arthur Bruton, a local attorney running against A. Bishop, won 252 to 212. However, this wasn't the end of activities initiated and promoted by his administration.

Judge/Mayor Arthur Bruton (Centralia Historical Society)

The Centralia Development Company that A. Bishop had been instrumental in starting and directing decided to dissolve after its oil-drilling failures. It was difficult to get a quorum but later it was dissolved with A. Bishop as president and O. B. Mayes as its secretary. This episode in the life of Centralia did not help either of these men's credibility. Those in the oil exploration business know all too well that numerous attempts are often necessary before a gusher is struck.

The progressiveness of Centralia in displaying the image and having the conveniences of a small city did not come without controversy and costs. Sidewalks were such an issue that were sponsored by Mayor A. Bishop Chance's administration and the March 1911 *Guard* continued to report concrete walk construction. A June 23 news item covered the Board of Aldermen touring the town to view the sidewalk progress and there was the hint they would authorize more. The rub was that residents had to put in these sidewalks on their properties and pay for them. But this can't be all laid on the progressive Aldermen's doorstep, four of who resigned by July, as early October saw a petition before the new city administration to have twelve residents build sidewalks. In the spring of 1912, there was talk of Mayor Arthur Bruton's machine politics and a report that some north residents were being forced to put in sidewalks. If not completed, then it would be bid out and the cost added to their tax bill. In the January 1923 paper there was an article on the bond-burning celebration in which Mayor Bruton suggested he was sometimes depicted as, "Bruton and his gang." So all was not rosy for Bruton's new administration. There would later be water pump issues and other demands that were like an endless parade requiring more local resources. In 1912 for instance, the Rodemyres (mirroring their father's support for such progress) reported the need for new city lights. The old arc-type streetlights from the 1897 period were not efficient in producing light or in economy of operation. They needed the new incandescent type. Now who in Centralia knew enough about electrical equipment to suggest such an innovation?

There are no signs that A. Bishop Chance had any desire to reenter local politics, but his endless chain of ideas charged life into new challenges that initiated changes. After his experiences as mayor, he could have withdrawn and not actively supported the community, however, in the July 10, 1914 *Guard*, A. B. Chance was found attempting to spin off enough electricity from the water plant to power the streetlights. He was working with a Mr. Rowland who obtained the contract a week later to complete this project. This would eliminate the electricity cost to the city of powering the streetlights now being paid to the Mexico Power Company. Also in 1914, there were those from the Commercial Club, such as A. Bishop Chance and Dr. Hope, who thought the city streets should be oiled to retard moisture and mud. The automobile people had also generated another problem, a parking dilemma, according to the *Guard* in April of 1915. Cars had to resort to parking in front of some stores, which sparked merchant hostilities, as they were driving away business from those still utilizing a horse and buggy!

Although the benefits of running water, sewers, and electricity would benefit residents, it was also needed to attract industry to Centralia. This would also benefit the existing Chance Manufacturing Company, but the evidence doesn't suggest this was the core motive of those city leaders who approved these programs. It was clear, however, that A. Bishop and his supporters saw Centralia and the need to progress differently than did the majority of residents. This opposition was mainly due to the financial impact of these programs to the indebtedness of the city. In the face of that, A. Bishop was still pushing for the grounds and a facility for his company. This compounded the problems that some area residents were already viewing with apprehension, in spite of the benefits these advancements brought. According to the May 26, 1911 *Guard*, a serious development arose when Mayor Bruton received a letter from a Mr. Rollins representing a Columbia resident, Mrs. Graves, who held a $3,000 note that she wanted paid. By June this had become a hot topic with Mayor Bruton, soon to be nicknamed the "Big Squeeze," asking why this hadn't been paid and using the June 9 local newspaper to put forth his opposition to the land purchase.

It was obvious in June 1911 that the new mayor and the Board of Aldermen did not see eye to eye. In a letter addressed to the Aldermen via the front page of the June 9 *Guard*, Mayor Arthur Bruton aired all the dirty linen, as he saw it. The Aldermen had passed Ordinance No. 516 at their last regular session, but without the mayor's approval. The ordinance that the mayor returned unapproved was the ratification of the purchase of fifty-eight acres (sometimes described as fifty-seven acres) between the railroads, earlier said to be west of town. He stated there was a $3,000 debt, which refers back to the Columbia lady who held the note discussed earlier. It is unclear who made the purchase, which apparently was already consummated, as the unpaid note suggests. It is not certain how far west this new property was located. As both railroad tracks leave Centralia they begin to spread out from each other, which produces more acreage as necessary to generate fifty-eight acres. This would suggest that this land mass could have extended to or beyond the current day highway overpass over both tracks. When the railroad wye was seriously contemplated in1909, across from and a little east of the current Wilson Street Plant, the *Guard* referred to the proposed location of the wye as "west of town." The second problem with a factory gaining control of all or a part of these fifty-eight acres will be understood after this situation is explained.

The mayor made two damning points, which undoubtedly stirred public opinion and supported his threat of bringing the issue before the people in an election. He was on pretty safe ground as his victory over A. Bishop, who represented the progressive group, suggested he already had majority support without bringing up details of the land transaction, which made its supporters look even more vulnerable. The first explanation in the ordinance suggested the land was needed to locate the septic tank, which according to the mayor only required one and not fifty-seven acres. Another purpose was for a cemetery, but the mayor thought that "is absurd, considering the location and surroundings of the tract of land." This may have been an and/or suggestion as the two certainly were not socially compatible. Another proposition was to donate it to a factory or other enterprises, which the mayor said was prohibited by the State Constitution. The only source to pay off the $3,000 was from the sewer fund, which apparently was already insolvent, according to the mayor.

Mayor Bruton's final thrust for the jugular, confirming that the St. Louis paper was daily transported by train to Centralia readers, involved an advertisement that ran in a St. Louis newspaper stating in the mayor's words: "This city had set aside this fifty-eight acres of land to be given away to factories and other enterprises, by the city, which was sent broadcast over the country in the face of the constitutional prohibition of such a procedure." The source of who instigated this was according to the mayor: "by some one's authority, I know not who's, paid for by some person's money, I know not whose." This was almost poetic, but it seals the purpose for the purchase of this land by these "unknowns." This whole episode ended in a whirlwind of activity. A few weeks later the June 30, 1911 *Guard* carried news of the defeat of a $6,500 bond issue to pay off outstanding accounts and debt of the water/sewer works. The paper suggested this was being pushed too hard by the progressive Aldermen and worked a hardship on the people. In the face of dwindling support, the Board of Aldermen approved the purchase of the fifty-eight acres for cemetery and septic tanks. On Monday, which appears to be their regular session, four Aldermen resigned. After the appointments of pro-Bruton representatives to fill these vacancies, the December 8, 1911 *Guard* reported that city Ordinance No. 544 was passed unanimously to sell the fifty-eight acres of city land. H. E. Stone, chairman of the selling committee, said a sale was made and money collected, as covered in the January 5, 1912 paper.

It was not until a decade later at a city meeting covered by the February 4, 1921 *Guard*, that it was revealed why Mayor Bruton and other residents were concerned about the city's indebtedness in 1911. Bruton told the members of the Board and assembled guests "of the $65,000 debt contracted in 1910 and how large it looked then." He said that back when the debt was placed on the city, it had a depleted general fund and "a citizen of the town had remarked that since the debt was saddled upon us there were children unborn who would not see the debt paid. . . ." The mayor suggested that when the debt was erased other improvements, such as the paving of the streets, could be anticipated. This makes clear why Mayor Bruton opposed certain city expenditures until the city got back on its financial feet. In a way, the aggressiveness of Mayor Chance could be applauded in bringing the many benefits of a modern water/sewer system. Mayor Bruton could likewise be applauded in a decade-long struggle in conservatively directing the affairs of Centralia that would bring closure to the burden for such a substantial debt. In November 1922 it was reported that Mayor Bruton realized a long-held dream by buying the last two outstanding bonds.

Dr. O. B. Mayes and A. Bishop Chance were very close as numerous news items indicate. In the July 21, 1911 paper they were found at Mexico along with Ed J. Schmidt on the previous Monday, attending a state highway meeting. On October 27, 1911, the *Guard* captured a trip that included Mrs. Chance: "Dr. O. B. Mayes is at home from Rochester,

Minnesota, where he went with A. B. Chance and wife to have an operation performed upon Mrs. Chance for tumor. He reports her condition to be most favorable since the operation and says she is getting along fine, which will be good news to her many friends here." Not only was Dr. Mayes a friend but also apparently the Chances' family doctor. His expert referral was deemed successful as this November 10, 1911 *Guard* item confirms: "A. Bishop Chance and wife returned from Rochester, Minn., Sunday, where they went to have an operation performed upon Mrs. Chance for tumor. Mrs. Chance's numerous friends will rejoice to learn that she is able to be about her household duties again. She was under the surgeon's knife for only thirty minutes during which time she underwent a most difficult operation, yet came out of it without so much as a shock to the nervous system, and immediately began to mend in health."

One would conclude with all the mayoral, personal, and family problems along with maintaining a telephone exchange and operating a factory, that A. Bishop had all the challenges he could handle. However, the paper reported on June 2, 1911, of a storm coming through Centralia and a telephone line coming down on an electric power line. This may have been a preview of a method needed to guy these lines that would be addressed early in 1912 when winter storms again raged through the area. What this mid-1911 problem proved was the incompatibility of mating telephone and power lines. The Electric Light and Power Company lines energized the grounded telephone circuit with such force that phones started ringing all along that circuit and the central switchboard was set on fire. The operators, Misses Nina Blanton and Lucy Marshall, bravely threw cups of water on the fire, in their excitement forgetting about the chemical fire extinguisher. They called A. Bishop who ran up to the office and extinguished the fire. It is fortunate no one was hurt such as J. Sam Brown whose phone was first burned out by the electrical power surge. He took his cane and pulled the offending wire off, a real hot stick operation that later relates to Chance Company's involvement in hot line tools.

Fortunately, in spite of all the diversions, A. Bishop continued to give attention to his factory operation and exposing Centralians to modern progress of the nonpolitical variety. In the same week as the hot telephone wire incident, the June 2, 1911 *Guard* carried this article:

> *A. B. Chance is making an electrical demonstration at his factory building on Sneed Street this week. The display consists of electric lights in carbon and tungsten and of varied form and design, electric sad irons, toasters, electric fans, motors for all purposes, vacuum cleaners, electric ventilating machines, etc. The last named article is a very recent invention and cools and freshens up a room making the air just like it is after a rain, fresh and sweet and free from dust. Mr. Chance will also have a number of experts from the factories to demonstrate the different devices.*

The word *demonstration* has an important meaning to the Chance Company and was a method of teaching, selling, and promoting Chance products that A. Bishop learned well and passed down. He had a keen interest and involvement in electrical items of the day, as this article suggests. Parts of this display were his items, but most would be displayed and promoted by other factory representatives. No doubt one of the most inquisitive and interested spectators would be the host himself.

Centralia was progressing with a report in June 1911 of a new ice plant coming on line using city water. To control congestion and wrecks on the main line, the Wabash built a second track for the Columbia branch trains to get to the east yard and to load and unload passengers. Local complaints about the condition of the Wabash's Centralia station was spurred on by the *Guard*, but the financially troubled railroad finally announced in April 1914 that the station wouldn't be built until after the current receivership had been dealt with. The automobile was becoming popular, although it may have been difficult to imagine it ever replacing passenger trains. What Centralia needed was an Auto Club to provide a place for members to store and to work on their cars. A. R. Toalson would have his old building on Hickman Street available soon for a lease of five years and a pit could be constructed to conveniently work under the cars. This was a time when there were few or no garages and service stations for normal automotive servicing. The Auto Club sounds like something that A. B. Chance and Dr. O. B. Mayes would organize. They did and were president and secretary respectively. They were to occupy the Toalson Building the following April, according to the paper on December 15, 1911. Mr. Toalson would have moved his concrete business to a new building on Railroad Street by that time.

As 1911 was ushered out and 1912 was welcomed in, several significant events would occur that coming year. The *Titantic* would make its first and last voyage and another controversial ship, the battleship *Maine*, would be refloated, inspected, and then towed to deep water and sunk again. A very important event in the Chance Company story would be introduced by a severe ice storm that would provide for the development of the Never-Creep® Anchor. But prior to those events, amid reports of thirty-six-below-zero temperatures and an ice storm, the *Guard* reported on January 4, 1873: "The prairie is covered with a solid sheet of ice, from the recent rain, and everything is fine condition for good sleighing." There were no electric or telephone lines to go down, because they were yet to come. So instead of worrying about loss of electric power or telephone service they didn't have, the hardy people of that time took it all in stride and resourcefully utilized what was given them and made the best of it. How much does a difference in time make? There was one, tucked away in his mother's womb awaiting the springtime day of May 26, 1873, who missed all these 1872–73 wintertime activities. Fast-forwarding some four decades, A. Bishop Chance was entering his middle years as 1912 dawned and again a winter sleet storm swept across the Centralia prairie. But modern times had brought change as those responsible for local telephone service couldn't consider taking time out to sleigh or ice-skate, a sacrifice born out of becoming a businessman and a servant to the public.

The *Guard* on January 12, 1912, concluded that it was already one of the coldest in years: "The mercury dropped to 9° below zero Saturday morning, 12° below at 8:00 p.m., and 24° below Sunday morning. 8° below was the warmest during Sunday. Saturday also brought a three-inch fall of snow and Monday a light snow. The weather the first of the week kept below zero." However, as the commonly told Never-Creep® story goes, we should be anticipating a huge sleet storm and as bad as the early January weather was, it did not appear to be weather that would develop sleet and take down telephone lines, like falling dominoes. February closed with little fanfare, which is a surprise because central Missourians have experienced some miserable Februaries. But March is another story, as convincing evidence suggests that during this month the telephone lines went down and the Chance anchor solution to get the lines back up and stabilized, was born.

Again the *Guard* comes to our rescue in their historical coverage, the first on March 1, 1912:

> *Sunday was a weather day and the weather man dosed out all sorts of weather. First it was rain, snow and sleet, then it was slush and rain and wind. At night it snowed about two inches and tried to freeze up. Monday the roads were all but impassable and but few farmers came to town.*

A 1912 scene of the sleet storm that took down power and telephone lines (*Chance Tips* March 29, 1939)

Another article immediately below said this:

> *S. J. Sturgeon tells us Lake Dutcher is full up to the dam the first time since last March, about 11 months ago. With all the heavy rains of this and last week the lake has never overflowed the dam. There are over forty acres covered by the water and the effects of the last dry year can readily be understood when one is apprised of the fact that with all the early Winter rains and recent snows and rains the lake is just about even full.*

But the weather conditions that played havoc with the power and communication lines began with this March 15 piece:

> *The biggest snow of the season fell in this section Sunday night and Monday morning and which was added to Tuesday morning, making a covering of about eight or nine inches on the level. March came in like a lamb but the weather man is cutting up some capers now. Thursday morning brought a rain and the temperature is about right for a freeze tonight, again.*

As this system continued to hammer central Missouri, the March 22, 1912 *Guard* even warned residents to stay away from downed lines:

> *LET THE LIVE WIRE ALONE. . . . If this sleet continues the wires will all be down over the city and it will be well for our people to keep a lookout for hanging wires. With the present high voltage running over the electric light wires and with the possibility of the telephone wires becoming crossed with them, it would be a good policy to consider all wires live ones. If a wire falls across your path go around it. Don't try to poke it with a stick, but let it alone and notify the company—they have experts who are capable of handling the live wires.*

The December 6, 1925 Sunday Magazine of the *St. Louis Post-Dispatch*, covered the story of A. Bishop Chance. Within the article were more details about the Never-Creep® Anchor development in March 1912:

> *The day after the storm a pressing need developed for a large number of anchors to save what poles were still standing. It was impossible to install 'dead men' fast enough to save the situation. A "dead man" is a log buried a short distance from a telephone pole as an anchorage for the guy wire which keep the pole from falling. My father suggested that a rush order be put in for a shipment of patented anchors of a type then being manufactured. However, there was no time to be lost in waiting on the shipment. We needed anchors at once and I set to work to devise some means of burying "dead men" faster than we had been doing. I worked nearly all night.*

The March weather was summarized as to snowfall totals in the March 29, 1912 *Guard*:

> *RECORD OF THE SNOWFALLS. . . . Snow records for many years have been broken by the winter of 1911–12, and especially for March. Twenty-two inches of snow have fallen in this section during this month. A total of 49.8 inches have fallen during the Winter. Saturday's snow was nearly seven inches deep here. This is the heaviest snowfall for March since 1881. Kansas City has had the biggest snow of the season which started at 6 o'clock, a.m. Saturday and continued until 6 a.m. Sunday. Twenty-five inches of snow fell and traffic was stopped. It cost the city $15,000 to clean the streets and the street railroad $10,000 to clear its roadway. Collapsing roofs caused $12,000 damage, and the fire loss was $50,000.*

Remarks by F. Gano Chance in a 1968 speech told how the development of the first Chance anchor took place.

> *The winter of 1912 was a disaster. A sleet storm loaded the telephone wires with ice and pulled down miles of line. Faced with the task of rebuilding the line, Bishop Chance had an inspiration. His biggest problem was anchoring the poles. Accepted practice of the day was to use a guy wire fastened to a log buried deep in the earth. But the Centralia telephone company had neither the time nor the manpower to dig big holes and bury logs in frozen ground. Bishop Chance had a new idea. Instead of digging a big pit, he bored a small hole at a 45 degree angle toward the base of the pole. Then he drove a rod into the ground angled to intersect the bored hole. The end of the rod was threaded. Next he drilled a hole in a small piece of firewood, reached down in the bored hole, slipped the wood over the end of the threaded rod and used a nut to hold it on. The guy wire he attached to the top end of the rod. At the time he thought of it as a temporary solution. He soon realized it could be a permanent one. Like many fine ideas, it was simple—almost as simple as the paper clip and the safety pin. The new anchors could be installed more easily and quickly and they held better because they pulled against solid, undisturbed earth. (1968)*

The Chance Company ties with the telephone industry is further strengthened in 1912 when the idea for the Never-Creep® Anchor was sparked by a telephone problem that had not been adequately addressed. Only later would it be used on other projects besides telephone poles. The anchor was neither the first nor even the second invention of A. Bishop Chance, although the anchor takes center stage and eventually added significant credibility and potential to Chance Company business activities. This clearly shows that not all inventions are destined to become commercial successes, whether due to mishandling of patents, promotional deficiencies, bad timing, and simply the lack of a need by the general public for the invention. What the earlier inventions had taught Mr. A. Bishop was invaluable and helped lead to his success with the anchor. The world is not always kind to the unaware and those not on their toes, as A. Bishop experienced with his bicycle handlebars idea. Simply trusting a person to do the right thing or relying on their handshake or their word, as was the custom in central Missouri, was not adequate out in the business world. The Chance father-and-son team would become very aggressive in protecting their product turf.

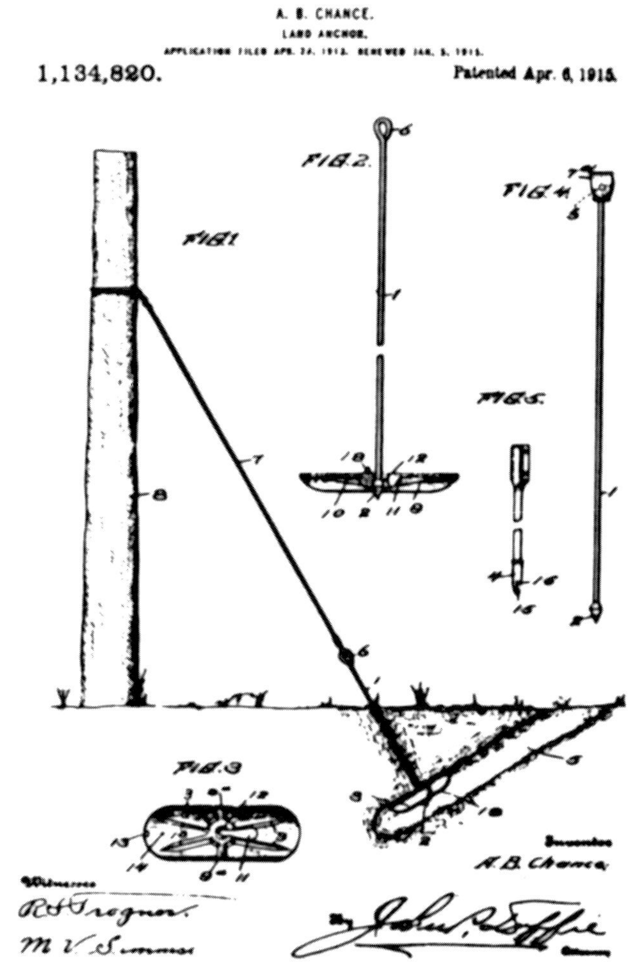

Pictured are drawings of a Land Anchor invented by A. Bishop Chance in 1913. It became Chance Patent No. 1,134,820 on April 6, 1915. Notice the anchor rod in Figure 4 was a part of this anchor patent, as suggested in the Thimbleye® section to follow. This appears to have delayed individual anchor rod patents until 1920. (A. B. Chance Co.)

What are the expectancies associated with going back and investigating the time when the Chance earth anchor was born? Can you see the headlines proclaiming the discovery of the Never-Creep® Anchor that restored telephone service after the destructive sleet storm inundated the area? The reality is that a significant invention or discovery often is not recognized immediately until it's developed further. The Chance solution for holding telephone lines up could have been lost locally like many other innovative ideas that A. Bishop Chance developed and frequently applied in solving problems. Just as important as the solution itself was the realization of its commercial possibilities, which led to product improvement, promotion, and eventual industry recognition. We may never know how this all transpired, but the successful application of this solution apparently opened his eyes to the fact that he had something special with great potential. His enthusiasm, problem solving, and inventive genius then took charge.

It's fortunate to have documentation of what gave birth to A. Bishop Chance's ingenious guying idea. One can contemplate all the efforts being made to bring the lines back up, but instead the *Guard* gave a preview of spring on April 5, with reports of the sun coming out and melting all the winter moisture away. Coverage of these events reflects the utter chaos that normally accompanies utility service interruptions. When this occurs even today, there is an all-out effort to get the utilities back up and running and A. Bishop didn't have the time to report his activities to *Guard* reporters. There may have also been a security issue with an unpatented idea and product being developed, but its importance likely wasn't understood until later. Certain local telephone people did become aware of A. Bishop's solution and took an interest in it. Some information suggests they were the initial encouragers for him to refine it, as it would also benefit them. This could also have resulted from later promotional efforts by A. Bishop, which appears to have begun in early 1913 after it was patented. The *Chance Tips* of 1939 did cover A. Bishop's recollections of that storm:

> *That night is still vivid in my memory. As manager of the local telephone company, superintendent of service, head linesman, and man-of-all work, I had been up all night trying to keep my lines clear. All of our efforts went for naught, however, and every line in the country was put out of commission. When morning came, we were faced with the enormous job of rebuilding the entire system. The temperature was below zero, there was snow and sleet galore. The ground was frozen, roads impassable, labor scarce and costly, and our subscribers naturally impatient to have their service resumed. New poles had to be put in, new wire strung and almost every pole straightened and re-anchored.*

As the construction of telephone and telegraph lines began to stretch across the landscape of America, the guying of the poles to handle loads exerted by the wire was without a practical solution. Such elements as wind and sleet intensified the problems in maintaining the upright integrity of these lines. Out on the bare Centralia prairies, natural objects used for guying, such as large rocks and trees, were almost nonexistent. So the need for a simple, man-made, mechanical kind of solution to guy lines on those open plains was even more of an acute requirement. How important a factor this setting was in driving the development of the Chance anchor idea can only be speculated, but the inventive genius of a mid-Missourian would provide what would eventually become a modern industrial solution in solving this problem everywhere. There was an interesting news item in the local paper in early February 1916 suggesting that another damaging sleet storm had hit the area. It said that the Mutual Group had lost many lines during the storms of the last month and being a competitor of the Home Telephone Group, one wonders if they were anxious to do business with Chance Manufacturing Company or vice-versa. No reports of downed Home Telephone lines were publicized, suggesting that the new Never-Creep® Anchor, which one could assume was used generously on Home lines, had solved or reduced the downed-line problem.

The origin of the Chance Never-Creep® Anchor was a log, according to F. Gano Chance. These kinds of natural objects weren't anything new. Farmers had used trees, rocks, and logs to hold down their fences in dips where wires in tension would tend to maintain an unacceptable straight and level position, which allowed room for an animal to get under. But early telegraph and telephone lines had a little different problem as the line was stretched to maintain a somewhat level configuration. Enough line tension was necessary to prevent excessive drooping so animals or vegetation would not get into it and break the line or ground it out, thus causing a "failure." Line loading was abnormally increased by ice, sleet, and snow from winter storms and this placed astounding weight on a line. In combination with strong winds, a line could be torn down when such loadings exceeded the mechanical capabilities of the line. The line wire could simply break or the insulator or fastening hardware fail. Another additional source of failure was in the pole due to side loading from crosswinds, collapse of rotted untreated early poles, improper pole installations, or changes in the direction of the line which could tend to make a line lean from a preferable, perfectly vertical position. These pressures intensified as the leaning increased, because of its height, tending to pull the pole and line onto the ground generating a failure. The anchor could counteract such forces and also be applied to prevent the pole from leaning toward the direction of the line tension and load. An illustration from the November 1985 Chance sales catalog, in the pole line hardware section, pictures the various components on or around a utility pole, including an anchor and guy line. Although it is a more modern setting, it pictures the basic products being discussed and the variety of components that would be developed.

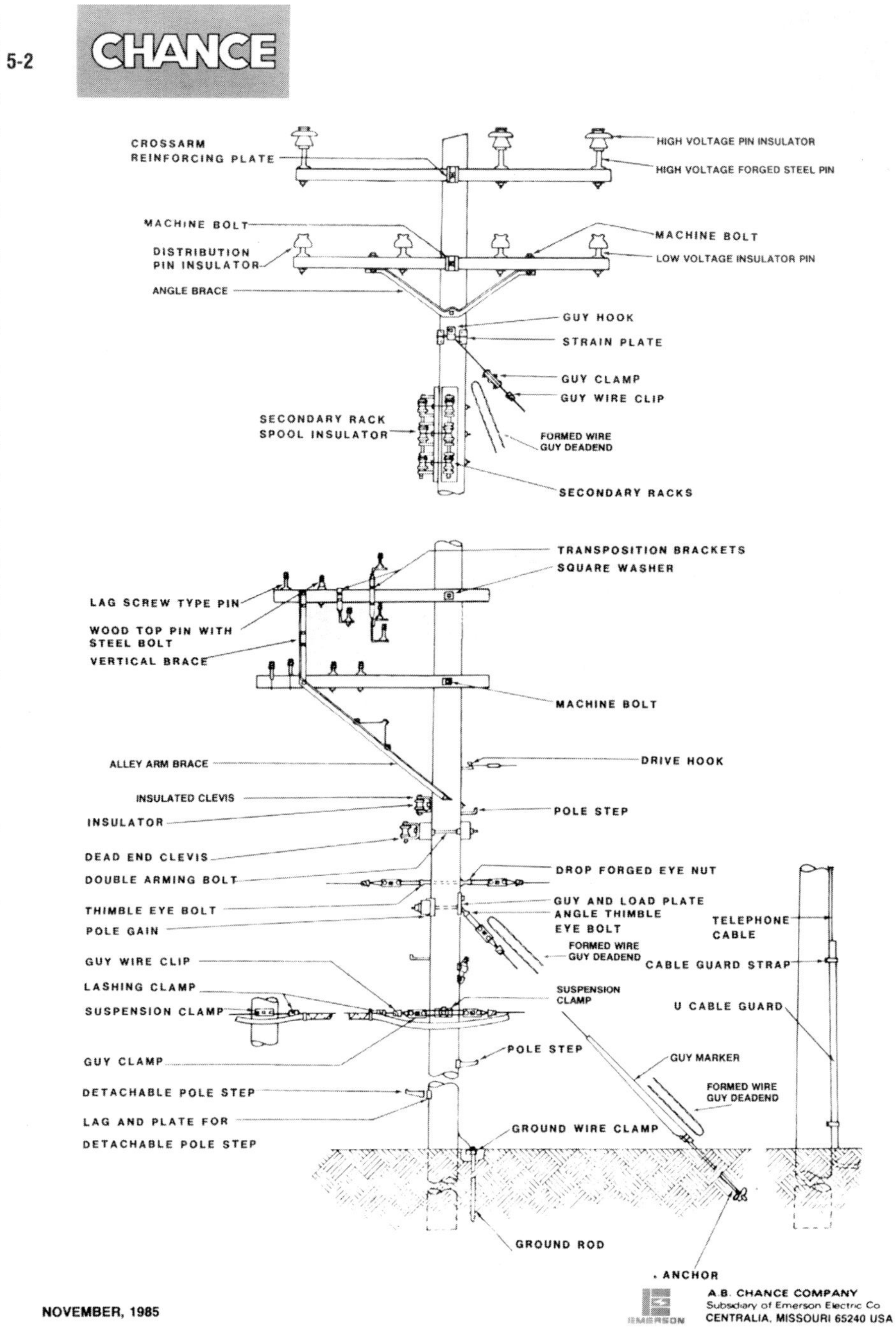

Hardware Items on pole (A. B. Chance Co.)

A. Bishop continued to remember the night his telephone lines went down in the 1939 Volume 1, Number 1, *Chance Tips*:

> *It looked like a long, mean job. The anchoring of the poles worried me most. Not only did they take the most time, but the strength and permanence of the pole depended upon the efficiency of the anchor, and so many of our log anchors had pulled out or slipped up that we were unable to use any of them. The usual method at that time consisted of burying a log 'dead-man' and cutting a channel for the rod attached and as the log was some four feet long and ten to fifteen inches in diameter, you can imagine the pit that had to be dug. Under normal conditions, one man could put in but two or three of these 'dead-men' a day. Here then, was the place to save time if a smaller log or smaller hole would do the*

work, even temporarily, it would be a life-saver for me. With this thought in mind, I dug a hole at right angles to the guy wire, instead of in line with it as formally. Then I drove the rod into the earth at right angles to the hole and in line with the guy wire, until the end projected into the hole. The log was then slipped down into the hole and anchored to the end of the rod. My new anchor was complete!

Notice that the dead man anchor was already being used and the driving of a metal rod down through undisturbed earth, thereby utilizing the weight of the soil instead of the anchor, may have been the most important concept discovered during the 1912 ice storm. The secret for success of the Chance anchoring system, in insuring line integrity, included another unique component between the anchor and a convenient out-of-ground point where a guy line could be attached to the pole. The general method being used at that time was simply wrapping a wire around the outside of some above-the-ground natural object or attaching a guy wire to a log that was placed in a hole dug for that purpose. What was needed was an item with special capabilities on both its ends to tie all this together and be able to be driven down into undisturbed soil for greater holding power. What eventually developed was one of Chance's most successful products, that couldn't be legally duplicated or improved upon to any significant extent by competitors. The device chosen by A. Bishop was a metal rod, which appears to have been the most developed component utilized during the 1912 sleet storm. It certainly appears more sophisticated than the log anchor that was simply a stick of wood with a hole bored in it, however, the secret was in how the log was prepared and placed in the ground. The innovative idea for the anchor rod helped give birth to the Never-Creep® Anchor, but of course the more common basic log anchor could have driven the development of the rod design because something was needed to connect the guy wire while making such a small hole that the anchor was holding against undisturbed earth. In the end, however, both these parts worked together as a team to make it all possible. Gano's understanding of these developments undoubtedly was influenced by his father's explanations about how this occurred:

Now the Thimbleye® came about, probably 7 or 8 years later. He called it the Thimbleye® but the standard rod, which he had (famous picture of Mr. A. Bishop with a replica of the first Never-Creep® Anchor) was started from the log anchor. He turned out a metal point that you could screw on this rod and drive the rod. Then he would reach down in the hole and unscrew the point and slipped his log on it. Then . . . bolts to the log. Now that was 1912. (1978)

Interpreting Gano's remarks along with information from other sources suggests that a hole was drilled or bored in the ground at a forty-five-degree angle from the base of the pole to be guyed. That is, the hole would be angled so it started a distance from the pole and would get closer as it was dug deeper. This forced the log (anchor) to lodge against undisturbed earth when the guying tension was placed on it. A log, which was more on the order of a large piece of fireplace wood, was placed in this hole. But first it received a hole in its center into which the metal shaft or rod would fit through. The point on this rod would allow it to more easily be driven into hard and/or frozen ground at a right angle to the hole that contained the anchoring log. It would also protect the threads on the rod end from becoming plugged with dirt. The trick was to line up and drive this rod in the proper direction of the small hole in the log. There were no laser beams available to precisely guide the rod. The need to be precise was driven by the same reason the log hole was angled away from the pole so stronger, undisturbed earth would provide adequate holding power to offset the stresses resulting from guying the pole. Of course valuable time could also be saved if the rod was lined up correctly to the anchor hole eliminating multiple rod drivings, but moving the log to line up with the rod was the likely solution. Holding power was always a concern in anchoring, especially with screw-type anchors, which loosened the soil as they were screwed into the ground. An anchor that "creeped" or released the tension on the guy line could be considered a failure in several respects including continuous maintenance. This was the importance and significance attached to the name Never-Creep®, implying that when installed, if the anchor was against undisturbed earth, it would not creep. The implications were that man-made anchors were considered suspect and even from the beginning the anchor, to be successfully marketed, was expected to be a product that line crews and management could forget after proper specing and installation.

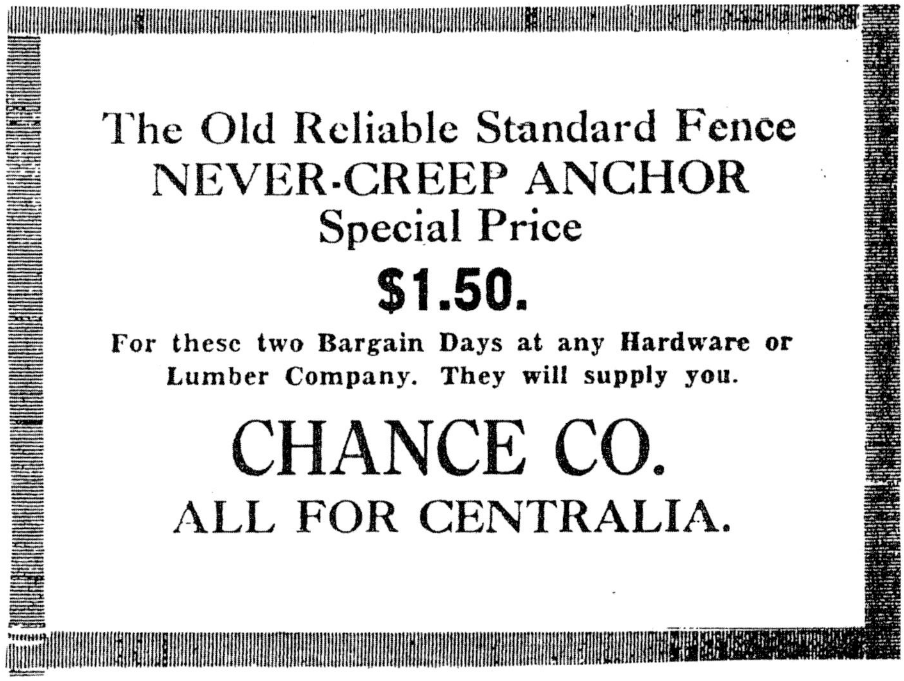

From the May 31, 1929 *Guard* comes a Chance ad for the Never-Creep® Fence Anchor. *(Guard)*

After the rod was driven down and through the log, a plate or washer was used along with a nut screwed onto the rod end, to hold the rod to the log. A. Bishop, interviewed in *Chance Tips* in 1939, recalled: "The big disadvantage, of course, was the difficulty of bolting the rod to the log after it had been sunk into the hole. The development of a metal plate was the natural result. This not only completely eliminated the difficulty but strengthened the anchor materially and made a much smaller hole necessary." This would appear to be a slotted plate so when the rod with a shoulder on its end was inserted it could be repositioned and "locked" into position. This would eliminate the need to reach down into the hole to screw the nut on and would allow a deeper hole increasing the anchors holding power. These items were the first metal parts, preceding the Never-Creep® all metal anchor. Gano explained this conversion from wood to metal. "Some time the last part of the year (1912). Just as soon as he got through with this job he realized he had something, so then he got a patent. . . . After he got the job done then he started to make an anchor." (1978) The thinner anchor could not only ease lining up the rod, but may have required a smaller hole in the hard frozen earth during cold Missouri winters. "The problem was he could not dig a pit in this hard ground in the winter time. So he conceived the idea of drilling a hole." (1978) This concept was quite different than simply attaching guy wires to an above-ground natural object, but instead encompassed many ideas that made man-made anchoring possible and some hundred years later A. Bishop's anchoring principles are still being utilized.

The promotion of the new Chance guying marvel began no later than a year after its development and employment in the 1912 sleet storm per this January 10, 1913 *Guard* news item: "A. B. Chance is in St. Louis this week attending a meeting of the Commercial salesmen and jobbers of electrical goods and supplies. Mr. Chance went to introduce his new 'dead man' a device he has just patented for holding guy wires in the ground." This is a very important piece as it suggests he is introducing his new product to the wholesalers and distributors of electrical apparatus. One could speculate that this included those who represented the telegraph and telephone industry as well as the electrical power people. Notice that A. Bishop had "just patented" his new product and that it coincides with the registration of the trademark Never-Creep® on January 1, 1913. Interestingly, the paper still used the generic term *dead man*, which described a whole range of stationary objects used in guying lines, fences and other unstable man-made items. Name recognition doesn't happen overnight, but has to be brought before the public through advertising and salesmanship. Another helpful aspect, especially for a new and unique product, is for the name to be descriptive and so identified with that item that it becomes more attractive, colorful, and easier to recall than the generic name. Although patent attorneys may have helped mold the names of the various Chance products, there is also a tinge of realism and product appeal that is identifiable to A. Bishop Chance. The idea of "never creep" is actually a negative connotation, representing the weaknesses associated with man-made guying, that this new name suggests had been overcome. Its originator knew this was a key element that could either make or break his new product.

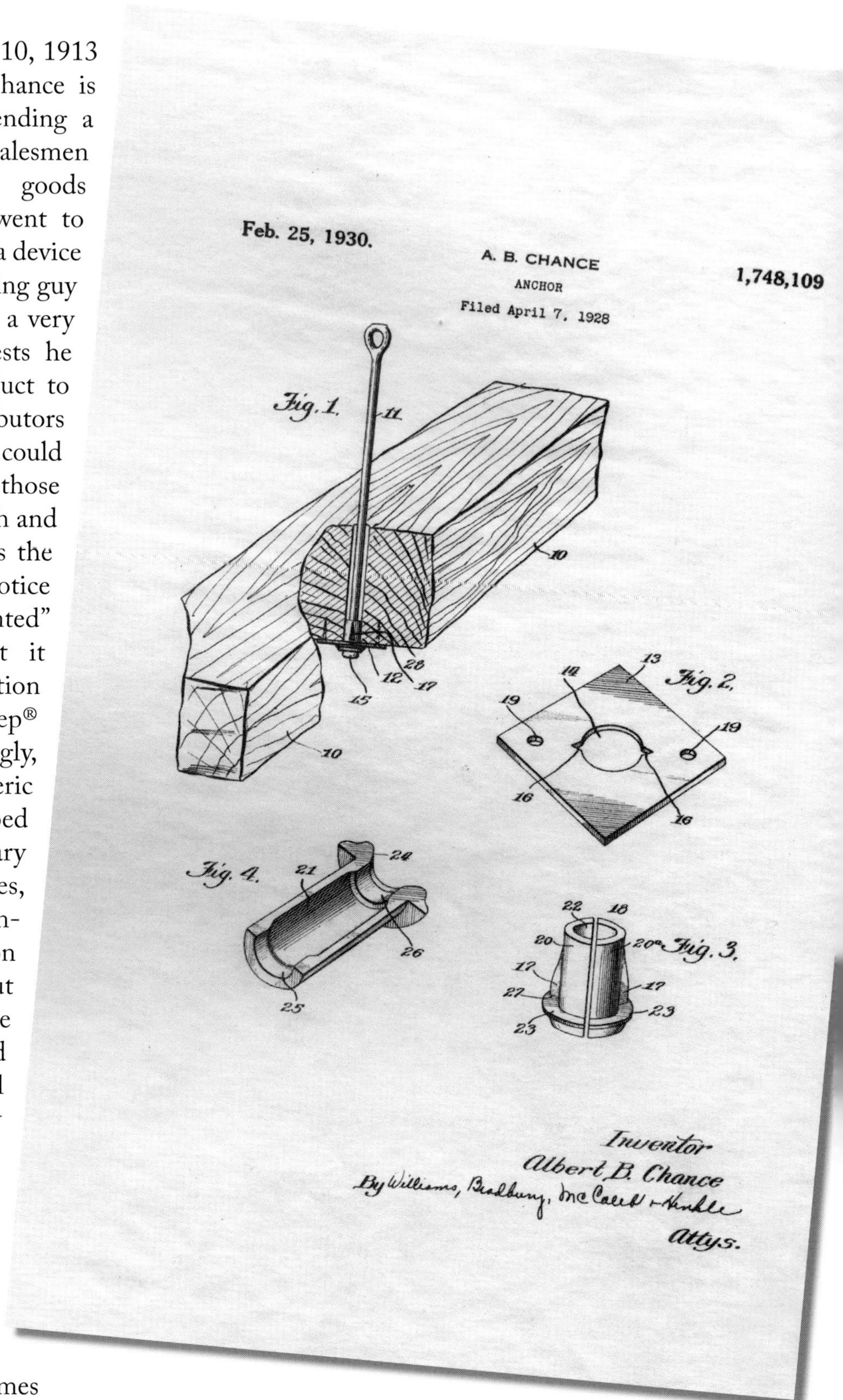

No. 1,748,109. (A. B. Chance Co.)

Surprisingly, some twenty years after its first use on Home Telephone lines, a patent file No. 1,748,109 revealed a new patent application was filed on April 7, 1928, and granted on February 25, 1930, for "Anchors." A handwritten note on a file copy, possibly A. Bishop's remarks, suggested it specifically was for "Dead Man, With N.C. Rod." The N.C. would refer to Never-Creep®. This patent raises the issue of why a seemingly obsolete anchoring arrangement that is distantly separated from metal anchor production, in which Chance Company was heavily involved by 1930, would warrant the pursuit for a patent. From an assortment of Chance patent information, A. Bishop, like other inventors and firms, tried to legally blanket a product from every angle that potentially would lead to a successful production item. In reality, the only way an inventor makes any revenue is if they sell their invention to someone or it becomes a sales success. With a concern of not relinquishing any advantage to a competitor firm or inventor, there was a constant challenge of reinventing the wheel kind of activity that was also fueled by A. Bishop's aggressive and protective nature. It is no accident that Chance Company became the leader in anchoring over such a long period. Another reason for this particular patent may indicate the "Dead Man" anchor, with little or no cost connected with it, was the choice of farmers who usually had a plentiful source of such natural objects. There was a constant struggle to bring on board new products to expand the hardware and anchor product lines to provide more sellable items in the Chance Utility Catalog. The attempt to gain new products to manufacture and sell was vigorously pursued, especially during the Great Depression. This 1930 Chance patent required an anchor rod and assorted hardware, which could bring increased business. This new product also led to a patent application in Canada, suggesting our northern neighbors had a patent program similar to the United States. Another revelation of this patent file is an explanation of "Dead Man." It suggested "a log, timber, metal body or other relatively fixed member which is usually termed a Dead Man. The Dead Man is usually buried in the earth and presents a relatively large surface for engaging earth above it and increasing the holding power of any member which may be secured thereto." Keep in mind the detail normally included in a patent filing was usually initiated by the inventor and then dressed up by a patent attorney. Preliminary patent application file copies with words inserted and suggestions made by A. Bishop suggest the process and the expertise and focus on detail that he impressively possessed. The following addition, which expanded the coverage of the "Dead Man" patent, is an example that is reflective of his thoroughness: "I desire it to be understood, however, that my invention may be used for attaching guy rods to other members besides those buried in the earth, such as trees, poles, and the like, whenever such procedure may be desirable. Consequently, the term *dead man* is employed throughout this specification and in the claims in the generic sense of including any relatively fixed member to which it may be desirable to attach a guy rod." The illustration of the Chance invention in the patent filing is reproduced on the previous page.

Even F. Gano Chance got into the act of anchor development with a better anchor: "I got the idea (I was seven years of age at the time) that as he was buying people out that he would have to buy me out so I designed an anchor. His anchor was good, it would never creep, but mine would never start to creep, so I called it 'Never Start'." (1978) The sharpness of Gano at such an early age shows his father was teaching him and Gano was displaying business skills that would later be put to the test.

As indicated, one of the first improvements in the Chance anchoring system was the substitution of metal over a wooden log to make the load-bearing object in the ground a true anchor. Although its function didn't change, this metal anchor became a single-purpose specialized object that could not be confused for anything else or used, for example, to stoke a fire on a cold winter night! This development shows the importance and the potential he placed on this man-made ingenious idea to support the dependability and reliability of electrical, telephone, and telegraph lines. At that time a national electrical system with its far-flung lines, which would rely heavily upon anchoring to insure line integrity and reliability, was hardly contemplated, although the few lines that existed were generating serious safety hazards when they went down, especially around all the telephone lines. The mass of telephone poles and lines upon America's landscape, that had developed almost overnight, were obviously requiring some form of guying mechanism. So the question could be asked as to which was more important, the initial idea or the continued refinement and promotion making it a more saleable, practical, higher quality, and more sought-after product. This was the genius of a Henry Ford or A. Bishop Chance. Henry Ford's first or even second auto is not what put America on wheels and neither was it to be Chance Company's destiny to sell wood log anchors. There are several activities previously mentioned that describe the phases and areas that required attention in putting Chance Company on the road to becoming the largest earth anchor producer in the world, and A. Bishop's talents were broad enough to cover them all.

Granny Ryman suggested that local people were involved in A. Bishop's enterprises:

> *I think that my Dad probably might have made some contributions to Mr. Chance and his effort to develop the anchor. Because he got together with people and invested in it. Everybody was for Boss Chance, to help him get the job done. He was a good promoter, don't think he wasn't. (1979)*

Local people were also involved in the actual operation of the Chance plant. As earlier covered, one of Home Telephone's former employees, N. Denton Palmer, whose marriage to Miss Iva Earl Day was covered in the April 18, 1913 *Guard*, also said "he is employed as expert workman at Chance Manufacturing Company plant." He was one of the longtime factory employees, very likely senior, who unveiled the bronze plaque of Mr. A. Bishop Chance in 1949! Granny also spoke about the early years of Chance Company and even his involvement in some of the manufacturing activities:

> *As I was saying, the telephone company was the original start of the A. B. Chance Company because of Mr. Chance's development of the Never-Creep® Anchor.* [Note: Granny is suggesting that telephone needs and use drove the anchor invention.] *He worked very long and diligently of getting this anchor accepted because it was something new. People would say that they couldn't very well afford to buy an anchor that way and then have to hire a Civil Engineer to put it in the ground. So, Mr. Chance would work and put these demonstrations on with it. Then time went on and he made some progress with it and along with it he developed an item known as the 400 companion coil lightening arrester. In those days you would have a crank telephone with a magneto on the wall and if you would be sitting there and a lightning storm would come along the lightning would play out across the ceiling of the house, etc. So he developed this lightning arrestor that consisted of two wafers of porcelain with a copper strip on each end. . . . Then he would wrap a alloy, lead zinc wire around it . . . at each end. (1979)*

The production machine that Granny said he ran several times was a sewing-machine-type apparatus that had wire instead of thread wound in its bobbin. The wires were then soldered on both ends of the assembly. As previously covered, the anchor came after the Chance Manufacturing Company start-up, which initially concentrated on electrical products like the arrester.

Patent No. 116,995 of 1939 is reflective of the original Never-Creep® Anchor design, described here as an Earth Anchor. Both the cover page and sheet one of two picturing the anchor was processed by patent attorneys Williams, Bradbury, McCaleb and Hinkle of Chicago. (A. B. Chance Co.)

Design

No. 116995

THE UNITED STATES OF AMERICA

TO ALL TO WHOM THESE PRESENTS SHALL COME:

Whereas ALBERT B. CHANCE, of Centralia, Missouri,

PRESENTED TO THE Commissioner of Patents A PETITION PRAYING FOR THE GRANT OF LETTERS PATENT FOR AN ALLEGED NEW AND ORNAMENTAL DESIGN FOR

AN EARTH ANCHOR,

A DESCRIPTION OF WHICH DESIGN IS CONTAINED IN THE SPECIFICATION OF WHICH A COPY IS HEREUNTO ANNEXED AND MADE A PART HEREOF, AND COMPLIED WITH THE VARIOUS REQUIREMENTS OF LAW IN SUCH CASES MADE AND PROVIDED, AND

Whereas UPON DUE EXAMINATION MADE THE SAID CLAIMANT is ADJUDGED TO BE JUSTLY ENTITLED TO A PATENT UNDER THE LAW.

NOW THEREFORE THESE Letters Patent ARE TO GRANT UNTO THE SAID Albert B. Chance, his heirs OR ASSIGNS FOR THE TERM OF FOURTEEN YEARS FROM THE DATE OF THIS GRANT THE EXCLUSIVE RIGHT TO MAKE, USE AND VEND THE SAID DESIGN THROUGHOUT THE UNITED STATES AND THE TERRITORIES THEREOF.

In testimony whereof I have hereunto set my hand and caused the seal of the Patent Office to be affixed at the City of Washington this third day of October, in the year of our Lord one thousand nine hundred and thirty-nine, and of the Independence of the United States of America the one hundred and sixty-fourth.

Attest:

H S Miller
Law Examiner.

Conway P. Coe
Commissioner of Patents.

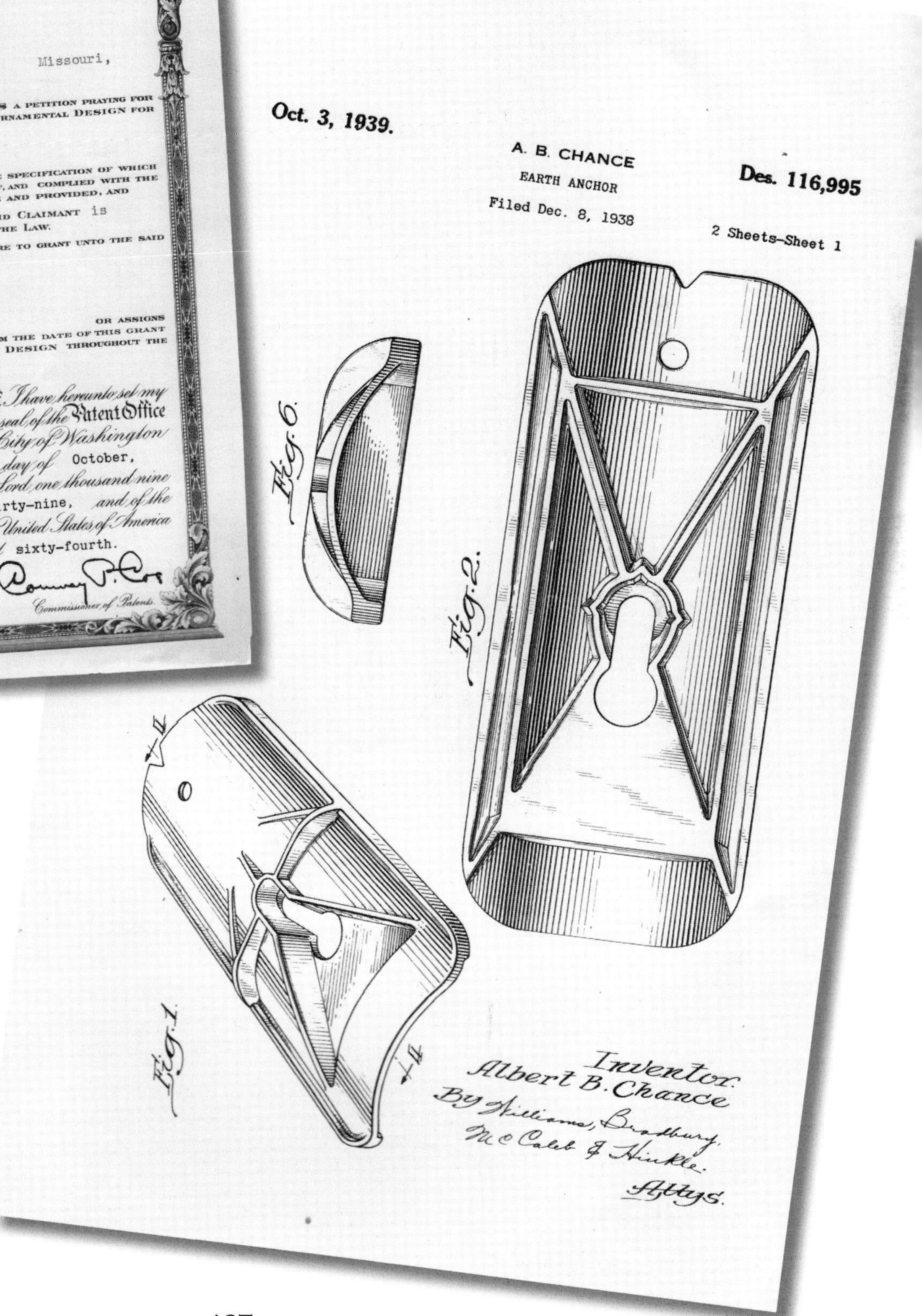

THIS SOUNDED LIKE A FISH STORY...

A Never-Creep anchor continues to hold during 6 years out of the ground.

Some time ago, when Bill Moss, the gent in the photograph, called on one of our South Carolina friends, he was pleased, but not surprised, when told of this company's fondness for the Never Creep Anchor. "In fact", continued the wire chief, "we have one Never Creep that has been out of the ground for about seven years and it's still holding perfectly."

This set Bill to scratching the old skull. His investigation revealed that several years ago, the Highway Department had cut a road through a hill and in doing so, completely exposed a Never-Creep Anchor. After seven years, the anchor is still exposed entirely and is still doing a perfect job.

From *Chance Tips* of June 22, 1939.

Gano in describing other Chance Company products in addition to the Sky-Rocket® said, "Well, the terminal bracket was another item . . . he put around his knee—rubber material knob . . . and he bent it around the knob and that is the way he made up the insulator tie wire. That was about the main thing he had when he was in Kansas City." (1978) Roy Hansen added, "So he was making lightning arresters in about 1910 and then along about 1912 or 13 he started making anchors." (1978) Gano agreed:

> *That is right. In the telephone business he had contracts with telephone companies and gradually with the power companies in which he made the original calls. He demonstrated the anchor and sold them. It was maybe 5 or 6 years before the supply houses decided—here is something we ought to have—but it is typical of a man starting. . . . In order to get a good patent, he had to buy out two other fellows that had similar ideas that were making application for a patent. (1978)*

Note the significance of this early acquisition or buy-out that would be repeated over and over at A. B. Chance Company. The anchor's possibilities and applications spread from that point due in large part to A. Bishop's early promotion and demonstration programs. A product's success often depends solely upon the ability to put it in front of the customer and to generate enough enthusiasm to produce a sale. One significant early customer was Western Electric, who according to Gano bought their plate anchor.

Along with obtaining a patent on the new Never-Creep® Anchor, the name was also registered as a trademark on January 1, 1913, according to U.S. trademark data. It would appear that the patent and trademark were applied for at about the same time in late 1912, which fits Gano's recollections. A. B. Chance Company, on January 4, 1954, renewed or reregistered this trademark. If not redesigned, a trademark can have continuous coverage. Then on November 9, 1974, a "1st Renewal" is recorded which coincides with Emerson's purchase of A. B. Chance Company, but this may or may not have been relevant. What is relevant is this famous Chance trademark, is now currently listed as dead apparently because it wasn't considered important, but obsolete with no future. Most firms today believe in streamlining and in eliminating a purchased firm's product offerings to a core group of high-volume items and don't concern themselves with the historical background of a product.

As the keeper of the item masters at Chance (1989–2001), one of this employee's functions was to code items obsoleted by Engineering with an "O" status code and apply what was called a "Y" code. This meant an item could have no inventory, customer or production orders, or usage in another item. Instructions were to keep count of these items after the new owners decided to reduce the product offering to a smaller core group. The result was something close to ten thousand parts, assemblies, and finished goods items were removed in a couple of years from the original thirty-thousand-plus items in the item master file. The loss of a part or all of a former firm's industrial heritage will hopefully generate a widespread concern of recording the history of U.S.A. manufacturing firms and products that represented a time when America's industrial self-sufficiency was evident.

As 1913 was coming to a close, the *Guard* ran a story on December 19 that could be labeled a Christmas Story, as it provides us the gift of insight into the magical world of one A. Bishop Chance. Many people during the Christmas holiday season visit the Plaza in Kansas City (considered the first modern shopping center) to shop and to see the lights and all the window displays. Those windows with miniature, motorized figures acting out some activity, represents the entertaining, sophisticated variety only a big city could offer. But then we weren't in Centralia in late December 1913 on a cold winter's night:

> *A. B. Chance has a very interesting and unique display in his window of the articles handled by the Chance Manufacturing Co., the display being lighted in a novel manner by a motor attachment. A small electric sign is lighted up advertising some special article and a small light is made to throw a glow upon that particular article in the display. This continues in progression until all the articles in the display have been pointed out, then the lights begin going on until they are all lighted then the display goes dark and begins all over again. The display is run on a small scale after the plan of the large motor signs seen in the larger cities and is a very creditable one indeed.*

Approaching its second anniversary, no doubt one article being lit up was the Never-Creep® Anchor! In addition to telephone line usage, as the development of electricity and the construction of electrical power lines progressed, the earth anchor was there to do its part. This was the beginning of a long and growing relationship with the electrical utility industry in which new Chance products were developed that not only solved guying problems in the ground around the pole but also elevated to hardware and electrical items on the pole and line as well. This is altogether appropriate as the first successful commercial product of Chance Company was the Sky-Rocket® Lightning Arrester that was an aboveground, more electrically oriented product. The practicality of a mechanical anchor would have to be sold over the inconsistent characteristics and even unavailability of natural anchors along the path of a line. The man-made anchor did become a success, due not only to its convenience and advantages, but also because of the untiring efforts of A. Bishop Chance It wasn't that much different than other industrialized solutions to everyday functions such as the lawn mower replacing goats, sheep, and shears designed for human hands. Once professional, specialized solutions were found superior enough to justify their cost, whole new industries sprang up to serve such needs. Today we see the electronic revolution driving and growing our economy in the same way and providing us with capabilities unthinkable a short time ago. We should recognize and appreciate these visionaries who continue in each American generation to provide the public with progressive technological advancements that we depend on every day.

The Teacher

To the many positions that A. Bishop Chance officially or functionally held, add teacher. Successful parents become effective teachers who help shape and mold their children. Gano wondered at times how many areas he got into on his own versus what his parents had introduced him to. He concluded that his father was a very clever man who had a strong influence on his early life.

> *He was able to influence people including myself. Looking back now and thinking of the things that he did, getting in business and in the holdings when I was quite young, my father must just have moved me into those directions. But he did it so cleverly that it felt it was really my idea. He used to give me lectures at night on electricity and I took notes as a kid. Occasionally we had parts of telephones, etc. My brother and I were very close together and he had a little more personality than I did but he didn't care anything particularly about science. (1981)*

Gano had learned as a young child how to take a telephone apart and make a crude telegraph, which was used by the neighborhood children to communicate with each other. This was one of Gano's first business ventures as he sold some of these components and services to friends, which mirrors his father's youthful scientific and commercial inclinations.

Chance Advertising Manager Art Houston addressed a questionnaire to Gano in 1941 to find out more about him. One area focused on was Gano's home environment while he was growing up. Excerpts from that completed document gives further insight into both father and son: "My father taught me the fundamentals of electricity, when I was 9 years old. I learned to make telegraph instruments at the age of 10 and made and sold many sets.... I read slowly—having read mostly science." In addition to the technical kind of science, there was also the political kind of science that Gano became emphatically associated with. His political science interest were begun or at least enhanced by the yearlong world tour. His father

Mrs. "Fancy" Chance with her two sons Buford and Gano. A small Gano in the center suggests this photo was taken around 1909. (A. B. "Buff" Chance IV Collection)

also taught Gano magic tricks at age 12, which became a lifelong hobby and interest. Although influences by Gano's mother aren't mentioned, her style and grace plus her presence in a childhood marked by her husband's frequent absences resulted in significant contributions. Gano's aptitude for education, plus his class and sophistication point to both his mother's and father's impact upon his life.

A. Bishop stirred not only the inquisitiveness of his son, but also other area children. It is difficult to comprehend just how far a person's influence stretches. The Centralia area was undoubtedly more progressive than similar communities, simply because A. Bishop was in tune with the latest developments, especially electronic devices. This would impact Chance Company because many people who were touched by the Chance family became influential employees. Granny Ryman is an example of one who knew Gano when he was a child:

> *Yes, when I was growing up he lived in the house just adjacent to the Stone home, which was on the corner of Jenkins and the old nursing home. Gano's father was always interested in everything that was new in electronics and radio, etc. So Gil Stone and I would go with Gano and listen to the radio and he had all these radios and all this sort of thing. We would just spend a lot of time with it. Gano was 2 years ahead of me in school. (1979)*

Parents hope positive influences and not negative ones will have an impact on their children. Substance abuse and addictions to drugs, alcohol, and tobacco can cause serious consequences to the afflicted and those around

Tomorrow: A full page Chance ad was run in the *Guard* on December 24,1937. Although it is a future piece and was directed toward responsible actions during union negotiation, it also connects with "The Teacher" text in the Chance beliefs it conveys. (*Guard*)

them. Mr. and Mrs. A. Bishop Chance taught their children not to involve themselves in such habits, as F. Gano's life demonstrated.

The fire that destroyed the Chance Snow Avenue home took place in early November of 1915, according to the November 12 issue of the *Centralia Fireside Guard.* The fire started in their neighbor R. C. Pool's woodshed. The Chances had time to save their car and some of their furniture, including their piano. The Fire Department was slow in reaching the scene and was only effective in preventing the fire from spreading to other homes. The new addition on the west or rear end of the house was nearly completed, and several contractors lost tools and supplies. The Chance loss of $5,000 was covered by $2,750 of insurance. The article added that Mr. Chance lost his "library of technical books, together with other valuable apparatus." The Pools, in addition to their home, lost a new Pathfinder automobile. In the excitement of waking up neighbors and fighting the fire, they forgot to remove their new car. Earlier that year in late June, the paper reported the painting of the A. B. Chance residence on Snow Avenue by Manny Berg. Before that, on May 7, 1915, Berg painted the front of the Chance "shop on the south side of the square." That "shop" would be the Chance factory building. It seemed that one of the Berg boys ran into some difficulties and this may be another early example of A. Bishop taking someone under his wing.

Gano recalled that his father built a two-story addition onto the back of their house. The second-floor area was being prepared to be a laboratory that Gano anticipated would be used by both himself and his father. It was so new that the exterior hadn't been painted and items such as a new x-ray machine hadn't yet been installed. There were also other pieces of special equipment such as a new vacuum tube radio that had replaced the earlier crystal set. Gano remembered the fire first struck the neighbors and quickly spread through some outbuildings to their new rear addition. At first it was thought that the fire wasn't a serious threat, but this quickly changed. Gano said he got out of bed and instead of trying to save toys proceeded to the laboratory where he took up the radio and x-ray equipment and removed them safely outside. It was frequently suggested that Gano seemed to know what was important, which was influenced by his father's teachings. The news article suggested even rugs and some fixtures were removed before the house was engulfed in flames, but many other items were lost. Fires were a greater threat than today because of their more frequent occurrences (flammable heating/lighting means) and because of a lack of high-pressure water or inadequate firefighting equipment. Although the Chance family may have been more affluent than the average Centralia family, this did not bring immunity from troubles and hardships.

It is significant that the Chances were involved in such advanced equipment of the day, in the mid-teens. Some of the earliest mentions of the x-ray's medical applications were found later, for example, in the 1921 newspaper. The *Guard* on April 4, 1924, also told of the new x-ray machine that A. B. Chance had purchased. This raises a question if this was a replacement or more advanced model of the original one. Gano had a lifelong fascination with radio equipment. He had a two-way radio in his car for security and communication purposes in the 1960s as this employee recalls. There were antennas on top of the main office that provided extended range for this equipment and for reception by his secretary at his office. Gano and his father would have been fascinated by the new electronic revolution including the cell phone of today. The early roots of Gano's radio interests can be traced back to before the 1915 fire. As the 1920s dawned, he had become much more sophisticated and advanced in this interest as the November 24, 1922 *Guard* revealed:

> *Gano Chance is one of our young radio experts and has fitted out a receiving set at his home, from which he has already tuned in on over 100 different stations scattered over practically over the United States and a station or so in the Dominion of Canada. From east to west he has been able to get the leading stations from the Atlantic to the Pacific Oceans and north and south from Winnipeg, Canada to the Gulf of Mexico. To the northeast he, a number of friends and members of the family have been enthused and thrilled by some of the great musical concerts in Schenectady, New York.*

Earlier in March 1922 there was a report of a radio station in Centralia at the residence of C. P. Burks, put in by Langenberg Brothers Grain Company of St. Louis. They sent daily grain quotations and set this equipment up reportedly wherever they had agents.

Gano's radio interests took on a new face in 1923 when ads in the *Guard* revealed a Drace and Chance business relationship. Ads from early 1923 indicate Drace ran the local music store, and was involved with Gano in radios. Along with complete sets, they carried a full line of radio parts that could be assembled by daring and capable individuals into a complete radio. Gano was active in this business as he installed a radio set for E. P. Stamper in early March 1923. A visit by F. M. Stamper, a prominent Moberly businessman, suggests the Centralia Stampers were surely his relatives. On March 16 the paper also covered Gano's installation of a radio set for O. G. Lease at Moberly. Mr. Drace was also involved in installations, as he set up equipment in the city park so residents could hear a political speaker, in late October

Drace & Chance Ad. (*Guard* January 19,1923)

1924. But the January 4, 1924 *Guard* indicated Drace Music had vacated its store east of the newspaper location, in the so-called Hubbard block (northeast corner of City Square). Gano was then at Central Methodist College in Fayette. There was an ad carried in the March 7, 1924 paper for Pruett Music Company that may have replaced or bought out Drace and Chance.

Gano was becoming a social figure as the January 16, 1923 *Guard* conveyed:

> *Gano Chance and Sidney Harley delightfully entertained a number of young ladies and young gentlemen with a DeMolay party Wednesday at the home of the former. Gano was 18 years old on the same day. Delicious refreshments were served and games were played to the delight of all present.*

The "DeMolay" party was a period social event and games were popular during those times and were enjoyed by Gano.

From the perspective of an older adult, Gano recalled a traveling show, the Chautauqua, that came to town in 1914, when he was only nine. The three-part play, entitled *It Pays to Advertise*, portrayed a son just out of college working in his father's plant. The first act found the son very critical of his father's backward ways but as the play progressed the son began to realize that he didn't have all the answers and his father was smarter than he thought. The father also comes to realize that his son had some very good ideas that should be tried. This influenced Gano in his relationship with his father and brought back remembrances because it was much like he and his father's arrangement and situation. Recalling this play at times, with a generation gap connotation, Gano said, "It kept me from being too far off base and kept us from a revolution!" (1978) That same play was repeated as the 1919 paper said this play was the most popular of the 1918 offerings. The Chautauqua programs were monitored because of Gano's connection to them, such as an early one in late August 1911. They had special speakers and various entertaining programs that ran several days annually. They continued to stop over in Centralia through 1927 and were held in late summer. The reports of the 1915 show told of a large tent that had been erected to house the events. A full page ad in the August, 1917 paper suggested that the Chautauqua was the people's university. The May 24, 1918 *Guard* spoke of the upcoming Chautauqua program in detail. The 1927 event occurred in late August, as customary, but the program apparently lost some $800. Those

Albert Bishop Chance was a well-attired man. This undated portrait is likely from around 1930. This is a very relaxed, natural pose that is also special because he is not wearing his glasses. (Jack Chance Collection)

Mrs. Frances (Fancy) Iglehart (Chanslor)(Gano) Chance married A. Bishop Chance on December 17, 1898. This undated photo could be from the 1920s. (Jack Chance Collection)

guaranteeing its success had to chip in $10 a piece and it was uncertain if it would be held in 1928. There has been a resurrection as in 2004 Chautauquas were held locally in Macon and Chillicothe. The 1914 *Guard* told of the many programs to be presented by the Chautauqua event that year. The comedy *It Pays to Advertise* was described as a modern drama and one in which a nine-year-old Centralian enjoyed and recalled over sixty years later!

Marketing and Sales Promotion

A. Bishop Chance could lay claim to being the first in holding many positions and in performing many different functions within Chance Company. Not only was he the first engineer, designer, product demonstrator, and president but he was a one-man show in the marketing and sales department as well. There isn't much information about his sales prowess, but it's obvious he became proficient in the art of selling. These marketing duties were inherited naturally because of the necessity of promoting his inventions and early products to insure their success and the company's economic survival. His outgoing personality, friendliness, and technical knowledge made him a natural for the sales job.

In addition to being an extrovert, another reason Mr. Chance was an excellent salesman is he knew his product. This was an advantage few other salespeople could claim with such a high degree of certainty. What better motivator could a salesman have than attempting to sell his own ideas that had been molded into a product. It wasn't a question of believing in his product as he understood it. If a weakness was found, the customer could delight in the fact that this salesman could immediately be transformed into a mechanic, designer, or inventor and fix it! It certainly was a situation in which there was no red tape or some middleman to deal with. Some labeled him as a tinkerer who would work with experimental or production items until they were transformed into an efficient and workable status. Maybe the word *practical* should also be added, as he was a man who used common sense to find practical solutions in solving problems.

The sale of the early cast-iron and malleable anchors was handled by A. Bishop, according to his son Gano:

> *He went out to the customer himself and made all kinds of anchor tests. He spent quite a few years just demonstrating the anchor to the utilities and then the jobber came. "Why can't we sell them for you?" they asked and that was the start of the real volume. The jobber does not create the demand, you always have to go out and start it and then the jobber sees the job and wants the business.*

Gano had learned firsthand the value of sales promotion from practical experience and also from the startup difficulties his father had encountered. The idea that customers will beat a path to the door of the inventor of even a marvelous product is not reality as the inventor usually will be the one wearing out shoe leather. Notice the reference to sales promotion by demonstrating how the product worked, which would continue throughout the history of the company. It did much to show the customer that they had a need for the product being demonstrated. Although A. Bishop Chance is commonly described as an inventor, recognition is overdue for his stamina, faith, and perseverance in promoting his inventions to the point that sales of his early products would sustain and grow his firm. Later, Gano again described the same point in a little different manner, which points to the importance that he placed on the value of sales promotion in the start-up of a new product. "He invents the thing and has to go out and create a demand to get distributors to take it on." (1978)

Another talent of Mr. Chance, according to Granny Ryman, was his selection of salesmen: "Mr. Chance got some pretty good salesmen. He had the ability, in my opinion, to be able to select and get good people that would go along with him and do almost anything to make whatever his ideas were a success." There are two significant talents that are reflected in Ryman's statement: the ability to select capable people and his charisma that generated an almost religious-like following of the people that worked with him. There developed a faith and confidence that what he was doing was important and would in the end be a success. Mr. Ryman believed even such people as his farmer father invested money in the development of the anchor. Those less convinced in the community, however, that couldn't visualize or see possibilities in Mr. Chance's business ventures, viewed his activities as suspect and believed his promotion activities exceeded his inventive genius and business capabilities.

Chance Company notes indicated that Mr. Montague of Kansas City was the first person in 1913 to hold the position of salesman, which Gano felt probably tied in with his father's connections to his Kansas City Company. Mr. J. P. Montague, an eighteen-year Chance Company veteran from Kansas City, according to the *Guard* of June 6, 1930, had just passed away. Personal papers of Mr. Chance suggest Mr. Montague was a Chance salesman from 1913 to 1928, or a fifteen-year veteran. Ill health led to his retirement. Gano continued to recall: "Father then hired Jack Milligan from Chicago, in 1916 before Mr. Montague died.

Jack Milligan was the second and Charley Clark was the third." (1978) L. C. Hansen in 1978 also recalled: "I remember Charley Clark. The night he retired was the day I came to the Company. We went to a party out at Buford Memorial Park, May 1, 1947." Charley Clark was a Chance salesman from 1919 to 1947. Chance family photographs included two taken at the Atlantic City, New Jersey Boardwalk of F. Gano Chance and Charley Clark. One of the photos is reproduced here.

F. Gano Chance and Charley Clark take a walk on the Atlantic City Boardwalk in what would appear to be the 1930s. Gano attended an electrical convention there in June 1931. Chance Salesman Charley Clark accumulated thirty years of service from 1917 to 1947. (Jack Chance Collection)

The sales promotion efforts of A. Bishop are noted by his recorded travels in the mid-teens. On March 13, 1914 the *Guard* found him in Mexico on business the previous Monday. Reflecting the influence of his travels on his family, on October 9 the paper revealed a trip by Mrs. A. B. Chance and her children to Warrenton, the guests there of friends and relatives. Early on January 29, 1915, the family had visitors from Liberty, according to the *Guard*. Mrs. G. A. Chanslor and daughter Margaret spent a week in Centralia. A year later, in early April 1916, Mr. and Mrs. A. B. Chance went to Liberty to attend Mrs. Chanslor's funeral. She was Mrs. Chance's aunt. Mrs. A. B. Chance was becoming active in Centralia society as on March 26 it was reported she entertained with a family dinner party on Sunday. The business world of A. Bishop seems to have entered a period when there were no sensational announcements, but routines were being established in which stable progress was made that would lead to the explosive events in 1917. In the meantime, the paper on May 14, 1915, found A. Bishop in Chicago for several days on business, likely nailing down relationships and product approvals that would stretch forth long into the future of Chance Company. He could have met with patent attorneys or Joslyn about his anchor rods that were under development. Joslyn was a competing utility manufacturing supplier who in the 1915 time frame had the forging capabilities

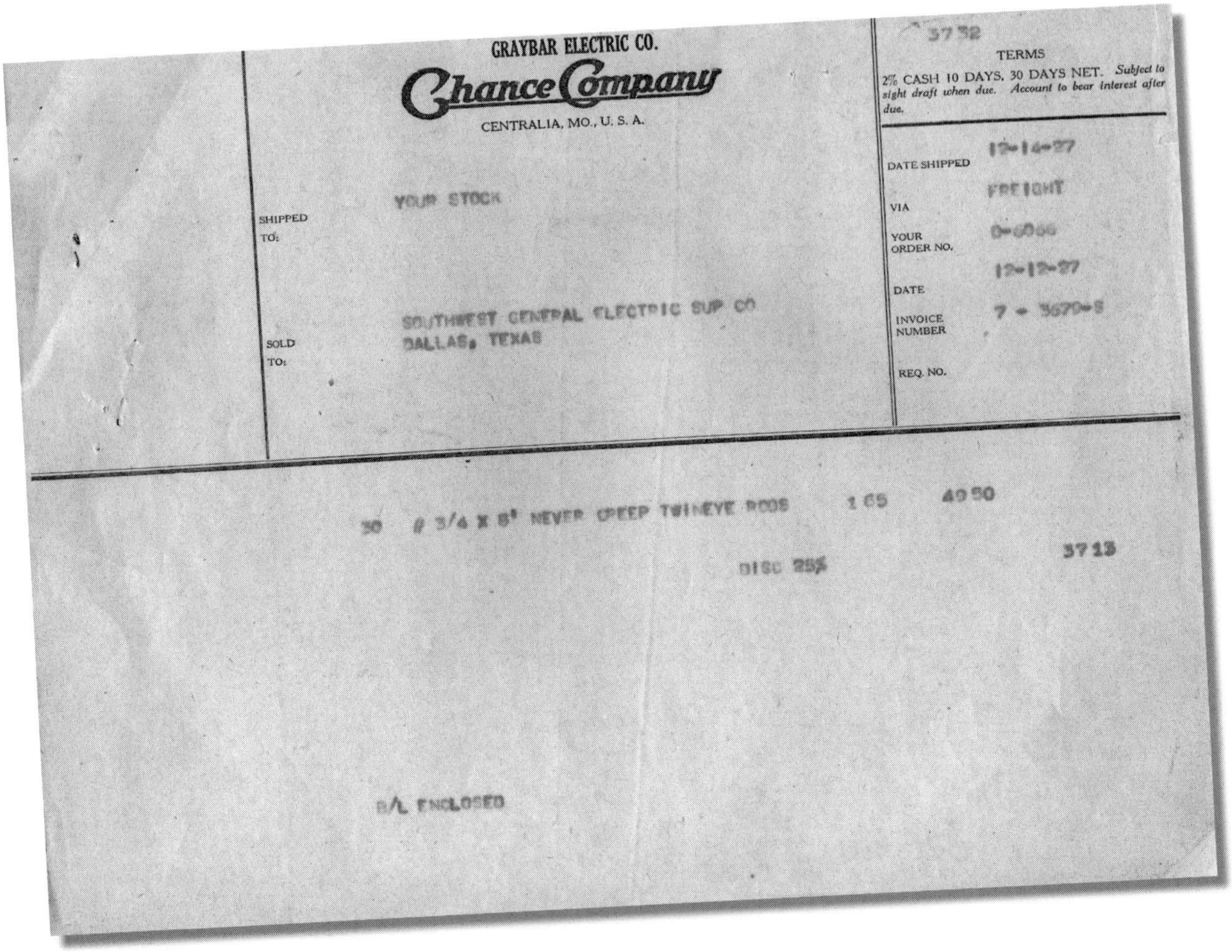

GRAYBAR ELECTRIC CO.

Chance Company

CENTRALIA, MO., U. S. A.

3732

TERMS

2% CASH 10 DAYS, 30 DAYS NET. *Subject to sight draft when due. Account to bear interest after due.*

DATE SHIPPED 12-14-27

VIA FREIGHT

YOUR ORDER NO. 0-6066

DATE 12-12-27

INVOICE NUMBER 7 - 3670-S

REQ. NO.

SHIPPED TO: YOUR STOCK

SOLD TO: SOUTHWEST GENERAL ELECTRIC SUP CO
DALLAS, TEXAS

30 # 3/4 X 8' NEVER CREEP TWINEYE RODS 1 65 49 50

DISC 25% 37 13

B/L ENCLOSED

Sales Order (A. B. Chance Co.)

to produce the Chance anchor rods, that will be extensively covered in the next Thimbleye® section. In late June the *Guard* brought news of the Saturday departure of A. B. Chance again to Chicago, Philadelphia, and other points, on a business trip. Keeping his bags packed, the paper on July 23, 1915, recorded an A. B. Chance departure on the previous Monday "for a trip to the gulf." It could have involved personal land holding and oil prospects or company business purposes. One safe conclusion was that Mrs. Frances (Fancy) Chance, besides maintaining the Chance household, had one ten-year-old and one teenage son to watch over during the absences of their father. The balance of the year was uneventful except for the November fire that transforms family activities into the difficult proposition of reestablishing the home base. The January 7, 1916 *Guard* ushered in the new year by recording a visit by Mrs. R. P. Greenlee of Kansas City. In February A. Bishop made another business trip to Chicago confirming that he spent a lot of time within the red passenger cars of the Chicago and Alton Railroad and the blue ones of the Wabash.

A. Bishop Chance's travels to distant cities—that others may have only dreamed of—became rather common. A *Guard* news report on September 22, 1916, is significant because it places the two sons together: "Mrs. A. B. Chance and sons, Buford and Gano and Miss Effie Faddis, motored down to Mexico Monday and visited with friends." It's not clear who Miss Faddis was, but the boys were a little old for a babysitter. In late October another article told of Mrs. A. B. Chance entertaining the Modern Priscilla Club. The wives of traveling businessmen had to deal with frequent absences and Mrs. Fancy Chance dealt with it in a positive manner by traveling with her family and maintained a social relationship with family and friends. This was a challenge that the families of salespeople also had to deal with, especially if they were younger adults with families, during the important years when their children were growing up.

The goal of any marketing effort is to generate a firm customer order. The sales order of December 14, 1927, reproduced on the previous page, is the result of all the hard work that went into what in reality is the lifeblood of any business firm. These three-quarter-by-eight-inch Twineye® Never-Creep® anchor rods at $1.65 each, less a 25 percent discount, is reflection of those times. It is very likely that Mr. Chance's capable assistant, Miss Ida Lou, made sure they shipped that very day!

The Thimbleye®

Anchoring's Silent Partner

Undoubtedly, the most enduring Chance Company product that's origin can be traced back to 1912, is the Chance anchor rod. It was developed, as was the anchor, to make the Chance anchoring concept a working reality. Neither the Never-Creep® Anchor or the Thimbleye® Anchor Rod existed in a perfected or final form in the winter of 1912. Their functions were instead performed by a stick of wood and a more sophisticated but early prototype metal rod with unusual and what would prove to be temporary fittings on both ends. The differences in both items became clear as A. Bishop was able to develop rather quickly a metal anchor that mirrored closely the functional aspects of the log anchor. The rod wasn't as simple as its more complicated and unique features would suffice for a short time as its operating and manufacturing deficiencies became clear. Engineering drawings of 1913 vintage indicated A. Bishop was also on the road to improving and perfecting a single purpose anchor drive rod. It too would quickly take on the signature of the Centralia Wizard with at least two important aspects of the rod that initially attracted his attention. The rod would appear to benefit by becoming a single part, followed by concentrating on reinforcing the eye area, making it truly a drivable product. Attention to the eye area, however, didn't stop there as by 1914 early Research Department information reveals an eye with a smooth, grooved design more common to a rope or cable pulley. Confirmation of the importance of this single feature is embedded and embellished in the name chosen to represent its famous trademark, Thimbleye®.

Different anchor types would come and go, their heyday reached and then they would slip into mediocrity and even extinction, as most other products life spans could attest. So what was the Thimbleye's® key to longevity? The function of both the anchor and rod didn't change all that much, but they were different in respect to each other. The anchor seemed to have more flexibility in how it could be designed allowing patentable variations. These various types, although each performed similar or the same job could be quite different in how they were designed, manufactured, and installed. But the Thimbleye® was on higher ground and its design elevated to an untouchable level. There are certain things that come along which can hardly be improved upon: Wheaties, Babe Ruth, Chrysler Mini-vans, the Bell System, renditions of a Russell, Benton, or Rockwell, and the Chance Thimbleye® Anchor Rod. When competitors are willing to pay for use of a registered name, then it is clear the developer has arrived with something very special.

The name of a commercial item can be important, as some get a colorful and unique title that even become famous, but others are assigned a name that simply is descriptive of what they are. How did a metal shaft come to be called a name associated with needlework? There is a connection between thread and rope as a rope's substantialness would allow it to be more in tune with guying. Both thread and rope are fibrous and somewhat fragile, yet strong. Utilization of these kinds of materials long predates metal wire or cable use, as there were many natural fibers that early man used to serve their everyday needs. Rope was commonly used for many functions besides sailboats, especially into the early 1900s when this Chance trademark name was applied. It is unthinkable to suggest a pre-dryer clothesline be made with anything but a small white, specially made rope, certainly not a slippery strand of metal wire. Understanding how rope functioned and had to be protected was interestingly applied to its metal counterparts, wire and cable that were destined for guying applications onto a metal anchor rod. Products do recycle and sometimes return in a different form. For example, the modern steel log chain used for securing cargo has been replaced in some applications by a nylon, ratchet-strap, tie-down. The new system is lighter, easier to operate, reduces vibrations, and has a softer, broader based surface, resulting in less damage to the area where the cargo is tied down. There is a weakness that if the strap is used against a sharp or rough surface, the nylon strap can be severed or severely damaged. A. Bishop found that where the guy line contacted the attaching point in the eye of the rod, there was potential for breakage of the guying line, whether it was fiber or steel. Incidentally, rope is a nonconductive material used by linemen in hoisting, hanging, and tying together or off various hot-line tool equipment and components.

Proceeding in how the trademark name came about, some remarks by F. Gano Chance are revealing: "A few years after the sleet storm he found that the thimbleye, the metal thimbleye that went up through the eye of the flat rod, was OK but why not have it actually a part of the rod." (1978) Gano is describing a separate part, a thimble-like object that ran through the eye of the early flat rod. It protected the guying line in a fashion more commonly associated with rope in static riggings, which resulted in the incorporation of the thimble shape into the eye or the upper part of the anchor rod. By making it thicker and then rounding or radiusing its shape so it wouldn't cut or stress the guying line, but instead spread the guying loading onto a smooth, rounded fitting, became a special feature of the Thimbleye®. Making a fist and noting the valley between the knuckles is reflective of A. Bishop's design in retaining the guying lead in the valley or groove in the rod eye. A rotating surface or a sheave (wheel having a groove) wasn't needed, as the guy line was fastened in a static position around the eye and didn't move once installed. So the choosing of the name *Thimble-eye*, which is listed in more technical and complete dictionaries, implies "thimble-shaped apertures in iron plates where sheaves aren't required—an eye in a plate through which a rope is woven without a sheave," was a logical choice.

Gano continued to speak of the Thimbleye® development

> *So he began experimenting with making up model wooden rods and developing this thimbleye on to the rod. That is where he got the name thimbleye and that is how he got into the rod business. He licensed all of his competitors, the pole line hardware companies, to make the rod. But the purpose of the ring or thimbleye was where you had a stranded cable and put tension on it, after it bends, why that ground cable winds up with some of the pieces of wire with more strings on them than the other. They did not stay in solidical form as they went through the rod and back up, so the way they found they could make the strand*

stand up and not break was to cover it in this cylindrical form. They used the thimbleye, it was called a thimbleye, it was a separate piece of metal but then father got the idea why not make a rod with it. I don't remember when he invented the thimbleye but I imagine it was about 1920. (1978)

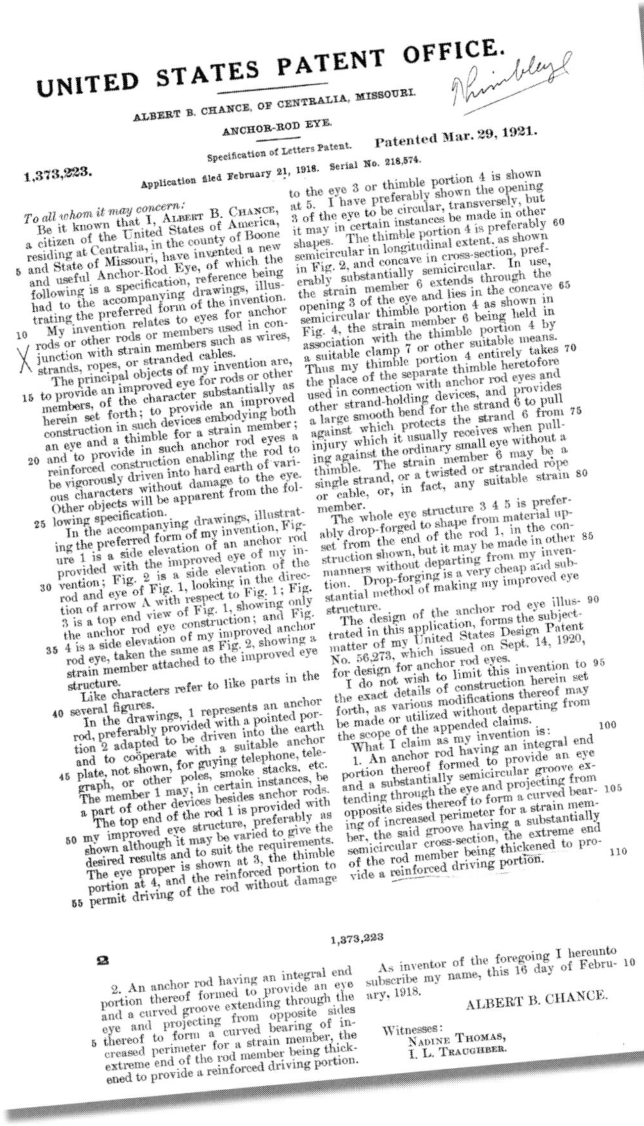

UNITED STATES PATENT OFFICE.

ALBERT B. CHANCE, OF CENTRALIA, MISSOURI.

ANCHOR-ROD EYE.

1,373,223. Specification of Letters Patent. Patented Mar. 29, 1921.

Application filed February 21, 1918. Serial No. 218,574.

To all whom it may concern:

Be it known that I, ALBERT B. CHANCE, a citizen of the United States of America, residing at Centralia, in the county of Boone and State of Missouri, have invented a new and useful Anchor-Rod Eye, of which the following is a specification, reference being had to the accompanying drawings, illustrating the preferred form of the invention.

My invention relates to eyes for anchor rods or other rods or members used in conjunction with strain members such as wires, strands, ropes, or stranded cables.

The principal objects of my invention are, to provide an improved eye for rods or other members, of the character substantially as herein set forth; to provide an improved construction in such devices embodying both an eye and a thimble for a strain member; and to provide in such anchor rod eyes a reinforced construction enabling the rod to be vigorously driven into hard earth of various characters without damage to the eye. Other objects will be apparent from the following specification.

In the accompanying drawings, illustrating the preferred form of my invention, Figure 1 is a side elevation of an anchor rod provided with the improved eye of my invention; Fig. 2 is a side elevation of the rod and eye of Fig. 1, looking in the direction of arrow A with respect to Fig. 1; Fig. 3 is a top end view of Fig. 1, showing only the anchor rod eye construction; and Fig. 4 is a side elevation of my improved anchor rod eye, taken the same as Fig. 2, showing a strain member attached to the improved eye structure.

Like characters refer to like parts in the several figures.

In the drawings, 1 represents an anchor rod, preferably provided with a pointed portion 2 adapted to be driven into the earth and to coöperate with a suitable anchor plate, not shown, for guying telephone, telegraph, or other poles, smoke stacks, etc. The member 1 may, in certain instances, be a part of other devices besides anchor rods.

The top end of the rod 1 is provided with my improved eye structure, preferably as shown although it may be varied to give the desired results and to suit the requirements. The eye proper is shown at 3, the thimble portion at 4, and the reinforced portion to permit driving of the rod without damage to the eye 3 or thimble portion 4 is shown at 5. I have preferably shown the opening 3 of the eye to be circular, transversely, but it may in certain instances be made in other shapes. The thimble portion 4 is preferably semicircular in longitudinal extent, as shown in Fig. 2, and concave in cross-section, preferably substantially semicircular. In use, the strain member 6 extends through the opening 3 of the eye and lies in the concave semicircular thimble portion 4 as shown in Fig. 4, the strain member 6 being held in association with the thimble portion 4 by a suitable clamp 7 or other suitable means. Thus my thimble portion 4 entirely takes the place of the separate thimble heretofore used in connection with anchor rod eyes and other strand-holding devices, and provides a large smooth bend for the strand 6 to pull against which protects the strand 6 from injury which it usually receives when pulling against the ordinary small eye without a thimble. The strain member 6 may be a single strand, or a twisted or stranded rope or cable, or, in fact, any suitable strain member.

The whole eye structure 3 4 5 is preferably drop-forged to shape from material upset from the end of the rod 1, in the construction shown, but it may be made in other manners without departing from my invention. Drop-forging is a very cheap and substantial method of making my improved eye structure.

The design of the anchor rod eye illustrated in this application, forms the subject-matter of my United States Design Patent No. 56,273, which issued on Sept. 14, 1920, for design for anchor rod eyes.

I do not wish to limit this invention to the exact details of construction herein set forth, as various modifications thereof may be made or utilized without departing from the scope of the appended claims.

What I claim as my invention is:

1. An anchor rod having an integral end portion thereof formed to provide an eye and a substantially semicircular groove extending through the eye and projecting from opposite sides thereof to form a curved bearing of increased perimeter for a strain member, the said groove having a substantially semicircular cross-section, the extreme end of the rod member being thickened to provide a reinforced driving portion.

2 1,373,223

2. An anchor rod having an integral end portion thereof formed to provide an eye and a curved groove extending through the eye and projecting from opposite sides thereof to form a curved bearing of increased perimeter for a strain member, the extreme end of the rod member being thickened to provide a reinforced driving portion.

As inventor of the foregoing I hereunto subscribe my name, this 16 day of February, 1918.

ALBERT B. CHANCE.

Witnesses:
NADINE THOMAS,
I. L. TRAUGHBER.

Patent No. 1,373,223 (A. B. Chance Co.)

Veteran Chance marketing executive, Roy Hansen added, "I know we were still collecting royalties on it in the '50s, but I think Joslyn paid us royalties based on the use of the trade name rather than the seventeen years of patent rights." (1978) Gano responded:

The original patent must have expired 17 years after it was patented, about 1930. But he had improved on the design a little bit then and had gotten an additional patent. Of course he got a copyright on the name Thimbleye®, then he licensed the different pole line hardware manufacturers to make it and pay him a royalty. Joslyn was his principal manufacturer and was paying this royalty. Well when the original patent expired [Chance got another patent by changing] *its shape a little bit. When Mr. Isbell was with us and Joslyn wanted us to lower our royalty, we had a contract with them and the packing numbers and everything was in the contract, etc. (1978)*

See Volume II concerning the confrontation between Joslyn and Chance Company concerning the royalty agreement.

The official U.S. Trademark record of the Thimbleye® confirms that it was first registered on September 15, 1921. During World War II, a filing date of February 8, 1943. was recorded, with a registration date confirmed to be June 15, 1943, revealing it took three to four months for it to be processed by the government. It was renewed in June of 1983 and surely in 1963. Using the twenty-year renewal time frame, the reregistration in 1943 seems to suggest that it had lapsed. The war could have generated a special holding ruling, etc., as it is unlikely that the company would have allowed such an important trademark to lapse. However, copyright law implies that constant usage with the use of the mark and the provision for a six-year renewal could allow the registration to be extended. As earlier stated, the trademark is alive and well today, reflecting an eighty-year life, which has no sign of ending. It has outlived its partner the Never-Creep® Anchor and has even given birth to similar products that have registered trademarks, such as the Twineye®, introduced about a year after the original

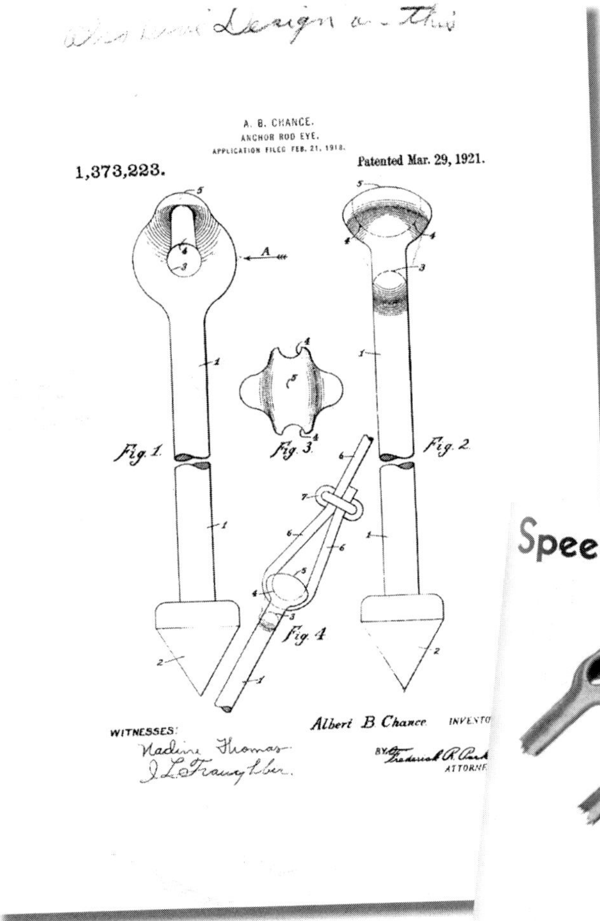

Patent No. 1,373,223 (A. B. Chance Co.)

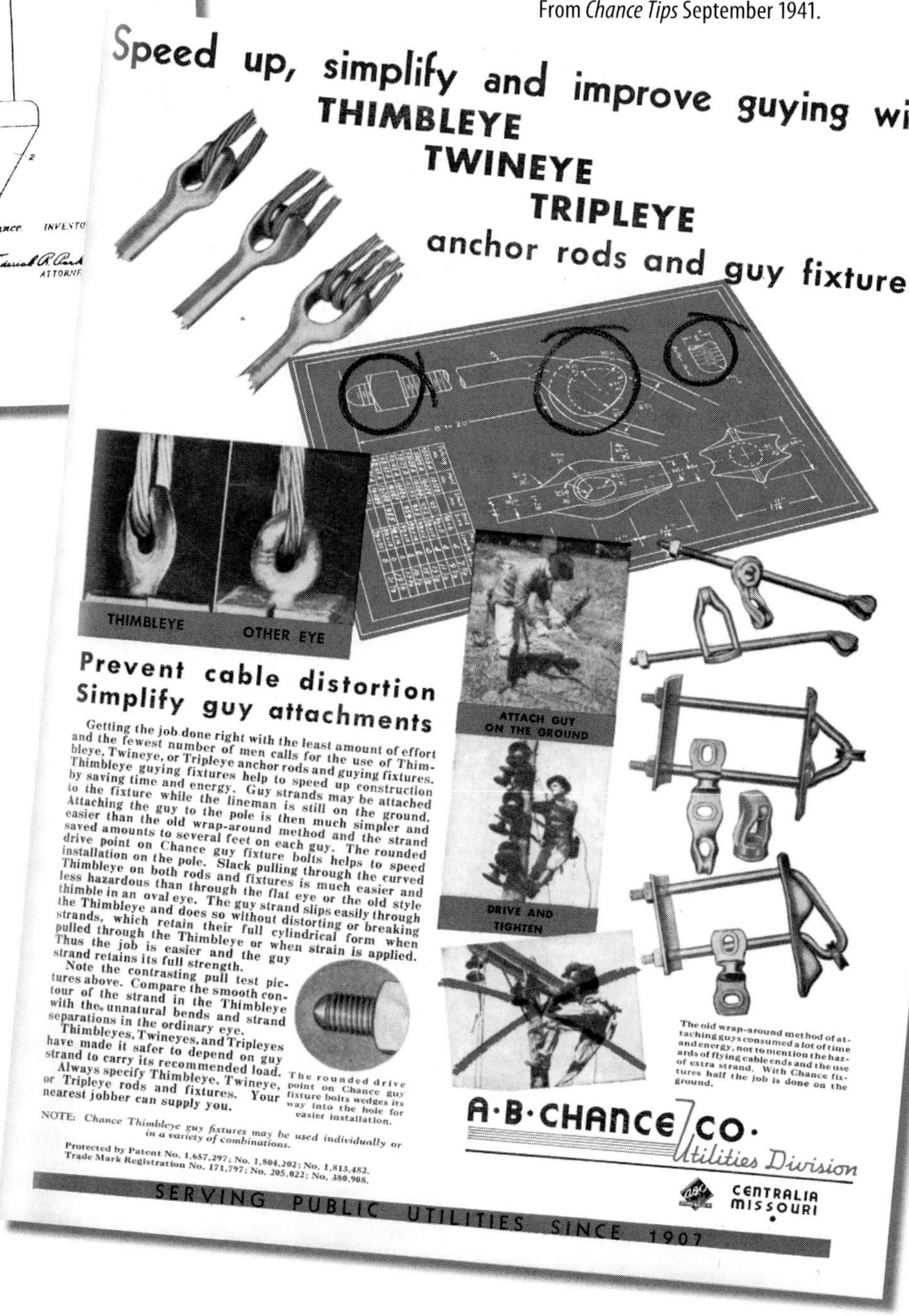

From *Chance Tips* September 1941.

on July 20, 1922. It's interesting that a filing date in renewing, of February 17, 1947, didn't result in a registration until July 19, 1949! The Tripleye® came much later on February 10, 1937, suggesting there was not a strong need earlier for a three-guy-wire rod or hardware eye. As alluded to, there were many different types of hardware items with these various trademark heads or eyes in addition to the common anchor rod. An example is shown in a Chance ad appearing in the September 1941 issue of *Chance Tips,* shown on the previous page, bottom right. Note the patent and trademark numbers included in this very detailed and quality Chance promotional piece.

Copies of two legal documents, reproduced on the right, surfaced concerning the Thimbleye® Trademark. The oldest of these (top), dated September 21, 1925, is the application, which does not coincide with the 1921 U.S. issuance reported in the previous paragraph. However, in noting the trademark certification of July 27, 1928 (bottom), in the top area is the word *Canada.* This would imply it is the paperwork confirming the request and certification of the Thimbleye® trademark in Canada, that followed the registration process in the United States.

It is beneficial to look anew at Chance Company history, especially the introduction of important products such as the anchor rod. It eventually became a separate product line from anchors that helped drive the growth of Chance Company. The opportunity to examine early company drawings and patent information will advance the knowledge of the development of the anchor rod by Albert B. Chance, as Patent Office communication/requests are addressed. As Gano alluded to, the first Chance anchor rod design didn't have the thimbleye feature. It is unclear if the rod used on the first anchor was of the off-the-shelf variety or a hardware item with Chance modifications. This could be why the anchor was patented in 1913 and the rod later. The four individual patent files indicate the first rod Patent No. 56,118, was patented on

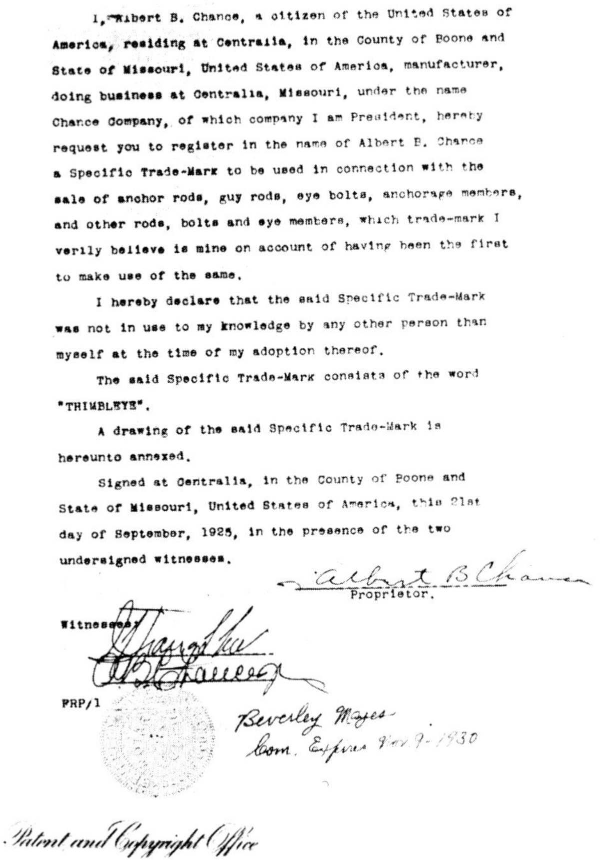

I, Albert B. Chance, a citizen of the United States of America, residing at Centralia, in the County of Boone and State of Missouri, United States of America, manufacturer, doing business at Centralia, Missouri, under the name Chance Company, of which company I am President, hereby request you to register in the name of Albert B. Chance a Specific Trade-Mark to be used in connection with the sale of anchor rods, guy rods, eye bolts, anchorage members, and other rods, bolts and eye members, which trade-mark I verily believe is mine on account of having been the first to make use of the same.

I hereby declare that the said Specific Trade-Mark was not in use to my knowledge by any other person than myself at the time of my adoption thereof.

The said Specific Trade-Mark consists of the word "THIMBLEYE".

A drawing of the said Specific Trade-Mark is hereunto annexed.

Signed at Centralia, in the County of Boone and State of Missouri, United States of America, this 21st day of September, 1925, in the presence of the two undersigned witnesses.

Albert B Chance
Proprietor.

Witnesses:

FRP/1

Beverley Mayes
Com. Expires Nov 9-1930

Patent and Copyright Office

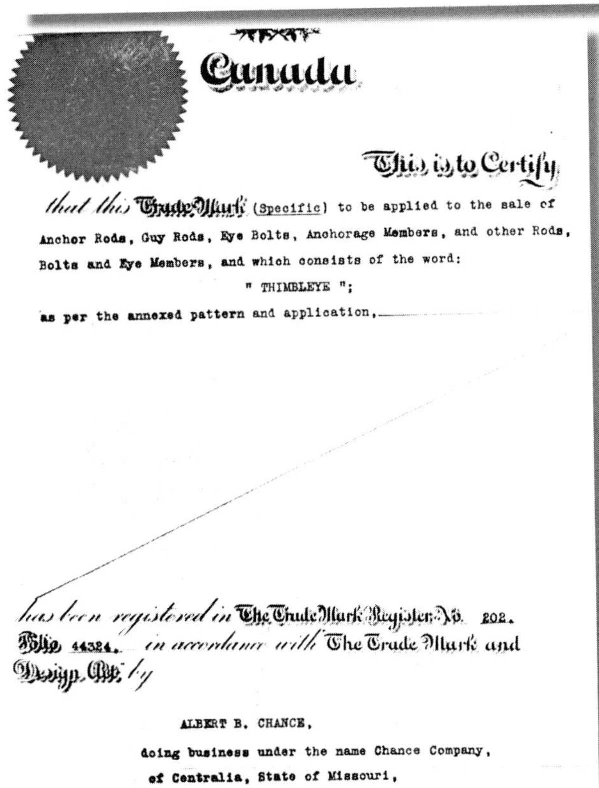

Canada

This is to Certify

that this Trade Mark (Specific) to be applied to the sale of Anchor Rods, Guy Rods, Eye Bolts, Anchorage Members, and other Rods, Bolts and Eye Members, and which consists of the word:
" THIMBLEYE ";
as per the annexed pattern and application,

has been registered in The Trade Mark Register No. 202. Folio 44324, in accordance with The Trade Mark and Design Act by

ALBERT B. CHANCE,
doing business under the name Chance Company,
of Centralia, State of Missouri,
United States of America,

on the 27th day of July, A.D. 1928

Trade-Mark Text and Trade-Mark Canada documents (A. B. Chance Co.)

August 24, 1920, for fourteen years implying a design-type patent. There is a further surprise to find this patent request: "For An Anchor-Rod" was not the Thimbleye®! The "Design" drawing does reflect the all-in-one-type rod construction with a point formed on one end and the eye formed on the other. It appears as a forged type item with a flat surface on the top of the eye that was "reinforced preferably as shown, to prevent injury to the eye portion in driving the rod." This was clearly a feature that separated it from a common-type rod with a small rounded loop or eye, as it was a rod designed to be driven into the ground. There may have been drive rods used on lightning rods, etc., which had a point like the Chance anchor rod. What the patent reflected was first having all these features in one component that made it special or patentable with an application specialized toward use with the Never-Creep® Anchor.

The early anchor rod story continues with the research file revealing a substantial number of engineering drawings. The 1914–15 group conveys a professional rendering and is a surprise given that engineer Bill Hollander suggested that when he arrived in 1929 Chance products were being manufactured without engineering drawings. However, to obtain a patent on a manufactured item, it seems logical that a mechanical drawing would be required to lock in specifications and reveal unique design features. The first sketches are dated August 30, 1913 (see below) and are professionally done, but are small, single-dimensional efforts. They do reveal, unlike the larger multidimensional drawings of 1914–15, that they are the property of "Chance Mfg. Co.," Centralia, Missouri, with the "OK" signature initials of ABC (A. Bishop Chance). This implies they were or became in-house documents. These types of sketches would appear to be beyond any capabilities within Chance Manufacturing Company and could have been outsourced. As the Chance Company story progresses, it will be revealed how this small firm grew into performing more and more in-house functions, although this didn't come easily without substantial internal efforts. The first three patent papers, several years forward, involving anchor rods and eyes reveal drawings with Attorney Frederick R. Parker as the source. He appears to be a patent attorney who became involved to make a professional presentation in the filings for the patents. However, it may have no connection to the origin of the previous engineering drawings, although A. Bishop's earlier experiences with electrical products (arresters, etc.) must have acquainted him with patent attorneys and contacts who could do engineering drawings. So it shouldn't be a surprise that A. B. Chance was beyond the level of development that historical sources tend to conservatively project for him, as he shows a remarkable sophistication for that time at his level of education.

Knowing the founder was working on the anchor rod in 1913, the same year the Never-Creep® Anchor was reportedly patented, raises

Anchor Rod August 30, 1913 (Book 1 of Research Dept.—1913-14) (A. B. Chance Co.)

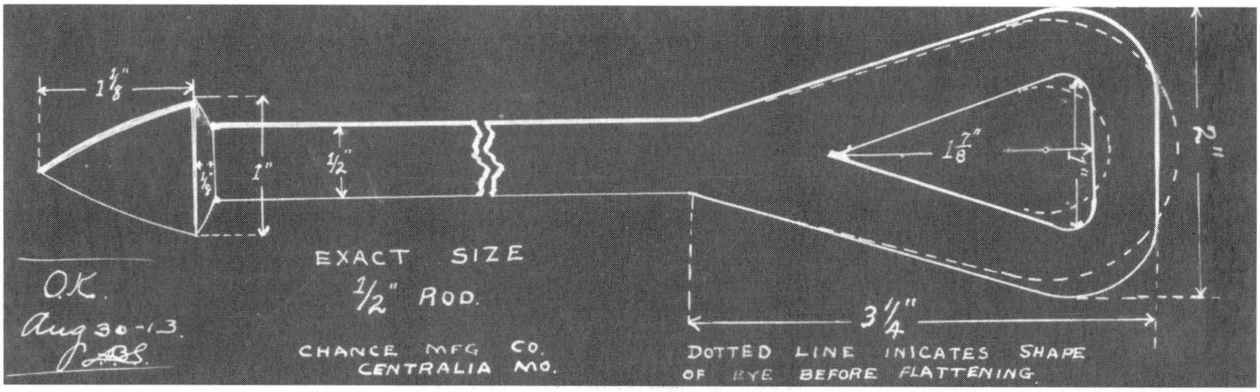

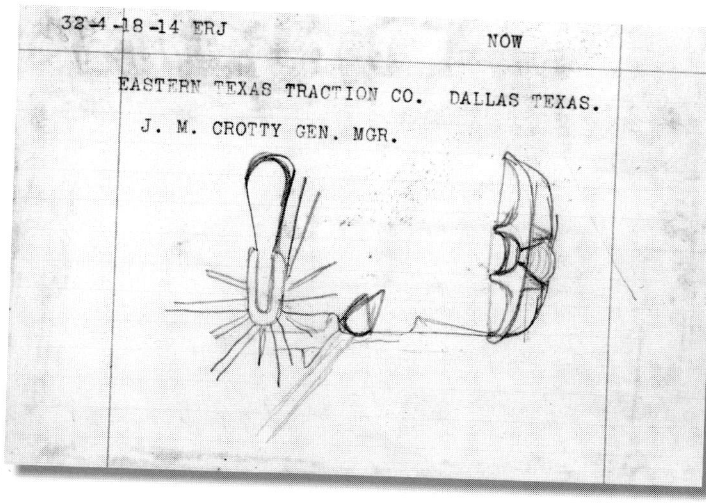

Design April 18, 1914 (Book 1 of Research Dept.—1913-14) (A. B. Chance Co.)

the issue of why the rod wasn't also patented. His anchor design was considered a special, patentable product and the rod may have been considered an integral part of the Never-Creep® patent. The rod may have also attracted much less attention and was a work-in-progress. The small August 1913 drawing pictures a rod with an elongated diamond (pear) shaped eye with a rounded top portion. This could reveal a rod with an eye that was formed by heating and bending the rod in a loop and bringing it back together. A later drawing has the word *weld* where the loop comes back together, suggesting a pre-forged construction method (Note: Later Chance patent-filing texts suggest the eye should be constructed by a drop-forge operation). Showing the advancing level of development, the rounded top was flattened, resulting in a flat surface on this August 1913 version. The drawing showed both the initial and final flattened dimensions that would allow the rod to be struck squarely on its top to drive it into the ground to intersect the anchor. This anchor rod had the point formed on the opposite end with a shoulder to ease driving and to slide into the anchor slot to complete the installation. These features seem refined and patentable, so why not submit this to the Patent Office? Of the four patents in the file concerning rods and eyes, the first, a forerunner to the Thimbleye® was not applied for until February 21, 1918! The basic anchor rod described in the 1913 drawings was not submitted until November 22, 1919. An obvious concern is the several years of the Chance anchor rod being exposed without any sign of legal protection. Later file information suggests no competitor or inventor conflicts. A series of sophisticated engineering drawings, dated September 11, 1914, reveal the 1913 design had been refined with a completely different oval eye that has the inverted-pear shape with the guying grooves in both sides of the upper head area. This is the Thimbleye® design! The only file material supporting progress of this design was a surplus customer index card! It was common for A. Bishop to use the back of surplus letters, etc.—utilizing everything until it provided no further utility. The index card has some sketches that surely represent the founder's efforts. The 1914 index card is shown above. What they appear to represent is sometime between April 18, 1914, and September 11, 1914, the development of the idea for the smooth rounded contours that would become the outstanding feature of the Thimbleye® was born. These early fall 1914 drawings are approved by A. B. C. on November 5, 1914, and also provide the four rod sizes concentrated on with a separate drawing for each. Drawing No. 1 is for the 1-inch rod, No. 2 was 3/4 inch (shown on page 146), No. 3 was 5/8 inch, and No. 4 was the 1/2 inch. Why would such an effort be made to get these drawings produced without attempting to get this unique design patented? This is out-of-character for A. Bishop Chance at this stage of his life, considering earlier experiences and actions. Adding to this drama was an apparent Joslyn drawing of February 20, 1917, described as "Drop Forged Eye For 5/8" Chance Anchors." This is a regular eye with a tighter and more compact reverse-pear-shaped eye opening with the flat top area for driving. Following was a March 19, 1917 drawing: "Joslyn Mfg. & Supply Co., Chicago Ill., *Selling Exclusively to Jobbers,* Chance Drop Forged Anchor Rods. Dwg. No-875." This drawing showed the four sizes

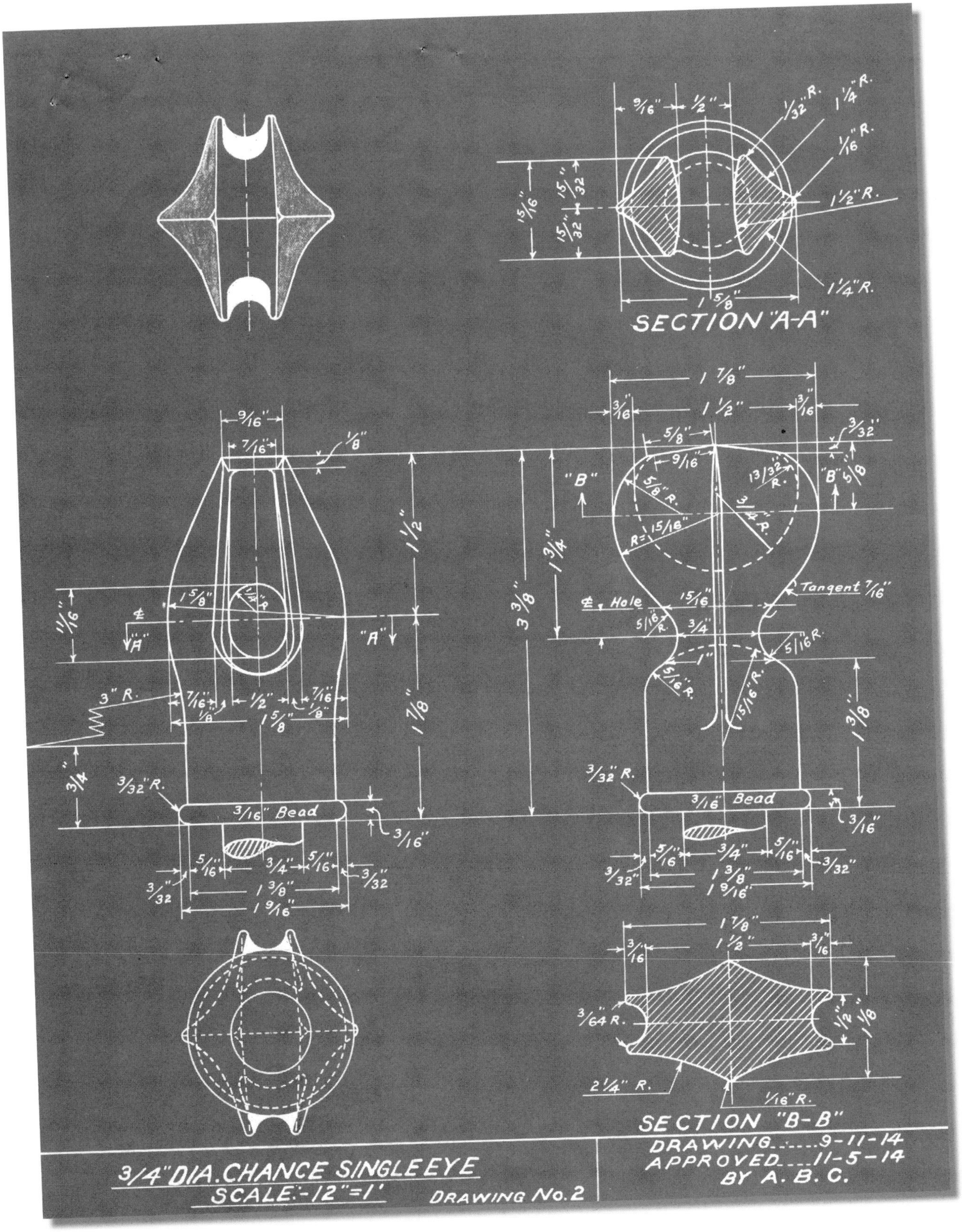

Drawing September 11, 1914 (Book 1 of Research Dept.—1913-14) (A. B. Chance Co.)

of rod diameters covered earlier. So we know that those outside of Centralia knew of the Chance designs and specifications. It is logical that they would not acknowledge these anchor rods as Chance unless there were patents or other protections in effect. Leading up to these Joslyn drawings are three Chance Company drawings for "Never-Creep® Rods—Hammer Face Drop Forged." The first is dated January 27, 1917, and the second with the additional description of "Oval Eye" is dated February 12, 1917. The four sizes shown on each of these two drawings focusing only on the eye end of the rods differ in that the first drawing depicts round eyes and the second has the reverse-pear shape or what is being designated as the Oval Eye®. Complicating this even further is another March 13, 1917 drawing with six different eyes. There is even a two-guy design that would later become the Twineye®. So there was a multitude of various designs and development work being done during the 1913–17 period, which suggests there is another part of the puzzle that hasn't yet been revealed.

The key to unlocking the mystery of the undocumented patented anchor rod designs appears to be driven by an association with other firms and one or more communication giants. There was the commonly circulated story of Western Electric influencing A. Bishop to get his Never-Creep® commercially accepted. After he convinced firms of the benefits of his anchor and rod combination, one eventual result was the standardization of these products. A later influence was R.E.A. and N.E.M.A. (National Electrical Manufacturers Association in which A. B. Chance Company was an active member). Marketing Services Department completed monthly/quarterly reports to N.E.M.A. for Chance sales activities in certain lines. But early on, besides Western Electric, another set of drawings in the patent file suggested the heavy influence of A.T. &T. (American Telephone & Telegraph) in setting specifications. One of the revelations of the description box in these drawings is the use of "Chance Company" instead of "Chance Mfg. Co.," possibly because of brevity. Many references of company products were simply described as "Chance®," further supporting this practice.

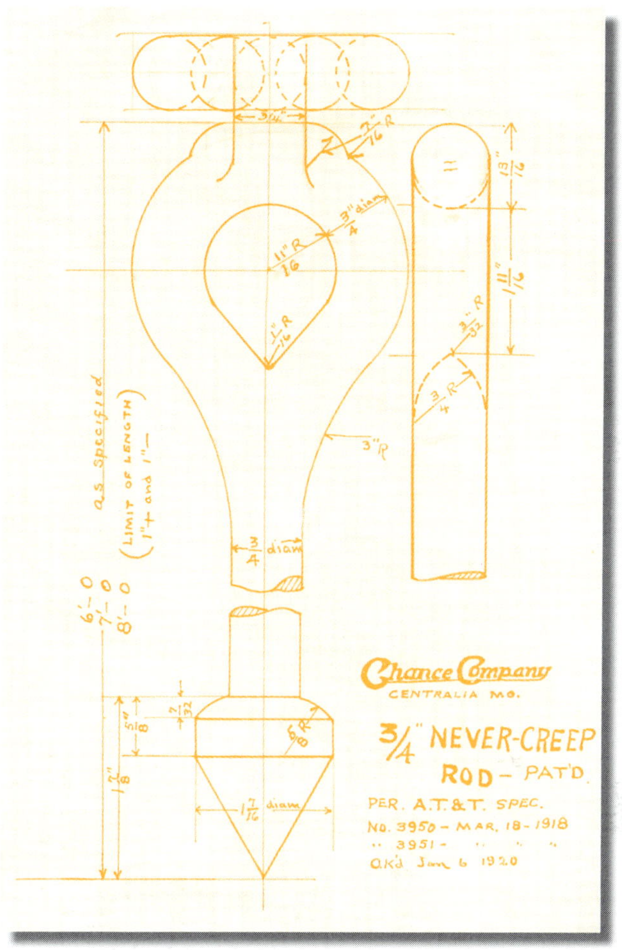

Rod March 18, 1918 (Book 2 of Research Dept.—1915-29) (A. B. Chance Co.)

One A.T.&T. oriented drawing reproduced here is titled "Never-Creep®, Rods. Pat'D per A.T.&T. Spec., No. 3950—March 18, 1918, No. 3951—March 18,1918." The two numbers 3950 and 3951 are not Chance patent numbers, but relate instead to 1845 patents with no obvious product relationship. These numbers are surely as they are described, as A.T.&T. specification numbers for Chance Never-Creep® oval eye anchor rods. The notation "Pat D" suggests they were patented, but were they covered under the Never-Creep® Anchor patent? As the rod developed, it may have become obvious that all the variations and unique designs like the

Thimbleye® would require more specific legal coverage. This is logical and reflective of A. Bishop's aggressive manner in legally protecting his later inventions and company products, especially in the expansion of hardware products within his increasingly more utility prominent firm. Note that the rod "Joslyn, Comm. Edison" (reproduced on page 149), and A.T.&T. were focused on was not the special Thimbleye® but the rod with the regular eye. Another noteworthy aspect of this information is the early anchor rods came under the title Never-Creep®, suggesting as suspected that they were closely associated with the more prominent and recognized anchor. There was also the advantage of name recognition and as mentioned the Never-Creep® patent may have also covered the regular-eye anchor rod. A. Bishop was wise enough to obtain outside help when necessary to provide material at a quality level he felt appropriate, and his approval had to reflect a level acceptable to large, more sophisticated firms. Of course his association with them, in providing designs and products not before available, also positively influenced his actions in becoming more professional. Chance Company success was not an accident but a result of continuously laboring to perfect processes, procedures, and products in a progressively expanding pattern. This was a key to continued company growth and the ability to weather the Great Depression.

In addition to the many designs of the eyes that were made an integral part of the guying end of the anchor rod, drawings dated March 2, 1915, revealed a removable eye that could be screwed onto a threaded rod. It was designated the Thimbleye Nut® as it had A. Bishop's grooved and smoothly contoured upper guy area feature with a broader base allowing adequate material that could be opened up to fit on a threaded rod. This would allow the installation of a heavy driving device on the rod, preventing damage to the upper eye. The early date of the threaded nut was a surprise. Of the many duplicated drawings in this series approved by A. B. C. on October 5, 1915, one revelation was they didn't include a 1-inch model. Little has been said about the anchor end of the rod having the bullet-shaped point. Some small sketches of each of the four rod sizes featuring the dimensions of these points may have been done locally on January 1, 1916. They led to a more refined drawing (undated) of "Never-Creep® Rod Drive Points." Besides the concern over ease of driving the rod into the ground was the important function of the shoulder area having the strength to hold the guying tension or load in the anchor slot. The base area was, for example, 1 3/4 inches wide on the 1-inch rod model, revealing the concern over adding the necessary bulk to prevent failure. Were all of these designs a result of the demands of customers like Bell and A.T. &T.? What this information suggests is the much earlier development (1914–15) of the sensational features of a design that would extend far into the future. This moves back the timetable of anchor rod development at least five years earlier than the 1920 time frame patents! It's exciting to consider that Centralia was the "Mecca" of such developments being orchestrated by the "Centralia Wizard!"

It is interesting that the original 1913 Chance rendering of what may be the first example of the single purpose Never-Creep® Anchor Rod, was the type that Joslyn and A.T. &T. were involved in. This assertion is based on the file information, which could be incomplete; however, this is what the two A.T. &T. Spec. numbers 3950 and 3951 appear to apply to. This rod has the later Oval-Eye® reverse-pear-shaped eye formed by a drop forge operation. This is the common variety drive rod with the flat surface on the eye without the more complicated Thimbleye® configuration. Another A.T.&T. drawing shows an "OK'd" date of January 6, 1920, which is puzzling unless these were later revisions. It seems like the anchor rod process was in slow motion at times, although it is clear A. Bishop's design developments were leading the legal and commercial recognition parade earlier than expected. The entrance of

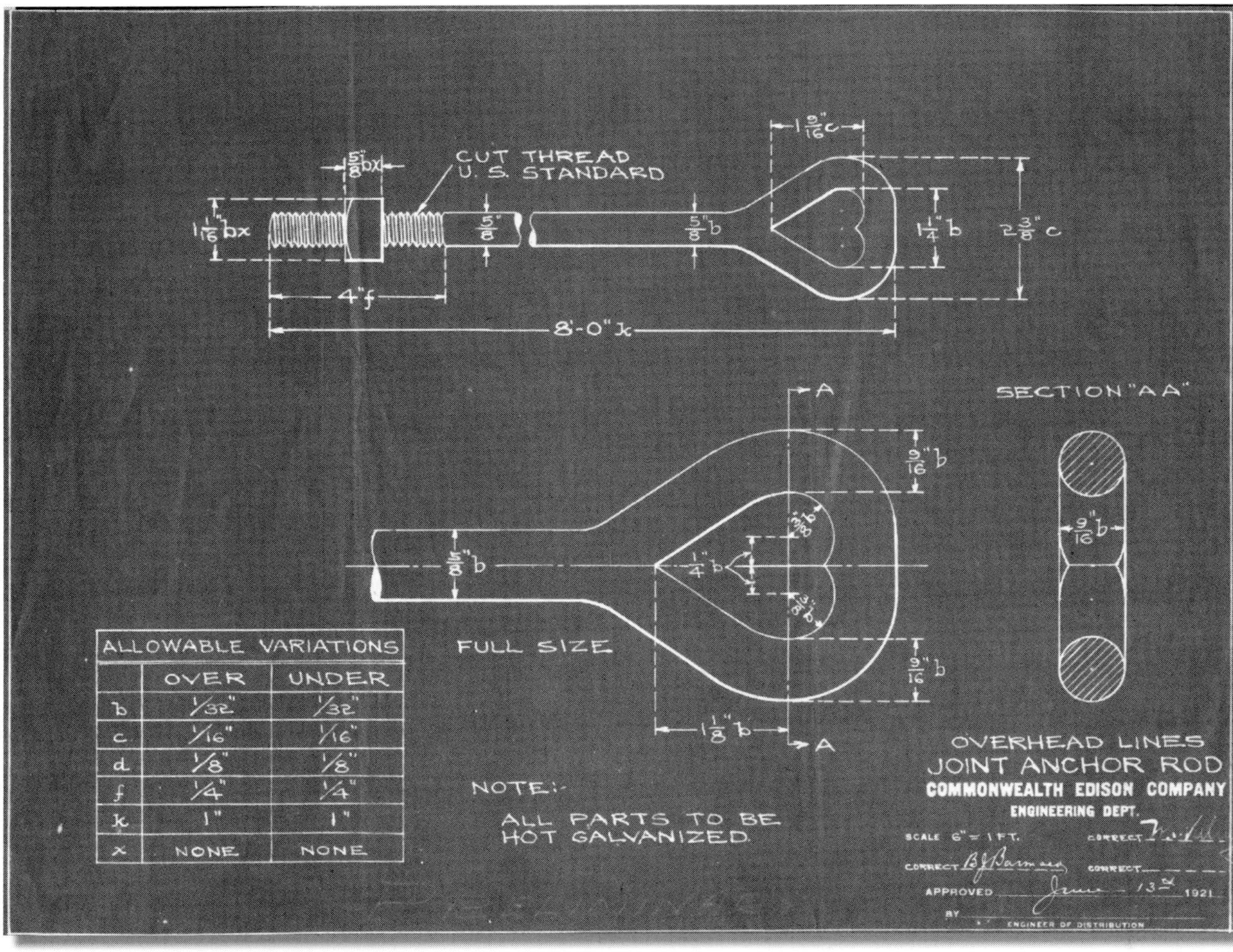

Drawing June 13, 1921 (Book 2 of Research Dept.—1915-29) (A. B. Chance Co.)

Commonwealth Edison Company into this process is confirmed by their drawing of June 13, 1921, reproduced here, which consisted of the twin-eye, Oval-Eye® discussed earlier. Their drawing title states "Overhead Lines, Joint Anchor Rod, Commonwealth Edison Company, Engineering Dept." There was an allowable variations table, a note reading "All Parts To Be Hot Galvanized" and a picture of the heart-shaped twin oval eye with threads and a nut on the anchor end of the rod! There are several important aspects of this drawing. Overall it reveals the increasingly more stringent expectations of a utility customer, which would be reflected in a short time by the entire industry. This is also an introduction to an old friend at Chances, what C. Edison labeled "Hot Galvanized," which translates into "Hot Dip Galvanizing" at Chance Company. However, this requirement was just one of several manufacturing processes Chance Company didn't yet have. To build what Mr. A. Bishop was designing, it had to be farmed out to other firms such as Joslyn. Bringing into perspective the time frame of 1913–21, is to reveal the pre– and post–World War I era. The majority of the population was still dealing with mud roads and streets, coal-oil lamps with no electricity, wood and coal heating sources, and no modern medicine like penicillin. Confrontations were still being generated between the horse and buggy and Ford's Model T. In the face of these conditions, the galvanizing requirement was another process that could be labeled as progress. It would insure that corrosion wouldn't prematurely cause product failures and a requirement of firms supplying future products to the communication and electrical utilities.

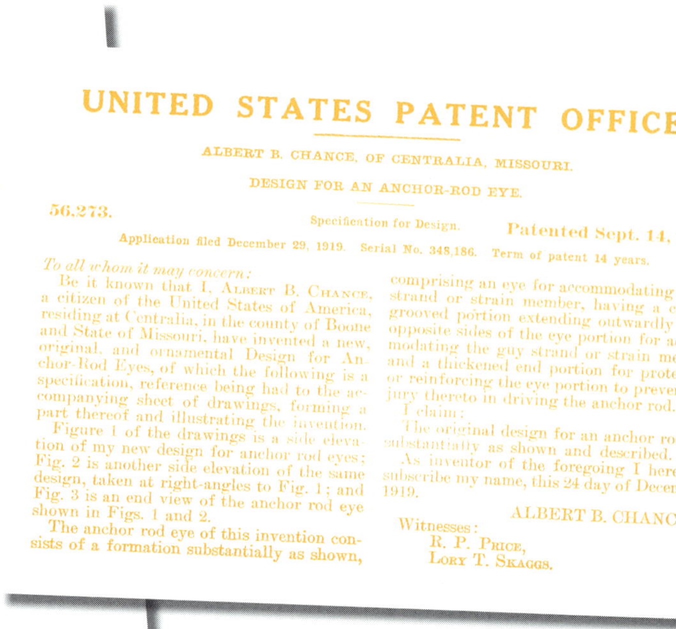

UNITED STATES PATENT OFFICE

ALBERT B. CHANCE, OF CENTRALIA, MISSOURI.

DESIGN FOR AN ANCHOR-ROD EYE.

56,273. Specification for Design. Patented Sept. 14,

Application filed December 29, 1919. Serial No. 348,186. Term of patent 14 years.

To all whom it may concern:

Be it known that I, ALBERT B. CHANCE, a citizen of the United States of America, residing at Centralia, in the county of Boone and State of Missouri, have invented a new, original, and ornamental Design for Anchor-Rod Eyes, of which the following is a specification, reference being had to the accompanying sheet of drawings, forming a part thereof and illustrating the invention.

Figure 1 of the drawings is a side elevation of my new design for anchor rod eyes; Fig. 2 is another side elevation of the same design, taken at right-angles to Fig. 1; and Fig. 3 is an end view of the anchor rod eye shown in Figs. 1 and 2.

The anchor rod eye of this invention consists of a formation substantially as shown, comprising an eye for accommodating strand or strain member, having a c grooved portion extending outwardly opposite sides of the eye portion for ac modating the guy strand or strain me and a thickened end portion for prote or reinforcing the eye portion to preven jury thereto in driving the anchor rod.

I claim:

The original design for an anchor ro substantially as shown and described.

As inventor of the foregoing I here subscribe my name, this 24 day of Decem 1919.

ALBERT B. CHANC

Witnesses:
R. P. PRICE,
LORY T. SKAGGS.

Rod illustration and Patent text
September 14, 1920 Patent #56273
(A. B. Chance Co.)

An Anchor-Rod Eye" (No. 56,273) required less than a year. Both featured the Thimbleye® design, with the first filing representing a complete anchor rod while the second attempt concentrated on the eye area of the rod, which likely resulted in its quick processing. The attempt to patent the entire anchor rod apparently raised broader issues that may reflect interference with other designs or the complexity of the overall rod area, resulting in a cautious approach taken by the Patent Office. The quick action on the Thimbleye® itself surely reflects on a design that was unique to a metal rod. Patent No. 56,273 documents are reproduced here and No. 1,373,223 is reproduced on pages 141 and 142.

Delving deeper into the early drawings file (1914–29), a series of Thimbleye® Rods was illustrated and approved on April 20, 1922, by A. Bishop Chance. They refer to patents issued on September 14, 1920, and March 29, 1921. This is another example of a post-dated approval of materials already patented. The September 1920 Patent No. 56,273 was submitted on December 29, 1919, and was in effect for fourteen years indicating a design-type patent. A March 1921 Patent No. 1,373,223 filed on February 21, 1918, followed. The first became the last as the "Anchor-Rod Eyes" (No. 1,373,223) took three years before issuance whereas the "For

In studying the patent request for "Anchor Rod Eyes," there was another mystery revealed. The text on line ninety of the February 21, 1918 request suddenly refers to the new patent No.

56,273 issued on September 14, 1920! This undoubtedly reflects a resubmission of the filing. It may have been common to prod the Patent Office with updated support material. Patent No. 1,657,297 of January 1928 refers in its text to "Serial No. D5225, filed February 20, 1923." The last paragraph, however, states: "As inventor of the foregoing I hereunto subscribe my name, this 28 day of April 1924. Albert B. Chance" This is yet another example of a later text supporting a patent submitted previously. The witnesses on No. 1,373,223 included Nadine Thomas who may have been a Chance employee or connected with patent attorney Frederick R. Parker. The other witness was I. L. Traughber who was known as Miss Ida Lou, one of the most prominent of early Chance Company employees.

There is one area that is not explained by these patent-era drawings or the patents themselves. How did the A.T.&T. Spec. Numbers 3950 and 3951, for instance, fit into the overall Chance Company picture. One reason Gano and others may have pointed to Thimbleye® patenting in the 1920 time frame, is because the patent file did not have copies of earlier patent activities. If there were patents pending, why would Mr. A. Bishop submit the 1918 and 1919 Thimbleye® designs for patents? Chance ads from the 1940 era suggest the 1920 patents were the initial ones. The last drawing in the file is dated June 28, 1929, and approved by A. B. C. on June 30, 1929. It shows the Thimbleye® Anchor Rod with the common-pointed opposite end. This appears to be an improved version of Patent No. 1,657,297 filed on February 20, 1923, or May 15, 1924, and issued on January 24, 1928 (four years!). Patent details suggest this design was a "beefed up" version of Patent No. 1,373,223 of March 1921 to sustain "extraordinary driving" for hard or rock-filled soil. A point Gano offered was that multiple requests were made by Mr. A. Bishop to insure that his patents remained in effect with slight changes. Of course the other reason was to correct weaknesses in his designs after being observed in actual service. Both of these reasons were undoubtedly in play, especially considering the competition intensity on the national level and the sometimes long length of time required to obtain a patent.

The earliest Chance Anchor Rod patent of the four in the research file was No. 56,118, dated August 24, 1920, submitted on November 22, 1919. However, this was almost two years before the filing of the Thimbleye® patent of March 1921. This may reflect a growing concern over obtaining the Thimbleye® patent and so Chance submitted a patent application for the common drive rod with the regular inverted-pear shape eye and pointed anchor end. This short request filed by the patent attorney did point to the flat, strengthened top of the eye so it could withstand driving blows, although it simply may have been an improved version of No. 3950-51. This was the basic original anchor rod that covers all of the requirements of the Never-Creep® Anchor's conversion from a stick of wood to a metal plate. The previous gulf in time that appeared to exist between the perfection of the anchor and rod, making up Mr. A. Bishop's original deadman anchoring principles, was much closer than the gap that appeared to have existed in issuances or filings of anchor rod patents.

A rare April 6, 1938 letter from Hubbard and Company to Gano concerning royalties on a Thimbleye® product is reproduced on the following page. The Chance files were arranged so expiring contracts generated a follow-up with the other party to keep these arrangements in an orderly and current status. Notice Gano filed this in the contract file. Two related letters were also sent to Hubbard with apparently two different secretaries typing Gano's letters. Another letter of September 19, 1938 concerning Tripleye® royalties was much more confrontational, as Gano pointed out to Hubbard the validity of Chance Company patents and Hubbard's failure to act in good faith in honoring valid and reasonable contractual agreements. This letter came on the heels of an

Letter April 6, 1938 (A. B. Chance Co.)

4

Hubbard and Company

ELECTRICAL MATERIALS DEPARTMENT

APR 8 1938

PLANTS:
PITTSBURGH, PA.
CHICAGO, ILL.
OAKLAND, CAL.

JOSEPH V. SMITH, VICE-PRESIDENT

Pittsburgh, Pa.

April 6, 1938

Mr. F. Gano Chance, Vice President
Chance Company
Centralia, Missouri

Dear Gano:

I have your letter of March 26, and it is correct we have not signed the supplementary contract on the large thimbleye for our steelwing anchor. However, I understand we have been paying royalties as per outlined. As you have requested, we have signed the three copies, retaining two in our files and are attaching one for you.

Yours very truly,

R. G. Robbins

Division Sales Manager

RGR:HV
Enclosure

filed in contract file
FGC

4
8/31/3

August 26, 1938

Mr. Joseph V. Smith, Vice Pres.
Hubbard and Company
6301 Butler Street
Pittsburgh, Pennsylvania

Subject: Royalties

Dear Mr. Smith:

The Accounting Department has called my attention to the fact that we have not received royalties checks from your company since March of this year.

We have received royalties from all of our other licensees, and we presume that you are manufacturing and selling items according to our contract, and that the payment of royalties is in order.

The royalties from your company are now four months past due, and knowing that it is the policy of your company to make prompt payment we have called this to your attention.

May we have your check at your earliest convenience?

Yours very truly,

F. Gano Chance
Vice President

FGC:BH

Letter August 26, 1938 (A. B. Chance Co.)

4
9/23/3

September 19, 1938

Hubbard & Company
6301 Butler
Pittsburgh, Pa.

Attention Mr. Carl Pierce, President

Subject: Tripleye Royalties

Dear Mr. Pierce:

Several months ago, my father and myself called on you and Mr. Smith in regard to a letter he had written us about the Thimbleye patent. We informed you that our patent attorneys had advised us that our patents are valid and that our contracts are good. Therefore, it was our intention to see that our patents were respected and our contracts lived up to.

The only other point of discussion was that Mr. Smith said he believed our royalties were a little high. We informed him that we did not believe this was true but we would investigate the matter and if we found they were high, we would consider the reduction.

In view of the inflation which has taken place the last two or three years and which is no doubt going to continue, we do not believe that the flat rate of so much per rod is out of line.

I wrote to Mr. Smith on August 26, copy of which I am attaching. To date, I have not received any answer to this letter, nor have we received our royalty check.

Since it has been six months since we received a royalty check from your company, we kindly ask that you give this your immediate attention.

Yours very truly,

F. Gano Chance, V.P.

FGC:V
enc

Letter September 9, 1938 (A. B. Chance Co.)

August 26, 1938 letter to a Hubbard vice president. Notice that the second September letter was addressed to the president implying it had become a more serious problem after his lower-level managers failed to respond. In spite of Hubbard's payment and response problems, Gano remained respectful and nonthreatening, but this wouldn't be allowed to continue.

CHANCO NEWS, CENTRALIA, MISSOURI 7

Five-eighths inch, Thimbleye Rods in the making! From four in the afternoon until one o'clock in the morning, Leo Sansbury, left, Sterling Adkisson, center, and Hershel Megown, right, toss rods from furnace to forge, and from forge to press at the rate of five rods per minute.

"WHAT GOES ON HERE"

(Continued from Page 6)

upset". These rods are inserted into a gas fired furnace situated to the right of the "upset" machine. This furnace, like the furnaces situated near the two drop hammers, is fired by natural gas and heats rods to a forging temperature of from 2100 to 2300 degrees F. Heating a one-half inch rod to this heat requires approximately thirty seconds while a one inch rod requires from a minute to a minute and a half. As soon as the rods reach the forging temperature, they are removed from the furnace and tossed to the upset machine operator. The operation known as "upsetting" is in reality the forming of a ball or knob on the end of the rod. This ball or knob is formed by a repeated number of blows being directed at the end of the hot rod. This particular function is the slowest one in the procedure of making rods, due to the way in which the machine must operate. For this reason, the upsetter is kept in use twenty-four hours a day, and two men form a third shift. It actually requires twenty-four hours of upsetting rods to keep enough finished for sixteen hours of forging.

After the rods come from the upset machine, they are placed in another furnace and reheated, removed and tossed to the drop hammer operator, where the eye is formed in a few rapid blows of the hammer, then the rod is tossed to the third man in the crew, who operated a press to perforate and trim the eye.

In forging with a drop hammer, the forging operation is performed by the machine lifting the ram to a certain heighth and then dropping it. The two drop hammers in our forge shop have rams weighing approximately 1500 and 2000 pounds. Thus, it is not 1500 nor 2000 pounds of pressure which is applied by these forges, but the pressure of those weights being dropped from a set heighth. Interesting to know is that in forging one-half inch rods, it requires four blows of the hammer. The weight of a one-half inch rod, five feet long, is three pounds. Forging the eye on a one and one-fourth inch tripleye rod requires nine blows of the hammer.

Although the blows of a drop hammer seem to fall in such rapid succession that there would be no time for the operator to make a move between licks, he actually lifts the rod from the die after each blow. This is done to keep the die block from absorbing too much of the heat from the forging. If the operator did not lift the forging, the die block would become too warm, thus shortening the life of the die, and the forging would cool too quickly, another thing which is extremely hard on the die. The life of a die is from 40,000 to 45,000 rods.

Threads are either rolled or cut on the rods, both operations requiring one man.

(Continued to Page 8)

The day shift crew who heat, forge, perforate and trim are Bob Daniels, left, furnace tender; Lester Anderson, center, perforate and trim; and Len Dailing, right, hammer operator. In this case they are forging five-eighths inch Twineye rods.

Forging Thimbleyes (*Chanco News* December 1946)

How the volume of licensed items was recorded is unclear. Was it simply up to the integrity of the licensed firm to report such activity? The 1938 period was a tough year compared to 1937, when business seemed to have blossomed only to fall back into the grips of the Depression. Businesses were stalling and trying to slow outgoing payments while accounts receivable at A. B. Chance Company suffered. *Chance Tips* magazine of June 1941 points out the importance of the Thimbleye® design and how widespread its sphere of influence was. Chance Thimbleye® guy fixtures were "Manufactured and distributed by licensees: Hubbard and Company, Pittsburg, Pa.., Joslyn Mfg. & Supply Co., Chicago, Ill., Line Material Co., Milwaukee, Wis., Oliver Iron & Steel Co., Pittsburg, Pa., Kortick Mfg. Co., San Francisco, Cal., Seyler Mfg. Co., Pittsburg, Pa., Copperweld Steel Co., Glassport, Pa., Utilities Service Co., Allentown, Pa., Electrical Mfg. Co., Battle Creek, Mich." This is a summary of who's who in pole line hardware manufacturing in the United States, now only memories of a bygone era.

A highlight of a visit to Chance Company was the opportunity to see the Drop Forge Department in operation. Tour guides always hoped that the anchor rod line was up and running as it was an impressive operation in action with huge hammers banging away upon the red hot steel rod ends, progressively forging them into a shape envisioned by the founder. The team work of heaters and machine operators pitching these hot rods to each other among the deafening noise of the hammering and the hissing of natural gas furnaces and the smell of nearly molten metal added to the drama of this unforgettable sight and experience. The faces and figures of untold men who brought life to this operation over decades have constantly changed and most forgotten, but the spirit of this operation lives on, a fitting tribute to Chance Company and the dedicated workers who represented it. It was an area where worker performance was demanded and bonuses expected to be earned for production above and beyond the rate and so it was manned by a select group. One reason for forging products is for strength as the compaction of the metal results in a stronger part, than a casting. Racing and high performance auto engines commonly used forged crankshafts and pistons instead of cast ones destined for regular production engines. Even though forged parts were more expensive it was the Chance belief that many parts that were to lead stressful lives in, around, and on utility power lines had to be of racing quality. For example, the tensile strength of a 1-inch Chance anchor rod was rated at thirty-six thousand pounds and the Chance catalog stated that "the drop-forge eye of each anchor rod is stronger than the rod itself." The wisdom of not having failure even in the face of such tremendous potential loadings speaks well—do it right as the quality will be remembered long after its cost has been forgotten. That is reflective of the wisdom of the founder and the high standards that Chance Company embraced, leaving a rich heritage for others to strive toward, long after this Mid-Missouri firm has ceased to exist.

Ladies and Gentlemen

THE TRAUGHBER, TOALSON, AND CHANCE FAMILIES

Women who worked at Chance Company are an intriguing subject because they made up an important part of the labor force. This appears contrary to what is voiced today about women's nonparticipation in the work force during the first half of the twentieth century. So either the critics are wrong or Chance Company was out of step with the normal hiring and working practices of the day. One could surmise, very early, that women locally seemed to be active and daring. A couple of brief references in the November 2 and the 16, 1894 *Guard* respectively, seem to support this: "There are said to be a number of school girls that are in the habit of smoking cigarettes" and "The lady drummer was here Monday."

There can be a special relationship and bond between longtime employees, regardless of gender, which can spawn positive working relationships. This became supportive to the work effort. Women that labored at Chance over the years, both in the factory and office, contributed to the success of the Company. L. C. "Roy" Hansen, during company interviews, was quick to recognize women's involvement as the interviews revealed some startling early revelations about the influence of women at Chance.

The early presence of women at A. B. Chance Company surely raises the question of how they were treated. Longtime employee Mrs. Betty Ridgway began work in 1952 for Brent Mayes in Accounting and later moved to Data Processing. She said the office jobs for women at Chance were eagerly sought over the years because they paid well, but she felt it was more difficult for women to be promoted. There are charges today that women were patronized by men in the earlier workplace and this employee's work experience at Chance covers the transition from business as usual until today when men have to be very careful in how they handle relationships with their female counterparts.

Mrs. Marjory Martin, who became secretary to F. Gano Chance and Nathan Toalson in the mid-century period, was an outsider who ruffled some feathers when she came into employment at Chance. The executive secretary position was highly pursued and when openings weren't filled from the pool of women within the company, this group let their feelings be known. Marjory's thoughts were:

> *Well naturally there is some complaint, not necessarily against Gano, but about the company—the favorites. Of course*

you naturally hear the gripes about Gano's secretary from the outside. Some of the secretaries were the favorite ones. You hear that in about every company. There are mistakes as you go into the pools. . . . You get someone from the outside because you really have to. You need the expertise. But those guys that have been there for 20 or 30 years, they didn't like it necessarily. (1981)

Marjory touched upon several issues, one being the idea of promoting from within versus going outside for new hires. Since Chance had the reputation of promoting from within, outside hires were viewed with suspicion by employees. There was also disappointment resulting in not having the opportunity for advancement. Employees could be passed over when their performance was not seen as up to snuff and going outside was a method of countering such problems and obtaining special qualifications. In addition, going outside the company helped quell charges of favoritism, thereby lending additional credence to those who were promoted from within. Management and people in general can be found to show favoritism for a variety of reasons. Some individuals are deserving of such on-the-job recognition while others are not. The executive secretarial position had to be filled by someone with whom the executive felt comfortable with and one who could fulfill the variety of job requirements including maintaining confidentiality.

There were additional discussions about the secretarial positions and promoting from within. Marjory continued:

Most of the secretaries, I think, had come from within. I know Marjorie Blanton, who was Mr. Hough's secretary, had come along with him. You kind of went with your boss. If he was promoted, you went along with him. I know there was resentment among some of the secretaries at the time. It didn't last long and probably 6 months or so after they got to know you, they didn't dislike you for it. But there is a strong feeling, they resent bringing people in. (1981)

Gano explained his position on hiring: "I favored the local people. I first support Centralia, then the county, the state, and national. I think it is a good philosophy and so when one does come in from outside it is rather an exception and it takes more for it to work out and to mellow." (1981) Gano didn't elaborate on why he had chosen Marjory from the outside, which was likely a personal issue. There had been charges over the years that certain boss/secretary relationships had been more than just business, and there is a sense that Gano was extra cautious in his secretary selection, not wanting to raise any issues of impropriety. This is not meant to belittle the home-grown employee as for a century they were overall a very talented and professionally developed group who likewise were expected to reflect this in their demeanor and appearance. From what has been learned from the early years and what has been observed since, they could match or better any other like group, especially in the quantity of quality work produced. Several women have spoken of the dignified appearance of the Chance office group in their long, stylish dresses and suits. They were expected to conform to an unwritten Company dress code that also covered male management personnel.

The early 1890s revealed a lot of socializing taking place in and around Centralia. Young men such as Bishop Chance were on the move, but young women were also calling on friends. The January 13, 1893 *Guard* had this account of such a visit: "Miss Ida and Myrtle Traughber were the guests of Miss Maude Chrisman in the country, Tuesday night." For reasons that will later become known, the Traughber name became one among many that was closely monitored in revealing the Chance Company story. It became obvious that the family had a certain prominence and high achievement level, which wasn't confined to the male members of the family. An October 13, 1899 piece in the *Guard* stated: "Miss Ida Traughber leaves today for Kirksville where she will spend a week with her brother, Dr. Frank, and attend the annual reception of the Osteopathy School." The following summer, on June 29, 1900, the *Guard* covered another Traughber visit to Kirksville. "Mrs. F. M. Traughber is in Kirksville attending commencement this week. She will be accompanied home by her son, Dr. Frank." This of course confirms that Francis M. (1848–1931) and Marnie (1848–1927) Traughber were also Misses Ida Lou and her sister Lizzie's parents. In July of 1900, Dr. Frank Traughber is identified as a recent graduate of the Still School of Osteopathy who opened a Centralia practice in August and became active in the state organization of that group.

Francis M. Traughber was a prominent Centralian who had become the president of the Centralia School Board and of the Centralia Building and Loan Association (established in 1892) in April of 1896. He maintained this banking position for some time; in April of 1896 and 1900 the *Guard* reported he had again been elected president of the Savings and Loan as he again did on February 26, 1904. He also became the Centralia postmaster in February of 1915. He served in that position faithfully, until he became ineligible for reappointment in August of 1923. He and Mrs. Traughber had begun spending the winter months in California, likely close to other family members. It was noted he had been in Centralia's public life as a businessperson for half a century. Additional details told of him being one of the founders of the Centralia mill. R. W. Traughber, the superintendent of schools at Hale, visited the F. M. Traughbers in July of 1916, indicating further the prominence of the family.

Miss Ida did not make the paper very often, which may indicate that as she entered her twenties she was still searching for her place in society. A year later on August 2, 1901, the *Guard* reported, "Miss Ida Traughber spent last week with friends at Saling." On May 15, June 12, and August 28, 1903, she came up from Mexico (Missouri) to visit her parents. This would suggest she was employed and residing in Mexico. Fast-forwarding to April 28 and July 21, 1905, seemingly little had changed as Miss Ida returned home to Mexico after a week visit on each occasion, with "home folks." But on February 23, 1906, was recorded a visit to her brother Dr. W. F. Traughber of Mexico. This suggests she may have returned to Centralia to live while Dr. Traughber had moved to Mexico. According to the *Guard*, the doctor's move had occurred prior to January 5, 1906. Since Dr. Traughber maintained doctor's offices in both Centralia (discontinued in February 1907) and Mexico, could Miss Ida Lou have lived in Mexico to cover the doctor's office of her brother? Besides a change in residences, Dr. Traughber had also married a Miss Nora Petree of Kirksville, which may have influenced the move to the larger community of Mexico. The marriage arrangement was an elaborate affair, at least in distance traveled, as the *Guard* on July 21, 1905, said they were to travel to the home of Miss Petree's brother in San Jose, California, where the wedding would occur in August. Veteran Chance employees will recall that San Jose was later a longtime Chance service-center (warehouse and sales office) location which was visited in the 1950s by a retired, veteran, and area resident, Miss Ida Lou Traughber! After her sister's death and her retirement from Chance Company in 1943, she relocated to the West Coast. Her brother Dr. W. F. Traughber and family had moved to Los Angeles during the summer of 1907.

The second son of Mr. and Mrs. F. M. Traughber, Leslie B., with a wife, young son and daughter died of typhoid fever at age twenty-eight, according to the December 1, 1905 *Guard*. Hardly any family was long immune from the hardships brought by those days. Later on June 8, 1906, there were two editorial pieces concerning the Traughber family. The first confirms Miss Ida Lou had returned to live

Meet the Treasurer

I. L. TRAUGHBER, TREASURER, A. B. CHANCE CO.

The world is full of surprises. Even "This guy Traughber", as she is erroneously referred to by men in the field who are not completely "in the know", is, "Well blow me down", a woman. But, we might add, a woman who can wear trousers whenever necessary and who can handle an executive position with masculine ability or better.

She is probably the most consistently industrious member of this organization, but occasionally finds time to join the fun and time to work with her trees and flowers. She has but one illusion - that an order received today should be shipped yesterday.

From *Chance Tips* of August 1939. Miss Ida Lou

in Centralia: "Miss Ida Traughber is at home from a pleasant visit with her brother, Dr. W. F. Traughber and wife, at Mexico. Miss Ida attended the reception at the Ringo Hotel Monday evening." There was also another son and brother, Charles Traughber, who was in the laundry business. Mrs. F. M. Traughber and son, Charles, visited in California with Miss Myrtle Traughber, who was teaching school on Cataline Island out in the Pacific Ocean, and other relatives. Miss Myrtle was to accompany them home. This reveals another California connection, besides the Petrees. In September of 1906 Miss Myrtle (sister to Miss Ida Lou) went to Washington, D.C., to attend National Seminary. Myrtle is not heard from again until news item in the June 7, 1912 *Guard* suggested she had married and was now living in California. "Mrs. Myrtle Traughber-Bryson, here from California for two weeks, the guest of her parents F. M. Traughber and wife, left for her home Tuesday afternoon." On September 21, 1906, the paper covered the move of Charles W. to Rolla where he attended the School of Mines. This agrees with interview material suggesting he later became an engineer. He was to graduate after his third year according to the September 3, 1909 *Guard*. Another article on December 31, 1915, reported Charley married Miss Georgia Powell at Memphis, Missouri. He was then managing a large California ranch for his brother-in-law, indicating he likely was not pursuing an engineering career.

On September 25, 1908, Miss Ida Traughber left for Oklahoma, according to the local paper. She returned in mid-February 1909 after an extended visit with family or friends. On August 13 her eighty-one-year-old grandmother died. Mrs. Traughber had come from Kentucky and her husband had died forty-two years earlier. She was the mother of F. M. Traughber and Mrs. Lucy Bryson. It was not until December 6, 1912, that the *Guard* revealed more Miss Ida Lou material: "Miss Ida Traughber left Wednesday for a three months visit in Los Angeles, California. Miss Ida's many friends know that she has earned a vacation and is entitled to a good time in the land of Winter flowers." Notice the play on winter, and in having so many family members in California this visit was not unexpected. However, what is surprising is a three-month leave of absence from her job at the Chance Company. Confirmation of her employment is her start date of February 7, 1910, stated by company material at her retirement in 1942 and 1943 (directorship). Her absence may have been possible in the early years, but as A. Bishop became more dependent upon her and her responsibilities grew, such an absence may have become more difficult. It is odd that her occupation, employer, or position isn't mentioned, just that her hard work had earned her a vacation? During the World War I period, the local newspaper did mention women with prominent positions more frequently.

The *Guard* of May 16, 1913, captured her spring return: "Miss Ida Traughber returned home from L. A. California, Tuesday, where she spent the Winter with her brother, W. F. Traughber and family." In May 1914 she was reported as a Mexico visitor on a Sunday, which suggested she was living in Centralia. In November, she ventured further to Kansas City and was said to have had a pleasant visit. In August 1915 the coverage of the comings and goings of locals continued, including Misses Ida and friend Fawn Stewart's trip to St. Louis to visit friends and relatives. April of 1916 found Miss Ida a guest of Mrs. G. S. Barger of Raymond, Illinois, for a week. Mrs. Barger was the former Miss Fawn Stewart so Miss Ida had lost a traveling companion and local friendly contact. But in August of 1916 she and a Daisey (Daisy?) Boone left for Colorado Springs to spend a month on their vacation, which suggests Miss Ida was working somewhere and had a new friend. From that point forward through 1920, there are no news items found of Miss Ida Lou. It is as if she had fallen off the face of the earth, but we know she is working at 210 North Allen Street in Centralia!

The paper's coverage of Miss Ida Lou's activities is given public exposure again on August 19, 1921:

> *Miss Ida Traughber was hostess to a line party at the Gem Theatre Monday evening, in honor of Miss Daisy Boone, of Mexico. Refreshments were served at Pruetts. There were twelve guests in the party. Out of town guests were Miss Boone, of Mexico; Mrs. Dr. Berger of Illinois and Miss Margaret Greenlee, of Kansas City; Mrs. J. R. Wiseman and daughter, Dorothy, of Texarkana, Arkansas.*

The Gem Theatre was in Centralia and in January 1922 it's name was changed to Pruetts Theatre as H. C. Pruett was the owner. Mrs. Dr. Berger was likely her longtime friend, the former Miss Fawn Stewart, although the *Guard* spelled her husband's name with an "a" (B*a*rger) in 1916. *Miss* Margaret Greenlee should have read *Mrs.* and would suggest that Miss Ida Lou's Chance Company relationship was now connecting up to the Chances in a social manner. It may not have been unusual for lengthy trips involving women during that time as in that same August 19 paper, it was reported that Mrs. A. B. Chance and Mrs. Greenlee had just returned from a four-week's trip to the Northwest, including a five-day trip to Yellowstone National Park. Mrs. Greenlee came to visit for two weeks in early 1922, which suggests she was a frequent companion of her sister and mother in Centralia. During that visit, Mesdames A. B. Chance, P. R. Diggs, and Margaret Greenlee spent several days in St. Louis. Early in 1924 she had moved from Kansas City to Columbia, and a March 18, 1927 news item stated she was living at the A.T.O. (Alpha Tau Omega at the University of Missouri) house, later confirmed as the chaperone.

By 1923 Miss Ida Lou must have become an important Chance Company office employee and frequent assistant to A. Bishop Chance. The ten-hour day and six-days-a-week work schedules had to take its toll. Employees made plans to get away occasionally, or at least Miss Ida Lou did. The October 26, 1923 *Guard* said she had just returned from a month trip to California where she visited her brother Dr. W. F. Traughber and family. The family seemed important, maybe more so because she was single, but likely just a natural inclination. In late November, the paper covered a visit she made to see friends in Mexico, Missouri. In April 1924 Miss Ida Lou was a Columbia visitor on a Thursday, possibly suggesting this was company or personal business related. Sooner or later it seemed that most local residents were threatened with influenza, as Miss Ida Traughber was according to the local paper on April 17, 1925. She recovered as the May 15 paper said she had returned home from a St. Louis visit. Her father returned in October from California where he had visited his son and family. This was a defining period in Miss Ida Lou's life at Chance Company in view of the maddening pace of activities there.

Early in 1926, Miss Ida Lou and Mrs. Baker were hostesses of the Modern Priscilli Club at Miss Traughber's home. Mrs. A. B. Chance was a member, likely one of the fourteen guests present. By the fall of 1926 activities intensified at the Chance Plant, because on September 17 A. Bishop left on the family's world cruise. In early December she entertained several guests with dinner. Taking care of financial obligations and her other duties in A. Bishop's absence undoubtedly kept Miss Ida Lou very busy at the Chance plant with many long-hour days. Of course it is not reasonable to suggest her duties were reduced upon his return. But with the march of time also comes losses and the July 8, 1927 *Guard* reported her seventy-nine-year-old mother, Mrs. Agnes (Bryson) Traughber (1848–1927), passed away. Born on a farm near Hallsville, she and Mr. Traughber had lived north of Centralia until 1926. She was a capable and well-respected lady. As fall 1927 approached, Miss Ida L. Traughber left to vacation at Hot Springs, Arkansas. Although

Miss Ida Lou and Nathan Toalson at Mr. A. Bishop's seventy-fifth birthday celebration in the spring of 1948. (*Chanco News* June 1948)

veteran employees referred to Miss Traughber as "Miss Ida Lou," the paper normally had referred to her by first and last names only with no reference to her middle name or initial.

The relationships between men and women have always been challenging. My wife's great-aunt, Ethel (Mozingo) Belcher (1899–2002), who lived to be 102, told us of her courtship in the early 1900s. Her beau, Ervin Belcher (1900–2001), made the mistake of hugging or embracing her before they were engaged, although there was an additional situation in which she caught Ervin dancing with another lady. She said, "If my papa had found out about this there would have been fireworks." She terminated the relationship and it took over a year of determined effort by Ervin to reestablish their courtship. They were married over seventy-five years! People from those times had a certain quality about them that can be admired today. This stricter code (in comparison to 2000) was revealed by an editorial comment in the March 15, 1895 *Guard*: "Centralia has some very forward girls that ought to be kept under restraint. Their bold flirtation with strangers is absolutely shocking, and gives a wrong impression regarding their purity."

Another story from this time comes from a lady who came to Centralia from Maryville, Missouri. as a schoolteacher in 1938. She was a recent graduate of Northwest State College of Maryville. By the way, Ethel and Ervin were also from Maryville, Missouri. On October 4, 1941, Miss Gara Williams became Mrs. N. A. Toalson. Nathan Augustus Toalson (1911–1985), a forty-six-year veteran who retired in 1975, eventually became the Chance Company CEO. Mr. and Mrs. Toalson were long active in the First Baptist Church of Centralia and were the parents of three children, Gara Ann, William Nathan, and Alice Bowman. The story that Gara told was from the 1910–1916 time frame when Chance Company was located in an old wooden, high-front building south of the Centralia square. A. Bishop Chance was a very personable and persuasive man and it was common for him to put his hand on the shoulder of those he associated with and talked to. Miss Ida Lou Traughber was A. Bishop Chance's secretary and one-person office staff. Miss Ida Lou was never married while in Centralia, and she lived with her mother less than a block away in a brick house just south of what is now Meador and Son Funeral Home. As the story goes, A. Bishop placed his hand on Miss Ida Lou's shoulder, which brought an immediate reaction. Miss Ida Lou quickly moved away and grabbed her purse and went home. Her mother asked her why she was home so early and she said Mr. Chance had put his hand on her shoulder. Mrs. Traughber said she didn't believe he meant anything by that and she thought Ida Lou should return to work. Miss Traughber did return and stayed for thirty-three years (1910–1943). This story paints a picture of the strength of character of Miss Ida Lou when she was a young lady, a strength that would endure and be demonstrated in her loyalty and long dedication to A. B. Chance Company.

This is Mrs. N. A. Toalson, the former Miss Gara Williams. (Jack Chance Collection)

In 1909 no other Chance family member could seemingly eclipse the accomplishments of a still young A. Bishop. But in a certain way his grandfather, G. H. (Garrison) Chance, had the credentials to unseat his grandson. The *Guard* on October 1, 1909, carried a story about a bridge, since replaced, 3.5 miles south of Centralia that spanned Silver's Fork (Cave Creek). It was just south of the Union Church. It was originally given the name Rotunda by the contractor who built it, Uncle Dan'l White. It was built in 1844 after White had completed what was called the "first structure" at the University of Missouri (M.U.). Garrison and his brother Albert Chance were working for White at that time and were involved in both construction projects. Garrison said he hauled off the first load of dirt from the foundation excavation and also hauled most of the lumber used to build this university building. At his ninety-fourth birthday celebration, Garrison told stories and mentioned many names of people in the Columbia area in 1840 when he moved there at age nineteen. At the time he hauled the first lumber for the State University and said there were only about 200 to 250 inhabitants in Columbia! An April 8, 1921 *Guard* article that was developed from an interview with Mr. Chance repeated this story and referred to his picture that had been hung "in the halls of the University of Missouri." An article in the February 22, 1924 *Guard* told of the first Missouri University building, built in 1840, being destroyed by fire in 1892. The pillars (six Gothic columns) were left standing and by 1895 they had become the subject of some hot debates as to their disposition. Today they still proudly stand, representing a universally recognized symbol of the University of Missouri. It is special that advancing four generations, Mr. Chance's great-grandson, F. Gano, would attend and graduate from M.U. Clearly Garrison Chance was a Boone County pioneer whose building skills were applied to some of the first buildings on the M.U. campus, in Centralia, and other county structures. This story was recorded just in time when Garrison was concluding a three-month visit and returning home to Adair, Oklahoma. An earlier visit mentioned in the October 11, 1907 paper of Charles Chance of Oklahoma visiting his brother John, would suggest Garrison was living near or with his son Charles. He later would remove to Tulsa to make that his new home when he was eighty-six years young. What a strong spirit he possessed.

The G. H. Chance legacy continued when the March 13, 1914 *Guard* carried a front-page piece about him getting run down in Oklahoma by a horse and buggy. This was apparently one of the first hit and runs as the woman sped away "escaping detection and has not yet been

The age of Buford in the forefront of this four-generation portrait suggests the 1905 period. From the right to left are Garrison, John, and A. Bishop. Note their heights: A. Bishop was six-foot two-inches. (Jack Chance Collection)

identified." Mr. Chance was then ninety-two years old and his serious injuries (unconscious for several hours) were complicated by his age. A letter to the *Guard* from in-law J. T. Roberts of Tulsa, published on April 9, 1915, said "Uncle Garrison was the oldest resident and doing well for his age." A 1916 letter said Tulsa had a population of about 50,000! Another letter from J. T. recorded in the March 24, 1916 *Guard* told of the celebration of Garrison's ninety-fourth birthday, but Garrison suggested it was his ninety-fifth. This is confirmed by a Tulsa letter the previous year (September 1915), which suggests he was then ninety-four. He was feeble and not able to do much. It also spoke of "Uncle Garrison's baby girl by his last wife."

Near the end of the second decade of the 1900s and on into the 1920s, there was much revealed about the early years of Centralia and its people in the local newspaper. This grew out of significant individuals reaching an age in which they began reminiscing about earlier times. What they had lived through was becoming recognized and important and this may have influenced those like Garrison to share their experiences with those around them. Schoolbooks were very impressionable and through pictures and words students came away with lifelong images of the various aspects of American history. One such image was prairie schooners pulled by oxen. They came from the East, heading to the West, across our great nation, transporting pioneers toward the Promised Land. Garrison Hardcastle Chance rode an ox-wagon across a large part of the eastern United States into Illinois. He was a pioneer like some of our early families. Before Mr. Chance passed away on February 20, 1922, in Oklahoma where he had lived the last fourteen years, he remembered some pertinent information that also found its

way into his obituary in the February 24, 1922 (No. 1) *Guard*. For organization and clarity purposes, these sources will be numbered. Another *Guard* article of April 8, 1921 (No. 2) will be referred to in summarizing some remaining aspects of his life. There are some glaring errors that will also be addressed. While reading the many letters written back to the Centralia newspaper, it was revealed why some moved to warmer and drier climates like Oklahoma to avoid the dangers of pneumonia and other life-threatening respiratory problems during the difficult winter months.

One of the areas where there was conflicting information was concerning Garrison's marriages and offspring. His first marriage to Miss Martha Ann Roberts (1827–1871), according to the Chance Bible, produced eight children: Cornelias Edward (C. E. or Ed) born in 1846; John Albert (John) born in 1849; William Eli born in 1851; Mary Elizabeth born in 1852; Elliott Butler born in 1856, Orson Jonas born in 1860, Sarah Ettie Francis born in 1864; and Charles Henry born in 1867. Garrison's interview with a Tulsa newspaper reporter on his ninety-ninth birthday (No. 2) reported seven were still living. Their honeymoon was taken in a one-horse cart along with all their belongings. Their first home "was a tiny cabin, and the furniture was rudely constructed from boards. . . . Later Mr. Chance married again three times." (No. 2) Martha passed away on August 15, 1871 (August 28 per tombstone) and Garrison married Miss Lucinda (Lurinda on tombstone) Wainscott (Winscot?) (1838–1885) in November of 1873 (No. 1). (Note: There were both Wainscotts and Winscots buried in the Red Top Christian Church Cemetery, even together, so it is not clear which is correct for Lurinda as her maiden name isn't on her tombstone.) Chance family research suggests the marriage took place on December 22, 1872. This marriage produced four sons according to the family Bible: Thomas Jefferson born in 1874; Aubra Bach (1876–1900); Bingharm Rollins (1878–1959); and Grover Cleveland born in 1884. The *Guard* reported Lurinda's death in 1885, as covered in Chapter One. This is also confirmed on her tombstone as being on December 5 (15?), 1885, at forty-seven years, eleven months, and seventeen days of age. Garrison married again in 1886 (No. 1) but at this time to an unknown lady, which apparently produced no children. She passed away in 1892 (No. 1). Information sourced by Linda Chance in May 2001 identifies Garrison's third wife as Justina McNeal (1832–1892). Although no hard evidence has been found confirming Garrison's fourth marriage to Miss Mattie Neal (No. 1) in 1893, the *Guard* on April 13, 1894, reported the birth of twins, a boy and girl. "Three children were born of this union, viz: William and Margaret, twins, and Afton." (No. 1) "His youngest child, a baby when her mother died, needed more care than her aged father could give her, and so an aunt adopted her, and it is her name and not her father's that she bears. She is Miss Afton Ames and is 22 years of age, and is now living with her adopted mother in Wyoming." (No. 2) Linda and Anna Lee Chance's research suggests Martha A. McNeal was a sister to Justina. There is no denying the hardships of childbirth during Garrison's lifetime, and it may not have been that uncommon for men to have multiple wives and families. But that was a reflection of those times and the lack of sophistication in medicines and precautionary measures. Of course not all female deaths could be attributed to childbirth alone as both male and female were constantly threatened by diseases and illnesses of various sorts plus work related dangers.

Garrison was a member of the Christian Church at Vinita, Oklahoma (No. 2). J. A. (John) Chance reportedly was not very active in the church, according to his grandson F. Gano, but he said his father A. Bishop was. It was confirmed John was a deacon in the Christian Church in 1881. It appears that John didn't have the flamboyant nature, as did his father and son, and his reserved nature tended to generate suggestions about him that may not be totally accurate. Reflecting a continuing influence in the Chance family, they were members of the Christian Church denomination. Garrison's obituary stated he was a man who loved his Bible, family, and enjoyed his many friends (No. 1) and who enjoyed his pipe and an occasional cigar (No. 2). He remembered when a farmer's plow was a piece of wood and "his ox-drawn wagon had sawed-off sycamore wheels and all of the clothes they wore as well as the food they ate, was produced on the farm." (No. 2) One thing that made a great impression on his life was when he saw the stars fall in 1833 (No. 2). "A devout student of the Bible, his careful study of prophecy has led him to believe that that astronomical event presaged the second coming of Christ and ever since then, to this day after living a life that spans a century, he has been looking to that coming." (No. 2) Mr. Chance said:

> *There was a shower of star fire just like snow. Everybody thought the world was coming to an end, and rushed out doors and fell on their knees and prayed. The stars were right bright, like they are in the sky, until they got "most down to the ground," then they went out. So while we thought they were on fire, there was no way of telling sure. (No. 2)*

Funeral services were conducted in Tulsa on February 21, 1922, for Mr. G. H. Chance with burial the next day at the Red Top Church, west of Hallsville. He was buried beside his first two wives. (No. 1) Incidentally, the September 29, 1922 *Guard* reported the one-hundredth anniversary of the Red Top Church. No tombstone has been found for Garrison, although he was reportedly buried beside his first two wives, whose graves are marked. According to Jack Chance (2005), the Red Top burial records were burned in a home fire of a person researching that information.

Chapter 4
Growth and Expansion
(1917–1924)

Rendezvous at 210 North Allen Street
Relocation and Refocus

The location of a new place of business is often limited to what is available and makes do, because too often such beginnings are short on resources and long on promises and the future is yet to reveal which is truer. An interesting question arises in special circumstances; what happens to a church building when it is no longer a church? In Nashville, Tennessee, Ryman Auditorium became the Grand Ole Opry and in Centralia the former Christian Church became associated with the beginnings of the Chance Manufacturing Company.

Start-ups aren't always clothed in stability and firmness of direction, but in 1917 A. Bishop Chance had decided that Centralia would represent a more permanent home for his small firm. How do we know that? A person's resolve is often reflected more purely in what they do. The convincing nature of this reasoning is reflected in the two-story brick structure that was erected in 1917 between the tracks of the two railroads that ran through Centralia. This new location would be the base of all Chance Company operations for the remainder of its long life. Although it would have been difficult to anticipate this at that time, it was a vision of the founder in adding to the continuance and permanence to that which he had earlier begun.

F. Gano Chance recalled his father's move from uptown to 210 North Allen Street:

> *He moved from here over to between the tracks in 1917 and the war was on and he couldn't use anything except the first floor. Right after it was built in the winter of 1917, he had a carnival to make money for the Red Cross, etc. for World War I. As soon as that carnival was over, it ran about a week, then Mr. Baker moved his chemical company upstairs. (1978)*

Gano also mentioned the production of automotive piston rings, which were to follow, and hitching straps. In another interview, Gano reveals additional information about this move. "I do remember some things during the war [World War I]. My father moved from his location off of the city square to over across the railroad tracks in 1917 and built a two-story building. He put in a machine shop on the first floor." (1981) In a 1968 speech, Gano gave

This 1910-era photo is of the between-the-tracks area where the Chance main factory/office building would be located in 1917. The building in the front center reads the "Centralia Coal Company, D. C. Pool Prop." This was replaced by J. K. Johnson and in the left background the tall two-story building has a D. ?. Kanatzar, Coal Office sign. One of the two-story buildings would be moved east of the Anthony Building and be rebuilt as a home. The angled structures with the rail spur appear to be coal bins. The city water tower looms in the background as does Crump and Humphrey (Centralia Milling Company) who can be identified by the Purina "checkerboard" signs. Ralston Purina was a St. Louis–based animal food giant who had many produce dealers in rural area towns. F. M. Stamper Company out of Moberly was a regional competitor. (Jack Chance Collection)

another perspective on the expansion of the Chance Company in the teens and early 1920s, commenting about the utilization of the new two-story, 210 North Allen Street building. "At first, the Chance Company just used the first floor of its fine new brick building. The second floor was rented to a man who made poultry feed. But Chance Company growth soon got us out of the chicken feed business and we began to use both floors and a rented building across the street." (1968)

There is little direct information of how financially solvent and successful A. Bishop Chance had become by 1917. Instead of looking at profit and loss statements or sales figures we don't have, there are other measures that can be used to determine how well someone is doing. For example, on July 29, 1921, the local paper ran a commercial ad: "Wanted: Woman for general housework. A. B. Chance." A pattern may be developing because when the proprietor has the resources to ask one of his factory employees to perform personal work, some might suggest he had arrived.

Car accidents were fairly common as more and more automobiles came into the hands of those more familiar with feeding oats to "ole Dobbin." A turnover recorded on the front page of the January 19, 1917 *Centralia Fireside Guard* involved Denton Palmer. As he followed the North Route, (designated the "Lewis-Clark Route") old Highway 22 north and west out of Centralia, he met Ed Riley's horse and buggy. Narrow, rut-slewed roads weren't supportive of precision maneuvers and when Denton politely pulled to the roadside he lost control and just as is common today, he overcorrected, ending up in the ditch where his car overturned, pinning him underneath. The lighter cars of that era allowed Mr. Riley, surely strengthened by hard physical farm work, to lift the car so Denton could escape, bruised but otherwise uninjured. He had been traveling to the Fountain Farm to pick up Mrs. Willis Fountain and her child. There were at least two Willis Fountains, a W. H. (1839–1918) and a W. T., who would appear to be the family member being described here. Members of the Fountain family were farmers in that section—including Stockton, a prominent livestock judge at the State Fair, and also a registered Hereford cattleman. Incidentally, Mrs. Willis Fountain was A. Bishop's sister and Denton Palmer was a Chance Company employee. Can you imagine the grief caused if she or her child had been injured or died in the accident? At Chance Company, it became common for company employees, especially prior to the union, to perform work at the Chance homes or farm. Early Chance employees appeared to be steadfast, disciplined, loyal, and hardworking individuals, who commonly became involved in a variety of jobs.

To better understand what Centralia represented in 1917 when A. Bishop decided to build a new plant, we turn once again to the *Guard*. A summary of the impressive features of the Prairie Queen on January 5, 1917, reveals a population of 2,860. It is obvious that

Main Chance factory and office building under construction at 210 North Allen Street in 1917. (Jack Chance Collection)

Chance Company didn't grow the town, as it already approached 3,000 before the Chance firm was a significant factor. In-town growth has since been stagnant when comparing to current figures of some 3,400 residents. Chance Company, however, has clearly influenced the stability of those numbers and impacted rural Centralia growth. The article also suggested:

> *Municipal, water light and power plant, with complete sewer systems. Three large horse and mule barns. Large poultry house and cold storage plant. Eight Rural Free Delivery routes. A league base ball park, broom factory, electric supply factory, cement block factory. Three school buildings [modern]. A fine 2-story brick city hall. Two extensive lumber yards. Two mills, one hitch yard. Forty-four business firms. Seven church buildings! Two telephone systems. Good Public Library. Three fraternal halls. Three good banks. Four auto garages. Two hotels. Located half way between St. Louis and Kansas City and in 21 miles of the State University and county seat. Railroads. Wabash, Chicago & Alton, Burlington-Alton and Centralia-Columbia Branch. 27 passenger trains daily.*

There were several items within this list that were attention getters. The description of "Three good banks" suggests that prior to the Great Depression there must have been some "bad" banks in existence or unstable ones attempting to stay afloat. The exclamation mark after the seven churches indicates the Centralia area was a strong and proud supporter of the Bible belt. At least one of the mule barn locations would eventually become a Chance plant site and the electric supply factory was obviously the Chance Manufacturing Company. Although this was an elaborate array of offerings, the most impressive was the twenty-seven daily passenger trains! To show how efficient, fast, and convenient travel on trains had become, a news piece in the February 13, 1920 *Guard* explains the advantages of having such rail service in Centralia. A reference to Glasgow meant John Chance's rails of choice was the high flying Alton, with their big, powerful, and fast steam locomotives:

> *J. A. Chance made his customary trip to Kansas City Tuesday and took treatment for the trouble in his lower limbs. Mr. Chance thinks he is improving nicely, which is good news. His trip Tuesday, incidentally was in a manner remarkable. He ate his breakfast at home, made the trip to Kansas City, received his treatment and made the return an entire distance of over three hundred miles, in time to eat a supper cooked by his wife and on his own table. Pretty fast moving for a seventy-year-old. In his younger days Mr. Chance would have thought himself swift if he could have made a trip from the farm to his trading point and back in time to enjoy the two meals at home.*

Adding to the routine of A. Bishop's business travels, the paper found him in St. Louis in early February 1917. From an inventive prospective, he was continuing to develop new products of the pole line hardware variety. A patent filed on February 17, 1917, for a Cross Arm Back Truss and Clamp, reaped a seventeen-year Patent No. 1,324,585 issued on December 9, 1919. The specialist assisting in procuring the patent was Frederick R. Parker of Chicago, a registered patent attorney with one witness being Frances Chance (Mrs. A. B. Chance). On March 9 there was verification that A. Bishop was in Iowa and had sent a "communication" to Mayor Bruton and the Board. It was read to the Board of Aldermen in their regular session on Tuesday evening, March 6, 1917:

> *Mr. Chance desired a location for a prospective factory building for his enterprise, and asked that the board give him a longtime gratuitous* [sic] *lease on the plot of ground vacated by the removal*

> *of the old city jail house, between the mill and Allen street. He also asked that the city furnish him electric current at not more than five cents a kilowatt. The matter was referred to the city grounds committee who will go over the matter thoroly [sic] with Mr. Chance before the board takes any action.*

The mysteries of the background and actual construction of the 210 North Allen plant are only revealed in bits and pieces. The struggle in gaining financial support for a site, which earlier included a factory, had fallen on deaf ears. The seemingly inaction by the community to romance and retain the Chance Manufacturing Company, in a sense is puzzling. That was one reason why coverage was given to all the political upheaval that Mayor Chance and his more progressive following had stirred up. Included was what could be labeled a short-lived industrial park for new industry, which could be construed to include the Chance firm. This and the slow growth in employment of A. Bishop's small operation may have dampened enthusiasm and raised issues that killed support. There were also a lot of sensational projects during this period, representing tremendous growth and industrialization, but there were also failures, some of which were outright scams. So there was a cautious, wait-and-see attitude prevailing, especially when there weren't enough dollars to go around. Although Mayor Chance's push to complete the water and sewer system was proven overwhelmingly beneficial to the local community, especially when a drought forced the railroad to use and pay for city water, the city struggled to pay off the debt. At one point total monthly volume out of the city's deep well was in the five-million-gallon range. Of course the deep well was indirectly the result of the failed efforts of oil speculators in Centralia, due in part to A. Bishop. If the capability had not been present to dig a deep well, the water system may have failed from an inadequate volume. The case could be made that because A. Bishop pushed hard for progress and the elimination of stagnation, Centralia benefited, but it may have turned on him and adversely affected his campaign to gather local financial support for his most important project, the Chance Manufacturing Company. Even though he had by this time become proficient in the art of selling, sometimes the hardest sale can be around those who know us as neighbor and friend. If he had depended on the support of the local community, the prospects for the dawning of his company would have been dim indeed. It instead seems to be driven by his enthusiasm, a spirit of never quit, and by a sound faith in his business endeavors.

In 1917 conservation of food and fuel was being enforced by the federal government because of World War I. This was the result of the demands abroad, or "Over There" as it was routinely referred to, and the railroads were controlled by the government with new equipment standardized for increased efficiency under U.S.R.A. Travel was curtailed and shortages of various materials made business difficult, especially becoming more intense in 1918. In local news the February 16 *Guard* reported the death of H. L. Ratliff by an electrical power line. This had a very important connection to the A. B. Chance Company of the future. It wasn't a downed power line that caused this accident but a combination of poor transmission line construction and a breaking of a safety rule by having a metal pipe near a power line. There was also local business news of a new soft-drink bottling company and a major addition to the Lyon Auto Garage taking place in Centralia. But the big news in the March 9, 1917 *Guard* continued to be concerns over the future relocation of the Chance Manufacturing Company. Because the paper's version varies and adds to the earlier Mayor Bruton quote, it is reproduced here:

> *ASK THE CITY FOR A SITE. A. B. Chance, head of the Chance Manufacturing Co., of this place, is asking the city to give his company the use of the ground, vacated*

by the old jail, between the two depots, for the erection of a building to house his enterprise. Mr. Chance asks the use of the ground gratuitously and he wants the city to furnish him current at a rate of not more than five cents per kilowatt, a little more than the city is paying Mexico for the current at present. It is said the contemplated buildings will be of brick and are planned for two stories in height. The site at this point is an ideal one for the location of a factory and can be placed arranged to have the switch track at the doors of the building. The board has the request under advisement and it is thought they will give the ground asked for, or a part of it.

The Chance Company story can be likened at times to thunder and lightning. After the passage of several years with no requests for assistance noted, all of a sudden this bombshell hits. Notice that the *Guard* piece, on the same front page as the city's official version, differs. The Rodemyres again appear to have the inside scoop on the details of the proposed Chance plant. How did they obtain that information, when A. Bishop was in Iowa making a last-minute emergency request to the city fathers back in Centralia? Could they have known of plans that had already taken shape or did A. Bishop make a phone call to the *Guard* office to insure additional newspaper coverage and support for his new plant request? There is another piece of the puzzle that hasn't been shared with the reader yet; A. Bishop wasn't the only party wanting the site where the old city jail had been located! It would be interesting to know who made the first request and if these two parties were initially aware of the other's request or how the second party became privy to the first request. According to the same March 9 newspaper, just above the *Guard*'s version of the Chance request was this item:

A MILL BUILDING IN PROSPECT. The proposition of Crump & Humphrey who desire to lease the land vacated by the old jail, is up before the city. The firm needs more room and a better location to get nearer to the switch tracks. The building in contemplation will be of brick and concrete, if built and would house the mill and elevator. The firm has long been considering the extension of the mill and the addition of exhaust elevators made of steel. A big mill and elevator would be of great benefit to the farmers and to the city. The present uncertain condition of the market precludes the possibility of erecting the building at the present, and the firm is asking an option to over a year on the ground. The board has the matter under advisement and no doubt will arrange to furnish them the land desired.

The location sought for the new plant generates questions, as did the delays in reaching that point. As a Christian, A. Bishop was likely familiar with the Scripture verse of asking for what you want: St. Matthew 7:7 (KJV): "Ask, and it shall be given you; seek, and ye shall find; knock, and it shall be opened unto you." One thing that hadn't changed was A. Bishop's insistence upon obtaining city grounds for his plant site. It doesn't seem logical that this would have held up construction of a new plant, which he apparently had accepted the financial obligation of personally constructing. The grounds would seem like a minor cost issue in comparison to the new building, but A. Bishop was a sharp businessman who hadn't given up hope in getting help from the city. This was now concentrated in obtaining a low-cost lease on the acreage and low electrical rates, a utility supplied by the city. There is substantial logic to suggest that what he was requesting was simply sound business judgment. This location offered water and sewer services, was close to the pump house and water tower, and was near where the Mexico electrical transmission line terminated, so he was close to the utilities necessary for manufacturing needs and for a facility with all the modern conveniences. In addition, this location could be served by either of two railroads. That was vital during that time, for the shipment of raw materials and finished products. The truck was just developing and many of the roads and highways were still dirt. The depots were close so express shipments could be pushed over to the stations by employees using "man-powered" wagons. So what can be construed as a request for city assistance may in fact be driven by the fact that this was prime real estate, if not the most perfect location for the Chance plant in Centralia! The shipping of the product was done quite differently in the 1920 time frame as described by Flynt Chandler:

> *Not much space when I started over there. As I said I done a little bit of everything, we were busy all of the time, we worked together, did electrical work and repaired machines and at that time most of the anchors were made out of gray iron and shipped in barrels. We would collect all of the sugar and salt barrels around town and put them in there and wired across the top of them. Had a cart with 2 big wheels in the center and on either end there was a little wheel and we would set those dudes on there and take them over to the old depot. That is when the old Wabash Depot was up in the corner. Hauled them over there and set them on the platform. Finally we got an old Model T truck and hauled them in that.*

Later during his thirtieth anniversary celebration in 1949, Flynt explained the reason why the anchors were placed in the barrels:

> *We didn't make anchors in those days. They were made out of cast iron, not like the steel ones we have now. We bought them from a foundry and then dipped 'em in black paint. They were liable to crack if they were thrown on and off freight cars so they had to be packed in barrels.*

Describing how the anchors were painted and set over a trough to catch dripping paint, he said: "We had a long bar about 3' wide and 4' long, a trough on either end and took just a common hay hook and dipped that anchor down in there and set it up on that rack." (1979) Have we lost the practicability and simplicity of the past? The reuse of the barrels points to a scavenging operation taking place at the rear of the stores of the local merchants. Flynt recalled those procurement missions that added another duty to the list of many jobs entrusted to him over the years. This also indicates how small an operation and dependent Chance Company was on unique methods to cut costs for survival. This is the kind of idea that one can picture A. Bishop or one of his ingenious employees coming up with. But the reuse of wood slats for floorboards from parts crates in the construction of the Model T Ford, suggests it wasn't all that unorthodox. For the man from Dearborn to use such an idea adds credibility to the methods of Centralia's most famous citizen. The salt and sugar contaminated barrels could generate an element of risk for more delicate products, but it's doubtful if these substantial hunks of iron, formed in the shape of an earth anchor, suffered premature rust-out!

A. Bishop Chance and Crump & Humphrey must have become anxious as to the outcome of their city requests. Focused before our very eyes in a rural, out-state, predominantly agricultural setting, is a struggle between this predominate farmer driven local economy versus the factory-industrial type of interests. Politically, this was a fragile situation and the outcome wasn't entirely clear in the spring of 1917. In a January 5, 1917 overview of Centralia's impressive offerings, the *Guard* stated that there were two mills located in town. The mill in question appears to have been located in the Rollins Street vicinity. According to Wabash depot information in 1923, Rollins did run across the tracks at that time to the rear of the original two-story Chance Plant. The Chance Plant expansion and the new 1925 Wabash Station would later make that impossible. A railroad stockyard blocked Collier Street between the tracks prior to 1925 but was torn down just before the new Wabash Railroad Station was built. This was deemed necessary as with the closure of Rollins another route was needed to cross the tracks east of Allen, giving access to the business concerns in that area. Collier continued to run across the tracks until a Chance Drop Forge expansion in the 1950s, closed it. In a 2003 visit, the sidewalk on the northwest side approaches the track as if no change had occurred fifty years earlier. To the west, Chance interests wanted to move from Allen eastward toward Collier and Crump and Humphrey wanted to move westward toward Allen. The contest, a clash of sorts, awaited the word of who would get all or a piece of this valuable real estate. Each was surely eyeing the other's "intrusion" and was concerned there wouldn't be enough room for future expansion. Ironically today, two elevators, one to the west (Wilson and Hickman) of the Anchor Assembly/Punch Press Building (Allen and Wilson Streets) and one (M.F.A.) further to the east on Jefferson, suggest to the serious observer, the struggles of those two interests in occupying this prime real estate. In a sense they have the former Chance Allen Street complex boxed in! This mill confinement would help result in the destruction of the Wabash Station in the 1990s, allowing the bulging expansion of Chance factory grounds to extend several feet southward.

So locked into the intrigue of who gets what, cautiously the two parties awaited the city's official decision. One characteristic of these two requests is neither guaranteed that a building or structure would be built. The mill request seems less sure, although that may have been a *Guard* slant. They left the image of what kind of structure they desired to build, but clearly the land was what was on both of their minds. The Chance request for electrical price concessions was a wise move, if only to suggest the seriousness and more thought out approach. Gaining the edge is important in any competitive contest and before the city's ruling in mid-May came more thunder and lightning. One was planned and one was not. Being a premier promoter, A. Bishop generated a lot of thunder. The March 16, 1917 *Guard* covered a business trip to St. Louis by H. E. Stone and A. B. Chance, but the big news in the newspaper, forced to shorten its pages because of war time paper shortages, was an announcement of the invention of a unique piston ring and the inventor's already established business willing to move to the new, proposed Chance plant. Such an announcement couldn't help but push the city to act on the Chance and mill requests for city land. The timing would suggest a power play as A. Bishop methodically played his cards to gain the upper hand.

The second unplanned incident involved Home Telephone presenting an ugly situation. At 1:45 a.m. on Saturday April 28, 1917, the Telephone Central, the switchboard nerve center of the local telephone system, caught fire. It was still upstairs in the W. O. Baker's Drug Store, which was located mid-block on the east side of downtown Allen Street, north of the City Square. Miss Blanton was on duty and sounded the alarm, but the fire spread rapidly, destroying the switchboard. Home Telephone moved to the Farmers and Merchants Bank building on the second floor in the southeast corner, across an alley from the *Guard* office. This building was located on the northwest corner of the City Square. In the meantime, to keep long distance calls in service, the long distance telephone office was transferred to the "Chance Manufacturing building" according to the *Guard*. This may have been a necessary move in A. Bishop's eyes not only to keep local long distance phone service in operation for Home Telephone, but also to ensure that Chance Company didn't lose any important business calls. One of the problems the lost equipment posed was wartime shortages and government regulations had made acquiring replacement equipment very difficult. This may have been a problem for the ordinary person, but not the much-traveled and well-connected Home Telephone "Electrician." A week later the paper reported that a switchboard had been found in Iowa, which brings up a possible connection to the Chances' old friend Theodore Gary who had extensive telephone operations in Iowa.

To put this incident into perspective, it came a lot later than expected. The previous idea that this fire caused the separation of the telephone from the Chance manufacturing activities doesn't appear to be the case. The telephone office, even in the expanded switchboard area of the Baker Building, was already a separate entity. There may have been a workbench or even a separate repair room, but there doesn't appear to have been a factory operation. The Chance factory had previously moved to the south side of the City Square in the old high-front Bush Building, several years before. The telephone operation had become a complex system as the installation of the new switchboard indicates. On May 11, 1917, amongst all this turmoil or possibly because of it, the *Guard* found A. Bishop in Chicago on business. That same issue announced the success of Home Telephone in locating a new switchboard. On June 1 the paper stated there were some five thousand estimated connections that had to be made to bring the new switchboard up into an operational mode. In that same issue, several miles to the south, the new Missouri State Capitol building was also hurriedly being completed for the July Fourth celebration. The June 29 *Guard* announced that the sixty-day out-of-service situation had ended and Home Telephone was back in operation. But the paper suggested this didn't come easy, without some magical modifications by the Centralia Wizard. The system was operated electrically, but could revert to storage cells in case of power outages.

> *A peculiar phenomena was discovered by Mr. Chance and his helpers in connecting the board and terminal rack. There were seven coils that came with the new board that were to be connected to the lines as "ballast" and when these were put on it was found that they would not operate on the current used. As a makeshift it was decided to try to test out with incandescent lights as the rewinding of these coils in the factory would require considerable time. A hunt for a lamp that was adapted to the use of "ballasting" the current was made with the result that a perfect "ballast" was secured which eliminates the induction troubles of a magnetic coil, while the enunciation of the voice is greatly improved both locally and on long distance. The phenomena of which we speak may not appeal greatly to the average reader, but it is this. In connecting the line for talking thru these lamps there is only 2 4-5* [sic] *volts used yet the filament in the wire glows to a very bright red despite the fact that so low a voltage is not sufficient to make a lamp of this type show any color at all. As the patrons are connected thru the switchboard and over the various trunk lines the impulse of their voices is carried thru this delicate filament in the lamp. A thoro* [sic] *test of the new arrangement has convinced Mr. Chance that he has a better arrangement than the magnetic coils and the service is improved by this discovery.*

This chain of events interrupted the drama surrounding the eventual location of the headquarters of Chance Company, by reverting to a local electrician's lowly efforts in bringing telephone service back up. But how could it more clearly be shown, the diversity of this man and the seemingly endless parade of challenges that he faced and conquered. This is A. Bishop Chance in his prime, one moment in a business suit heading to Chicago, Iowa, or St. Louis in his selling or procurement posture, and next in charge of connecting up hundreds of switchboard terminals and connections, but at the last minute finding that this new switchboard wasn't working properly. Normally this would generate serious delays in getting a factory person to town in solving the problem. But not in Centralia, as A. Bishop changes his focus for a brief moment to inventor and problem solver and magically all ends well because the Centralia Wizard was there, taking care of business!

On April 13 the *Guard* reported that no action had been taken by the city fathers earlier in the week during their meeting. The intrigue continued as to whether the city would grant a factory or a mill the grounds to erect new plants. At this point there may have been a growing concern that the city had tabled the issue and would make no decision. The central area between the tracks in the Rollins and Collier Street areas was where the municipal water tower, wells, and power plant facilities were located. Many Chance employees surely considered this an intrusion on Chance or railroad property, but now we know this was a faulty conclusion and the shoe was instead on the other foot. The grasp for a portion of these grounds came about when the old city jail apparently fell down and or was removed. The city could have looked out for their interests and said no, "because we need to have room for a new jail and potential expansion of our municipal plant!"

More Eggs (*Guard* March 9, 1917)

The dominating news stories weren't the city's quest for more concrete sidewalks, or six alleys and two street crossings concreted in mid-year 1917, but instead the faraway European War that now included speculations in April of 1917 about the United States entering it. There were already food shortages in America from all the shipments made abroad. The July paper confirmed that American troops were now in France and soon a young captain from Independence, along with his unit, would see action. The Chances, were still traveling about as brother Bingham and wife from Los Angeles visited with J. A. of Centralia, Ed of St. Louis, and then on to Oklahoma and other places. It had been nineteen years since he departed Centralia. The local and former area residents could reminisce about old times when the Centralia area was a frontier region. The June 29, 1917 *Guard* said of Bingham: "When he left the old neighborhood southwest of Centralia there was over a thousand acres of virgin timber land thru which he said he was often afraid to go when he was small. Today the same land is dotted with houses and farms." Wood from what appeared to be an endless supply in central Missouri, was used for fuel, building materials and even wastefully on plank roads. The huge beams in many early buildings speak to the size of these virgin timbers where today either brush or small trees grow. Frontiersmen like Daniel Boone found traveling through woods with these towering trees easier than today as they impeded the growth of underbrush and other tangling vegetation. The woodlands of the area had changed like the prairie around Centralia.

By the time the May 18 newspaper hit the Centralia streets, the word was surely out concerning what the city had decided to do with their highly sought after grounds. It is not clear how big a story this was or even if it registered highly amongst the local topics of conversation. Nor is there a sense of any controversy arising because of these two requests, but we were not there. Of course the issue of who got these grounds could generate such a question, but that was not to be. The city wisely decided to grant both parties the land to build their plants! This may not have been viewed by the two parties as the paramount judgment that could have been rendered in view of future expansion possibilities, but one would suspect that the overwhelming attitude at this juncture was one of relief over being given the opportunity to locate their plants in such an excellent location. Because there will be future requests and questions surrounding this issue, the May 18, 1917 *Guard* article is reproduced here:

> *BOARD GIVES GROUND TO ERECT BUILDINGS. The board of aldermen met in special called session Monday evening to act on two new ordinances just prepared. One of these laws was for granting the use of the city's ground between the Wabash and C. & A. stations to the Chance Manufacturing Co., for a factory site, and to Crump & Humphrey for the erection of a mill and elevator. The ordinance gives these concerns the use of the ground for a period of thirty years from date and specifies that buildings must be erected within two years and if vacated at any time before the lease is up, the buildings are to revert to the city.*
>
> *No rental is charged by the city but the lessees are obligated by the contract, to use the city light, power and water at the regular rates. The city will continue to use the ground in any manner they may desire until the lessees build upon it.*
>
> *The Chance Co., wants to erect a two-story brick building 50* [60] *x 150 feet to manufacture their products and to operate a foundry for casting some of the parts of their inventions. They contemplate the erection of this building at once, or as soon as arrangements can be made with the railroads for proper roadways to get to the building on the north and south sides.*
>
> *The milling company contemplates putting up a two story mill and a three story elevator of brick and concrete. They ask two years time in which to get the building up, because of the high cost of materials just now.*
>
> *The old coal sheds and shacks will be moved off of the city's ground.*
>
> *The other ordinance authorized the preparation and printing of a new set of city ordinances to cost the city not more that [sic] $500.00. Work on preparing the ordinances will be commenced at once and then the printing bids will be asked for. The sum of money prescribed will have to cover the cost of revising and writing the ordinances, which is the largest end of the work.*

The second ordinance to organize and print out the city's prior ordinances and to establish an accurate and accessible account is wise, especially when considering the importance of the ordinance details concerning the Chance and mill proposition and agreement. It was important to legally document what had been done so no future questions would arise. This reflects upon Mayor Bruton, a lawyer, who didn't want any controversy surrounding city actions as was noted several years earlier on the city land deal.

Being a brick and mortar person who appreciates the skills of fabricators and builders, this same enthusiasm doesn't seem to be shared by those in search of news items. So with a thirst to find anything suggesting that construction of the new Chance plant had begun, Ordinance No. 593 issued by the Centralia Board of Aldermen is welcomed. This is from the July 20, 1917 *Guard* and the quote here represents a portion of that article:

> *Ordinance No. 593, a bill to create and establish sewer district No. 14, was read and passed. This district comprises the territory from the Globe Hotel north to the north side of the C. & A. railroad and east to the water plant, line, then south to Railroad street and thence west to the point of beginning. This sewer will serve the Wabash and Alton Railroads, the mill, the Illinois Oil Co., and the new Chance building. The expense will be taxed against these parties and the city will not pay any of the costs except to advertise for bids for the construction. The sewer line runs north 544 feet, thence east 470 feet, thence south 544 feet and back west 470 feet. Ordinance No. 594 specifying the lines of sewers and providing tax bills for the cost of same, was read and passed.*

To summarize certain details of what occurred, the posture of the City of Centralia was clear in not shouldering any costs involved with this sewer construction. The proposed Chance plant wasn't given any utility concessions, and not unlike sidewalk costs being assessed to property owners, the water and sewer work in this block area (Allen–Collier) were going to be charged to those concerns located there. Keep this in mind in later discussions concerning the mill situation as they may have gotten more than they bargained for to quickly be assessed such costs. Neither was this a welcome additional expense of erecting and putting into operation the Chance plant. As earlier mentioned, there weren't a great deal of incentives to locate in Centralia although the land was important. Today there is such a hunger for new industry that there are all sorts of concessions to induce prospects. Tax concessions could also be visited, but in this 1917 situation the reference to taxes suggests they would also be transferred to the occupiers. Mayor Bruton ran a tight ship and apparently wasn't going to allow any questionable situations or unanswered questions to persist. Note the Ordinance No. 593 remark: "the city will not pay any of the costs except to advertise for bids for the construction."

The idea that Chance Company had 100 percent local support and acceptance with no reservations, is simply not true. Neither did a very capable and community oriented F. Gano Chance find an always smooth road in dealing with the local community, even in spite of the many financial contributions. It was always a "work in progress" although there was a high level of public acceptance, appreciation, and realization of what a special situation existed and how the company positively impacted Centralia and a large area surrounding it. Power, position, and financial clout does not necessarily equate to public approval, but certain firms today minimize or ignore it entirely. Although certain things the Chances did had multiple purposes, there was a naturally pure desire to make their world a better place and to be accepted in the process. Acceptance was never an automatic thing and they learned that truth, regardless of their perceived position and success.

To conclude the documentation of city action and other sources concerning the Chance plant building at 210 North Allen, on a Tuesday evening of August 21, 1917, the Board of Aldermen received bids for the sewer construction in District No. 14. This was the block area where the Chance plant was to be located. Only one bid was submitted by a J. W. Robinson of $441.10, which was approved. The *Guard* on August 24, 1917, stated: "The cost of the lateral will be paid by the city and in turn will be collected in tax bills from

The completed two-story 210 North Allen Street Chance Manufacturing Company plant is from the 1918–25 period. Note the siren on the roof that warned of bad weather or working hours during the day. (Centralia Historical Society)

the Wabash, Chicago & Alton, the Chance Company, the Centralia Mill and the Illinois Oil company." Construction probably began no later than mid-year 1917, on the new Chance factory at 210 North Allen. Evidence that it was being constructed in 1917, is confirmed by the construction photo reproduced on page 167, and the year 1917 over the south door entrance that A. Bishop chiseled into the stone. This confirms all or part of the walls were up in 1917. Comments by a then young Gano also suggested it was completed near year's end. It can be assumed, but not confirmed, that the chemical company Gano referred to in describing how the new building was first utilized was the Centralia Manufacturing Company. An article in the August 8, 1919 *Guard* said: "A third story is also (in addition to 100 foot eastward addition) contemplated for the plant, which would give the Centralia Manufacturing Co., which leases the upper part of the Chance Co.'s plant, some much needed room." Flynt Chandler in 1949 described the second floor renter as W. O. Baker (a Centralia Manufacturing Company official), who manufactured a powder for dusting chickens.

What was the year 1917 like, especially since it represented an important milestone for Chance Company? It was a year that was ushering in more modern times and World War I was in progress. Tobacco ads for Chesterfield, Lucky Strike, and Prince Albert were appearing more frequently in newspapers, magazines, and on billboards. Moviegoers were introduced to Mary Pickford. At Sagamore Hill, Oyster Bay, New York, former President Theodore Roosevelt, sixty-one, passed away at his home on January 6. In Centralia, 1917 could be considered a year of troubles for the community. After the April Baker Building/Home Telephone fire, a cyclone (tornado) came through killing seven local people and several other area residents. It also destroyed many buildings and did other damage in early June 1917. The week before John Chance had injured a finger severely in a sharpening accident and Mayor Bruton was fined for assault/battery when he struck a Mr. Waters with his open hand. They were having a strong argument when Mayor Bruton had heard one too many complaints and strong words from a local patron. If the embarrassment of this wasn't enough, papers across the country and in Canada picked this story up. Letters came to Centralia and one religious leader severely criticized the mayor. His absences at the city Board meetings reflect his dilemma, but it apparently all blew over as he was reelected and continued as mayor.

The war hadn't stopped travels of Centralians as A. Bishop Chance and his sons Buford and Gano, were St. Louis visitors in mid-July. Mrs. A. R. Toalson and son Nathan were at Browns Station the previous Sunday attending a birthday dinner at the home of Charles Turner. An item on July 24 reported Mrs. A. R. Toalson and daughter Miss Anna Lee and son Nathan spent Thursday with the Joe Hardins of Hallsville. On September 7, 1917, the *Guard* carried a story about the marriage of a cousin of Centralians Mesdames P. R. Diggs and A. B. Chance, a Miss Margaret Chanslor, who had frequently visited in Centralia. There was increasing war news of troop and military supply movement through Centralia by rail. Food, fuel conservation, and shortages were frequently mentioned in the paper and druggist R. L. Hope became the local government fuel administrator.

When a "black-out" hit Centralia, as reported by the October 5, 1917 *Guard*, one might incorrectly consider it another result of war. In 1917 the technology, skill, and training weren't present officially to work on an electrical power line without shutting off the current. However, when it shut down the presses at the *Guard* as they were rushing on Wednesday and Thursday to get out the Friday paper, this was an unfortunate situation for the power company as it would get coverage in the paper's edition completed during the nighttime hours. Their complaint was "we deplore the thoughtlessness of the department in not notifying power users." As people and firms depended more and more upon electricity, its reliability became an important issue. The whole episode was caused by transmission lines into the city's power plant that were too low, said to have caused the fatality recently covered. The power company shut down the power to raise these wires after the impending lawsuit. The irony is this work occurred a couple of hundred feet (just a few feet from the new 1948 Switch and Tool plant) from where the Mecca of world hot line tool production and development would be centered by A. B. Chance Company in 1938. Assurances of its eventual worldwide value was confirmed in Centralia twenty-one years earlier!

Some information reveals how small an operation Chance Company really was. It naturally would be assumed that the Company was bursting at the seams, which it could have been in its previous city square setting. With the construction of the new, two-story factory building, one might think that it was a much larger operation with seventy-five to one hundred people, but all of this building wasn't being utilized by Chance Company immediately after its construction. One of the more prominent family names around Chance Company in the early years was Chandler. Flynt and his brother Ollie began in 1919 and March 1, 1921, respectively. A marriage license was issued to Ollie Chandler and Miss Eula Hicks, both of Hallsville, according to the February 5, 1915 Guard. By December 1921 Flynt was also married according to news items in the paper concerning the death of his grandmother in Hallsville. Flynt Chandler, in recalling his childhood, said his family came to Centralia around 1915 when his father, a rural mail carrier, was transferred from Hallsville. It didn't take long for Flynt to get to know A. Bishop.

> *I knew him when he was over there in this little building, just back of Dr. LaChance's now. I was just a kid then and I knew him when I saw him. He remembered me being*

> *a kid around. He had just been over here* [210 North Allen] *maybe a couple or 3 years when I went to work for him. (1979)*

Flynt went to work at Chance Company on Monday, August 25, 1919 which was easy for him to recall since he was hired the Friday before on the twenty-second, his twenty-first birthday. When he came to work there weren't many full-time employees and when asked what he first did, he replied:

> *My mother said, "what have you been doing?" I said, "I've been working in the office today." She said, "Office, what have you been doing in the office?" I said, "mopping the floor." That is the way I started out, just anything that came up, that is what I did. Mr. Chance was generally right with me, we worked together. When I started there were just three men; Earl* [Irl] *Hicks was a machinist and a fellow by the name of Whittaker, he was kind of a flunky. The first job I done, there was the extension* [words lost] *and Mr. Chance and I had a* [words lost] *with long handles on it and he would get on one side and me on the other and we would turn that* [words lost]. *(1979)*

This appears to be the same job that Bill Hollander witnessed in 1929 when Ollie was involved in screwing screw anchors into the ground by putting a bar or shaft onto the anchor and walking around in a circle screwing the anchor into the ground. It would appear A. Bishop was testing screw anchors very early indeed.

Apparently Flynt considered himself the third employee along with Hicks and Whittaker, although early accounts in the *Guard* included Denton Palmer as a Chance Manufacturing employee. This *Guard* article of June 3, 1921, suggests Denton was a telephone employee:

> *How many people know the different human elements that go to make up the complex telephone system of Centralia? You pick up the receiver and yell into the transmitter, or wind the magneto with the little crank that is no less cranky possibly than you, and you are familiar with the little box or the desk set, and you may know the manager, J. A. Chance, thru business associations or O. G. and Denton Palmer, who magic hands can take all the muleishness out of the telephone box and make it sing like a bird, but the heart of the telephone is the central and the heart of the central is the central girls.*

The Palmers had apparently been taught to take care of A. Bishop's earlier electrician duties at Home Telephone as his company's demands increased. Although Denton was then a Home Telephone employee, it is possible he was still at this late date, a "swing" employee who worked in both places. Later when Home Telephone was sold, he again moved his employment back to Chance Company, as later employment confirmations suggest.

The group picture on the following page, circa 1920 described by then one-year veteran Flynt Chandler, was taken in the fall of 1920 by another employee, Christine Moberg (Hicks). The other employee not shown was Miss Ida Lou Traughber. Flynt in 1949 commented about the ladies in the picture and the interest their attire generated. "Everybody who sees this picture always gets a kick out of the overall suits the girls are wearing. They needed them, though because they were running piston ring machines." Flynt also said the picture was taken at the front of the 210 North Allen factory with a cracked window temporarily taped. They were standing in front of the double doors in the middle front of that building which would later become the main entrance. In 1920 it was the entrance to the factory and "the storeroom where anchors and rods were stocked." The offices were on the southwest corner, supplied from another source.

Group picture. Front row: Jewel Brooks (Neiman), Ruth Noel (Gilmore), Irl Hicks, Sr., and "Boss" Chance. Second row: Harold Mosely, H. L. Carnahan, and Flynt Chandler (*Chanco News* September 1949)

The number of employees in the 1919 and 1926 periods are revealed by their time card numbers. Flynt said his number was No. 7 and Les Reynolds said:

> *Mine was No. 41. We had them for years and years, while you were in the same department. If you changed departments you might get another number. . . . As I was going to say, there were just forty working when I went to work in 1926. You got your clock number, just after the next one* [last hire]." *(1979) Flynt said, "Yea, I had #7 for years over there." (1979)*

The peaks and valleys of the labor requirements for a small, pre-union firm also had an ingenious twist. When the company had a load of steel coming in, according to veteran sales employee Jack Curtright, they would put a call out to local people for part-time laborers to help unload it. Jack said it was "heavy" work and he "knew there was a better way." This undoubtedly influenced Jack to go to college and escape this type of work.

As the year 1917 came to a close, the new Chance two-story factory building at 210 North Allen was said to have been completed. Besides the machine shop or the small anchor and arrester operation on the first floor, A. Bishop had a renter for the second floor. It was the other firm that would steal the immediate spotlight in 1918 and for a few short years. The next section will describe their rise and fall and the growth of Chance Company and their need for more space to expand their manufacturing complex.

Exploding Commercialism Between the Tracks

1918–1924

The word manufacturing in Centralia, generally without exception, brings to mind the Chance factory operation. Even the mention of chemical, to those knowledgeable of Chance Company history, will bring thoughts of the Chance Big C auto line. However, in early 1918 there was a prominent non-Chance factory operation in Centralia, involved in the manufacturing of chemicals. Research without curiosity is life without spirit!

There are at least two reasons to examine this "other" Centralia Manufacturing Company. There was much intrigue over this chemical concern that temporarily occupied the second floor of the Chance building. This was eventually where piston rings were boxed, many office employees worked, and where Chance executive offices were located for decades. Most of the time Chance Company, in its continual growth mode, was bursting at the seams and needed all its space for company activities, but this could have also been the destiny of this new firm. The *Guard* carried a story of its organization in its February 15, 1918 issue:

> *ORGANIZED COMPANY. CENTRALIA BUSINESS MEN BEHIND NEW MANUFACTURING ENTERPRISE. The new Centralia Manufacturing Company, which was organized to make poultry remedies, held its first meeting Monday evening and elected officers. W. O. Baker was chosen temporary chairman of the meeting and R. L. Hope secretary. The assembled stockholders then proceeded to elect the officers of the company for the ensuing year, and the following were chosen. O. B. Wilson, president. H. E. Stone, first vice-president. R. A. Fountain, second vice-president. H. F. J. Lake, secretary. John M. Waller, assistant sec'y. W. O. Baker, treasurer. Mr. Baker was also chosen general manager of the company.*

There was frequently an ad in the *Guard* of the O. G. Wilson Remedy Company of Centralia, advertising his "Distemper Remedy" and in fact one such ad appeared in a later March 22, 1918 paper. It is unclear if O. G. and O. B. were related as they are two separate individuals. The Baker-Lake Chemical Company letterhead confirms O. B. was associated with that firm. Instead of getting clearer, however, this subject becomes more complicated per the December 5, 1919 *Guard* article of "Want Consolidation" revealing there was also a Wolf and Webb Chemical "companies of this place." A new company was being incorporated to absorb the "companies" of Centralia which appears to include the Centralia Manufacturing group. The new company " will likely be Baker-Lake Chemical Company." (Note letterhead on page 193) Noticing the officers of the Centralia Manufacturing Group covered earlier would

O. B. Wilson (Centralia Historical Society)

Plant from 1917 to 1925 (*Chanco News* June 1957)

suggest an already established web between these two concerns. Also note that Gano referred to this chemical firm as Baker's. Apparently a lot of the output of the Centralia plant went to Iowa, which explains why the advertising and sales offices would be located in Waterloo, Iowa. One of the reasons cited for the additional capital that this move would hopefully generate, was to establish warehousing with enough stock to take care of customer demands and to get away from the current jobber distribution arrangement. Demand for the Centralia plant's production was consistent, according to the article, and increased activity could be gained with increased work force and equipment to take advantage of potential business that reportedly had been turned down in the past.

There appears to be a lot of wheeling and dealing taking place as it was in earlier times when empires were built and financial giants were born. One can surely imagine the interest of a certain Centralian in all these goings on, although A. Bishop Chance's direct involvement can not be ascertained. Life can give us an education and you can be assured A. Bishop was an observant pupil. He did get involved indirectly in a quest to get more ground from the city, as outlined in an August 8, 1919 *Guard* article. It suggested the need for expansion of the Chance facility which had reached a crisis point, even in the contemplation of adding a third story to the existing two-story structure. The later part of the August 8 article stated: "A third story is also contemplated for the plant, which would give the Centralia Manufacturing Co., which leases the upper part of the Chance Co.'s plant, some much needed room." It would not be a very popular stance to stymy the growth of two prospering Centralia firms, which boosted the chances for success of A. Bishop's request for more city land. It also revealed that the third story, added in 1925, was being considered as early as 1919 from what appears to be non-Chance business activity.

While this introduction has reflected on the existence of the chemical group and how such local business actively supported and fostered the growth in the grounds and plant of Chance Manufacturing Company, the second part deals with the seeming disparity in local support for Centralia firms. Returning to the February 15, 1918 *Guard* article on the organization of the

Centralia Manufacturing Company the last half of that article reads:

> *The new company has been incorporated under the laws of Missouri in the sum of $25,000.00. Stock in the enterprise has been sold to Centralia business men and farmers at $10.00 a share, and nearly all of it has already been taken. There are already sixty Centralia holders of stock and the number is increasing daily, so there is little doubt that the stock will soon be disposed of. The company will manufacture a number of poultry remedies that are made up of the same materials as used in making two well established brands, that are handled by twelve thousand dealers and backed by a two hundred thousand dollar advertising campaign. The Centralia company will manufacture the remedies for these companies, insuring the success of the home company from the start. It appears to the Guard that this company will be a great thing for Centralia and the fact that so many of our business men have taken stock should insure its success. Mr. Baker is still selling stock for the company while it lasts.*

Although it appears that the sometimes reluctant Centralians were lining up to buy stock in this new firm, including farmers who use such products, it did have initial capitalization of $25,000 which is much smaller than the ill-fated Chance venture of a decade earlier. This reason alone may prove the difference between success and failure. It surely wouldn't have the financial risk of a half-million dollar proposition and apparently its performance and potential were more in line with its financing than the earlier Chance proposal. The flamboyance, boldness, and possible overextension of an optimistic and young A. Bishop Chance may have been his early undoing as much as any lack of local support since his business was growing much slower than he originally contemplated. However, there seemed to be a group of Centralians that were always prepared to invest in some venture, as the local paper confirmed again in 1918.

The events of that time help explain the establishment of what could be labeled the second generation Chance Company. A window through which plant expansion can be viewed, is the continuing *Guard* articles on Chance requests for more city land. There appears to be an exciting drama unfolding between Chance and the mill (Crump & Humphrey) interests. In the same issue as the articles on the initial requests for the between the tracks city land (March 9, 1917), a mill ad (see page 174) revealed additional information. First, the names Crump & Humphrey are familiar names to the Chance employee list. Bill Crump was a longtime employee, a capable mechanically oriented person, and Don Humphrey was a key Industrial Relations figure whose employment began in the World War II era. It is probable that the firm Crump and Humphrey, also known as Centralia Milling Company, was operated by relatives of these two men. They incidentally were the local Purina dealer, a St. Louis based food giant. The question seemed to be if chicken, eggs, and hog feed could win over Never-Creep® Earth Anchors and Sky-Rocket® Lightning Arresters. In 1917 the agricultural business was a significant part of our nation's economy and its importance even greater as our involvement in World War I accelerated.

To kick off the New Year in style, city officials stated in the January 4, 1918 *Guard* that the city was in the best shape in years. This had to be a relief to A. Bishop who had taken some heat for pushing the construction of city utilities. The dominating coverage in newspapers, however, was war news, and letters from servicemen began to fill the paper by 1918, along with draft news of young men being sent to Fort Riley, Kansas. Agriculture production was almost at full capacity, and the government began to regulate and operate many areas including telegraph, telephone, and the radio industries.

The government put out a call in July for twenty-five thousand young women to become Army nurses. Contact between the Chance families continued as Mr. and Mrs. Charles Chance of Tulsa came by car in mid-1918 to visit his brothers John of Centralia and Ed of St. Louis. On June 21, 1918, the Diggs had a dinner for the A. B. Chance family and Kansas City guests Sara Gorin and Margaret Greenlee.

Government intervention continued to impact everyday life. As earlier noted the government's regulations of the communication industry became very real when Union Mutual was merged into Home Telephone, consolidating their local phone networks. This occurred on September 2, 1918 "at the insistence of the Government" according to the *Guard* article. Owing to the foresight of Home's "Electrician," the switchboard at the new central would have enough additional "drops" to accommodate both systems. The government regulatory agency had sent notice to increase the cost of new phone installations and were not encouraging, as the paper put it "telephone talking." A. Bishop Chance, representing Home Telephone, concluded the transfer at a meeting with Union Mutual managers. He seemed to be handling a lot of duties that his aging father may have had responsibility for. In that same September article, the government's long arm also impacted the Centralia railroads as it had the telephone business. It too was sort of a merger of the C & A and Wabash, if only in connecting their tracks. The Mexico to Clark trackage was to be shared and become directional and the depot/freight facilities at Mexico was combined into the Alton. But the big news, from our perspective, was how a proposed connecting line would play havoc with the between-the-tracks real estate in Centralia. One purpose of the new crossover was to allow the Alton to also serve the Columbia branch. Because there are certain aspects of the specifics of the proposed line that are confusing, it will prove beneficial to quote the first paragraph of the early September article:

> *R. R. CONSOLIDATION: C & A and Wabash Consolidate Lines Here for Period of War. Government surveyors were in Centralia last week running a connecting line from the C. and A. to the Wabash. The connection as laid out runs from west of the Wabash depot northeast to the Alton just west of that depot and cuts thru the Chance building lease. The Chance company are arranging to erect a foundry building on the spot where the line crosses, and they will fight the construction over that point. The new line practically hems the mill in on three sides, making it bad for teaming.*

This article is both informational and confusing. One can picture this track slicing across the couple of hundred feet space, in close proximity to where the old mill was located. Beginning to the west of the Wabash Depot, is clear. This could be the actual location of the siding into the Punch Press building today, which also crossed Allen to serve the Chance shipping warehouse and circled back south to complete the loop to the main line. This dictated the "stair step" shape of the 1948 Tool building. The Wabash Depot was located on the east side of Allen Street, along the Wabash tracks. This is the pre-1925 depot, as the big station was built to the east of it, or somewhat centered between Allen and Collier. The proposed track would then run northeast toward the Alton tracks slicing behind the original two-story Chance building, thus spoiling A. Bishop's plan to extend the plant eastward from the back of his new factory. The idea that the track would run west of the Alton depot is confusing. This suggests the Alton Depot was on Rollins for the track to come out to the west of it as it switched into the Alton yards. The Alton depot sat on Allen as did the Wabash Depot, and when torn down the parking lot to the north of the original Chance plant was developed. A photograph reproduced in Chapter Five clearly shows its Allen Street location. Another explanation could be that the Alton had a freight station near Rollins and it was not identified correctly.

A very early photo of the Anthony Building yet to be rebuilt by Chance Company. The Chapter One Globe Hotel photo (see page 29) refers to the current Railroad Street as Wabash Street and reveals the Anthony Building across Rollins in the background, confirming this is the future home of Crump and Humphrey, Honey Ford, and the Tips Tool Division of A. B. Chance Company in 1938. (Centralia Historical Society)

There are several issues that this news item raises. Government action, including the just recorded telephone merger, may have marked the beginning of a questionable view of government involvement in private business, by the Chances. This includes A. Bishop though most other sources would dwell on his son F. Gano's private activities and beliefs concerning the government's role in our lives. Of course, A. Bishop's resistance confronted that which would interrupt the growth, expansion, and the efficiency of his centrally located and concentrated plant complex. However, he was not the only one complaining as it can be assumed that the reporter writing this article interviewed both A. Bishop and Crump/Humphrey. In certain respects this was a more serious situation for the mill, as it had an immediate affect of boxing them in. The statement "making it bad for teaming" is significant as the danger of runaway teams of horses being frightened by an approaching snarling and snorting steam locomotive a few feet away, was bad for business. This may have been the determining factor in the mill's departure and the Chance plant eventually obtaining the grounds from Allen to Jefferson. So in a way it may have quickly and decisively achieved what could have resulted in a long and bitter struggle and even a different Chance plant layout than exists today.

Although what may have occurred because of the cross track is logical, there is no confirmation that it was ever completed. What we do know is the war had become a serious proposition. For example, the September 13, 1918 *Guard* told of the Ford Plant shutdown. It said car production (Model T) had stopped Friday and they would only build war products. Henry Ford was frequently criticized for his views on various things, including war, but when it got down to it he supported his country. The same would be true in World War II. If this is not enough to convince us that this connecting track would soon be in operation, the following November 18, 1918 *Guard* article might:

> *WILL MOVE CENTRALIA MILL. Crump & Humphrey are repainting and refitting the Anthony building on Railroad and Rollins streets and will modernize the building for a home for the Centralia Milling Co. Machinery from the mill between the railroads will be moved to the Anthony building and installed and new machines are to be added to make this an up-to-date mill. The old mill will not be abandoned but will be used mostly for storage purposes for this growing institution. They are painting the outside of the new mill site red, which is probably one of Jim Humphrey's ideas, and it will loom up like a candy egg on a wedding cake when the job is done.*

Sometimes more questions arise than answers are found. The mill partnership is apparently represented by a Jim Humphrey, but where is the Anthony Building? It appears to be what is known today as the Narraganset or old Chance Tool Building. The Railroad and Rollins Street location would fit although this could also mean the corner to the west. There was a piece of information that said the Narraganset Building was once the location of a mill and a 1925 newspaper article confirms that it and the Anthony Building were one and the same.

The modern Galvanizing Department building of 1950–51 is where the proposed track would have sliced across on the north side, which defined the west front of most of the Drop Forge. It was built on grounds that came under Chance control during or after the expansion in the early 1920s. It was built as close to the old Alton tracks as legally or physically possible. During nighttime adventures in both departments in 1961–62, on the second trick (shift), passing trains on the then GM&O tracks were close enough that it felt like you could reach out and touch them. According to the January 31, 1919 *Guard,* it didn't look very promising that A. Bishop would gain these grounds. The mill had apparently been abandoned, relating to its announced move the previous November. "The board leased the plot of ground occupied by the old mill, and between the railroads to Porter Wade for a period of five years at a rental price of $20.00 per year." So instead of smooth sailing, A. Bishop now had another between the tracks rival. Porter Wade's territorial aspirations, however, are unknown, but it doesn't appear to have been in the best interests of the Chance firm, which had also become boxed in.

As the summer of 1918 ended, the war tide was turning in favor of the Allies. By November, peace was in sight and by months end the Army was coming home at a rate of five hundred a day. Touring the country during that time was an Army tank that was promoting Victory Liberty Loan notes. When Centralia's turn came in April, some area residents were disappointed at the tank's small size, as it rumbled up and down Allen Street after backing off its rail car on the government train. Of course people wanted to see the real "McCoy" that helped win the war, but a few would have only been happy if it had shot off a couple of rounds.

The paper on April 4, 1919, carried the news that Mayor Bruton was elected for another term. The people must have loved this tobacco-chewing mayor who brought slower progress with stability. In late April, Mayor Bruton and his aldermen had some tough issues before them, including the paving of streets with potential costs being charged to property owners on either side. Some wanted it while others did not and some could pay while others could not. Estimates were too fluid for those affected to accept, especially multiple-lot property owners so the controversy raged through July. During this time a lot of local business travel was to St.

Louis, including A. Bishop on April 11, 1919. On June 20 the A. Bishop Chances entertained a Mr. and Mrs. J. F. Gorin of Hope, Arkansas, with the P. R. Diggs as guest.

The Water Plant building and equipment was destroyed by fire on July 15, affecting the whole community. The city's electrical generator and a water pump survived which allowed electrical and water service to be reactivated. The Centralia electrical power situation was not 99+ percent reliable as it is today, but it drove a substantial portion of Chance plant machinery and A. Bishop clearly became more knowledgeable about it as time progressed. He undoubtedly influenced his son F. Gano of its importance, which would later dramatically change the direction of the company. But back in 1919, the day-to-day problems of running a firm as efficiently and economically as possible was on the Centralia entrepreneur's mind. The *Guard* of August 8, 1919, in covering the city's business earlier in the week, found a familiar party before them once again:

> *A. B. Chance was before the board asking that he be given a power rate on the small motors now connected on the light service wires. He is now paying the 11 cent rate for this current because the motors can not be connected with the three-phase current. A committee was appointed to arrange a satisfactory adjustment. There is a prospect that the Chance company will also be put on a separate line, so that there will be no interruption of service as is now the case. Mr. Chance stated he was ready to pay the cost of such a line and connection, which was estimated at $185.00. The Chance company also asked for more ground back of their building so that they could build a 100 foot addition at an early date. This matter was also put up to the committee.*

The source of power was still out of Mexico but $185 seems inadequate to establish a separate line from Mexico to Centralia, even using the existing power line poles. In the midst of all these operational challenges the company was expanding to the point that it needed a hundred-foot factory addition, which meant more city real estate. The requests for more grounds are intriguing as it does reveal growth and conditions within the company. His current expansion situation was a continuing saga, developed a decade earlier when he began an active quest for a plant and grounds. But now it was driven by a more realistic need. As is customary in the *Guard*, they would repeat the request for more land as discussed before the city, but in their own words, possibly with the help of a local resident as they did on August 8, 1919:

> *CHANCE COMPANY TO BUILD: The Chance Manufacturing company have asked the city for an additional fifty feet of ground east of the present building so that they can extend the plant a full hundred feet more. The city property committee is working on the proposition now. There is only eighty feet between the mill and the Chance building and it is probably all that can be given them. This will cover a thirty-foot passageway west of the mill. A third story is also contemplated for the plant, which would give the Centralia Manufacturing Co., which lease the upper part of the Chance Co.'s plant, some much needed room.*

The dimensions of these requests are interesting. How could a 100-foot plant be built with only 50 feet of additional length? Referring back to city action of May 18, 1917, taken on the original request, instead of asking for 100 feet, A. Bishop had instead asked for space to build a 50-by-150-foot building, but it turned out to be only 100 feet long, but 60 feet wide. The length of 150 feet appears to be what was given by the city, as with the current 1919 request of 50 feet, a total length of the Chance plant could now total 200 feet or allow a new 100-foot addition. A practice of asking for more than used might have been approved once, but possibly not a second time. Someone like Mayor Bruton may have reminded A. Bishop that he had originally planned to build a 150-foot building that instead turned out to be 50 feet short of its mark. To sell and convince the public and city officials that his demands were genuine, a notation was added about the possibilities of adding a third story, or a skyscraper of sorts in Centralia! To gain further impetus, the needs of the chemical producers, Centralia Manufacturing Company was also aired.

One of the reasons there was a cap on additional grounds to the east was interference with the old mill. The later article stated that there was only eighty feet separating the mill and the Chance building. This would put all or part of the mill behind the east end of the Chance plant. So it appears that A. Bishop's request would intrude onto mill property or into the thirty-foot passageway. The city's leasing of the mill property earlier in January to Wade decreased the possibilities of a Chance infringement upon it and apparently this was too difficult a situation to resolve as no further action on the Chance request is found. It is not clear what the frustration level of A. Bishop may have been, regardless an additional trait he possessed was that of persistence.

The chemical firm Wolf and Webb Companies or a part of it sometimes referred to as the Centralia Manufacturing Company in its reorganization described in the December 5, 1919 *Guard*, was to remain in Centralia. One of the reasons to do so was the convenience of utilizing space in the Chance plant. The economical impact it had on A. Bishop's enterprise is unclear, but it did provide funds to help pay for his new building. However, A. Bishop requests for more city property and his references of adding to his plant suggest his growing firm was also needing extra space. The closing remarks of the above referenced article suggests the chemical firm's position and challenges:

> *It is now up to the Wolf and Webb Companies to accept the offer of the new company. If we understand the matter they propose to exchange the new stock for the old on an equitable basis or buy it after the new company gets on its producing feet. The main point is that the factory is to remain in Centralia and is to be enlarged, as it is our understanding that the Chance Company have agreed to put another story on top of their building to hold the new equipment and materials of the chemical company. This will admit of the installation of the new automatic machinery which will enable the factory to put out a car load of any of these popular articles in a short time.*

As 1920 dawned, it would represent a decade in which the maturing of America's industrialization might would blossom. The march of progress was detectable in the local paper with ads and articles on cigarettes, washing machines, farm tractors, self-starters for Ford cars, oil stoves, Victrolas and Victor Records, airplanes, auctions, and movies. Closure of the war and the exiting from it as a world power could no better be represented by an American from the central Missouri area, General John J. Pershing. He was in Centralia en route to Columbia to receive an honorary degree according to the local newspaper of April 23, 1920. Also in a leadership role, A. Bishop was installed as one of the directors of the reborn Commercial Club in July 1920. Continuing community spirit was demonstrated by the August installation of an electric siren at the Chance plant, which is visible on the roof of the factory picture on page 177 of this chapter. The *Guard* titled the news item "WEATHER SIGNALS" but explained in the article that it would also daily signal the plant work related times of 7, 12, 1 and 6 o'clock, in addition to fire alarms. Notice this indicated a ten-hour workday, confirmed elsewhere in early interview material. The same August 6, 1920 paper also covered a visit by Buford and Gano into St.

Louis to visit relatives, but the big news item was another request by A. Bishop for more city grounds. It was a revival of sorts of the 1919 request, but with a couple of new twists. Instead of repeating the space needs of the Centralia Manufacturing Company and the general needs of the Chance firm, A. Bishop got specific by announcing plans for a blast furnace! This was pretty sensational for Centralia and sounded more like a "Pittsburgh" sort of expansion project. This change was similar to the tactics of an infantry commander, if you can't go through them, then go around them. Since A. Bishop found little success in an earlier attempt to proceed uninhibited toward Collier Street, he instead simultaneously headed north and south toward an even more defined barrier, the two railroads. The key point, however, would appear to be his acknowledgment of the mill's need for adequate lands, only allowing the availability of eighty feet to the east behind his plant. The city couldn't compromise on their earlier ordinance but a ninety-nine-year lease on grounds not occupied was a wise and necessary move. Surprisingly the city seemed anxious in calling a special session to address the latest Chance request, per the September 24, 1920 *Guard*:

> *FOR A BLAST FURNACE. The board of aldermen met in special called session Tuesday evening, to act on the proposition to lease a plot of ground to the Chance company and to take up the matter of marking the streets for the free delivery of mail in the city.*
>
> *A. B. Chance made a talk before the board, stating that his company had been handicapped in securing castings for his manufactured specialties, and he had under consideration the putting in of a blast furnace in Centralia. He needed a long time lease on the 160 x 80 feet belonging to the city and lying east of his present plant. The building he expected to erect would cover this space and would be one story in height. The blast furnaces, which would have a capacity of two tons a day, would occupy the north end of the structure e[a]nd would employ 20 men. It was expected to operate the furnace every other day. The furnace would be fired by oil and the consumption would cost about $250 a month. A 70-foot stack over the furnace, which carries an intense heat will, he said, practically eliminate a smoke nuisance.*
>
> *Mr. Chance asked for a ninety-nine year lease on the ground, as he had to have a long time lease to secure necessary financial aid in putting in the plant.*
>
> *The cost, outside of the building, will be about $75,000.00.*
>
> *Besides the city ground asked for, sand pits will be put in on Wabash ground, south of the Chance building, just over the spur tracks running to the water plant.*
>
> *The board instructed Mayor Bruton to draw up a lease, reserving the water and sewer main rights, which run under the proposed building, at an annual rental of $25.00. The plot of ground to be leased is 80 x 160 feet from the Chance building to the oil station, with space reserved for a walk along the south side, to be put in by the lessee.*
>
> *The ordinance will be ready in a few days and the lease made so Mr. Chance can go ahead with his project. Work will be started, he says, as soon as the loan can be floated.*

By this time, the Chance manufacturing firm had over a decade of small but successful operation under its belt, and now its potential and increased activities were becoming obvious to the public. The new plant and the additional boost in activity by utility and automotive products propelled it beyond its origin as a tiny job-shop operation. His latest request was further explained per the *Guard*'s coverage of the city's board meeting on October 8, 1920:

> *BOARD MEETING. The board of aldermen met in regular monthly session Tuesday evening. The bill for the leasing of the city ground to the Chance Co., for the erection of an addition to their present plant was presented to the board and read. It provides for a lease of ground 80,360 feet* [sic], *which includes the present building, for a period of 99 years at a rental of $25.00 a year. The proposition was held up after the first reading as the board stated they had understood that the lease asked for was only for the ground needed east of the present plant. The city gave the company a lease on the ground now used, for a period of thirty years, free of charge, four years ago. After thirty years the building was to revert to the city. The proposition will be taken up again at a called meeting, when Mr. Chance will be asked to come before the board and explain why the change is asked. This will probably come up next week.*

A. Bishop had a sense of direction that continually surprised those around him. An example was a proposal for a ninety-nine-year lease on all the grounds he held and the additional acreage contained in his latest request. Of significance is how it relates to today. Little that we do during a lifetime has very long-range legal affects, but a ninety-nine-year agreement would run to 2019! Will the city renew it at that time? In certain respects, the 1920 plan was as important as the initial 1917 one. This would allow for a compact, efficient, and centralized arrangement that would remain the center core of the Chance Company of the future, regardless of additional expansions. Clearly the momentum had stretched in A. Bishop's favor as the November 12, 1920 *Guard* told of the city board meeting and the passage of the ordinance leasing the city's plot of ground between the railroads to the Chance Company for ninety-nine years. The original ordinance was amended to read that the property was to revert back to the city in case the plant was destroyed by fire, or if the buildings were discontinued for company use. Another unrelated ordinance divided the city up in corners with directional identifications, the dividing lines being Rollins and Singleton Streets. In addition, a four-course dinner was served by the Chances, symbolic of what positive winds of fortune were coming their way, as recorded in the positive ruling by the city: "A. Bishop and wife gave a four-course dinner Monday afternoon. Plates were laid for twelve. Mrs. Mabel Gooch-Price of Phoenix, Arizona, was guest of honor."

The crack varnish of the Alton represented locally by their "Hummer" continued to scorch the ribbons of steel around Centralia, especially out on the open prairie where seemingly endless straight track allowed engineers to crack wide the throttles and make up lost time. This generated crossing problems where life hadn't advanced enough to contend with unheard-of speeds of sixty- and eighty-mile-per-hour passenger trains. A Columbia Branch accident in 2004 suggests the crossing accidents continued to occur, but the irony is the autos on Highway 63 were now traveling seventy mile per hour and the COLT coal train twenty or so! Unfortunately in 1920, daring individuals were also trying to extend auto speeds in the same manner with tragic results on predominantly unapproved dirt roads, as turnovers became common. For example, the *Guard* recorded three in the last half of 1920. Within Centralia there was also frustration over lack of progress on the paving of city streets, now extending out from the downtown district. Those living along Sneed, for example, led by A. Bishop Chance, orchestrated a charge to get it hard surfaced. However, poor roads didn't deter some hardy souls to travel cross-country from a Michigan factory with a new car. An Oklahomian, a former Boone County resident, Charley Chance returning from Flint (longtime home of Buick), after taking factory delivery of a new auto stopped over in Centralia to visit his brother, J. A. Chance, before continuing on to Tulsa.

The dawning of 1921 brought one of the prettiest Januarys some locals had ever seen. The winter months of January/February in North Missouri are usually the most trying and are viewed with apprehension. This was also a rich period in the generation of material that relates to our story. The continual Chance Company expansion obviously had an impact both inside the plant complex and within Centralia, with the influx of new employees. Street names and house numbers were assigned, which was convenient in getting mail and pinpointing where people lived. Aristos Flour was being advertised by Crump & Humphrey and four-course dinner parties were now in vogue. The struggle over state highways raged on and the *Guard* was promoting the "North Route," the one through Centralia and touted as the only one crossing all rivers by bridges instead of ferries. Theodore Gary, however, would bring logic to all the fighting over who was going to get the St. Louis/Kansas City state-supported route. Additional support for the North Route was a new power line from Kansas City that primarily followed the Wabash north of the Missouri River, east toward St. Louis. The Chase Electric Line had been surveyed in 1906 but they were now putting up lines as reported by the January 21 *Guard*. This was good news to A. Bishop and the city who were depending more and more on electricity, especially in hopes of the more reliable variety.

The cold weather was momentarily interrupted with some hot air from the proposed new blast furnace described in the January 14, 1921 *Guard*:

> *Architect Ed Chamberlain is now busy drafting the plans for the new furnace and molding room to be added to the Chance Mfg. Co. building between the railroads: The new building will be brick, glass and steel construction. Much glass will be used to give needed light, and steel trusses are being figured on to bridge the big roof. The building will be only one story high but will cover a plot of ground 80 x 240 leased by the city, and just east of the present plant and adjoining it. The blast furnace will occupy part of the new building and the molding sand and stock rooms the balance. The company intends to do its own casting of its specialties.*

The accumulation of information about the Chance operation and plant became more substantial with each additional newspaper account. The proposed plant design with a multitude of windows was common for that period in providing light as mentioned and also for ventilation purposes during the hot summer months. The building described is the Foundry addition of 1925 and long referred to as Hardware Assembly. Down its center was a wide walkway and forklift traffic lane used by employees in reaching other areas of the Chance complex. It would appear from this news article, that the construction of this building was planned a few years prior to its erection. On January 28, 1921, the paper said the "new blast furnace at the Chance plant was fired up last week and Friday some experimental casts were made in the temporary building erected to house the furnace. Miss Maurine Chamberlain lighted the first fire made in the new furnace." Maurine is likely one of two daughters of local architect Ed Chamberlain. She was or became a Chance office employee. This suggests that a foundry was in operation several years prior to some of the remembrances of early employees although it was housed in a temporary structure. It is obvious the foundry equipment was something A. Bishop needed to generate production casting, regardless if he didn't have a permanent roof to put over it. One worry that could be ignored was a heating system in January, as the foundry was always a place in the Chance factory where salt tablets were consumed the year around!

Architect Ed Chamberlain (1854–1941), celebrated his seventy-third birthday in late May of 1927. He had come to Centralia in 1867 and eventually would be responsible for the design of most business buildings in Centralia, including city and Chance Company structures. He also designed houses especially along what was called Chamberlain Row on Lease Avenue. Like A. Bishop Chance, Ed Chamberlain's positive influences can still be seen and are utilized even today.

Addresses can become synonymous with the name of a place, like 10 Downing Street or 1600 Pennsylvania Avenue. In Centralia, the most famous address was 210 North Allen and to the many secretaries who mailed in thousands of letters or the utility purchasing departments who sent in countless customer orders from around the globe, this address became firmly etched in their minds. The January 1921 *Centralia Guard* explained the source of the Allen Street name, plus the street names of Rollins and Barr. The Rollins name was a familiar one to local historians and could be credited with having a widespread influence in area towns. Major Rollins was said to have considered himself a politician, but his talents and accomplishments exceeded that narrow description. It appears his sale of some of the first lots in Centralia led to his prominence in the Prairie Queen. However, even the *Guard* seemed unsure of what early individuals were responsible for the street names of Barr and Allen. Harvey Hulen, an early area pioneer, wrote the paper after they had recently, in his opinion, pointed to the wrong persons. Mr. Hulen said Allen was named for the early pastor of the Allen Street Church of Christ. Mr. Hulen suggested the pastor may have conveniently been later ignored as his antislavery stand led to his loss of local support after the Civil War. Incidentally, Mr. Hulen was a Confederate. Barr Street, two blocks west of Allen, was named after the son of Robert Barr Sr. who was born and raised in Boone County. He was the brother of Colonel Singleton's wife. Barr helped build the railroad and was with the Corps of Engineers. He may have also been the original surveyor of Centralia. The *Guard* appeared to concede that Mr. Hulen was correct or out of respect dared not question his contentions.

The February 4, 1921 paper covered the city's regular meeting and a surprise announcement by Mayor Bruton. He predicted that an end could be seen to the water and sewer indebtedness. The mayor felt this was a key accomplishment and that the burden of his office could surely now be shouldered by someone else. The February 18 *Guard* captured an interview with Mayor Bruton in which he said he was searching for a candidate for mayor. He said "he had held the office long enough to get all the cussings he can assimilate." Like A. Bishop Chance before him, he had found that not everyone agreed with all his actions. In the last several elections he had chosen not to be a candidate, but continued to allow his name to be printed on the ballot, when no other candidate stepped forward. In spite of his efforts the April 8, 1921 paper told of Mayor Bruton's reelection. A month later on May 5, a former Centralia mayor (1887–88), C. E. Chance passed away in St. Louis. He was seventy-four. Of the Boone County Chances, he registered many firsts. The *Guard* in their May 13 obituary said he " was a progressive and forceful man and made friends readily." Cornelius Edward Chance was born August 8, 1846, the eldest child of Garrison and Martha Ann (Roberts) Chance and had three children, with two married daughters surviving. He married Miss Mary E. Green in Scott County, Illinois, in 1873. The passing of another area pioneer, James Grant Bruton (1845–1922), was reported in February 1922. He was Mayor Arthur Bruton's father who came to the area in 1853 as a farmer and later was in the livery/stock feed business. The *Guard* said he was a moral man who took consistent stands and was affiliated with the Church of Christ denomination. Both these men surely influenced Mayor Bruton and former Mayor A. Bishop Chance. There was also a Dr. Frank Bruton that Harvey Hulen remembered in a letter to the *Guard* in August of 1922. Harvey left Centralia in 1874 and was living in Grace, Mississippi, in 1921. He knew Dr. Bruton in 1859 as a teacher and leader of the singing school in church. He also was an auctioneer and became a doctor. Dr. Bruton and Dr. Sneed had most of the local Centralia medical business in those early years according to Mr. Hulen.

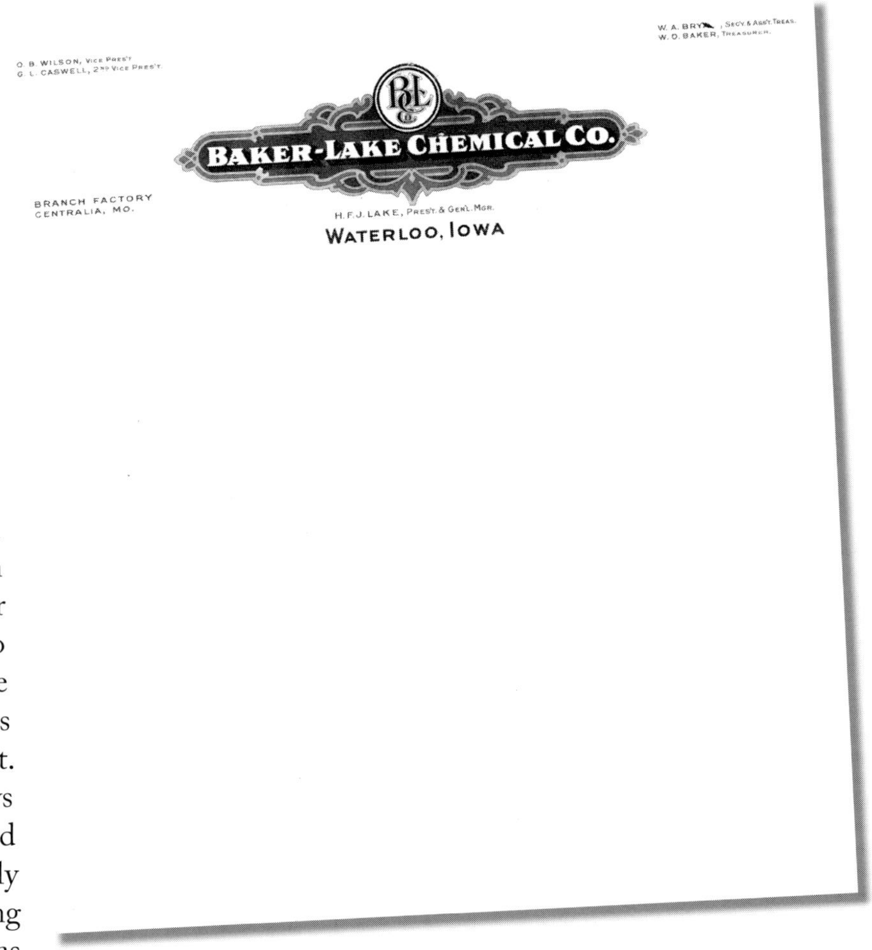

Baker-Lake Letterhead (Chance Foundation Files)

The lighter moments were alive and well as visits, dinners, and other social events continued, as it was a time when people found entertainment in their contact with each other. This was prior to the electronic age a half-century future although a young F. Gano Chance was installing and selling radio equipment. In the summer of 1921, bad news developed concerning the Webb and Wolf chemical companies that directly impacted the Chance Manufacturing Company. The first sign of trouble was sheriff sales notices posted in Centralia, directed at the three chemical firms. The chicken and bug remedies providers were in financial trouble as the August 5, 1921 *Guard* recorded:

> *Wm. R. Nelson estate of Kansas City has levied on chattels of the Wolf company and the Audit and Bond Co., on those of the Webb concern. A. S. Barada & Co. has levied on the equipment of the Centralia Manufacturing Co., and Sheriff Fred C. Brown will sell all or enough of the levied goods to pay the bills held by the parties above mentioned, on next Friday.*
>
> *The three concerns mentioned occupied the upper part of the Chance Manufacturing company's building between the main line depots on North Allen Street.*
>
> *The first mentioned two firms formerly were located at St. Louis and Quincy, Illinois and the Centralia company was formed here to handle the supplies needed in the manufacturing of the remedies of the former. They had been running here six or seven years and when first started appeared to be moneymakers. A large number of Centralians invested their money in stock of the concerns, some put their savings in and some invested considerable money. We are told one small dividend was paid the stockholders.*
>
> *Two or three years ago the stockholders had a meeting and the name of the three concerns was merged into that of the Baker-Lake company with offices at Des Moines, Iowa. Last year is alleged something went wrong at the head office, where it is whispered some one misappropriated the funds of the company and it was on the verge of going under, when it was re-organized here and*

local men put more money into the thing to get it going. They were doing a fairly good business, it is said, but accumulated bills of the old firm kept coming in and swamped the new company.

No prosecution has followed the alleged misapplication of the funds, and we are not apprised of their nature nor if there is any truth in the rumors.

Besides the many Centralians who have lost heavily in the business, there are quite a few stockholders in the country and in nearby cities.

Several hundred newspapers of Iowa, Illinois and Missouri, whose managers accepted stock in the Webb and Wolf companies in payment for advertising space, we understand are in the same boat as the balance.

The business was organized in this city by W. O. Baker, who was in the drug business and he was the main promoter of the several enterprises here. Mr. Baker, we understand, is working in the John N. Taylor establishment in Moberly at present, although his residence and family are still in Centralia.

We regret the failure of this company and its kindred enterprises, not only because it takes another enterprise out of the city and renders idle several men and women, but because such failures are bad for future enterprises which might have to depend upon local capital.

This may have very quickly given A. Bishop some unexpected plant capacity as there are no indications, through 1922, that suggests these firms were revived. On March 23, 1923, there was an article in the *Guard* stating that the Wolf and Webb poultry products would be put back on the market by new owners, Sames and Sons of Centralia. It was not said if the Chance Plant would be the location of the manufacturing and warehousing of these products.

O. G. Wilson (Centralia Historical Society)

A competitor to the Hope Drug Store was O. G. Wilson Drug Company. He frequently advertised his chicken medical remedies and was said to be a thorough chemist. The *Guard* of September 16, 1921, carried news of Mr. Wilson contemplating establishing a "Gardite Company" to make and market his products, suggests his operation may have remained separated from the Webb and Wolf concerns. He went to St. Louis to further investigate the possibilities of such an enterprise, but the recent failures of other chemical ventures surely did not help such a vision.

The activities of the city found favor with the local newspaper, which is advantageous since Chance Company developments were frequently associated with city actions. These were not always clean and noncontroversial, but they were at times colorful. On August 5, 1921, the *Guard* carried a request from another source for a tract of land eight hundred by fourteen hundred feet to build an airport for the Chicago/Denver mail and passenger air service. The reliability of early planes was apparently supplemented by the building of landing strips every fifteen miles or so along the route. The city wisely turned the request over to the Commercial Club, stating they had no land to offer. Although the later part of 1921 found no Chance Company news of any significance, the silence was broken at the instigation of the city as covered in the January 6, 1922 *Guard*, in its routine coverage of the regular city meeting:

> *The city clerk was asked to notify the Chance Co. to empty the road oiler of six hundred gallons of road oil so the machine could be put in out of the weather. The oil had been bought from the city but was never used, owing to some reason. The clerk was also instructed to present the Chance company a bill for several hundred loads of dirt used from the city dump grounds to fill up their new storage room floor.*

Controversies were never too far away as the billing for the dirt generated some sparks according to the February 10, 1922 *Guard* in its coverage of the following month's city meeting:

> *The city clerk informed the board that he had asked the Chance company to make a settlement for dirt used from the city's ground west of town, and found that a little over 300 loads were used. Mr. Chance told him that as most of this dirt was used in grading the city's streets around his factory and that the city had never paid its third of the cost of the electric siren as agreed, he thought it right that the cost of the dirt be butted against the cost of the siren and both debts canceled. The city had agreed to pay one-third of the cost of the siren which was to be approximately $375 and as the price of the dirt was about $65, or about half of what the city agreed to pay, the board decided to call the matter settled.*

As the spring of 1922 took root, things also began to warm up at the Chance plant. A news item in the *Guard* of April 21, 1922, told of a roof fire on the "Chance storage plant," ignited by sparks from the foundry. Employees used a garden hose, among other things, to extinguish the blaze. Another item in the May 5, 1922 *Guard* was a report of a tour through the Chance plant.

> *A GROWING FACTORY. We were taken thru the Chance Manufacturing plant Monday and saw the making of the various articles made there from the molding to the finished product, packed ready for shipment. The plant is now employing 47 men and women each of whom draws a handsome weekly wage which they spend in Centralia.*
>
> *The plant is now turning out five thousand Gastitle* [sic] *piston rings daily and will have to increase the output to seventy-five hundred to take care of the increasing demand.*
>
> *Mr. Chance says the plant needs more room and we understand it is to be enlarged at an early date, taking in a fifteen foot strip a block long just south of the present plant and north of the big storage obse* [sic].
>
> *It is a busy bunch of fellows in the molding and finishing rooms as they were at that time getting ready for a night molding run.*
>
> *The working quarters of the women upstairs have been arranged to give them every comfort, even to a cozy lounge room, with easy chairs, tables and reading*

material, and also with individual lockers for their clothes.

This was the first account in a while of what the Chance Manufacturing Company had become and how piston ring production had grown the company. These newfound manufacturing capabilities would also impact the company's ability to make important utility oriented products. Flynt Chandler's 1919 report of only three employees had quickly expanded to forty-seven. What was emerging was a picture of Chance Company as it was entering into an almost two-decade period that could be labeled the Piston Ring Era. This generated a volume of plant activity and increased employment that helped set the base from which a sole entrance into the electrical utility market in the late 1930s, could be successfully orchestrated.

The livestock feed and grain business was one of volume in Centralia during the 1920 period. Sometimes several rail car loads of grain were shipped out each week from Producers Grain Company. They seemed to have that business pretty much sewed up in town. They were located across Allen between Wilson and the Alton tracks and eventually became the M.F.A. elevator before Chance later obtained control of those grounds. The success of Centralia firms was a mixed bag, however, with a December 9, 1921 report of the broom factory across the Alton tracks from Producers having shut down, but it would later be revived. The start-up of the Hamilton Brown Shoe factory (parts initially), on the west side of the park, signaled new employment opportunities for some fifteen employees, many of whom were women. A year later on December 15, 1922, the local paper told of the demise of yet another small business, Crump & Humphrey. They had assigned the business to R. P. Price, citing inadequate profits. The owners were identified as Luther Crump and Jim Humphrey. Five years earlier, when they went head to head with A. Bishop for city land between the tracks, the future looked a lot brighter. This would make vacant their building on the corner of Railroad and Rollins, known later as the Narragansett or Anthony Building. This apparently resulted in a new business, Crump and Dawson, who were local dealers for the Boone County Elevator Company of Columbia, per a January 26, 1923 ad in the local paper. Another new business, a shirt factory was checking facilities out in Centralia, especially the Anthony Building recently vacated by Crump & Humphrey. The dimensions of fifty by one hundred feet of this Railroad Street building tends to confirm it as the Anthony structure. An added bit of information resulting from the shirt factory overtures was their need to "add a 20', 2nd addition on the south, to fill the lot."

Some Rodemyre humor was always welcome, as it was in the February 4, 1921 *Guard*. This local news item concerned the death of a hog on the Wabash. They concluded it was a clear case of suicide "as the Wabash hasn't got a train on the west end that could run a hog down on level grade." The Wabash was usually a struggling railroad financially and their track and equipment suffered as a result. Strength problems with the original St. Charles Missouri River Bridge, threatened with heavier trains, was being repaired in March 1921, causing Wabash detours over M.K.T. (Katy) tracks. In late September there was a report of Kansas City Power and Light possibly extending their line east from Glasgow to Mexico, which would also include the Moberly and Columbia areas. Local Centralia service was considered poor, from a reliability standpoint and also from inadequate voltage at times hardly able to keep the lights bright. Of course, the *Guard* was hot about this issue, as outages shut down their very time sensitive presses. Their critical cadence moved to a higher plateau in mid-March 1922 when a power failure shut down factories and shops in town. The *Guard* responded by calling to Mexico to get a straight answer of why power went down. They were infuriated with the double talk and inability to get an answer. The paper's editors were likely concerned about later in the week when their paper went to press. Obviously when factories such as Chance and Brown Shoe went down, this had a financial impact on the community. One of the functions of a newspaper in those earlier times, as it should be today, was to put a voice to and pressure on getting community problems solved. In the December 30, 1921 paper, the editors even questioned the city on why they did not fire up the city generator to furnish temporary power. When the power went down, it isn't very difficult to imagine A. Bishop pacing his plant's floor and his fertile mind considering the nearby power plant's capabilities.

Sometimes it seems that progress can bring more concerns and problems. My dad once spoke of a farmer west of Duncan's Bridge that had a gas-powered generator that would illuminate his barn while doing chores. Incidentally, these Delco-type generators were being advertised in the *Guard* in the 1920 period. That's about as close to electricity, except for a bolt of lightning, that most rural people experienced in the pre–World War II period. So like other conveniences, you hardly miss it if you've never had it. Because of several forward thinking Centralians such as Adam Rodemyre and A. Bishop Chance, Centralia became exposed to electricity early according to Sam Locke. Mr. Locke, associated with the Mexico Power Company, the source of Centralia's electricity, said during city bond-burning ceremonies in January 1923, that Centralia was one of the early pioneers in the use of electricity. The transmission line from Mexico to Centralia was only the second one of its type in the state of Missouri! A. Bishop was at home sick, so he was unable to attend those ceremonies. The remarks by Mr. Locke must have made Centralians feel better, especially a short time later as in the March 30, 1923 *Guard* an article spoke of the immediate rebuilding of the just mentioned transmission line. However, to replace the frequently troublesome line would take further sacrifices as hot line maintenance and construction work had not been perfected or even become widely known. Service would be interrupted during the day when linemen were rebuilding the line. To overcome this, the city generator would be brought up during the day, but it couldn't handle all the demands, including the Chance plant. Arrangements were made to run the Chance operations at night after a test run was made with only the Chance plant connected. A 25 percent overload resulted! The Chance factory was said to use "a prodigious quantity of current." It was hoped the line work could be completed in two weeks. Surprisingly, this line also supplied power for Hallsville, Sturgeon, Thompson and later Clark. This power was bought by the city of Centralia and resold to local users. Conditions were apparently deteriorating as power had been going off more frequently per the June 15, 1923 *Guard*. They concluded that you can't do without it as electrical machinery runs with it and profanity won't substitute! In mid-July 1923 the Missouri Utility Company of Mexico, owned by E. R. Locke and W. F. Corl, was sold to Illinois Power and Light.

A twentieth-century icon Will Rogers was emerging in the early 1922 newspapers as was famed western author Zane Grey. Ads for Kellogg Corn Flakes and Oshkosh B'Gosh overalls were also hitting the big time. In Centralia the oiling of streets issue was a longtime discussed and an apparent frequent activity with the lack of hard surfacing taking place. Another result of the dawning of the automobile age was a local call for a tourist camp. These so-called auto courts were becoming a popular stopover, overnight facility for long-distance travelers and were being built in other towns. This likely had some major highway location ramifications. Theodore Gary, chairman of the State Highway Commission, was trying to take care of the state's roads already built, by suggesting the progressive idea of load limits in the July 1922 paper. Another automotive item was the purchase of the Lyon Auto Company by a St. Louisian, J. F. Honey. This was the exclusive Ford dealership in Centralia that also offered Fordson tractors.

It is doubtful if there was an off-spring of Garrison Chance who showed more resiliency and adventurous spirit than did Charley Chance. This must also speak of his strong feelings for family, as the July 21, 1922 paper covered another visit by car with his wife to see his brother John. They had recently lost their older brother, C. E., so this may have made the visits even more important to Charley. They incidentally found the roads in fair shape most of the way from Oklahoma. Another longtime Centralia resident, A. R. Toalson, was quoted in the March 3, 1922 *Guard* as having lived fifty years on the southeast corner of Sneed and Hickman Streets. There were few current citizens who were even residents of Centralia in 1872. Along with new milestones, the passage of time also brought departures. A James A. Hudson (1853–1922) of Columbia may have appeared to have no Centralia connections, but he did. One similarity with the Chances was his telephone association as head of the Columbia Telephone Company. He was formerly from Macon and once owned at least one and possibly both newspapers there (*Macon Times* and *Macon Examiner*), plus the *Chariton Courier*. He also became associated with Theodore Gary and in Gary's 1907 book (*Independent Telephony*) he is listed as a contributor and wrote a letter of recommendation of his intimate association with Mr. Gary over the last twenty-five years. His death at sixty-eight in mid-1922 may have allowed the Gary group to gain full control of the Columbia operation, which would lead in a short time to the purchase of the Centralia Home exchange. However, this may not have been a factor as it appeared Mr. Gary and Mr. Hudson worked closely together and may have originally co-purchased the Columbia exchange.

The initiation of a long and frequently talked about Company tradition was apparently begun in 1923. Hard-charging people like A. Bishop, who would soon reach the half-century age mark, occasionally needed a reprieve from the daily grind. The March 16, 1923 *Guard* reveals what occurred:

> *The employes* [sic] *of the Chance Manufacturing Co., will give Mr. Chance a banquet tonight, Thursday, as a mark of their appreciation, before his departure for the West. All the employes, with Mr. Chance and wife, will attend the picture show and after the show they will assemble in the girls waiting room of the Chance plant, where they will enjoy a banquet. Mr. Chance is expected to make an address to the employes, after which those who desire will participate in a dance in the upper stock room. Mr. Chance will leave Saturday for California, where he will spend a month recouperating* [sic] *his health.*

As earlier stated, A. Bishop was ill during January and apparently hadn't made a full recovery. His practice of meeting with employees just prior to his extended absences appears to have been established in an official manner in early 1923 with this company sponsored social event. It is unclear if he had returned by April 15 when his eldest son Buford, quickly got married! But undoubtedly he was back when his Red Snappers baseball team was defeated by Clark 39 to 0 in June! This shows further refinement of the Chance operation, although it's highly questionable that his son F. Gano would have later approved of such a name for a company-backed extracurricular activity. The promotion of his products made travel a frequent companion, as was noted in a late November 1923 business trip to St. Louis. Like the period Wrigley Juicy-Fruit gum ads, Chance utility products were growing in sales and becoming industrially recognized.

Part of the tradition of Company get-togethers was expanded due in part to the large building that sat on the southeast corner of Rollins and Railroad Streets, known as the Anthony Building. The financial success of A. Bishop Chance was not only being applied to plant expansions, but also in local real-estate purchases. According to an October 31, 1924 article in the *Guard* of the Railroad Street Garage, this property was now a Chance-owned building. This answers a lot of questions especially because it was utilized by A. B. Chance Company over several decades. But the most interesting reflection of this site was its many faces, including a recent apartment complex rejection and its gradual demise today into a condition that blinds one to its magnificent rebirth in late 1924–early 1925. A. Bishop Chance worked hard to make Centralia a showplace of progress, modernity and impressive and well-maintained properties. A revelation of the late fall 1924 article was the condition and size of the original Anthony structure was changing in the face of remodeling and new construction activities being orchestrated by A. Bishop. Although it may not have been referred to as the Chance Building, this would appear appropriate in the face of its resurrection in the mid-1920s. Immediately one would assume all the expenditures and attention being given this building would be for Chance Company use, but one suitor J. Ford Honey of local Ford dealership fame was already lined up to occupy a portion the new Anthony as soon as the refurbishment had been completed. Before covering the details of the changes taking place at Railroad and Rollins, further analysis of the owner's intentions and reasons for this venture can be very revealing. Suddenly the real A. Bishop Chance is emerging, although in reality it's just another dimension of this bold, daring, and driven entrepreneur. As the just discussed farewell event may have been a first, the Anthony appears to be another first in a major activity outside of the manufacturing arena. The similarities with the later Rock Garden and Farm will become obvious. The idea of taking care of family and employees also gets a boost as the benefit of this building in many ways was in support of local residents. There was a lot of talk over the years of having a large auditorium to hold various events, but the city fathers had decided it wasn't the place for the city to provide such a facility in the new City Hall of 1923–24. An ad later in the February 5, 1926 *Guard* for the "Chance Auditorium" spoke of "the Greatest Musicale Centralia ever had, and the best, cleanest dances." It suggested people should support these "wonderful accommodations" as "it cannot exist as a free gymnasium alone." There were reports the Centralia High School had basketball games in this second-story gym. Although the Chance Company would later utilize the building, it appears Mr. A. Bishop's earlier intention was to provide a facility that was also beneficial to the local community and the "Club Room" as the large second floor gym later was called would hold countless local and company events.

Another revelation of the October 31 Railroad Street Garage article was the significant changes taking place in late 1924 on the Anthony Building. The north and west sides were replaced giving a modern front with many windows including four large ones of about eight by twelve feet on the west side, which would be utilized by Honey as auto showrooms. The most significant change, among construction cost reports of between $25,000 and $30,000, was an addition of fifty feet to the east. Consider that this substantial cost was at a time when a new Ford cost under $500! Also included was a fifty-by-twenty-foot basement in the rear that housed the furnace, coal and other supplies. The front included what was described as "massive steel girders and posts" which represented the strength being built into this two-story structure. One of the reasons A. Bishop was earlier trying to get Rollins and Railroad Streets paved, with portions of it around this building already completed in late 1924, was to enhance the convenience and appearance of this location. It also suggested the importance of advertising Centralia to the multitude of rail travelers coming through town, as the Anthony Building was one of the prime focal points of Wabash passengers. This was confirmed at a Kiwanis meeting while viewing a film that included or was about Harry S. Truman's famous Whistle Stop Campaign of 1948. Of the many movie sequences taken from the rear vestibule of his train where President Truman would stand and speak or wave to the thousands who turned out when he came through town, was one of Centralia. It was a camera shot from the Allen Street crossing area east as the train departed, clearly showing the west side of the Anthony Building. Historians, however, will have a problem with this as the 1948 Truman Campaign train didn't come through Centralia. Instead this appears to be footage taken in 1950 when President Truman addressed the graduating class at the University of Missouri and passed through Centralia on the Columbia Branch. A 1950 picture of Mr. Truman waving at the crowd of Centralians, which included the Chance plant employees let off work for the occasion, will be included in Volume II. Incidentally, his late returning train didn't stop in Centralia, but did briefly in Mexico. Mr. Truman usually took the Missouri Pacific from St. Louis through Jefferson City and Sedalia and on to Independence during normal travel to and from Washington, D.C., and home.

Progress was being made on the Anthony Building according to the February 20, 1925 *Guard*. "The Chance building on Rollins and Railroad Streets is now ready for the sheathing of the roof." The steel trusses were up and ready for the carpenters. The roof on the east was to be rounded with the west side overhanging the roof of the old building being guttered to carry the water off to the rear. Early pictures suggest the corner building may have sat beside another building to the east with a space between them. The smokestack was finished with additional brickwork near completion. According to a September 19, 1924 article, Centralia's architect Ed Chamberlain drew up the plans for this new and revised building. At that time it was said, "A. B. Chance has rented the building on Railroad and Rollins streets known as the Anthony building to J. Ford Honey, for a garage." The addition on the east was to be used as a garage and the many changes, it was suggested by the owner who wasn't going to "stint money nor pains to make an up-to-the-minute building."

The March 13, 1925 paper stated that the Anthony building was now the largest single building under one roof in Centralia. That the Anthony had more square footage than the original two-story Chance Plant at 210 N. Allen is a surprise, although it was yet to receive its third story and rear addition. As April 17 rolled around Honey Ford was preparing the Anthony Building for its move there and the big news was a large electric sign that would grace the front corner. This elaborate sign would be a first for Centralia, weighing in at some nine hundred pounds. The word FORD at the top was eight feet long, and the bottom section twenty-four feet from top to bottom with various alternating lighting effects. It was stated by the May 1, 1925 *Guard*: "It will be a showy place and will attract much attention to passengers going through on trains as well as the stranger on our streets." Some of its additional features were modern restrooms with lounges. Other areas were described plus above the new east room on the second floor was the paint and workroom, "where cars are taken up on the big 8 x 18 foot elevator which will lift the heaviest car or truck." The filling station was located on Railroad Street although cars ran under the cut-off northwest corner. It was suggested on May 1 that it still was not ready for occupancy. A week later the paper said the Anthony Building was to be extended to the east with the new ironworks on the ground ready for that extension. It appeared that the construction had already been completed, but apparently not.

It is no small thing to rename a Centralia landmark or to rediscover its more glorious past. A book can perform such magic, like reviving a flower that has faded and wilted and Mr. Chance would have liked that. The name could just as easily been relabeled the A. B. C. Building: it was *A*nthony *B*efore *C*hance! But as 1924 comes to a close our focus again shifts across the street to the northwest at 210 North Allen where the nucleus of A. Bishop Chance's world is centered. After the advent of the new building in 1917, there began the usage of a term "Up Front" at Chance Company. This terminology shares more than one meaning. An initial response might suggest it means to be open and honest with no hidden agenda. This is appropriate for Missourians who have carried forth the heritage of being forthright and to "tell it like it is." President Harry S. Truman reflected this in a very powerful way and the Chance family and Company also displayed this practice. Another meaning of "Up Front" speaks of physical location. The front door of a home or business is appropriately up front. A guest or visitor is not expected to come to the side door or the rear utility entrance. This was no different at the new 210 North Allen Street address of the Chance Manufacturing Company as the offices were up front, although one or the main entrance was on the south side. These offices initially only took up about the first twenty feet. Although there may have been visions of expansion there is little reason to believe that eventually the entire brick structure would be offices, especially after its two stories became three in 1925. It is not clear which came first, the need for more office space, production expansion or both somewhat simultaneously. When one was called up front for conference, direction, or routine procedure such as the shipping foreman's several daily trips, it was to the nerve center where power and control resided. In spite of this, however, A. Bishop never allowed the factory to become totally separated from the office. Many companies have progressed to a policy where contact between employees is frowned upon and a cold, impersonal atmosphere is enforced. This is made easier for many headquarters people who are often located remotely, a long distance away from the operating divisions of the corporation. So the connotation of up front has changed to mean some far off place or a separated and fortified position unlike A. Bishop Chance who was as comfortable in the plant as in the office and contact with employees was of paramount importance.

New developments were coming at a quickening pace during the early 1920s and some were negatives such as the national Teapot Dome scandal. But there were a lot of improved items being introduced by 1924, or more extensively promoted, such as Thor and Maytag washing machines, Bayer aspirins, and Murine eye drops. In Centralia, Mayor Bruton ordered the cornerstone for the new City Hall and the red exterior, rough-surfaced brick were arriving. The mayor's mother, Mrs. J. G. Bruton, seventy-five (1849–1924) passed away in late February. The rapid pace of one A. Bishop Chance was reflected in a trip to the St. Louis auto show promoting the Big C auto line, getting the vitrified telephone underground cable installed in downtown Centralia, becoming chairman of a citizens' group promoting street paving in June 1924, and returning home in late July 1924 from a week-long business trip to New York City. In spite of this, A. Bishop found time to visit with his father and mother. The A. B. Chances entertained the J. A. Chance and his sister's family the Willis Fountains in early February. In late June 1924, Mr. and Mrs. A. B. Chance motored to Mexico with his father and mother on a Saturday to procure some plants and attend the carnival there. Mrs. A. B. Chance was also trying to make their new daughter-in-law, Mrs. Buford Chance, feel welcome by inviting her to her tea club meeting in early July. The J. A. Chances must have enjoyed their grandchildren as they entertained their granddaughter Lucille Fountain in late August.

As late October 1924 rolled around, town folks including the staff of the *Guard* must have kept a curious eye focused on the area between the tracks, as it was a hotspot of activity. If nothing was visibly occurring, one then became suspicious of what A. Bishop was up to next. If he had been an athlete, which he was in durability in the pace he maintained, he was at the top of his game. Financial success had finally caught up with his experienced abilities and visions and all facets of his life seem to blossom. This was also a time when the residential area grew and progressed. Leading a parade of locals concerned about street conditions, who wanted more street paving and eyesores removed, was the A. Bishop Chance family. At times it may have only been a parade of one, but his enthusiasm and position made it appear as a crowd! In fact, he and J. R. Brown in July 1923 presented a petition signed by eight to the city for street paving of Sneed, while an opposition proposal for no paving had five signers. He wanted Sneed from the new City Hall to Jefferson paved with the cost shared by property owners. This may have generated as much opposition as support because not everyone had the resources he did, although there are some substantial homes along east Sneed. Regardless, A. Bishop had the street at his residence paved and curbing put in. Apparently "driveway" hadn't come into vogue as the use of street ensued instead, although this could have been the short street to the east of his home. This confirms his family was residing in the current Centralia Museum, Garden House or the last A. B. Chance family home. The oiling of streets wasn't very popular because of the smell and mess it generated, plus the repeated applications didn't prove very cost effective. The *Guard* pointed out that its founder Adam Rodemyre had long been in support of street paving and believed he would see it take place before his death, but of course that hadn't yet occurred almost two decades later. But petroleum was popular for heating purposes as back at 210 North Allen three new oil tanks were added to the current two for the plant furnaces. These were large as it was estimated it would take a dozen carloads to fill them. Electricity, another important source of power for the Chance plants, would be enhanced with the nearing completion of the new power lines from the Mexico powerhouse.

Along with the just mentioned paving at the A. B. Chance residence in 1923, portions of which can be seen today as it approaches the century mark, is a discussion about an "Elm" branch or waterway running near or on the Chance and J. R. Brown properties. This suggests that the J. R. Brown home may have been located across Sneed to the south. There is a rose garden where a church was located across the street east from the Chance home according to native Centralia resident Bill Miller. The gully was described by the July 20, 1923 *Guard* as "the unsightly and unsanitary ditch." The proposed solution for eliminating this problem was formulated with a proposition to the city by Brown and Chance to cover Elm Branch. They would put in concrete walls and a concrete cover over this waterway along their properties if the city would cover their share of street costs and cover the entire work with twelve or eighteen inches of dirt. Mr. Chance's engineer recommended the covered area be made three feet high and six feet wide. This would later prove beneficial as the east border of the Rock Garden. The existence of this short street running close and parallel to Jefferson was somewhat of a past mystery, but the provisions to handle a city drainage problem may have led to its existence. At that time there was a home behind the Garden House where the Rock Garden is located. Over a year later, in September 1924, the *Guard* covered the purchase of the Jeff Jerman residence on East Singleton Street for $1,250 by A. B. Chance. This would appear to be the Rock Garden location. He was going to remodel the house vacated by the Jerman family who moved to North Barr Street.

Electricity had become an important part of life in towns and cities across America. By 1924 it was also a vital source of industrial power for such firms as Chance Manufacturing Company. Not only was reliability becoming important, especially in view of Chance Company's later contribution to worldwide electrical efficiency, but the cost of this power was also an issue. Chance Company was competing against large-city-based factories, which had lower operating overheads and the founder made several requests to get reductions in power rates, because of the large quantity his firm used. This is today referred to as a volume discount. One such occasion was in early January of 1924, in

The former Elm Branch has now become a street beside the Chance Garden House, Rock Garden, and the Chance Bungalow on the right looking north. The paving of the driveway entrances and the street is in progress. (A. B. "Buff" Chance IV Collection)

which Centralia City Superintendent Gillam stated that the Chance rate was close to the city's cost and the Mexico Power Company was working on further reductions in rates from their source. A special aspect of the Centralia community was its exposure to the outside world. New ideas, fads, and methods were brought back to town by residents like Adam Rodemyre and later the personnel from the Chance firm. A. Bishop's frequent travels also brought the latest developments to Centralia, as he was constantly attempting to keep abreast of what was occurring in other manufacturing firms. He also traveled for leisure, although it is hard to imagine his total commitment to that end, without maintaining a watchful and curious eye toward technological and mechanical trends of the day. However, his focus on leisure and human relationships, whether it be a company get-together, rocks or flowers for his Rock Garden, or world cruises, suggest he considered that important and it gave him the opportunity to land and occasionally get away from work.

The Chance enterprise wasn't the only growing business and employer in town as the broom factory, which routinely employed about twenty-five workers, was again going strong. In March 1924 it was even busier as the Maher Brothers Mexico plant had recently burned. Progress was also on Theodore Gary's mind as he was trying to get Missourians out of the mud by asking for more state money for paved roads by increasing registration fees. In another historical tour around the City Square and down Allen to the Alton Station, A. Bishop was vehicular host to *Guard* officials in pointing out the poor conditions of city streets. The resulting article in the paper, on March 28, 1924, asked the question: "To Pave Or Oil?" Holes in the streets were up to "hub" deep. Fourteen years of oiling Allen Street seemed wasted, according to the article, as A. Bishop had cindered Allen Street in front of the Chance plant, and routinely had his factory men fill new holes with fresh cinders. It was suggested the heavy traffic carried by Allen Street made maintenance of it even more acute and more frequent. Reflecting like conditions out on major highways they also had to be separately marked to prevent travelers from getting lost. The North highway through Missouri was to be specially marked by Shriners in view of their upcoming Kansas City convention. Spring floods in April brought conditions similar to the early 1990s flood at Glasgow, as the "Big Muddy" preferred to straighten its channel at the bend a mile west of town and render the new million-dollar Alton Railroad Bridge useless. It was considered one of the strongest bridges in the country and it still carries untold tons of railroad traffic today. A new highway bridge, or wagon bridge as it was called, costing $400,000 was halted by the contractor in the face of the flood. Some refer to these as Theodore Gary bridges, many of which have already been replaced today. Unlike the railroad bridges serving a much more mature industry, the automobile and truck traffic grew in size and weight, eventually obsoleting those bridges.

A series of names included in the mid-May 1924 school graduations would be familiar to most veteran Chance Company employees. Two Centralia High graduates, Anna Lee Toalson, who would enter college in Columbia that fall and future Purchasing Agent Harold Cox, who would attend business college at Quincy, Illinois, were noted. Two longtime fellow employees at Chance, Jack Curtright and Nathan Toalson, were also eighth grade schoolmates who would be initiated into Centralia High that fall. The Chandler brothers, Flynt and Ollie, were becoming well-known locals whose trips to Hallsville and Columbia to visit family and friends were becoming routine. They were products of the stability generated by local residents working at the Chance factory. In September, Ollie was reported to be walking with a limp after his right foot was injured while helping move machinery at the plant. To A. Bishop Chance's delight, the concrete street proposition passed and bids were opened in September, ushering 1924 out in grand style.

As the winter of 1924–25 came, how little some things have changed at 210 North Allen. Moving ahead some seventy-five years, on January 4, 1999, early morning hours brought a thirty-degree below zero chill index and snow piled high over Chance finished goods and raw materials stored outside in the elements. There has, since Chance Company's inception in 1907, been an effort to get everything under roof, but continued expansion makes that a hope and not a reality to those in the materials handling area this cold winter day. Neither had the prairie conditions changed much with few natural barriers in the winter to slow the charge of swift frigid winds that sweeps through Centralia with the urgency of an express train. Only the tall three-storied main office building seems to stand in silent protest. As one approaches it this January morning, you could expect to be buffeted and swept around the north side or pushed in the front door. Even in the stillest summer days, there always seemed to be air currents shooting around it as if to remind one that the principles on which it was built are worthy and will somehow survive, resisting the winds of change in new and strange directions.

The Big C Automotive Line

What is a dreamer? Is it someone that can see what others cannot? Could Native Americans visualize in nature a spirituality that their White brothers did not? Is a dreamer one who lives in a fantasy world of make-believe, like Walt Disney or P. T. Barnum, who brought reality to their dreams? A. Bishop Chance was occasionally called a dreamer, apparently because of what he did or said he was going to do. It would seem that a test of whether the title of dreamer carries a positive or negative connotation is if one's dreams come true. A compliment that the Chances earned is what they became involved in, they completed. They were no quitters. It's amazing to follow the decisions and examine the many things the Chances participated and became involved in. It wasn't an atmosphere in which the faint hearted would ever feel comfortable.

The second story of Chance Company's new 1917 factory building is the key in unlocking a mystery of sorts. The question arises, if new products were developed and purchased to fill this plant, or if production of former products were expanded to the extent that all of this extra space was needed. There is an expansion spirit that Americans share in that whatever we build, it is never quite large enough. It's not a certainty that Randolph Hearst shared this opinion after completing his massive castle but most of us have a two-car garage and we are still building on, as the Waylon Jennings song goes. A. Bishop had a positive attitude toward life, as he was always building for the future whether it was in human relationships, his community, or in the plant and grounds of his company. Gano remembered their use of the ground floor but the company hadn't yet grown into the second story. All of those doubting Thomas's who called A. Bishop a dreamer, in such instances, once again would have to conveniently forget what they had said because in a year or so he would again prove them wrong. Gano, in talking about the new building said:

> *At that time we were not using the second floor. There were quite a few young business men and they set up this carnival to make some money to give to the Red Cross for the War. They had all kinds of carnival acts and they ran that about a week in the second floor of our Company. Shortly after that we rented it to another party who used it for about a year. Then we decided we needed it for ourselves. (1981)*

The question still exists as to why A. Bishop built a considerably larger building than he needed. How did he finance it or sell the idea to the bank? Lower construction costs, greater utilization of limited grounds, and expansion possibilities made a second story logical and many brick buildings of that era had an upper floor. Were there visions of expansion and growth, especially after the war, as they were about to enter the Roaring Twenties. Was there a plan to acquire another product line, even if it wasn't remotely connected to the telephone or utility industry and did he encounter a party with a piston ring design that generated an interest in the automotive area and his optimistic and aggressive manner took it from there? The last scenario would appear to explain how Chance Company got into the manufacturing and supplying of auto-related parts and accessories and decided that they needed the second floor to handle this new product line. This piston ring contact had been made in 1917, prior to the construction of the new building at 210 North Allen.

Clarification of what actually occurred came via the *Centralia Fireside Guard* in the March 16, 1917 issue:

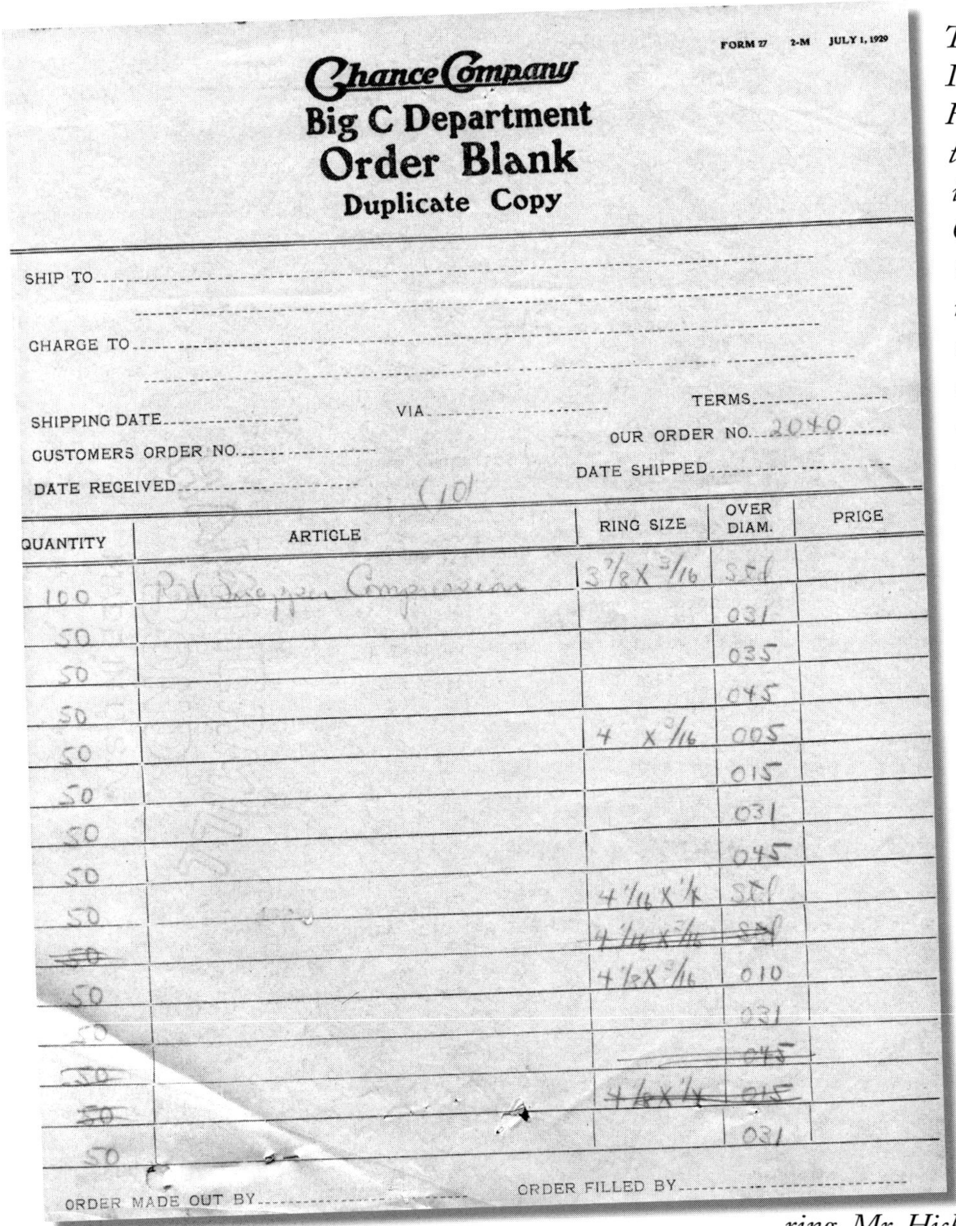

FORM 27 2-M JULY 1, 1929

Chance Company
Big C Department
Order Blank
Duplicate Copy

SHIP TO

CHARGE TO

SHIPPING DATE VIA TERMS

CUSTOMERS ORDER NO. OUR ORDER NO. 2040

DATE RECEIVED DATE SHIPPED

(10)

QUANTITY	ARTICLE	RING SIZE	OVER DIAM.	PRICE
100	[illegible]	3 7/8 x 3/16	Std	
50	"		031	
50			035	
50			045	
50		4 x 3/16	005	
50			015	
50			031	
50			045	
50		4 1/16 x 1/4	Std	
~~50~~		~~4 1/16 x 3/16~~	~~Std~~	
50		4 1/2 x 3/16	010	
50			031	
~~50~~			~~045~~	
~~50~~		~~4 1/2 x 1/4~~	~~015~~	
50			031	

ORDER MADE OUT BY ORDER FILLED BY

The Big C Order Blank is the first document found that describes the Chance Company auto line as the Big C. Note that it is a subheading under the company name, meaning it was subordinate. This is further confirmed by it being described as a "Department." It is possible the position of the Chance auto line varied as the information in the text suggests and department in this instance represented more of a division between it and the utility area. This is revealing of a lesser position than usually thought. It can be dated because the form number in the upper-right-hand corner of July 1, 1929, means it couldn't be earlier. Because it was a surplus paper, reused on October 30, 1929 correspondence, it could not be older. (A. B. Chance Co.)

TO MAKE DEVICE IN CENTRALIA. Irl R. Hicks, who recently moved to Centralia from Moberly is now connected with the Chance Manufacturing Company and is placing his new "Gas-tite" piston ring on the market. These rings are being manufactured by a concern near Chicago but Mr. Hicks intends to begin making them here just as soon as the necessary machinery can be installed. The "Gas-Tite" rings are not made like the ordinary cheap rings on the market but are made from individual castings and ground (not turned) to an absolutely true circle to exact size while in the closed position to insure a perfect fit when installed in the cylinder of an internal combustion engine. An illustration is included to show the contrivance and its utility in making a perfectly gas-tight piston ring. Mr. Hicks is the inventor and owner of this device, which is only one of his many inventions, some of which are now being used by large manufacturers. Mr. Hicks' invention is being used and recommended by leading engine manufacturers and mechanical engineers, and it is a tested device. Centralia is fortunate in securing this enterprise.

The timing issue was covered earlier in the previous section about the building of the new 210 North Allen facility. This article clearly reveals that Mr. Hicks had already developed his piston ring when he became associated with Chance Manufacturing Company, prior to the construction of the new two-story plant. This association and its growth possibilities appears to have added credibility to Chance Company's proposed costly expansion project. This would have also aided financing and some of A. Bishop's business trips to Moberly or Chicago may have included talks with Mr. Hicks. However, limited tooling and machinery and lack of foundry capacity at the small Chance operation resulted in a delay of immediately taking advantage of this business opportunity. A sales outlet for the product and maybe even the transfer of the Hicks operation, could have also slowed the start-up process. This envisioned slow beginning led to the flexibility of renting out the second floor of the Chance plant to a chemical concern. In reality this was advantageous, as it drove rental revenue and borrowed time to get the ring manufacturing operation up and running.

It is unclear how Irl R. Hicks and A. Bishop Chance reconnected with each other, assuming Irl was an early employee and had previously left Chance employment. However, they both were from the local area and the March 16 article suggests Irl was not a Chance employee in early 1917. They may have been looking for what each brought to the table, as together they could satisfy each other's needs. Irl Hicks appears to have wanted a local factory to produce his piston rings, among other automotive items and A. Bishop wanted to build and operate a larger plant. As earlier covered, there seems to be a clear connection between their early relationship and some type of agreement or contract. The potential of this business agreement seems to have helped drive the construction of the new plant at 210 North Allen, although there was A. Bishop's expansion of his existing business and other possibilities. The utility business, from a Chance perspective, was slower in growth but the automotive area was a sensational growth industry due to Henry Ford's insistence upon everyone being able to afford a new car. His manufacturing complexes ensured that unheard of production levels would allow this to occur. This established a need for splinter companies that supplied parts to these maintenance hungry machines and this is where the Hicks and Chance potential business laid. A. Bishop was an opportunist who went with the flow and was ready to take advantage of new possibilities, which is reflected in the company until about 1940. Clearly one of the reasons for continued success was an ability to not only get into business with potential, but to get out of it and move on to something else, when it dried up.

The Hicks story generated considerable personal intrigue after the 1917 announcement that Irl R. Hicks was responsible for the piston ring design that was to be produced by Chance Manufacturing Company. Earl Hicks, an early employee mentioned by Flynt Chandler would appear to be Irl R. and because Irl may have been pronounced like Earl, the error developed from interview material. It would appear he was a part of Ollie Chandler's wife's family of Hallsville and the Irl Hicks, Sr. shown in the photo on page 180 appears to be one and the same person. Because Irl Hicks was from Moberly in early 1917, however, this suggested a possible connection to the Moberly or east of Cairo Hicks family but there appears to be no connection. Irl Jr., who is Irl's son, was discharged from the Navy in the Spring of 1946 and was mentioned occasionally in the *Guard* when he was growing up in Centralia.

What has accumulated since 1917 is several newspaper accounts of the Hicks families in Centralia through the year 1922. A May 14, 1920 *Guard* ad placed by Irl Hicks, covering a non-piston ring product, is reproduced. Another report was found of the Hicks in a May 6, 1921 news article of Mrs. Irl Hicks undergoing surgery at Kirksville and convalescing at her mothers, Mrs. N. W. Howell of Madison. There was an ad in the February 25, 1921 paper and in other issues of Mrs. Hicks selling lady's hats in Centralia. The floodgates then seem to open. On July 29 the paper said Mr. Hicks had been offered a fine position at Fort Madison, Iowa, in a tractor factory. The *Guard* indicates they would regret to lose Mr. Hicks and wife from their city. On September 2, 1921, an article announced they were leaving with additional valuable information about his Chance Company involvement:

> *Irl R. Hicks is leaving next week for Fort Madison, Iowa, to join his wife. Mr. Hicks has resigned his position as foundry superintendent with the Chance Co., and expects to do electrical work in Fort Madison. Mr. Hicks is a mechanical genius. He originated the "Gas-Tite" piston ring that is made here and is the inventor of other mechanical appliances that are being manufactured and sold. In fact, all of his inventions made good because of his practical, hard-headedness. We regret to lose him and his good wife from our city but we can recommend them to the good people of Fort Madison.*

From a company standpoint this indicates Chance Company had a foundry operation by

Piston Ring Machining. Pictured from the left: Cleo Palmer, Harold Mosely and H. L. Carnahan, factory superintendent. (*Chanco News* June 1957)

1921 and Irl's expertise was being utilized in an important position. The business relationship between these two men is still unclear although it may have been like other Chance product buy-outs or license to manufacture, such as the early arresters. Irl's departure would not halt the manufacturing of the Hicks piston rings and other of his products by Chance Company.

There were departures at Chance Company over the years, some permanent and some only temporary. The company seemed to be generous in regards to reemployment if there initially had been no problems and the employee offered beneficial contributions to Chance Company. There later was a rule that a year had to transpire before reemployment, but evidently not in 1921. The September 16, 1921 *Guard* reported that Irl R. Hicks was "back on the job at the Chance plant." This was only two weeks after his departure! In the September 1, 1922 paper Miss Rebecca Howell, formerly Mrs. Ira Hicks, was in town from Fort Madison where she was a saleslady in a large store there. Her return to Centralia may have been due to a fire that destroyed her millinery stock and business. This would indicate that the Hicks were now divorced. The question of what Hicks family Irl was a part of seems to have been cleared up in the following March 16, 1923 *Guard*:

> *Irl R. Hicks who has been at the bedside of his mother near Hallsville, is at home in this city again. Mr. Hicks has been promoted to the position of chief electrician in the Chance plant, a position which he is eminently qualified to fill, and which brings him a larger remuneration than his old position of superintendent of the foundry.*

What this suggests is the Hallsville Hicks family had a prominent presence in the early Chance Manufacturing Company. The other fascinating part of the article is how it ties the electrical talents and expertise of Irl with A. Bishop. What was the impact of this and how did it further influence what was occurring within the walls of the Chance plant?

A March 30, 1928 *Guard* announcement of the death of M. F. Hicks (1855–1928), father of Irl R. from his second marriage, confirms the family's Boone County roots. He was a member of the Christian Church and a well-known area person who as a child witnessed portions of the Centralia Massacre. The July 6 paper reported the death from gastroenteritis of Ella Lee Hicks (1925–1928), the daughter of Mr. and Mrs. Irl R. Hicks. She had one sister Mary Ann and a brother Irl R. Jr. This suggests that Irl had remarried.

Why did Chance Company get involved in automotive products? There is no doubt that Henry Ford, by the late teens, had put the world on notice of the auto's popularity and his company's success. It was a phenomenon rarely seen in the industrialized world. Of course, to the owner of the first auto in Centralia neither was this a strange area for the progressive-minded A. Bishop whose eldest son Buford also became infatuated with the auto. There was a tremendous potential, from a business point of view, especially since the early auto had a healthy appetite for repairs and maintenance parts. Gano spoke of the automotive venture in a 1968 speech:

> *Through the twenties, the little Centralia Company grew slowly and steadily. A foundry was added in the middle of the decade, soon after the Company started on one of what you might call side trips—ventures into unrelated lines of business. A Missourian patented an improved piston ring and my father began to manufacture it. He moved further into the automotive supply business by adding a line of polishes, top dressing and radiator sealers–chemical products. We never made much money in that line–nor any others outside the utility business, but we kept on making piston rings until the start of World War II. (1968)*

L. C. "Roy" Hansen in 1950 period. (Jack Chance Collection)

As a farm boy in Iowa, L. C. "Roy" Hansen, longtime Chance marketing executive and key figure in the Chance employee interviews, witnessed the constant maintenance required on early motorized machines. "I remember when I was a kid living on a farm up north, every summer somebody in that area was grinding the valves or putting in new piston rings in a car. You never see anybody doing that any more." (1979) An employee who worked in the piston ring department in the 1930s at Chance Company, N. A. (Nathan) Toalson, responded.

> *No I guess it's all done in garages now but that used to be a big business. Model T Fords, that was one of the big sizes of piston rings as I remember 3-3/4 diameter x 1/4" thick that went to Western Auto. They only had about three outlets but they were all practically sold under their trade name. I remember them buying cartons and cartons of piston rings. (1979)*

Western Auto stores were in many towns in the Midwest selling everything from bicycles to refrigerators, and of course carrying a full line of automobile parts. At the turn of this century they seemed to have fled the public scene like many other former period business firms.

An additional reason to get involved in automotive products was the extra plant capacity and a quality local work force available to man such an operation. There were also a couple of other reasons: an employee who developed a piston ring and other related products and a salesman who came through town and sold A. Bishop on other auto items. These other auto-related items were purchased to fill the line initiated by piston rings. According to Gano:

> *A fellow was with us by the name of Bud Hicks and he got an idea for a piston ring. So we started making piston rings and at that time some man came through here with a lot of formulas for automotive products and sold it to Dad: valve grinding compounds, radiator cement, polish, tire patch, etc. So he started in the automotive parts business. That was about 1920 and the trademark was the Big C. (1979)*

Shinever Polish ad (*Guard* December 3, 1927)

Gano mentioned that they had kept some Big C plates to which his son Jack responded, "I have a full set of them at the museum." (1979) These were printing or press plates that embodied the Big C trademark symbol.

Piecing this information together reveals a partial picture of what occurred. Irl R. Hicks apparently became known as Bud around the plant. He was an employee as Gano said, but there appears to have been a lapse in his employment. He had already developed several products and was producing the piston rings elsewhere when production was moved to Centralia. Irl apparently wanted to get his business closer to home. It is not clear if production overlapped and how long it continued in other locations. According to Gano, there were people coming through trying to sell various products. It is possible that Hicks developed these auto products while working earlier at Chance. What hasn't been said is the possibility that Chance Manufacturing Company could have become an entirely different firm if the auto division had continued its growth. The prospects were very bright in contrast to comments made concerning this auto activity years later, after it failed because of the stiff competition of established firms with superior marketing/distributor clout. Another important revelation is an early established Chance operating practice of line expansion. For example, in August 1928 or earlier, an association with a Mr. Van Orsdel brought the Jiffy Rim Tool into the fold. Trying to get other auto related items to broaden their offerings, was a philosophy being put into practice by A. Bishop, although he had already applied the full-line approach to his utility products.

By the early 1920s the manufacturing of piston rings at Chance had become a very important part of the company in sales volume and the number of employees dedicated to ring production, although its profit contributions were generally lower than hoped for. It was a highly competitive market in which Chance

An ad for Protexun from the May 14, 1920 Centralia *Guard* appears to be an independent effort by the inventor of the "Gas-Tite" piston ring, Irl R. Hicks. It is promoting another of several products he developed. Irl was already a Chance Manufacturing Company employee whose piston ring design became the impetus for the establishment of the Chance Big C automotive line. *(Guard)*

was pitted against producers across the country. Part of the plant expansion and need for such processes as a foundry operations, was driven by piston ring needs for lower cost in-house production, although the Chance utility products were also being benefited. In early December 1923 there was a plant shutdown for a complete overhaul and upgrading of the factory facilities. Only about ten employees were left working in the office and factory. The *Guard* suggested the plant would open up the first of the coming year with the capacity for a larger work force. There is a continuing question of what the true business relationship was between Hicks and Chance. This unknown situation becomes even more unclear with a December 21, 1923 *Guard* article entitled "The Piston Ring Merger." After all the business upheavals witnessed in Centralia over the last two decades, this article sort of reminded one of such turmoil. A. Bishop is now a spokesman shaping the destiny of the Hicks' Gas-Tite piston rings as "he informed the *Guard* that a merger has been effected of the piston ring manufactures of the leading industries centers of California and the Eastern states." This article revealed that the piston ring department at Chance had been shut down recently, "owing to the fluctuations in prices of all foundry supplies." This may suggest how close a profit margin there was on piston rings. A comment about the continuance of company utility products production "owned solely" by Mr. Chance, helps clarify that he wasn't in 100 percent control of piston rings as he was with his anchors and other utility items. "The "Never-Creep" anchor and the manufactured electrical supplies department of which Mr. Chance is the sole owner, are still running at full capacity." The new merged company would become, "the C-K Company of San Francisco, California, Cleveland, Ohio, and other large cities of the East." The new management hadn't decided that ring and other related product production would remain in Centralia, but A. Bishop said he was trying to positively influence that. The last paragraph is puzzling in describing what might occur, for example, if production remained in Centralia, manufacturing activities would begin early in the new year of 1924. Even in view of this questionable start there was already a search for a new superintendent. There appears to have been personnel difficulties that generated problems as had high raw material and manufacturing costs.

Buford Chance would become involved in the auto products in the mid to late 1920s, before his untimely death. Bill Hollander talked about what he had been told about Buford and also revealed further details of the Chance auto line.

> *Yes, Buford at the time he rolled his car, he was the manager of what the folks up here called the Chemical Building. It was the automotive supply building. He had charge of the piston rings and they had bought a whole bunch of odds and ends for automobiles. There were boots, tube patches, pills you dropped in your gasoline tank that was supposed to stretch your gas mileage and a gauge to hang on a Model T Ford so you did not have to lift the seat cushion to stick the gas to see how much gas you had. Oh, some little outfits that were called spark intensifiers and a whole bunch of little odds and ends that were used on the old Model T and the earlier automobiles. It was the Automotive Division but they used to mix up oil enamel which was about the same thing as black tar, but it never did dry, it just stayed sticky. They also got rubber cement in barrel lots and they put it in small tubes for tire patches and the like. The people up here decided they were chemicals so they always called it the Chemical Department but it was the automotive supply and that was what Buford worked with. (1979)*

The oil enamel or black tar Bill was referring to was a substance used for top dressing. As the automobile evolved from the early open to closed all-weather bodies, they continued to make the top of the bodies out of fabric, wood spars, and screen wire. It wasn't until the 1930s that most closed cars got a steel roof or what marketing people labeled as a "steel turret" top or roof. In actuality, it was simply the use of sheet metal to "skin" the exterior surface of the top of the body. *Hemmings Classic Car Magazine* of June 2005, doing a profile on Chile Harold Wills who held the patent on the all steel body until 1937, suggested that Ford/G.M. held out until then to avoid royalty payments. Prior to this, the tar paper tops would dry out from the sun and would crack and leak. A quick fix was to use a tar-type substance that would freely flow into cracks and crevices and seal it up, much like a building roof could be repaired. The other items mentioned, including tire patches, were also important to carry in the car as a flat per mile was not an abnormal situation with the fragility of early pneumatic tires and the bad roads. Some of the items mentioned are what is sometimes labeled as "mouse milk," which are items with questionable benefits. Other items were accessories the manufacturer left off, but most of the Big C line was maintenance and repair products. Thus, early car owners became mechanics of sorts and with each automobile sold came a demand for a package of repair items, many of which were carried in the car to handle on the road breakdowns. As a car owner, A. Bishop realized this and saw the potential of the automotive area and apparently after hearing Irl R. Hicks' sales pitch, decisively entered it.

The "Chem" building, as later employees called it, was a two-story brick structure with a basement, that at one time sat by itself on the east side of Collier Street, between the tracks. The brickwork for the construction of the Chance Chemical building was begun in late April of 1925. When Collier Street was closed with the 1950s expansion of the Drop Forge Department, the Chemical building became

Chemical Building (*Chanco News* June 1957)

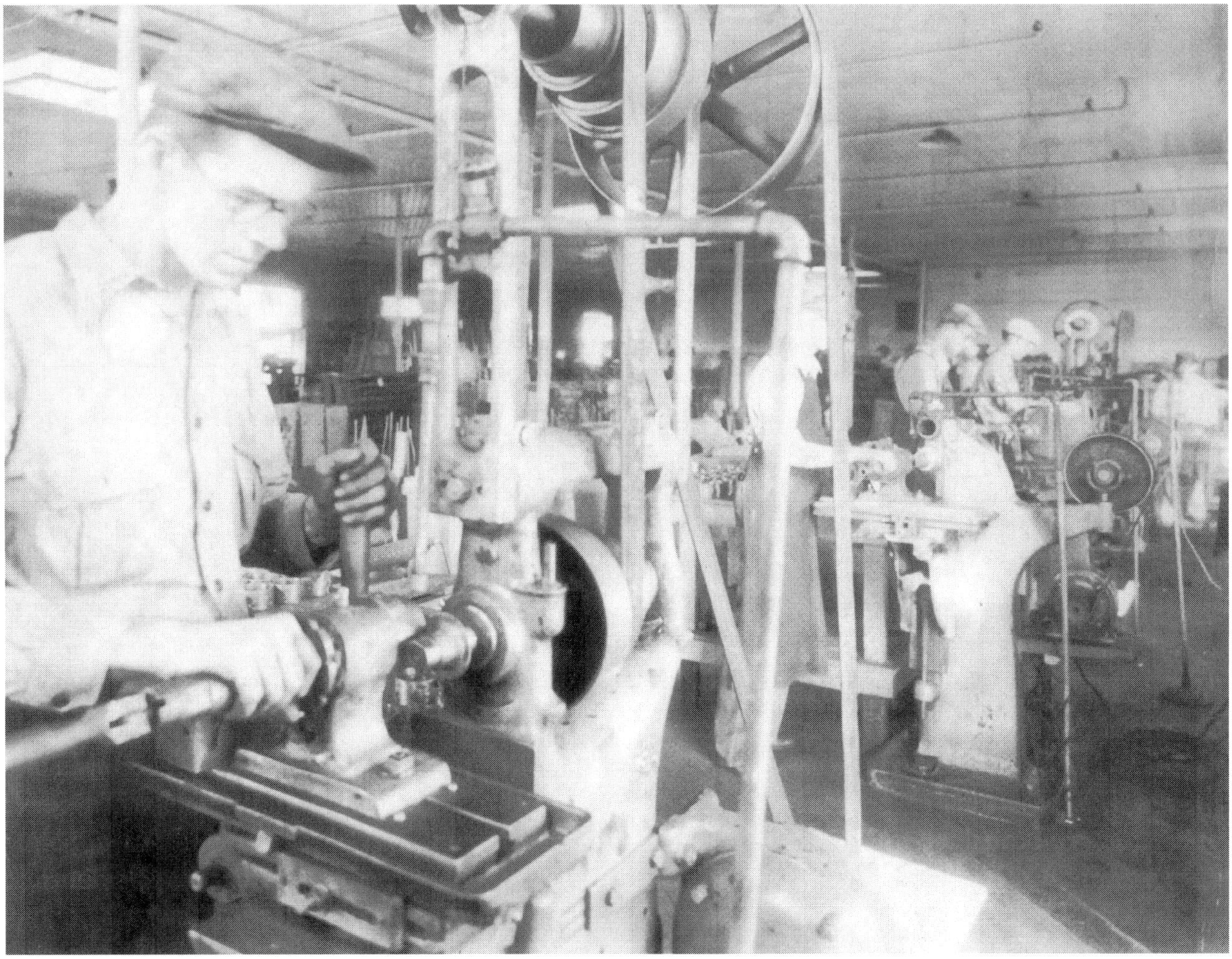

No. 27 Operation 11: Employees of the Big C department at Chance Company are performing various operations in developing a finished product that could end up on a Western Auto retail store shelf or shipped express overnight to some local garage. (Centralia Historical Society)

its southeast corner, although the Drop Forge building narrowed and extended on further east on the north side. The ground-level floor of the Chemical building was used to house machines for several decades and the upper floor was used as a storage area for overflowing office records. According to longtime Maintenance Supervisor Don Smith, it also had a basement with an elevator. Engineering or an operating part of that department was also located there prior to the completion of the new Research Center in 1961. Numerous trips were made to check out travel expense reports of Sales and Marketing, although it was a more common, but not aggressively pursued hangout for the Accounting Department. There were rows of filing cabinets filled with older records and correspondence. It was said that Don Miller was charged with the responsibility of destroying older records before the second sale of the company, but dated, selected materials were routinely destroyed because the volume would become overwhelming. Because of its stark environment and the pounding vibration from the big hammers around the Drop Forge, trips to it were kept infrequent, short and to the point. It was torn down in the mid-1990s when the southeast end of the Drop Forge building was again rebuilt and expanded.

Sales and Marketing efforts in the early years of the company were directed by A. Bishop Chance. Promotion of one's products is a challenging part of any successful business. Once a company becomes established and known in its business circle, it can become easier to produce sales if

Gentlemen:-

The car owner of today is "motor wise." Show him a piston
ring that is a positive performer -- that absolutely and permanently
prevents the escape of gas -- and he'll buy it without argument. Thousands
of dealers are proving this every day with GAS-TITE PISTON RINGS.

Yes, sir! GAS-TITE dealers are winning the lion's share of
the piston ring business in their communities -- getting this profitable
trade speedily and easily -- because the superiority of the GAS-TITE
PISTON RING isn't a matter of experiment -- it's been proved and
recognized.

Here's a piston ring with a score of talking points -- and one
that will back up your promises with actual performance. And our
guarantee protects both you and your customer absolutely. Every set
you sell means a booster for GAS-TITE PISTON RINGS and for you --
brings you business you don't have to spend a cent to get.

When you take on the sale of GAS-TITE PISTON RINGS, we'll
advertise your name persistently direct to every man in your local-
ity who owns a motor car. You give us a list of the names you want
to reach. We'll do the rest -- keep sending them persuasive liter-
ature, bearing your name, until they come in and buy. We'll
bring the business to you -- all you've got to do is hand out
the GAS-TITE PISTON RINGS and pocket the profits.

Isn't that a proposition worth investigating? Mail
the handy card for all the particulars. It's the key to
profits -- so do it now.

Yours very truly,

CHANCE COMPANY

A. Bishop Chance

A promotional piece that could be directed at garages, dealers, and distributors (customers) owed its existence to the fact its backside was reused in some patent correspondence dated December 19, 1927. So it was made-up and utilized before that date. The 1926–27 time frame is reasonable. After reading its flowery word and phrase selection, one can assume it is the work of a professional salesman. It could be the efforts of the new director of sales of the Piston Ring Department, Frank H. Dukesmith, who was employed on February 5, 1926. This theory is not offered to suggest A. Bishop was not capable of such material, as examples of his composition skills truly reveal. (A. B. Chance Co.)

the products and company generate a positive image. Trying to break into new markets and be price competitive can be difficult, as A. B. Chance Company later found, for example, in promoting small anchors, house trailer anchor kits, and foundation anchors. The automotive area of Chance Company was such an orphan, but for two decades it was a big volume stepchild. The customer base was entirely different than the utility market where A. Bishop had for decades concentrated so much of his inventive genius and sales promotion efforts. When an opportunity to employ a sales expert presented itself, the Founder surmised that to allow such a person to take over the automotive marketing end of the business would be a wise and beneficial move. No one within the company had apparently reached a level of competency that he approved of and he didn't have the personal time to commit, as his utility products and sales were expanding, among his many other activities.

As if to suggest that the piston ring area was totally revamped after the recent reorganization, the *Guard* on February 5, 1926, announced the Chance Manufacturing Company employment of a Frank H. Dukesmith as director of sales of the Piston Ring Department. He was charged with organizing a national sales force to sell the rings direct to car owners. He was first going to work through Missouri, Arkansas, and Texas, organizing state crews in the cities he visited. He had come to Centralia some three months earlier and had a local city directory published, which had gained local notoriety for him. In spite of his heavy traveling schedule in his new position, he would make his headquarters and home in Centralia. This article explained his background as beginning the first school of salesmanship known as the Chicago School of Salesmanship. He believed that salesmanship could be taught and he had sold his institution to A. F. Sheldon. He began a magazine called *Salesmanship* that he also sold. He had various other industrial experiences that he brought to his new job, but salesmanship was his focus and subject of study. The association of Frank Dukesmith with Chance Company is significant. One can sense that A. Bishop was all eyes and ears, learning any and all sales secrets that his new sales director could offer. It was a longtime practice at Chance Company to learn from those brought in from the outside. Dukesmith may have been the first professional sales manager on the premises. The impact of his Chance tour is difficult to assess at this point, but undoubtedly an accumulated body of knowledge, procedures, and practices were handed down developing a more sophisticated sales program at Chance Company. So it's with caution that credit be given to later officials of progress or the origination of certain company programs, when it instead may have been influenced or begun by A. Bishop and those like Frank Dukesmith and Jack Isbell from the outside who were employed at Chances.

The brand of promotion tactics reflected by Chance Manufacturing Company's new piston ring sales czar was captured by the *Guard* editor in a February 1926 article: "The Chance Piston Ring has exceptional qualities. It is a perfect circle due to its design and also to the manner of manufacture. It will not pump oil in its operation and has other good features. It has a great future and needs only the proper selling force to make it go on the market." Time would reveal if Mr. Dukesmith's textbook type posture would be successful in the real business world.

The industry's approach to manufacturing piston rings varied and A. Bishop wasn't satisfied with the methods being currently utilized, as he illuminated in his patent filing of April 12, 1928. This clearly suggested he was unhappy with the current system in the Big C plant. His new proposal was "a novel method of manufacture of split rings and novel apparatus for carrying out the method." His approach would provide a "more simple method of manufacture whereby split rings of more uniform quality and size may be manufactured more economically and more quickly." This new method would help address the Big C pricing and competitive problems by speeding up ring production while adding to the quality of the final product. The patent application was described as a "Method and Apparatus for Manufacture of Split Rings." By January 1929, certain Chance patent claims had generated conflicts with Johnson, Bartlett, and other patents pending in the piston ring manufacturing area. There was a Chance decision made to not file for foreign applications. One ongoing question between the patent examiner in Washington, D.C., and the Chance patent attorneys, Williams, Bradbury, McCaleb and Hinkle of Chicago (W.B.M.&H.), was that the Johnson system did not hold both sides of the ring so it came out noncircular or noncylindrical as only one side could be machined properly while the other side was untrue. The Chance method claimed to hold the ring in position so both sides could be machined true, making it fit the cylinder more perfectly.

A letter of April 7, 1930 from W.B.M.& H. to Chance Company confirmed that Chance Case 8 had been formally allowed with two claims. Still awaiting a patent, Mr. Johnson, the president of the Piston Ring Company of Muskegon Heights, Michigan, had his attorneys communicate that there was interference between his 1923 Patent No. 1,471,030 and the Chance filing. Legal action was suggested. A. Bishop was now intruding upon the "big boys" in Detroit and this likely wouldn't go down easily. Mr. Chance's attorneys saw the Chance claim as very specific and saw no Johnson interference and suggested they were bluffing and may have even progressed to the Chance method, but had failed to cover it legally. Johnson's patents consisted of thirty-nine assignments to his company and his coverage was broad, but the Chance claim had a lock on a specific part of ring manufacturing. They suggested a compromise might be reached in offsetting license agreements to cover both areas. Mr. Chance was actually using the Edelmann system which he considered superior and even had them manufacture some ring equipment to apparently make rings incorporating the new Chance ideas, but he suggested that working out the Johnson situation would be beneficial. Johnson correspondence stated they weren't interested and would start a proceeding like one they currently had in court. In the meantime, Patent No. 1,780,003 was issued to Albert B. Chance on October 28, 1930. Three years later the *Johnson vs. Chance* suit had been pending for two years, suggesting Johnson didn't have a strong case. Dismissal was discussed with a $50 charge for legal services to be the defenders burden. All agreed and the case was closed in Jefferson City, Missouri. The Johnson group stated new methods had been established and they weren't interested in the Chance system. This could reflect the quick progress in ring manufacturing technology or that Mr. Chance's lawyers and the Centralia Wizard had called their bluff!

Right in the middle of boom times in mid-1928, an unexpected event occurred: a layoff! This wasn't a headline event in the *Guard* of May 11, but followed in the article of another injury accident at the plant. The Molding and Piston Ring Department of Chance Company was shut down with twenty-two employees laid off, adding further to the confusion in understanding the piston ring arrangement at Chance. Apparently an eastern concern could make the "Red Snapper" rings cheaper than the Centralia operation, with its "present equipment." However, these Big C rings would remain on the market. The comment about machinery may suggest that other companies had invested in more efficient and productive machinery thus reducing their costs to the point that producers like the Big C Line couldn't remain profitable and competitive. The just discussed changes in manufacturing the piston rings may have led to the shut-down to change over to the new methods and the use of the new name "Red Snapper" may suggest that.

As the month of July 1928 closed, there was an announcement of a record customer order for a million piston rings. Foreman [*sic*] Bryan Fenton was in Chicago preparing in some way for the filling of this order. The Drop Forge Department was closed down temporarily to await his return, likely in preparation for this huge order as the July 27 *Guard* article suggested "filling this order as rapidly as possible." This likely meant the normal long ten- or twelve-hour day, six-days-a-week work schedule would be expanded even more, as one of the conditions of getting the order apparently was a quick delivery. This up-and down atmosphere in the piston ring area seemed to be a characteristic of this business, or of the Chance auto operations. A final note said the Drop Forge Department was starting up again to make the new Jiffy tire tool, as the cutting dies had been completed, which may explain its role in this large order as rings were cast in the Foundry.

Automotive technology advancements were coming at a quickened pace if the changes in the Big C piston rings were any indicator. A September 21, 1928 *Guard* article announced a new type of ring being made at the Chance plant which had small vents through which lubrication oil could flow. This became a standard feature of later rings for reductions in wear and heat. It appeared to be more of an improvement to existing rings than an entirely new design. An October news item called it the "ventilating ring" and it was reported being demanded by customers in growing volume.

Automotive piston rings became an important part of Chance Company, volume wise, which kept a lot of people employed, especially during the Great Depression. It was not a big moneymaker because of the tough competitive climate surrounding it, distribution problems,

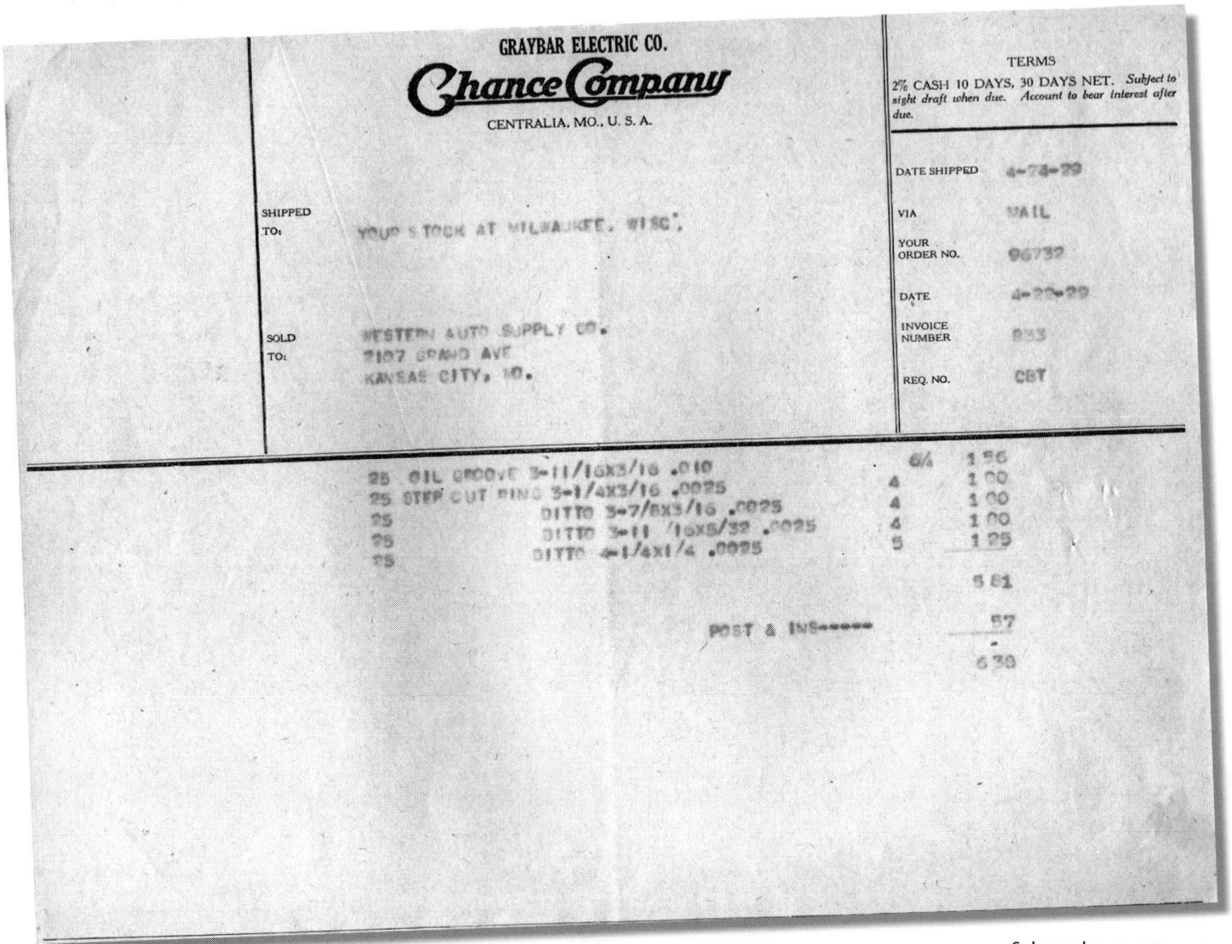
GRAYBAR ELECTRIC CO.

Chance Company

CENTRALIA, MO., U. S. A.

TERMS

2% CASH 10 DAYS, 30 DAYS NET. *Subject to sight draft when due. Account to bear interest after due.*

SHIPPED TO: YOUR STOCK AT MILWAUKEE, WISC.

SOLD TO: WESTERN AUTO SUPPLY CO.
2107 GRAND AVE
KANSAS CITY, MO.

DATE SHIPPED 4-24-29

VIA MAIL

YOUR ORDER NO. 96732

DATE 4-22-29

INVOICE NUMBER 833

REQ. NO. CBT

25	OIL GROOVE 3-11/16X3/16 .010	6¼	1 56
25	STEP CUT RING 3-1/4X3/16 .0025	4	1 00
25	DITTO 3-7/8X3/16 .0025	4	1 00
25	DITTO 3-11/16X5/32 .0025	4	1 00
25	DITTO 4-1/4X1/4 .0025	5	1 25
			5 81
	POST & INS-----		57
			6 38

Sales order (A. B. Chance Co.)

and the other big players in that market. As Bill Hollander recalled from post-1929 times: "At that time we made piston rings under our own name called Big C lines. We made them for Western Auto, J & R, and several other different auto suppliers." Making piston rings for other distributors under their name, is not unlike the manufacturing of anchors for Joslyn. As the keeper of anchor inventory and the processor of customer orders, the packaging of Chance expanding anchors into Joslyn boxes is recalled. Instead of the Chance "C" numbers (C88115) we labeled them "J" numbers (J88115, etc.). Sears and Wal-Mart have personified this situation best by making nothing, but selling everything.

The reuse of surplus letter backs was revealing of how ring orders were processed. An April 24, 1929 shipping date on an order from Western Auto Supply Company of Kansas City, Missouri, shown above, suggests their headquarters may have been in Kansas City. If so, this is another reason for A. Bishop's travels there besides the idea he liked the site of the stepping-off point of America's great western trails. This particular order was being shipped direct to Western Auto's Milwaukee warehouse. A similar order to another customer was filled "from KC Stock." This apparently was an important stocking center for Chance rings, but it may suggest there were other stocking locations. Further details show the quantity, type (oil, compression), the dimensions, the oversize (.010–ten thousands, etc.), and the price. Twenty-five rings for $1.56 was a good deal in comparison to today's prices. The abbreviated name, Chance Company, appears to be officially used in the late 1920s, early 1930s on company documents, including patent papers. The reference to electrical distributor, Graybar Electric Company, appears to show a close Chance relationship as early as 1929.

Many Chance managers and executives, such as Charles P. "Cab" Atkins and N. A. (Nathan) Toalson were employed in the Piston Ring Department and many employees got started there. Nathan Toalson told a story involving piston rings and A. Bishop's ability to teach employees better work habits without uttering a word. This shows not only his understanding of people and their psychological makeup, but it also displays a seriousness he felt about the work being done in the plant. Not only did he expect employees to work hard in an organized atmosphere, but to work smart as well.

> *Well, I remember one thing, when I was working in the Piston Ring Department on the third floor of the present office building. Piston rings would have a habit of getting on the floor and didn't anybody pick them up. You would drop one and put it in the machine and drop one and there were unfinished piston rings laying on the floor around every machine. I remember one day, he [Mr. A. Bishop] came up on the third floor, walking around inspecting and looking. He had a whole pocket full of dimes and he started throwing dimes around on the floor. Everybody started looking at him and asked what is going on. Well, what he was doing was telling everybody that those piston rings laying around on the floor was just like leaving a dime laying on the floor. (1979)*

Roy West, who began his Chance work one summer in the mid-1930s in the Piston Ring Department, mentioned an interesting prelude to World War II when the Japanese were preparing for war in the late 1930s–1940s. "I've always heard that Japan bought quite a sizable order of piston rings. Western Auto pretty well claimed the Midwest, with all of those stores. Along in the end, they [Chance] had a pretty good ring, a pre-fit ring." (1979) There was additional information about the Japanese connection from F. Gano Chance, surprisingly in a discussion about the world tour that the Chances' took in the late 1920s. From Honolulu, the ship set sail for Japan:

> *. . . to Yokohoma, that is the seaport of Tokyo. I took movies and I was signed up with Fox News to send them the negatives but they wouldn't give me a copy so I quit taking pictures for Fox News. Coming into Tokyo, I was taking pictures through the porthole of the harbor and when World War II came along I advised the government. They were asking for pictures of foreign countries and I sent them and they looked them over and made copies. I was very much impressed with the Japanese. The Chance Company was making piston rings and we had a piston ring distributor over there, a Japanese. He was so nice and he met us when the ship came in. He took the three of us around over the city and explained to us about the city. We were very much impressed with Tokyo. The people were very friendly. The University was there and students became personal guides for our students. There was one Japanese student whose name was Sudiki, which is about as common in Japan as Smith is in this country. For three days he came down to the ship and went with us. When we left to go to Osaki by ship, we had become so acquainted that he cried and was in tears when we left. So we found the Japanese people wonderful, very clean people, and little things meant a great deal to them. I enjoyed that country more than any country we visited because of the people. The Japanese are wonderful people and when we got into World War II it was my opinion it was a case of high politics, not the will of the people but a few powerful people decided to bring on the war. (1981)*

Jack Curtright's perspective on piston rings was enriched by a stint with GM and their Chevrolet Division. He said he was working at the H. A. Jones Motor Company, the Chevrolet garage in town, and that is how he first became familiar with piston rings and got a job with Chevrolet in St. Louis. Jack suggested that the production of piston rings was a catalyst for the eventual Chance Company involvement in Cone Anchors.

Nathan Toalson began work in 1930 sorting through and organizing the items that the Chances had accumulated on their around the world cruise. He later was sent to work in the Chance Lab after working alongside Paul Sturgeon in the Piston Ring Department. Both were running milling machines. The Lab's primary function was, in Nathan's words, "Analyzing the iron every day. In other words, the pull irons today would be analyzed the next day to see whether or not you made the right formulation." (1979) This iron was what Nathan called, Gray iron, primarily for piston rings and the mixture of the iron had to be pretty well controlled. Because Chance automotive products originated a decade or so before Nathan came upon the scene, he didn't recall its origin within Chance Company.

> *Well, they were in the piston ring business well before 1932 but I don't know how they got into it. As I recall, Buford was more or less in charge of all of the automotive operations and they had gotten into tire patches and a lot of other things besides the piston ring for automotive trade. The so-called, Chemical Building, was pictured as the old building that is now the tail end of the forge shop. It was a three storied brick building that is still there* [in 1979]. *The forge shop has been built around it but that was what they called the Chemical Building for years. That is where they had all of the tire patching compounds and other items for the automotive trade. The piston rings were never down there. They were manufactured first in the foundry and then on the second and third floors of the present Office Building. The third floor was where they machine piston rings and the second floor was where they boxed, dipped them in oil, and boxed them for shipping. (1979)*

When asked when the laboratory was constructed, Nathan responded:

> *That was about 1940 and I think it was completed in 1941 or 42. The* [piston ring] *machinery was sold. Some of it was sold to McGraw Morrison in St. Louis and some of it was sold to Ramsey in St. Louis. We got out of the piston ring business and sold them the inventory, etc. We didn't sell it as a business but just got out of it. (1979)*

The chemical lab was established on the second floor of the main plant in 1928. It is assumed this is what Nathan is referring to, although an electrical type lab was established on the third floor after the Tips Tool purchase in 1937. Nathan's comments about the sell off of the Big C line suggested it was a piece-meal type of sale and understandable in view of the lack of viability that A.B. Chance Company represented at that time in the highly competitive auto replacement business. A veteran employee, Les Reynolds remembered the piston-ring sale, as one or more involved in the Chance operation were employed by the new owners in St. Louis. Incidentally, his daughter Mary Dell Young was also a longtime Chance employee.

> *When they moved the ring business to St. Louis, Herman Sent, he went down and came back with his boss and wanted me to go down there. I said, I have two kids and that is no place to raise kids in St. Louis. I am just going to stay up here and I have always been mighty glad I stayed. (1979)*

Although Nathan didn't mention his involvement in the sale of the Chance piston ring business, his knowledge of whom it was sold to is a clue. When asked about the Chance piston ring business, F. Gano Chance's sharp recall again comes into focus and gives an overview of the activity in that area over the years and details of their departure from it.

> *When we moved from our location over at the City Square to the brick building* [210 N. Allen] *we had the second floor which we still couldn't use. We rented that out to a fellow by the name of Baker who was in the chemical materials business for the poultry business. The first floor had this little machine shop and we had storage for anchors. We had a fellow working for us who got a patent on a piston ring and it was called the Gas Tite and it looked like a good possibility. So we started making piston rings and we had not advertised but we were making contact with companies like Western Auto, Oklahoma Power and Supply, etc. So my father was selling piston rings and we got into the manufacturing of piston rings and found, however, that the original piston ring wasn't what we thought it was and it was difficult to make. So we changed the design. We went along until about the time of World War II, about 1940, in which we had developed a company and sales of several million piston rings mostly to Western Auto. Of course the problem was, most of the business was going to one company who owned the name because we put their name on it. So all they had to do was go to another piston ring company and ask, "can you make this and what would you change?" So it wasn't too stable a situation and so it looked as though it was the wrong business to be in unless we would really go out and sell it under our name. So we called an advertising agency and started pushing the piston rings under our own name. We came to the conclusion that we were just a little bit too late in the piston ring business. There were quite a few others that did a good job of selling. So I said to Nathan, he was with us then, "Nathan, I think we had better get rid of that company and I think a company in St. Louis might be interested." So he made the contacts and worked out the sale and I went to Arizona. When I came back we were out of the piston ring business. I used to laugh when somebody would ask, "what business are you in Mr. Chance?" I would say, "when I was home last, we were in the anchor business and so I am not sure. I will have to check." Nathan did a nice job and got us out of the piston ring business. (1978)*

The minutes of an A. B. Chance Company Board meeting explains what occurred. Apparently the total proceeds for all parts of the business was $50,000. "At a meeting in January, 1941, it was agreed to sell the piston ring business including the inventory, machinery, etc. for $50,000." (1941)

Rediscovering Mr. A. Bishop Chance

It is important historically and satisfying to reveal material not before released or that which may never again be available, but some of the best material is that made public so long ago that it has been forgotten or is unknown by later generations.

On July 30, 1998, rainy weather postponed the usual noonday rounds of hiking west up Wilson Street or south down Hickman to loosen muscle and joints tightened by another four-hour round of "deskitis." Recalling the special material entrusted with the day before and anxiously sorting through it into the late hours of the previous evening, a new adventure emerged. One discovery concerned the first-aid door area. My tenure was sufficient to know the location of this door and where office employees had once parked and entered on the south side of the main office building. However, the expansion of the warehouse in 1965 had changed all of that.

From the inside of the main office building at 210 North Allen, the old first-aid doorway no longer existed. But from the warehouse side the stark reality of bygone days could still be viewed although some south windows and other openings had been bricked in. Circling back through the office and then south into the warehouse, one is immediately confronted by the roar of forklifts busily running back and forth with material destined for some faraway place. This could have been the same scene thirty-three years ago, little seemingly had changed. Galvanized anchor rods would soon feel the stress of tension in holding a power line vertical in some remote location, but this wasn't new as it was A. Bishop Chance's idea back in 1912 on telephone lines. Of course Walter Turpin now occupied the foreman's hut instead of Roy West or Ollie Chandler and the faces of the forklift drivers had changed but the continual rush to get material out the door to customers worldwide remained the same.

Comforted by this, it became more promising that the object of the search was intact. Focusing on the tall brick wall, once exposed but now under roof, along with the detail of original windows and doors, generated sort of an eerie feeling, not unlike what one feels when a picture of the sunken mass of a *Titanic* or *Yorktown* are viewed. Inching westward closer to the corner where a new wall now hid vehicles passing near by on Allen Street, metal racks stacked with anchor rods and extensions had to be overcome. Getting closer to the door area, it now appeared surprisingly intact. It was too dirty and lifeless to visualize one's entrance through it today, but it had survived. Although the story could end here, the door wasn't the main object, only a roadmap indicating the whereabouts of a stone with the inscription of 1917.

Searching immediately around the door frame brought nothing of significance but moving upward, a small stone block appeared with 1917 carved in it. The significance of this date block is it notes this building's construction date. This was a common practice on brick buildings of this era, in carrying their building date high on their fronts or on their sides, in this instance. Although this is important, it's even more valuable to know that A. Bishop was the inventor turned temporary stonemason, who cut the year into that stone with hammer and stone chisel eighty-one years ago! Although the future isn't always kind to the past, the integrity of the small stone block, a monument of sorts to a man considered a hero in his time, is now sheltered from the elements. At the same time it also has been hidden from public view and there may come a time when its survival is threatened by obscurity or viewed as nothing of worth by current owners. May future generations be vigilant in preserving such valuable artifacts and let this be a reminder that the torch of responsibility has passed to them.

Years earlier the Centralia mayor had turned his attention to the architectural area. The June 17, 1910 *Guard* had this to say: "Mayor A. B. Chance is an architect of no mean ability and this week he had an opportunity to exercise his ability along this line when it came to running the circular walks around the city hall and making the job symmetrical. How well he succeeded one can see when the forms are taken away from the walks."

Another surviving artifact of a more fragile nature seldom generated today in our computerized age, was temporarily entrusted to my care and examination. In this case there is built-in credibility because it was an article dated March 1, 1931, from a quality and widely circulated publication of that time, *Forbes* magazine. This article is also valuable because it looked at Chance Company development from its beginning up to 1931 and the many other activities of A. Bishop Chance. Since its in the first-person perspective of the founder, it becomes a keeper worthy of being reilluminated especially since it provides a personal glimpse of Mr. Chance by one outside of Centralia. Although it takes place a few years beyond our current chapter, "A Chance-Built Industry" captures the "Growth and Expansion" theme of the mid-1920s Chance Company.

"A Chance-Built Industry"
by Thomas W. Parry, Jr.

One of the most severe blizzards in a decade set in about an hour before the main line flyer arrived in the little city of Centralia, Missouri. I had to change to the branch line there for the twenty-mile trip to my destination, but the conductor told me the train would be at least two hours late. So I walked across the street to the Century Café and asked the waitress for a cup of coffee and some rolls. A cursory glance of the surroundings assured me there was little to be seen in Centralia, so I opened a magazine and leisurely sipped my coffee.

Presently a wiry man whose keen eyes literally danced as he spoke walked in, ordered the plate luncheon and was gone again before I had scarcely realized his presence.

"That Man," I commented to the proprietor, "seems to be in a great hurry about something."

"That Man," replied the proprietor, "is always in a hurry. That is A. Bishop Chance. His factory just beyond the railroad station is about the biggest thing this town has ever seen. If you're waiting for the local to pull out, you might spend a profitable hour over there."

And that is how I happened to meet Bishop Chance and to learn the story of his patent anchor and automotive industry—a business which he has elevated from a tiny machine shop in an abandoned church building to an industry covering four square blocks and doing a business that can well be envied by companies far greater in size.

After being directed to the printing plant, the galvanizing plant and half a dozen other departments, I finally found Mr. Chance in the pattern shop. Within two hours, during which time I must have walked five miles following him about, he had given me the story of his career.

The Chance Company ships its electric and automotive products throughout the United States and to many foreign countries and its patented "Never-Creep®" anchors are utilized by the largest telephone and telegraph companies in every corner of the country. It was that anchor which finally started this inventor-industrialist on the way to success. And success came after a series of setbacks which might have proved the undoing of a man less determined.

At the age of seventeen Chance, after leaving the Centralia High School, secured a job in a drug store. "After six months," he told me, "when I had received one large orange as salary, I decided to seek other work. There followed two years' employment in a clothing store, and then I procured a job as night clerk in the Globe Hotel. At night when everyone had left the lobby I had an opportunity to study numerous volumes on electricity, and shortly I had perfected several devices."

"My first invention, at the age of eighteen, was a bicycle handle bar. I sent it to a manufacturing concern in Chicago, but repeated letters failed to bring a reply. Several months later I learned the company had obtained a patent and placed the product on the market."

"But," he digressed, "you don't want to hear about my life; you came to make a trip through the factory."

"I'd rather hear the story of your career," I urged him.

"Very well," he acquiesced, "but I'd much prefer to occupy the background in this picture. I do not deserve so much credit for building this business. The thing simply grew."

"Did you invent anything else after the unfortunate handle bar episode?" I suggested.

"Yes," he replied, "and I lost no time obtaining a patent on it. It was a hotel signal-calling device, but it proved too advanced for the times, for there were so few telephone exchanges then it did not pay to place this invention on the market. It was several years later that I installed the first telephone in Centralia and shortly after that, at the age of twenty-four, I made my first money in the telephone business. I bought up parts from various manufacturers and made telephones and installed the first switchboard and lot of "phones in Centralia."

"Not until 1907, however, did I launch the business that eventually grew into the factory you see here now. It was a small manufacturing shop in a discarded frame church structure. There I manufactured and placed on the market about fifteen inventions, all of them pole line hardware, including telephone lightning arresters, arm trusses and similar articles. Because of lack of funds with which to place them on the market adequately, I met with only partial success with these inventions."

"My first big order came from the Canadian government for six thousand lightning arresters. Incidentally, this customer is still buying the same product from us. During the winter of 1912 the ice first broke for me and certainly necessity was then the mother of invention. A severe sleet storm had mangled the telephone system of Centralia and the surrounding community. There was imminent danger of all the heavy leads going down. It was necessary to take some sort of preventive measure immediately. My father urged me to telegraph a metropolitan concern for some patent anchors, but I believed the time was too limited and doubted if those anchors would hold the heavy load."

"Throughout that night I remained in my workshop, designing a device which I hoped would hold the load. The following morning, not without misgiving, I asked my men to make up some rough anchors from wooden slugs, tying wire to them. The idea worked so well that I obtained a patent. The result is the 'never creep' anchor which we are shipping throughout the United States today. They are exported regularly to Mexico, Canada, Cuba and many other foreign countries. The anchor is an appliance to hold telephone and telegraph lines, derricks, silos and fences in place."

The principle of this anchor is ridiculously simple, yet it remained to Chance to develop it. It is simply this; you bore a hole, drive the rod into the hole, pass the plate down into the hole and hang it over the rod and the work is done. The pull is entirely against solid earth; all of the disturbed earth is behind the anchor plate.

"Since then," Mr. Chance continued, "the business has grown continually, and the anchors no longer constitute our only stock in trade. Sales offices have been established in New York, Pittsburgh, Chicago, Atlanta, St. Louis, Kansas City and Oakland."

Today the Chance plant is one of the largest of its kind in the country. It is complete and modern as any industrial giant in a metropolitan area. When the automotive and electric specialties had increased to more than two score products, it was evident that factory additions must be made and Mr. Chance put his profits back into the business until he had a well-rounded plant. He has gone so far as to install a printing plant where all labels and cartons, as well as advertising material, are handled.

The gray iron and semi-steel and nonferrous foundries represent the ultimate in modernity. So do the drop-forging, the punch press and stamping departments, the galvanizing plant, machine shops and every other department of this growing enterprise in a city of less than 3,000 persons. The entire plant is driven by more than 300 individual electric motors.

"This plant," Mr. Chance explained, "is small in comparison to many factories in the United States, for it covers only about four square blocks. But those who have visited it have been of the opinion that it is one of the most complete of its kind to be found. Our business now extends virtually throughout the world. Chance piston rings are used in automobiles in many countries. Japan, Java, India and the Straits Settlements are exceedingly good markets."

Mr Chance's constant use of "we" or "our" in referring to the management of the plant prompted me to inquire of the other owners or directors of the Chance Company. I learned that there are no directors, no other owners. This is a one-man business.
"To what do you attribute your success," I asked.

"We have endeavored to supply fairly the needs of the persons we serve," Mr. Chance replied. "It is our policy to better the quality of our products and to sell them at a price that profits the user. We believe—and we teach our employees to believe—that we have no claim to existence unless we can produce better goods at a lower price or at least, equal quality goods at a lower price than our competitors."

It took imagination to build the business of Chance, to visualize the potentialities which this rural community held. It took determination which could not be dimmed. It took courage to face a host of discouraging bankers who insisted that their notes be paid. It took understanding to face those same bankers with a smile after they had flatly refused to carry him longer.

Such a man is A. Bishop Chance. He was forced to switch his account to three different banks before he finally found one that would carry him long enough to give him an opportunity to develop his business. Until four years ago, he had not indulged in a vacation of any duration. With his family, he toured Europe for nearly a year, but since then he has hardly been away from Centralia.

At fifty-six, Bishop Chance has lost none of his youthful imagination. He is forever looking forward, seeking new markets and better products. His love for his business, which brings him to his office at 7 o'clock every morning, has gathered momentum with the years. He will perpetually be fired with all the romantic vigor of a Beowulf—that is the secret of his success. (This Chance Company source was reprinted through the courtesy of Forbes magazine, March 1, 1931.)

Delays, interruptions, and detours in our lives are not to be wasted and many so-called successful people have learned to reap where there appears to be no grain to harvest. The railroad brought such opportunities to Centralia. Another person en route to Columbia via the Wabash's Columbia branch line also encountered a delay, but being the resourceful and opportunistic man that Jack Isbell was, he was rewarded with a lifelong career. The reader is also rewarded by T. W. Parry Jr. who brings a fresh perspective of A. Bishop Chance. It may never be known, at this late juncture, why Mr. Parry was en route to Columbia but it may have been associated with the University of Missouri, Stephens College, or an up and coming organization like the M.F.A. (Missouri Farmers Association) led by another charismatic Central Missourian, William Hirth. Columbia's growth and significance spilled over and benefitted Centralia and other surrounding communities, as it does some 70+ years after Mr. Parry's interrupted visit there. However, the timing is interesting and potentially revealing, as the East Coast reeled with the devastating results of the stock market crash in 1929. F. Gano Chance commented that it took a couple of years before the dark days of the Depression hit with its full fury upon Chance Company. So an unusual thing may have occurred, to find pockets of optimism and success meant those based on the East Coast had to venture to the Midwestern frontier. There they could develop an understanding of why such giants as Mark Twain, John J. Pershing, Omar N. Bradley, Harry S. Truman, and Walt Disney were produced from this area. Values such as faith, optimism, and those who still dreamed of possibilities that would become realities, resided there. There was a mental and physical toughness that could still be found in the heartland in a time when the hearts of many retreated in fear as they confronted the gloom falling over our nation. But one individual, as the local bankers suggested, in the face of certain disaster began to run instead of walk and grinned even broader instead of frowning. So Mr. Parry had found a gem in all the business rubble being slewed across America, just before a famous leader from New York would utter a cry to regroup and take a firm grip on hope as the only thing we had to fear, was fear itself. President Franklin D. Roosevelt and A. Bishop Chance did no time-sharing with fear as they both knew to counter fear and failure one had to take strong actions and perform worthy deeds in their respective roles. This they both did.

What was the extent of influence that the *Forbes* article provided in inspiring others to visit Centralia? How many of the multitude of daily travelers on passenger trains coming through Centralia became more aware of the business district lying to the south and the Chance factories hovering over the tracks to the north and south? While rail passenger service was alive and well into the mid-century period, Centralia profited from its exposure and travelers continued to seek out A. Bishop Chance. It's natural for people to first look at the results of one's success and then begin to look more closely at the individual, searching for the secrets of their success. Mr. Parry may have been one of the first nonlocal professionals to expose Mr. Chance nationally in such a powerful manner.

There were two points contained within the article that could be considered questionable. The first is a reference to the Chances world tour, suggesting Europe was the limit of their travels. Secondly, Mr. Parry's lead-in was difficult to initially understand. Weren't the Chance factories near the station and across the tracks from the restaurant impressive enough to attract Mr. Parry? It is doubtful that it had no impact, as even in later years many visitors have found it special to see such a factory and company headquarters complex in such a small agricultural town, which Mr. Parry respectively labeled a city. But during that time factories were being idled and so those still being operated pointed to the importance that leadership and the strength of the human element provided within those plants. Mr. Parry saw something in Mr. Chance that caught his interest and attention as he realized there was something special about this man.

THE MENNINGER SANITARIUM TOPEKA, KANSAS

October 29, 1943

Mr. A. B. Chase
Centralia, Missouri

My dear Mr. Chase:

It was very kind indeed of you to take the time to show Mrs. Menninger and myself your beautiful offices and a part of your plant there at Centralia just before our train left for Kansas City last week.

I was deeply impressed with the idealism and consistent purpose with which you had carried through your life's activities. In my work as a psychiatrist I am daily obliged to confer with people who have made failures or who are about to fail, and I once helped to write a book on WHY MEN FAIL. As I grow older, I am more and more interested in why men succeed, and some day I hope I can write a book on that subject. But the longer I live, the more complicated I think that problem is.

I wish you would turn it over in your mind if you feel like it, and when you have time write me a letter and tell me what psychological factors in yourself seem to you to be most important in the success you have achieved. Sometime, also, I hope you will get as far west as Topeka and spend a day or so with us. I should like to talk with you for a long time.

Sincerely yours,

Karl Menninger, M.D.
CHIEF OF STAFF

KM:AD

Menniger Letter (Chance Foundation Files)

Were there other encounters of travelers stopping in Centralia to visit with Mr. A. Bishop and/or to tour his plant? Knowing his extroverted, friendly nature and the convenience of two main rail lines having stations there for the arrival and departure of train passengers and the Columbia branch line, it is a certainty that this occurred. The *Forbes* exposure undoubtedly gave a degree of notoriety to the Chance name. As you may have guessed, there is a record of another significant visitor to Centralia, hosted by a locally prominent tour guide, A. Bishop Chance. People like to be around those who are successful, especially those with the social grace and the charisma of a senior ambassador like Mr. Chance who welcomed the opportunity of showing off his plant once again. The public even pays to hear such people explain their methods and how they too can become rich and famous. The secrets of an individual's success may more clearly be revealed by looking inside that person instead of analyzing their accomplishments. This may sound like something a psychiatrist would be involved in by trying to determine what drives those we label "a successful person". Most of the focus of such professionals, however, is on those who have problems and are failing. This is why we can personally associate wisdom with Dr. Karl Menninger after his 1943 visit with Mr. Chance. The positive impression that A. Bishop left with this world-famous doctor gave impetus for his further interest in people like Mr. Chance for material in doing a book on success instead of the failure theme.

Dr. Menninger revealed his thoughts on this subject in his October 29, 1943 letter to "A. B. Chase" reproduced on the previous page. Apparently A. Bishop didn't have a business card to give his visitor and so the slight mix-up in the uncommon name, Chance. The Chase name was associated with an electrical power concern in the Kansas City area. Note the three underlined words that were corrections penned in, that match the signature style and type/color of ink of Dr. Menninger's signature. It is impressive for a famous man to still remain humble and to display honesty and sincerity. Apparently he did not feel that is was necessary for his secretary (AD) to make those corrections and retype the letter. Maybe the obvious impression this leaves, possibly even purposely, was on the order of honesty and sincerity he recognized in Mr. Chance. An interesting aspect of Dr. Menninger's remarks is his recognition of the importance of man's ability to lead a normal, successful life. He had come to the conclusion, however, that such a study was more complicated than it appeared on the surface and the more he investigated this area, the more difficult it became. How these two men made contact and arranged the Centralia visit isn't discussed, except that it occurred just before the Menninger train left for Kansas City. The Menningers could have been in Columbia and since there was at times, a substantial wait between the Columbia branch trains and the mainline Centralia trains, Dr. Menninger could have been intrigued by the Chance Plant and was put in touch with its founder. Regardless of how the visit occurred, this contact with A. Bishop seems to have generated a renewed interest in Dr. Menninger's new study.

Dr. Menninger concluded his letter by inviting A. Bishop to write a letter "and tell me what psychological factors in yourself seem to you to be most important in the success you have achieved. Sometime, also, I hope you will get as far west as Topeka and spend a day or so with us. I should like to talk with you for a long time." (1943) This appears to be an invitation for A. Bishop to get on the couch for one of the most famous of all psychologists. A. Bishop responded to the correspondence invitation with a November 7 letter to Dr. Menninger at the Menninger Sanitarium. A side benefit is it holds importance to the story of Chance Company and is reproduced to the right.

The immediate response is a characteristic of A. Bishop and is indicative of a personal habit that was passed down within Chance Company.

A·B·CHANCE CO·
Utilities Division

CHANCE GUY ANCHORS
THIMBLEYE ANCHOR RODS
THIMBLEYE GUY FIXTURES
CABLE RINGS AND CARS
CONSTRUCTION TOOLS
TIPS LIVE LINE
MAINTENANCE TOOLS
TIPS TAP CLAMPS
TIPS GROUNDING EQUIPMENT
TIPS FUSED TAPPERS

CENTRALIA
MISSOURI

OFFICES
NEW YORK
CHICAGO
DALLAS
KANSAS CITY

Nov. 7th. 1943

Mr. Karl Menninger, M.D., Chief Of Staff,

The Menninger Sanitarium,

Topeka Kas.

My Dear Doctor Menninger:

It is truly a pleasure to receive your letter of the 29th., ult. and to note the many kindly expressions in my behalf. I enjoyed, very much, the privilege in talking with you and Mrs. Menninger and the courtesy you extended in making a rapid visit in our plant. I regret the time was so short -- and Sunday-- when the plant is devoid of animation, it is much more interesting when alive.

I note, with much interest, your work as a psychiatrist which I should have easily guessed. I can think of no profession I could consider more interesting.

Truly you have me"stumped" in your request,for a letter giving psychological factors,I consider most important in the success achieved. I'll have to stop and analyze the meaning of success. Was the results obtained born of pyschological factors - or- just like Topsy "just grew"?

Most all men have a philosophy of success --probably an unconscience one-- never the less they follow it. Success, accomplishment, achievement-- or -- what ever it may be called, are all born of the same parents. Feltham said " The greatest results in life are usually attained by simple means and the exercise of ordinary qualities. These may for the most part be summed in these two - common sense and perseverance." I have always endeavored to use psychological factors (just common sense,)for that is the best definition of psydology I know) in dealing with men and general business pursuits.

I believe that merit is the recipient of highest rewards of remuneration. No business hcan hope to succeed that does not render outstanding service in some form. Equal quality at less price - or - better quality at equal price.

"THERE'S NOTHING TOO GOOD WHERE LIFE IS AT STAKE"

SERVING PUBLIC UTILITIES SINCE 1907

2.

I believe that the real reason for the greater number of failures is "doing to day as they did yesterday". Doing today as you did yesterday is not advancing. Advancing only comes through improvement. Experiencees of yesterday can only be used as a basis for improvement to-morrow.

I appreciate the courtesy of your invitation to visit you when in Topeka and desire to extend you and Mrs. Menninger to visit us whenever you happen to be near Centralia.

Yours very truly

A.Bishop Chance
Chairman of Board

Chance reply to Menniger Letter
(Chance Foundation Files)

Employees were taught that answering correspondence quickly was a good business practice and an expected courtesy to those being dealt with. Dr. Menninger must have also adhered to this policy, as he referred to the Sunday visit of a week earlier as the reason for writing. In A. Bishop's response there is a mixture of business and personal philosophy along with a quote from Feltham that makes it more impressive: "I have always endeavored to use psychological factors (just common sense, for that is the best definition of psychology I know) in dealing with men and general business pursuits."(1943) Common sense is a phrase that Mark Twain could have done wonders with, but one of our Founding Fathers, Benjamin Franklin, had already become deeply associated with it. The dictionary describes it as sound and prudent judgement. The common man relishes an association with it, suggesting that it's even on the order of a poor man's college degree. Like formal education, to gain common sense took time and was a learning process. Few youngsters were ever entrusted with its full honors. Early in A. Bishop's business life there surely were those who accused him of having no common sense. Starting a successful business doesn't protect a person from mistakes and ridicule, but the experiences encountered on the way can allow one to attain a higher degree of wisdom. The reference to common sense by A. Bishop was from the positive and not negative point of view. In fact, he admitted his utilization of common sense in attaining success. It's routinely suggested that this is the path of a self-made man. In explaining to Dr. Menninger about his make-up, A. Bishop had accumulated a body of knowledge that had served him well in determining what to do and not to do in his association with people, both in his personal and business life. He simply labeled this as the common sense approach. In Chapter Six is coverage of a highlight listing generated by A. Bishop in about 1936. His listing in the year 1933 is adorned with his usage of "common sense." Its significance suggests that sound judgment again ruled at Chance Company and they were on solid footing heading up and not down. There is a right and wrong way of doing things and one of the foundations of his success was centered in his honed abilities in the human relations area. There is little doubt that Dr. Menninger recognized that in this man and wanted to know more about what made him tick.

Perseverance, the other quality mentioned in the Feltham quote, was a very strong characteristic of A. Bishop. Although he was considered a people-oriented person who could charm and influence people, one of his strongest traits was his "stick-to-it, never-give-in" stance. The 1931 *Forbes* article also spelled out the business philosophy that had insured Chance Company's success. A. Bishop had been able to capture these principles in only three sentences, as follows.

> *We have endeavored to supply fairly the needs of the persons we serve. . . . It is our policy to better the quality of our products and to sell them at a price that profits the user. We believe—and we teach our employees to believe—that we have no claim to existence unless we can produce better goods at a lower price or at least, equal quality goods at a lower price than our competitors. (1931)*

A. Bishop's 1943 response to Dr. Menninger was similar to the *Forbes* quote, but with the addition of the word services. It will become obvious that the Chances had an intense concern over satisfying the customer. This intensity was passed down to the employee by requesting, for example, that they work harder and smarter. There was never the idea that the company had it made and employees could slack off. The surviving Chance response letter appears to be the next to the last draft as penciled-in corrections have been made to this original typed copy and it isn't signed.

A. Bishop's 1943 Menninger reply letter also had some additional advice. The complete quote is:

I believe that the real reason for the greater number of failures is "doing today as they did yesterday." Doing today as you did yesterday is not advancing. Advancement only comes through improvement. Experiences of yesterday can only be used as a basis for improvement tomorrow. (1943)

Caught up in the middle of World War II, one couldn't help but be impressed by all the technical developments and the tremendous production in war materials which had blown the Depression away, or at least temporarily so. To win the war or be a business success A. Bishop was declaring to press on. Even though he was in the winter period of his life he was filled with hope as he viewed all the awesome changes taking place right before his eyes. Can you imagine the enthusiasm of an optimistic, extroverted, inventive-type person who had supported such advancements throughout his life and the satisfaction he must have felt in examining all the new technologies being developed during this period. The secret process of heat-treating aluminum and the x-raying of castings for aircraft, which A. Bishop was responsible for by his purchase of the x-ray machines many years before, were taking place right in the plants at 210 North Allen Street, Centralia, Missouri, U.S.A.

An inventor has to be a positive person as they have to visualize something that doesn't exist. Enthusiasm is contagious and Dr. Menninger surely noted it in A. Bishop Chance, even though he was seventy-some years old in 1943. His son, F. Gano Chance, also captured some of his father's ideas and positive points and proceeded to catalog various aspects of Chance Company, one of which was the overall operating philosophy Gano tagged, "The Ideas That Guide Us." Although its coverage will come later, it is important to trace any connection and verify if there was a common theme that permeated company policies that can be traced from Chance Company's origin to its end. Even though there are certain portions of the later Gano material that are new, there are connections, for example, in the more specific business principles. There is the idea of profit sharing, covered in Volume II that will note a special relationship between the two Chance generations. The following quote from a Gano speech in St. Louis in 1968 entitled "The Ideas That Guide Us," clearly shows a common theme of serving customers with quality products and services effectively and efficiently, in which Chance Company and the customer could both profit.

We prosper by providing useful, high-quality products and reliable service to the electrical power and communications industries and other customers. Through these products and services, we fill a need of our society. By performing our functions effectively and efficiently, we make a profit and help those with whom we do business to profit.

Parking Then and Now

In the early 1900s, the horse required a hitching rack/post and stables. The locations of these facilities were usually in-behind or out-back because of obvious sanitary reasons. As the automobile roared upon the scene, the parking space or lot became a necessity, allowing the public the opportunity to get closer to the front door, a challenge that survives and

Young A. B. Chance III (Little A. B.), pictured in front of the Chance Garden House garage. He is riding an early motor scooter. (A. B. "Buff" Chance IV Collection)

Parking autos around the City Square was a start but now main street parking became an issue. The city decided to permit cars to park in the center of downtown streets because of congestion and the need for more parking spaces. They didn't like curb parking as it took up too much room and blocked traffic. Also local merchants were not yet prepared to drive off horse/buggy customers. Finally, after all these years, it's obvious why certain local towns had such a wide main street. It wasn't for a four-lane right-of-way but instead for angled, center parking. The new Centralia policy of center street parking became official May 13, 1921, a Friday no less. One of the benefits cited was allowing more autos to be parked in the business section.

flourishes today. The transition from horses to automobiles was one of those areas that were most difficult because until one eliminated the other, a way had to be found to satisfy the needs of both. One of the most common complaints of early auto owners was poor streets and roads. The auto couldn't handle rough, muddy and poorly maintained roads as well as horses or reach their potential of greater speed and range. An example of the lack of progress in Centralia of hard-surfacing city streets, was noted in the February 18, 1921 *Guard* which suggested city streets were in a worse condition than county roads. Later that spring a volley of complaints reached the city officials about parking problems. Back on April 11, 1919, the *Guard* had called attention to progress for auto parking in Centralia:

> *Owing to the great demand for auto parking space the street commissioner was instructed to remove the few remaining hitching posts on the south side of the square. This gives the entire four sides of the square to park machines. The Thomas Lumber Co. was given the old chains and will erect a line of free hitch racks along the east side of their yard. This gives more racks than the city ever had at any one time.*

As the number of cars grew, companies converted from hitching posts to parking lots. At some point someone got the idea that a way to compensate employees with position was to designate a parking space for them and place their name on it. When the executive was out of town, their secretary could park in their spot, which was one of the benefits of the prestigious position of executive secretary at Chance Company. For several decades the main office executive parking lot was on the north side of the main office. Holes in the brick wall are all that remain where name placards listing F. Gano Chance, N. A. Toalson, D. R. Miller, A. H. Knepper, A. H. Allen, J. P. Mullen, J. M. Arndt, L. C. Hansen, P. D. Sturgeon, C. C. Hough, and A. B. Chance III once were located. One of the eventual changes brought by Emerson was the demise of the executive parking lot, which became a storage area for anchors. Hopefully this was done entirely for practical reasons but it signaled an end of an era when Centralia-based executives grew fewer in number with lesser positions. In mid-

1999 the anchors were removed from this area, returning it to a parking lot, but for visitors, as executives had become as extinct as mule barns and hitching racks.

In the 1960s a humorous situation occurred, when a young lady had gotten a new job at the Research Center. Pulling into the parking lot she had the usual new job jitters. Finding a parking space near the entrance, however, seemed like a good omen so hopefully everything would go well. Later that day Skeets LeVee, the Vice President of Engineering, came into the office somewhat red-faced and obviously irritated and asked who had parked in his parking space. As it turns out, Judy didn't like the job and left the company, but not before she was teased about what the consequences were of parking in the vice president's parking space!

Standing beside the big Packard in front of the Chance Garden House garage is Porter Williams, the A. Bishop Chance's chauffeur. Porter had earlier been associated with the Oliver Carnival. Across the street is the "Bungalow" that reportedly the Chances built for their son Buford and where other family members and company officials later resided. (A. B. "Buff" Chance IV Collection)

In August of 1925, the *Guard* carried a story of a Ford runabout being stolen from Chance Company grounds. A window was broken and the "storage house" near the Alton railroad was entered. They then let down "the big bars and opened the big doors." It is unclear if this was the first company car or a personal vehicle parked at the plant. This may be the car A. Bishop challenged a Wabash boxcar with! The building it was stored in was on the north side of the original Chance plant and may have been the executive garage before it was located in the 1929 Punch Press building.

Although many executives were extended designated parking privileges over the years, Mr. A. Bishop and Miss Ida Lou could boast of a parking garage. In a discussion about factory additions, Paul Sturgeon stated that:

> *We had occasion in the late 40s or early 50s to do some modifications on that building. Bill Decker was the Plant Engineer at that time when we decided to close up the two garages that were in the southeast corner. Mr. A. Bishop and Miss Ida had stored their automobiles there. (1979)*

These changes were announced in December of 1946 in a *Chanco News* article concerning substantial changes being made all over that building. The building being referred to is the Punch Press Plant across the street from the Main Office. Examination of the southeast corner, which became the Foreman office area, revealed those 1946 modifications. The narrowness of the stair-stepped sections on each side doesn't appear wide enough to welcome two vehicles at the same time, however, garage door construction during the 1920s were tall and narrow which reflects Model T Fords and other period vehicles. These doors are shown in the photo of the Punch Press building on page 234. Another key is noticing the long

The two doors on the left side of the Punch Press Building, across Allen Street from the Main Office, were where the parking garages for Mr. Chance and Miss Ida Lou were located. (Jack Chance Collection)

white masonry stones extending the full length of the front of the south section. Structurally this supports bricks above openings such as doors and windows. Closer examination shows telltale signs of newer construction and two narrow but tall passageways can be detected. Trying to visualize Mr. A. Bishop and Miss Ida Lou's use of this garage raises the question of which side was designated to each. Regardless, it provided the founder with auto accessibility for his probable frequent usage during the workday and when returning home on the midnight train from some far away place. Miss Ida Lou can be seen crossing Allen Street facing the rising sun each workday and at the end of the day her shadow casting back toward the Main Office as she faced the setting sun. That is the story of dedicated working people as they spend the better part of each day, or a lifetime, within the boundaries of the work place. As we recognize the discipline and contribution of countless millions who have labored, may we realize as President Calvin Coolidge said in 1925: "The chief business of the American people is business." *(Speech to the American Society of Newspaper Editors*, January 17, 1925) So another of the colorful sagas is played out at a factory in a small rural town in out state Missouri–where the common daily occurrences were anything but ordinary. In today's world we are led to believe that progress doesn't come naturally but is gained by force and confrontation. But there was an earlier time when there was class and a quiet dignity, when a lady parked alongside the president!

Chapter 5
Success and Hardships
(1925–1930)

The One-Hour Lunch Break
Activities In and Out of the Plant

The origin of many things we do can be traced back to an earlier time. When working days were longer, more physically challenging, and fast food was a cold sandwich, the logical length of lunchtime was one hour. That worked especially well in Centralia where most people went home for lunch and enjoyed an important lengthy break. Today we think little of driving five, twenty, or forty miles to work and many of us have become temporary transplants during the workday and noonday meal.

Many employees of A. B. Chance Company, especially management, were required to live in or near Centralia and found the one-hour lunch break to their liking. The out-of-towners either went to a restaurant or brown-bagged it. This one-hour period could be a lengthy one and employees developed various routines. You could play cards or horseshoes, gossip, read, attend service club meetings, shop, eat at Squires, and if you weren't a very ingenious individual you could always work. This employee's routine over several later decades became one of reading Scripture, walking, and napping. In the mid-1960s, a gathering of the male variety routinely took place in a large office to the north of the elevator on the first floor of the main office. The spokesman for the group was a one-man show by the name of Charlie Stone. Like Sam Sneed, he would soon have everyone roaring from laughter. At Pitman, a later division of Chance, Dick Roth, George Diddie, and others had some enjoyable pitch games, and groups over the years played cards at the various locations. Near the Research Center there was a horseshoe-pitching area for engineering personnel to try and make some "ringers" during their noonday sessions. The Chance Company also provided lunch tables in various locations, and inside break areas where employees could congregate. Over the last quarter of the twentieth century, the makeup of employees from outside Centralia grew rapidly, especially after the company was sold. The purpose for the long lunch periods was no longer based on employee needs as much as company considerations. This meant office employees could be at work for nine hours instead of the normal eight and one-half, providing longer coverage for customers in other time zones.

Undoubtedly, one of the most memorable stories of lunchtime activities involved the Chance Company founder. Not every early employee had a car and as A. Bishop left for lunch some of his employees would ride with him. Neither was it out of character for him to stop and pick others up. Those walking were spoken to or waved at as he passed. This establishes an image of a man who was much more possessed by an interest in the people around him than anything in front of him. There are several Flynt Chandler stories scattered throughout this book that shows the diversity of his thought process and the phenomenal recall of this man who either lived or worked through the entire existence of Chance Company! This is a favorite:

> *I don't think there was a better fellow than A. B. Chance. I never saw him mad and he always had something to say to you. Before he bought that Model A* [Ford] *he had an old Model T* [Ford] *with a top on it. Every day at noon he would take us all home for lunch. Mrs. Sires worked with me and she lived down in the same part of town. We were coming back one day at noon and there was a freight train over there and Boss was driving. I'll be damn if he didn't run right into it! The brakeman came over and he said: "What the hell are you fellows trying to do here?" I told him "You better get that damn train out of the way!" (1979)*

With no electric signals, multiple tracks, extensive switching operations, and the many freight and passenger trains passing through Centralia in the early 1900s, it was a dangerous situation. There were two examples in late 1921 and early 1922 of vehicles running into trains, both occurring at the Allen Street crossing. The first on October 15, 1921, involved a Mrs. J. A. Marshall and her small son, who were bringing cream into town. The car's brakes failed and she hit a slow-moving Wabash freight train. Only the front of the car was damaged. On January 13, 1922, the *Guard* told of another head-on into an Alton train switching by the Allen Street crossing at the Alton station. The badly damaged Ford coupe, owned by Junior Lyon, but driven and occupied by "Lory Skaggs and Hugh Howell in company with Miss Katherine Paxton," met with glass cuts and bruises.

For nearly a century the one-hour lunch break has endured for the office staff of what was Chance Company. Although many traditions have tumbled by the way side, there are some things that even time seems to have no dominion over. In spite of fast food, computers, quick this and that, at noon in Centralia during a normal workweek, time stands still for an hour and the basic elements of life again become more important to employees than the machinery of business. A. Bishop Chance and Flynt Chandler would surely approve.

The GM&O acquired the Alton and this post-1940 photo pictures a Kansas City–bound passenger train making a stop at the former Alton Depot. This rare picture reveals the station that would later be razed to make room for a Chance Company parking lot. (Centralia Historical Society)

The mid-to-late 1920s was a time of success for the A. Bishop Chance family, but at the same time it would also bring heartache. Hardships were no stranger to this family or other people of those times as life was not protected from the elements as well as it is today, although adversarial conditions face every generation. Ironically, the very mechanical machines that were contrived to reduce the hardships of everyday life would at times generate a backlash that fueled serious mishaps. This would occur even to those like A. Bishop who was an avid promoter of the industrial revolution and a personal contributor.

Before entering into the last half of the 1920s, a summary of conclusions reached concerning the Chance family follows. One of the first surprises was the discovery of the pioneer influences of Garrison and his brother Albert, representing the first generation of Chances who lived in Boone County. Their University of Missouri connection is significant and the fact Garrison remained in Boone County during the Civil War adds to the family's strong presence. The Chance family involvement in the hotel business is more complex than commonly reported. Another surprise is the expanded coverage of A. Bishop's local telephone influence and of his technical skills and electrical knowledge that were applied to various projects. His very early business negotiations, acquisitions, and product development, patent activities, and monitoring of his competitors is impressive and reflective of the future. Home Telephone came much later than expected and wasn't owned solely by J. A. and A. B. Chance, although they did manage and appear to be in control of its day-to-day operations. The aggressive business policies of A. Bishop and later F. Gano Chance generated capital shortages in the face of continual expansion. This was a reality of grassroots people who started and grew substantial businesses. By the mid-1920s, A. Bishop has become financially successful, confirmed by the tremendous building and expansion of the Chance plants and facilities, extensive travels, and local property and real-estate purchases and refurbishment. To become successful at the level he envisioned, he became self-taught, learned quickly, remained inquisitive, maintained a positive and enthusiastic attitude, and was toughened by his experiences. He became well traveled and aggressively pursued knowledge of innovative products, procedures, and technologies of the day. He felt it was important to be honest, aggressive, self-sufficient, responsible, responsive, and service-minded in his business dealings and his organizational and personnel skills also served the company well. His explosive pace in patent filings and continued inventive-type production are beyond anything earlier imagined. However, A. Bishop wasn't the first mayor or a very popular one, as his push for progress was considered extravagant, but after the water and sewer bonds were paid off in the 1920s, the public recognized the benefits of these modern conveniences instead of the costs. He then began pushing for street paving, although as a private citizen. The negative experience of being an elected official appears to have dampened A. Bishop's and F. Gano's enthusiasm for seeking future political positions. A. Bishop Chance's compassion toward employees, friends, and community-betterment projects grew as he became more successful.

Developing on the Fast Track

Progress Between The Tracks

The Roaring Twenties is appropriate in describing the growth of Chance Company during the mid-1920s. The normal way of finding out what's occurring within a firm is to look at their books or sales records, but when they are not available there are less-sophisticated barometers of growth. A census of automobiles in the parking lot or viewing a firm's plant to see if any new construction has taken place, can help determine a firm's health and if something positive is occurring. A case in point is how many Americans believe it becomes a necessity when they reap prominent success, to build a mansion. Of course plants are built to utilize while mansions are built to impress. However, both indicate an increased level of success as it did at the Chance Manufacturing Company.

The importance of America's railroads in the 1920s is reflected by it reaching its peak in rail mileage and development during that decade. Neither can its convenience be overstated as it brought the Interstate Highway or Lambert Field within a couple hundred feet of the Chance Company headquarters. The importance of company officials traveling by rail has been documented, and steel and other manufacturing materials were transported in by rail and finished goods were sent back out in a boxcar or by Railroad Express on a crack Alton or Wabash passenger train. The existence and location of Centralia directly relates to the railroad and geography is also important to the Chance story. So if the railroad was so important, especially with all the Columbia Branch traffic, why was the Wabash's Centralia facility so poorly represented? The old station was an eyesore and not well kept according to local news items, but both it and the Alton's Depot appeared to be similar.

The poorer Wabash is one reason the seemingly more financially healthy Alton was popular, but of course it didn't have the heavily trafficked Columbia Branch although some of that activity was siphoned off by the Alton into and out of Centralia. The trackage of the Alton clearly made it the preferred route to and from Chicago. The St. Louis service was the Wabash's domain early before the Mexico cutoff was utilized by the Alton. Kansas City traffic favored the Alton, although the Wabash was also popular, especially for Iowa or Omaha traffic. The condition and appearance of the Wabash Station apparently reflected a poor image for Centralia and nearby Chance Company. The *Guard* had sent out cries since 1912, of how badly Centralia needed a new Wabash Station, but few if any envisioned what eventually resulted. However, a 1912-era picture of the Wabash Station reproduced below was surprisingly more fit than what is being described and of a size commensurate with those days.

This postcard from a Mr. Sturgeon of Centralia to his daughter, a Miss Pauline Sturgeon of Mexico, was postmarked March 26, 1912. It is in color. This is the pre-1925 Depot, located on the northeast corner of Allen and Wabash tracks. (Darrell Darst Collection)

Progressing through the early 1900s, rumor after rumor surfaced of Wabash plans to build a new Centralia Station. One such front-page story graced the *Guard* on February 23, 1923, but it appeared this one was different. This movement was suggested to be like "Banquo's ghost," it just wouldn't go away. On this occasion Wabash engineers actually came to town and drove stakes, thus destroying the rumor that the new station would be located out west on the Columbia wye. The east stakes were near their water tank and a small icehouse and such a location would close Rollins across the tracks. The positive consequences of this to Chance Company's eastward expansion aspirations are obvious. Blockage of main streets by passenger trains would happily be reduced by centrally locating a new station between Allen and Collier Streets, which at that time was closed (Collier) to across the tracks traffic by a stockyards. Another reason this initiative appeared legitimate was what actually occurred. The new facility was to be like the Macon Station, it was stated, yet another connection with their northwest neighbors. However, in spite of local excitement over a 1923 closure to this issue, it was not to be.

In contrast, the proposed new City Hall was spurred on by a poorly designed city facility which was threatening to collapse. An April 20, 1923 *Guard* article said the roof was spreading the walls out and a new one should be built outside the City Park (Square) leaving it clear of any buildings. The Settle Livery Stables and Early lots east of the City Square were purchased for the new City Hall. The old stables were moved and the lots filled and leveled. The September 14, 1923 *Guard* reported the foundation for the new city structure was being poured and it would be allowed to season and cure over the winter. An April 11, 1924 article covered A. Bishop Chance donating a brass box to be used in the cornerstone of the new City Hall. There was also talk of street paving later in the summer as Mayor Bruton in April of 1923 was reelected, unopposed for another term. Centralia, with 2,860 residents, still had a broom factory, electric supply factory, an iron foundry, and a proposed new Post Office, City Hall, and Wabash Station. In spite of high hopes all is quiet along the Wabash, but numerous Centralia city activities may have grabbed the spotlight and the station situation just faded into obscurity for another year. However, the competing Alton was building a new water tank north of Producers Elevator. This successful and longtime grain-and-feed dealer was located on the Allen and Wilson corner lot that would become the M.F.A. elevator and later Chance Company property. A. Bishop was also busy in real estate and in support of street paving, however, another year passed before he and city officials got the Wabash involved in paving and city street extension issues. In the September 26, 1924 *Guard* was an excellent article entitled "Opening Up Street" that explains the Rollins and Collier Street situation. This article suggests that the new Chance Foundry Building, described as an addition to its iron foundry, would be built across the existing Rollins Street between the railroads, thus forcing its closure across the tracks. There appears to be no resistance by the city to this proposition and instead of the future new Wabash Station forcing this issue, it instead was the expansion-oriented Chance Company enterprise. However, this would work right into the proposed location of the new station and the railroad readily cooperated in beginning the building of a new crossing east at Collier and Railroad Streets. An answer to another lingering question about these two streets is the statement that Collier "has never been opened thru the railroads." To complete it would require some twenty carloads of cinders, relocating a switch, eighteen-inch tiling, plus the tearing out of the stock pens blocking the street. This was to be provided by the Wabash with the city performing the street grading. The object of this as stated was to do away with the very dangerous crossing at the Wabash steam engine water tank on Rollins, but it's clear Chance and Wabash aspirations for facilities expansion was at its heart.

It is interesting how the railroad had gotten into the business of street construction. Because a portion of the grounds on the north side of its railroad tracks were Wabash property, a portion of Collier became Wabash responsibility. This later was true of the brick street or driveway constructed by the Wabash from Allen to Collier, north of their new 1925 station. The ashes from coal-fired locomotives were a high demand item, especially on railroad roadbeds

This photo was likely taken from the Anthony building in about 1928 or later. It features the Wabash Depot and three distinctive signs painted on two non-Chance structures. The first on the large Wabash water tank is promoting the Red Snapper Piston Rings. The second on the city water tower tank is for Never-Creep® Anchors. On the extreme right is the Chemical Building with the trademark symbol of the Big C Line painted on its south side. These were all positioned to catch the eye of Wabash train passengers or those traveling Railroad or Collier Streets. (Jack Chance Collection)

and as fill in other applications. Noting the earlier Chance and city discussions about payment for dirt that was used as fill between the tracks, it is clear the raised elevation of the railroad roadbeds through the wet Centralia area generated lowlands between the tracks that acted like a lake during heavy rains. Consequently, the between-the-tracks area was built up close to the railroad level. This is one reason the Chance complex is more imposing as it sits at a higher level than the approaches on either side of the tracks.

The best-laid plans of mice and men don't always progress as envisioned by logical thinkers. The October 3 paper had a small news item about the city dragging its feet in not removing an "old frame building out of the street." This is sort of a mystery structure, but it may be the laundry building mentioned later in this paragraph. The railroad said it didn't need the crossing and so it was up to the city to do its part and thus an impasse of sorts had developed. A week later, in the October 10, 1924 *Guard*, the complexity of this project is expanded further when it's noted that the "Chance plant is tearing down part of its storage house just south of the foundry building in order to let the railroad move its switch track several feet to the south. It is understood that Mr. Chance will build to the east of his present factory at an early date." The switch track per 1950-era overhead photographs, show its southerly migration to the main line as it progressed east along the south side of the Chance Tool Building (1948 era) beyond Collier Street. The Wabash official was also asking the city when they were going to relocate the poles in the street which were likely electrical power line poles in the vicinity of the power plant. The Wabash hadn't begun street construction and the city must have thought they were dragging their feet as Mayor Bruton suggested they would do their part after they saw some action from the railroad. The city said it had moved the "old laundry building out of the way and was ready to grade the city street along the city's property." So there was progress, but with three parties involved it was slower than expected.

The focus on streets moved to heavily traveled Allen Street between the Globe Hotel and the Alton Railroad on the north. The Wabash, in October, requested no paving in this area, as contemplated track relocations would destroy new pavement. Although the spur track may have been a reason, changes in the track near the proposed new Wabash Station, information of which was held confidentially, were surely driving concerns by the railroad. Amongst all the activities occurring in what could be described as the Between-the-Tracks Industrial

Park area, one of the prime instigators was none other than A. Bishop Chance. What he didn't stir up, others would shoulder him with the responsibility of carrying out. The April 24, 1925 paper said he wanted Railroad Street from Rollins to Collier paved. He felt the Wabash should share half the cost and eight-foot sidewalks were also mentioned. This was important not only to his Anthony property but for access to the east end of his plants including the new Chemical Building, especially with the imminent closure of Rollins across the tracks. In the struggle to get Collier opened, with the Wabash dragging their feet, the city fathers had presented A. Bishop with its costs! In reading the city's position Mayor Bruton in November of 1924 said the city had nothing to do with the proposition other than approve its opening. It was suggested that A. B. Chance be contacted as to the cost of filling and grading to open Collier Street. It is entertaining from our isolated perspective of the back and forth negotiating taking place or more accurately, "passing the buck!" However, a sense of change has taken place as 1925 progresses with Chance Plant activity intensifying into a maddening pace. The success of the Chance operations had registered, and to get in its way was like daringly standing in front of a steamroller. The city's position had become sort of you can do it as long as you pay for it, vintage Mayor Bruton.

As early March 1925 arrived, University of Missouri students F. Gano Chance and classmate Robert Austene likely traveled on the Wabash on their weekend visit back home. But in spite of the many travelers coming to and through Centralia, the railroad station situation appeared to have become a dead issue. But after all those years of pestering the Wabash, the *Guard* wasn't about to give up on its new depot campaign. Their reliable source suggested that the future Wabash Station would be of red brick with a red tile roof (correct) and be a 60-by-70-foot structure (incorrect). The dimensions were more correctly amended in the April 10 paper to a 60-by-150-foot structure. Suddenly the May 8, 1925 *Guard* headlines stated the Wabash Station project had a clear track and local Wabash Supervisor J. J. Prewitt had been given orders to clear away the stock pens and begin moving track. A passing track to the north and east of the old station serving the livestock yards would be removed. A new second track would be extended from Collier eastward a half-mile to near the pump house on Lake Deutcher. The space between the two tracks would be widened to 18 feet and concreted by the station for passenger and freight access. A new steel water tank east of Collier (incorrect) would replace the wooden one on Rollins with several supply pipes on steel cranes so water could be taken without spotting the engine at the tank. No contract had been let, but it was expected bids for the new station would be opened soon. A Chance storage house on railroad property along the north switch track, would be removed to clear the way for the new station.

A revelation of the changes necessary to construct the large Wabash Station indicates the heavily concentrated array of buildings, businesses, and various structures then located between the tracks. A more official article in the June 12, 1925 *Guard* provided additional information:

> *WABASH STATION. A. B. Chance has been notified to move his large galvanized warehouse immediately north of the Wabash depot, by the road officials.*
>
> *Skaggs & Esteppe have also been notified to move their ice storage house, located on the north side of the railroad's water tank.*
>
> *These are the first moves toward clearing a way for the construction of the new station here.*
>
> *The bids for the construction of the new station were closed Monday and were to be opened Tuesday. We understand the Bell Brothers of Moberly former Centralia boys,*

were the only contractors inside of the state who submitted a bid on the entire building and here's hoping they may get it.

The Wabash Railroad and A. Bishop had not come to an agreement on the street paving issue, according to the June 26 paper. The railroad wanted to hold off until the station was completed, but A. Bishop suggested a mud street to the new station wasn't a very attractive situation. That same week the Wabash released information about a proposed new brick-paved street north of the new station, which would give station access to highway-type vehicles. On July 10 the *Guard* confirmed the removal of the sidetracks and the selection of the depot construction firm, J. J. Wuellner and Son of Alton, Illinois. Erection of the 178-by-40-foot station was to be carefully monitored by D. O. Graham, a Wabash bridge-and-building engineer. A week later, the evacuation for the foundation was begun by men and teams with the dirt taken to the east. An optimistic prediction of completion in one hundred days seemed a bit quick. A lengthy *Guard* article of August 21, 1925, spoke of a former resident, Elgin T. Cash of St. Louis, who helped get the new station by writing a letter to Wabash's H. E. Watts in January. He wrote of the old station that he could remember back to 1878, which he felt was disgraceful in view of Columbia students from probably half the states laying over there. He suggested in his wide travels across America that the Centralia Wabash Station was one of the shabbiest, unfit for decent people facilities, existing! Mr. Cash headed a Commission Company at the National Stock Yards of Illinois. Surprisingly, the *Guard* suggested that no other party or organization besides Mr. Cash and themselves, had registered an official request with the Wabash for new facilities.

What followed in the *Guard* was an almost weekly update on the construction of the new Wabash Station. In contrast, few erection details were found when the new Chance Plant was built in 1917. The August 21 paper said the foundation work was being completed and the three different grades and colors of brick were being unloaded. Some of these bricks were brought home when the station was being torn down in 1992 and photos taken of interior and exterior detail. The light yellow interior brick was contrasted with "a smooth dark chocolate and a mat surface chocolate." The smooth chocolate is a beautiful brick with brownish, rust-colored impediments that along with the light yellow brick made a beautiful interior, especially in the large waiting room. The exterior dark red brick had a rough texture as used in other period buildings. Large marble slabs would mark the floor entrance of the main doors and materials awaiting use were covering the entire depot area in late August of 1925. A well was dug for the new stock pens west at the wye and even local farmers had been hired in helping to grade and fill the new track line, which must have been referring to the second track. As the walls were being erected, in spite of delaying rains, the significance of the size and beauty of the station were beginning to register its architectural and construction excellence. Later comments by Burlington Railroad (CB&Q) officials coming through town, described their amazement at the huge station for a town Centralia's size. The Columbia Branch traffic wasn't always obvious and neither was what Chance Company and Centralia might become. The paper called it a station designed for the future and like Moberly, in many ways Centralia was becoming a "magic city!" What wasn't known was the approaching Great Depression would force a focus on survival instead of growth, which curtailed the tremendous expansion of America in the 1920s. However, the framework for our industrial might was being laid and as World War II was entered, the facilities and capabilities planned and built in the 1920s would help carry the Allied powers to victory. As the fall of 1925 rapidly approached, the *Guard* was full of optimism with the new wider railroad bridges and double-track work being completed west and east of town, which in their opinion was a preview of the eventual double-tracking of the Wabash to St. Louis. This kind of progress certainly dampened any lingering ideas of the Burlington and Santa Fe constructing double-tracked expressways through the area. In many ways the new Wabash Station would become the center of all area activities. It was a beautiful and modern facility that grew and boosted Centralia's positive image in a substantial way, and along with Chance Company's progress established a confidence that reflected community success, pride, and promise.

As the work progressed, the station began to take shape. The inside was wainscoted up about four feet with a light buff brick per the August 28 *Guard*. The south track or Columbia branch and siding, was moved to the south three feet. The 125,000-gallon steel water tank was being constructed north of the station and near the Chance plant with locomotive outlets east and west of the station to facilitate taking on water while waiting at the station for passenger and freight activities. The water source was still from Lake Dutcher east of town. Window and door frames were being installed as were heating pipes under the floor level and in the basement. The walls went up slowly because of the "difficult manner in which the ties have to be made in the inside and outside finish work." A week later the September 4 paper noted further progress. The between the track concrete platform would be 8 by 470 feet to allow passenger and depot trucks movement. The main concrete platform was 12 by 495 feet except on the south side of the depot where the width would be 18 feet. The north road would be 30 by 596 feet and run along side the north passing track or what could be described as the Chance plant spur. A street-type paving machine was being utilized for the concrete paving work. The concrete pillars for the new water tank were also put in place and the *Guard* described this location. "This steel water tower is on a line about 20 feet north of the west line of the former site of the stock pens, and about 85 feet east of the center of Rollins Street." Other activities included work on the two-mile-long south track. Eight brick masons were in charge of the brickwork, with most of the west section up. Wiring concealed inside the brick walls plus water pipes were being installed. The east brickwork would begin in a few days.

An extensive article outlining the completion cycle of the depot work was run in the October 23, 1925 *Guard*. The roof sheathing to hold the tile roof was nearly done and the major brickwork had been finished three weeks prior. The sand pipes and water cranes for the traction and water hungry Wabash steam engines were placed; the west one in front of the old station near Allen and the east one just west of the new Collier Street crossing. Work had now begun on assembling the materials that had arrived for the 150,000-gallon water tank, to be located on the concrete foundations. The concrete curbing was in for the new north side roadway that would be completed with brick. This would benefit the loading and unloading of freight, with Chance Company being a major customer and benefactor of such out-of-the-mud advancements. Five new tracks, side by side to the east, suggest the new Centralia Yards east of Jefferson. One of the new tracks "curves to the north and goes around the station joining the other tracks on further west." This is describing the spur track that began up in the Hickman Street area and ended east of Collier, that served the Chance factory complex. The east half of the old depot would be moved to become a new freight house just east of Collier.

A week later, the October 30 paper focused on the paving controversies with four major parties involved: the Alton, the Wabash, the City, and A. Bishop Chance. The main concerns were Allen and Railroad Streets. The Wabash was responsible for Allen Street from their south right-of-way to the south side of the Chance factory. The Alton, per the *Guard*: "would inherit one hundred feet of Allen Street from the north side of the Chance garage to their right-of-way" on the north. A longtime

This is a south view of the 1925 Wabash Railroad Station looking across Railroad Street, formerly Wabash Street. The freight and Railroad Express area was on the east end or to the right in this picture The station office and operator area was within the center cupolo area and the large visiting room to the left with the west end being the large open porch. The debris along the front is the platform area that was being removed along the main line track nearest the station, indicating this picture was taken during the period of its eventual destruction. This debris was dumped in the Columbia wye island to the west. A local photographer, Mrs. Perry Roberts, provides us a final view of this large Centralia landmark. (Centralia Historical Society)

bone of contention was the paving of Railroad Street from Rollins to Collier with the Wabash responsible for fifteen feet and A. Bishop the other twenty feet. The *Guard* pushing for a paving solution spoke of a bad mud hole at the front doors of the new station and the much greater costs if the city was forced to use the tax bill plan. This may mean between Allen and Rollins on Railroad Street.

The passage of two more weeks brought further progress as the November 6 *Guard* told of eighty-five Wabash track workers raising the track eight inches through town. The east and west station sections, with its new red tile roof had a "pretty effect." A problem, however, halted the completion of the center section, caused by faulty iron beams rejected by the superintendent. These were actually huge wood timbers with metal plates holding them together. There were four of these positioned over the waiting room. The interior floors were being concreted and the water tower erection was proceeding slowly. The foundation for the new freight station, fifty-five feet east of the center of Collier Street, was being laid. On November 13 the paper again spoke of the heavy timbers for supporting the center section of the roof were rejected. "They were large and lengthy timbers and had to be shipped from the mill creating a months delay." There seems to be some confusion as to what was rejected, the metal parts or the wood timbers. They may have been constructed in such a way that to reject one part was to reject the entire assembly. These exposed red beams played an important role in the beauty of the waiting area and were wisely retained by Mr. Bill Reinhardt whose company got the contract for the station's removal. These, including a portion of the roof tile were still in Mr. Reinhardt's possession in late 2003, although he said he was anxious that they be restored and utilized in an appropriate structure. In 2005 a new Wabash shelter house in the City Park on North Allen Street suggests this material had finally found a home a couple of blocks from where it had resided for nearly seventy-five years.

No party was more anxious to get the new station completed than the St. Louis based Wabash Railroad. The November 27 paper said the roof was on and the water tank completed. The tile portion of the roof was about done on December 4 and the windows and doors were in. The December 11 paper reported the work was begun to install the tile floors. The December 18, 1925 *Guard* told of the work on the new north driveway that would give passengers uninhibited access to the new station. Passengers could enter through the north doors into the large waiting room and tickets secured at the ticket window in the east end of the waiting area. Patrons already having tickets could go directly to the train by going under the open, long west shed to the platform. The large waiting room was for both ladies and gentlemen with the ladies restrooms in the northwest corner and the men's in the northeast corner. After Christmas and the New Year were celebrated, the January 8, 1926 *Guard* told of the final preparations for the new station's opening. New seats of the double-row type, each holding ten to twelve people on either side, had arrived. They had a walnut finish with tall curved backs and would occupy the center of the waiting room. On the west side were some special seating for ladies. The new tile floor with individual tiles of about eight inches of a slate color with white pebbles imbedded in them, were being laid, but cold weather was holding up the concrete driveway completion. The opening of the new, long-awaited station was only a week or so away, according to the local paper. But even that prediction was optimistic as the official opening announcement wasn't forthcoming until the February 19 edition reported it occurring the previous Monday.

Although the street-paving issue was still undone in early 1926, the Wabash was busy completing the depot project. A new fence was placed along Railroad Street in front of the new station to keep people from crossing the tracks in that busy and dangerous area. They also were doing landscaping work around the station and grounds in March of 1926. It's understandable why the Wabash, without financially deep pockets and a need to complete paving to their new station, wasn't focused on Railroad Street. In early May they were just completing sidewalks on Allen in front of their new station. The long brick driveway that extended from Allen to Rollins on the north side, allowing passengers and freight to come into or be shipped out of the new depot, was also being laid. In May, Mayor Bruton and A. Bishop traveled to St. Louis to talk with the Wabash further about street-paving issues.

These four 1992 photographs reveal the interior detail of the 1925 Wabash Depot waiting room area. The first picture with the rounded doorway in the east side with the ticket counter on the right. The four roof beams had been removed and stacked on the floor. The second picture is of the west end leading to the rest room areas from the waiting room. The third picture is looking north toward the Chance plant showing the exquisite detail of the brick walls. The fourth photo shows the roof beam detail on the east Operators wall area of the 1925 Wabash Depot. (Darst Collection)

There is a moment in time when there is a balance in a setting that speaks to a correctness in appearance and practicability. It had not completely arrived yet, but the new Wabash Station was certainly an important part of an image that would register on tens of thousands of local residents and visitors that would pass through this area of Centralia over the next several decades. As dump trucks passed by this employee's northwest corner office at 210 North Allen in 1992, with previously precious building materials that had now become station debris, it signaled the closing of an era that had begun by the sale of A. B. Chance Company in 1974. Remembrances of the Centralia Station weren't from usage of it, but in knowing that its silhouette had existed during my lifetime and Chance work experience. A most vivid station memory was going to lunch uptown on a sunny day in about 1963, with a crowd of Chance office workers. As we headed south up Allen, a Wabash passenger train destined for St. Louis pulled up to the station, blocking the crossing. Through the car windows, the passengers were looking at us as curiously as we were at them. At that time the death knoll was already sounding, as the decline in rail travel raised the thought of why anyone would be riding the train. Certainly not anyone that had something more important to do! So amidst such negativism even among those who appreciate architectural excellence, why the imposing Wabash Station stood long after its original purposes had vanished, is a credit to its substantialness and its place in the history of the Centralia area and even A. B. Chance Company.

Returning to other news, as 1925 dawned one of the first orders of business was to find out what everyone got for Christmas. Not only did the Chance family long have an eye for the latest merchandise of the day, they now had the resources to buy it. The local paper in its initial 1925 issue carried a news item about Mrs. A. B. Chance's new cabinet radio, a Christmas gift that allowed the family to receive programs from " all over the continent." In late February, a report in the paper told of the new "Chance chemical building" being planned. It was going to be three stories high and located east of the present iron foundry and near or on the site of the Illinois Oil Company, which was moving west to Hickman Street. Those familiar with the former "Chem Building" know it was a two-story brick structure with a basement. Prior to that announcement in early February 1925, A. B. Chance and wife were home from a two-week trip to New York, Philadelphia, Boston, Washington, D. C., and other "important places on the near Atlantic seaboard."

During the early Hansen and Chance employee interviews, the Chance Development and Investment Company was discussed. It was incorporated about the same time as the Chance Manufacturing Company in 1932. This special development and investment organization held A. Bishop's real-estate holdings. The license for other manufacturers to fabricate and sell his patented and trademarked products appear to have been handled by Chance Company and A. Bishop appropriately compensated as an official of the same. One of the comments made by Roy Hansen was the impressive amount of Centralia real estate he owned at one time. As he became increasingly affluent, he purchased more Centralia properties and real estate. One such transaction was recorded in the February 6, 1925 *Guard* in their column of sales or trades. Property transactions had become so prevalent that the paper was covering such occurrences along with new construction resulting from the building boom that permeated this period in Centralia. A Mr. Jarred Pryor, who was "colored," was living in a dilapidated shack, "just east of the garage he is constructing." It was called the most unsightly property on Railroad Street. This information suggests it was to the east of the Anthony Building. Before Social Security was ushered in by F.D.R.'s New Deal programs in the early 1930s, older people were often at the mercy of their church, family, or friends to help them sustain an existence. Many older people hadn't accumulated resources to the extent of being self-sufficient and hard physical work of that day often left them incapable of adequately caring for themselves. This is the picture that comes into focus of Mr. Pryor. Such hardships were born equally by Americans of all races. Mr. Chance was said to be fixing up a house on a lot east of the Doc Burnham place and "will turn it over to Jarred as a part of the deal." To be fair and balanced, A. Bishop and other Centralians were continually trying to make Centralia a showplace, especially those properties along Railroad Street and other areas visible to thousands of Wabash railroad passengers stopping and going through town or exposed to other visitors and local residents. In addition, this eyesore right next to his newly expanded and modernized Anthony Building just didn't work well at all, especially with the many community events and the Honey Ford Auto Dealership to locate there. So before we give Mr. Chance the humanitarian award for 1925, it is clear that there were other factors in this situation. However, the idea of making sure that Mr. Pryor had a new home is evidence that Mr. Chance was concerned about his welfare.

To provide more detail about A. Bishop's involvement in local real estate, the Pryor story continues into early May of 1925. The Pryor house had been torn down and the old city hall and jail building moved onto the lot. Local architect Ed Chamberlain generated plans to convert this old frame structure into a modern two-story house and a basement dwelling. The outside was to be stuccoed and the house rented when completed. This could be the house on the corner of Railroad and Collier Streets, occupied for a time by the Fentons, Husteds, and with the move of Tips Tool and many of its employees to Centralia in the late 1930s, the Kalips would also reside in that home. In mid-May 1925 the *Guard* told of A. B. Chance's involvement in building a "new modern bungalow on the site of the old J. M. Nichols place north of the Alton Railroad and when it is completed Mrs. Luna Eaton and family will occupy the house." Undoubtedly A. Bishop had a vision of making Centralia a prettier and well-kept town by replacing some of Centralia's earliest structures, which were in a poor state of repair and condition, with new or rebuilt modern homes. But there may be another story here. Notice that a Mrs. Luna Eaton and family were to occupy his latest project. This could suggest that she was a widow and that wasn't a station in life that offered a very promising future. Combining Mr. Pryor and Mrs. Eaton's stories suggest A. Bishop desired to help the less fortunate in the community.

In the same paper, to illustrate the numerous activities A. Bishop was involved in during this period, another project on Bradford Street was discussed in the paper. This was a five-and-a-half-block street running from Singleton to Sneed Street and three blocks south. Property along the first two blocks was said to be owned by A. Bishop and was being paved by him. The Fry Brothers Construction Company was doing the street paving. This was another improvement project closer to his home neighborhood and a street he may have used to go back and forth to the plant. This is another example of one man's desire to make his community a better place to live and visit. The extent of his community betterment projects has been a pleasing surprise. It indicates an important transition from trying to get it done as an elected official, to a current stance of personally having it done. In 1925 he opened his wallet and simply financed such improvements himself. This shows how big a man he was as hardly anyone could charge that he had some secret motive in performing such acts.

One of the new pieces of equipment that the Chances had been using for some time was a movie camera and it could be assumed there were home movies made of family activities. Documentation of the Chance movie camera being utilized in promoting the street-paving issue was given coverage in the local paper. Later in this chapter a reference is made about the filming of an anchor test. This equipment was bulky, apparently not very reliable and was not something the average family would have, but this isn't unusual for the A. Bishop Chance family. One of the earliest corporations that used movies and established a motion picture department for business applications was International Harvester who first produced a movie for business advertising in 1911. Other firms such as Ford Motor Company later copied this, along with standardization of parts and other industrial firsts. The application of the movie camera for business purposes at Chance was of course just a matter of time, with A. Bishop at the helm. Like the x-ray, this movie equipment would be another personal item later utilized within the company. It would give birth to an extensive Chance movie library and program to educate customers on the safe utilization of their products. In serving their customers it also became an ingenious method utilized in promoting Chance product sales. Fortunately a *Guard* news item of February 6, 1925, confirmed the possible existence of an early Chance Company advertising film. The occasion and the setting are unexpected, but that uniqueness likely resulted in this information being recorded at all. The location of this event was the new City Hall. "The Centralia Commercial club met at the new municipal building on Tuesday night and held election of officers and had a lengthy session witnessing a motion picture advertising film projected by A. B. Chance."

As this book began, one initial goal was to unlock a perceived mystery surrounding A. Bishop Chance and enough has been learned about him to significantly add to the image of his most famous and most frequently encountered portrait. A picture is often how we remember someone, like the stately one of George Washington or Abraham Lincoln that commonly hung in schools and other public places. These staged pictures, however, usually fail to capture the energy, spirit, and life of an individual, although that is often the image that is established of them and not a battlefield scene or surviving the winter at Valley Forge. A picture that hung in the reception area at 210 North Allen, or at least in the 1980 to 2000 period, although it likely was put up in the Chance headquarters after his death in 1949, was one of the founder A. Bishop Chance with a replica of his Never-Creep® Anchor invention. This is a picture of an elderly, dignified, tranquil-appearing gentleman with his crowning achievement. It suggests his highest achievement plateau, that when reached it wasn't bettered, and nothing else before or after really mattered. This Kodak rendition isn't a fake, but is a true image caught on film for all times, but it falls short of representing the actions, enthusiasm, aggressiveness, daringness, business prowess, wide and constant travels and all his technical/inventive accomplishments earlier and later. So if there is an uneasiness about a portrait or a picture as if it's suggesting there is an untold story, look closely at the sparkle in this gentleman's eyes as it hints of a time when he was a shaker and mover who had an impact on each of our lives well beyond the impact of the primitive anchor and rod appearing in his aging hands.

The expansion of the Chance plant was on the verge of going ballistic in 1925 although increased activity had prevailed since the 1917 move to 210 North Allen. Another development in the unwelcome category was a few injury accidents reported at the Chance plants, some of a serious nature. The activities at the plant were being actively reported in the *Guard* and this seemingly open-door policy revealed much about the company. The Chance plant was a work in progress, as there were problems arising with expansion and growth that were difficult to anticipate and combat. Many of the operations at Chance were some of the most dangerous and extreme of manufacturing processes and in the early stages of these quickly established capacities there were enough unknowns to generate safety issues.

One of the impressions gathered from the pages of the *Fireside Guard*, from 1876 to the late 1920s, was the seemingly lack of safety rules and provisions to counter the frequent serious injuries that resulted. It was painful to read of horrible injuries resulting from most everything from horses to the iron horse. The railroad's safety record, for instance, was terrible from frequent run-togethers of trains and the loss of hands, arms, and legs by switchmen caught between uncoupled cars. There was even a fascination by the public of such occurrences. It impressed people during the dawning of modern times, especially those caused by all the new apparatuses of the day, most of which wouldn't receive any safety awards. In almost every facet of life there were continuing stories of death and injury even in jobs and areas one would today consider a safe haven. Ladies with long dresses in the home would sometimes get too close to stoves and open fireplaces and die from severe burns. There seemed to be no concerns for safety and sometimes there was an overwhelming desire to somehow reach out and ask those people to question why such dangerous practices were tolerated. But it was a rougher, tougher, and less affluent time and many modern conveniences like advanced heating systems, machinery safe guards, strict operating procedures, safety devices, wide variety of medicines, and instantaneous communication systems were not present to help protect and serve people. There was also the long development process of technology in safety items like safety tip shoes, earplugs, and protective eyewear just as there was in manufacturing techniques and equipment. First there had to be a realization that health issues such as hearing loss was occurring before solutions could be developed to address it, as all new technological developments always have certain unknowns to deal with. There was also the numbing effects to such occurrences hardened by years of experiencing the destructive results of seemingly uncontrollable ills such as disease and other things bringing hurt, injury, and death. A positive attitude seems to have developed to just survive and prosper, as life didn't necessarily offer smooth sailing or safe passage, so acceptance and toleration seems to have become the practice of the day.

The personal habit of wearing safety glasses to protect eyes during cutting and grinding operations in metal fabrication, is a safety practice that was learned at A. B. Chance Company. When the federal government came down hard on auto safety such as requiring seat belts, a cry erupted from those charging too much government intervention. It's difficult to understand the resistance to wearing seat belts then and even today. Safety was always a personal concern for our Chance Field Sales force who drove on the average of thirty to forty thousand miles per year in reaching their customers. The Kaiser and Frazier did promote such passenger protection items as padded dashes in the 1950 time frame, followed by the mid-1950s Fords promoting many safety features including seat belts, but they didn't sell well especially when there was a price tag attached. People treated such movements as something to be ignored, and like cancer, it would go away. So businessmen, corporations, or the government weren't entirely to blame as it was difficult to gain acceptance of safety items and procedures/rules. It became obvious that accidents and injuries were costly and to generate a more positive, productive, and constant environment, more attention to safety was needed. Not always received with enthusiasm, the use of safety glasses in the plant and hardhats in certain areas were just good common sense practices and surely no one enjoyed seeing people get hurt. In addition, like profit sharing, a later developed Chance employee program, safety didn't really cost in view of its benefits, and it was the responsible and compassionate thing to do. But before this would become policy and gotten under control all across America, there were casualties. This story will end with a personal tragedy within the Chance family, which appears to have resulted because of a disregard for the proper and safe operation of an auto. This appears to have had a profound impact on the founder and it must have further influenced the stringent attention to safety in the Chance plants.

Surprisingly two accidents in the mid-1920s involved the Chandler brothers who were prominent members of the Chance Company family. Occurring in September 1924, Ollie got his right foot caught while helping move some machinery at the Chance Building on a Saturday morning. Another accident reported in the January 16, 1925 *Guard* involved Flynt, who fell twelve feet from a ladder onto a pile of bricks. A bad scalp wound requiring three or four stitches and a serious hip injury resulted. When he didn't improve under local care, A. Bishop took Flynt to the Mayo Clinic. This shows Mr. Chance was concerned about his employees. The accident appears to have occurred at the Anthony Building.

For two and a half years no accidents at the plant were found or noted, which is impressive with the increased activity and number of employees. In July of 1927, Chance employee Harry Reimers received a serious injury to his left eye when a piece of steel flew from a drill while working on an air compressor. He was immediately taken to a doctor, which may have indicated the company was handling his medical needs. In September 1927 there was an injury at the Chance Foundry. While Berry Benton was rolling a seven-hundred-pound wheel in the factory storeroom, the wheel fell upon his leg requiring seven stitches to close the wound. It was estimated he would be laid up about a month. In mid-October, Roy Witmot got a piece of steel in one of his eyes while at work in the Chance plant. He was taken to the Audrain County Hospital in Mexico for removal of the metal and was reported improving. Even with a mandatory safety glass policy, which came later, there was a frequent need for medical attention to remove foreign material from factory employee's eyes. This eventually led to a full-time nurse during the day to handle minor medical problems and to assist employees who were more seriously ill or hurt or were being transported to a hospital. This shows that the management of Chance Company was serious about medical issues concerning their employees and made adjustments to serve their needs. Ann Davis capably filled the company nurse position for several later decades.

Health issues were important to A. Bishop. One example takes us back to the July 1925 *Guard* in which an article suggests he might build a hospital in Centralia. The local paper was in favor of such a proposition indicating the advantages of having a local medical center that would keep such expenditures in town. In June of 1927 the *Guard* outlined some special equipment that the Chance Company had installed in their Department of Physiotherapy. Since it is two years removed from the hospital proposal there would seem to be no direct connection except it was another example of A. Bishop taking matters into his own hands, in this case to benefit not only his employees and company, but the local community as well. The article stated: "With this department in operation people of Centralia or especially the employees of the Chance Company will have as much advantage as a modern hospital would give them as far as electrotherapy is concerned." This type of treatment must have been popular in that day. There were three special machines installed in the plant under Gano Chance who was currently a college student at M.U. and a summer student employee of his father's company. His involvement is further explained in a September 9, 1927 *Guard* news piece stating: "Gano Chance is home from Kansas City, where he has been studying x-ray and physiotherapy." The October 14 paper confirmed that he had gone back to school. Although Gano's involvement in the x-ray machine is documented in his involvement in soybean research for the University of Missouri and during World War II at the Chance Plant (See Volume II), the rest of this information is new and not discovered elsewhere. Notice the more common reference to the Chance firm as Chance Company.

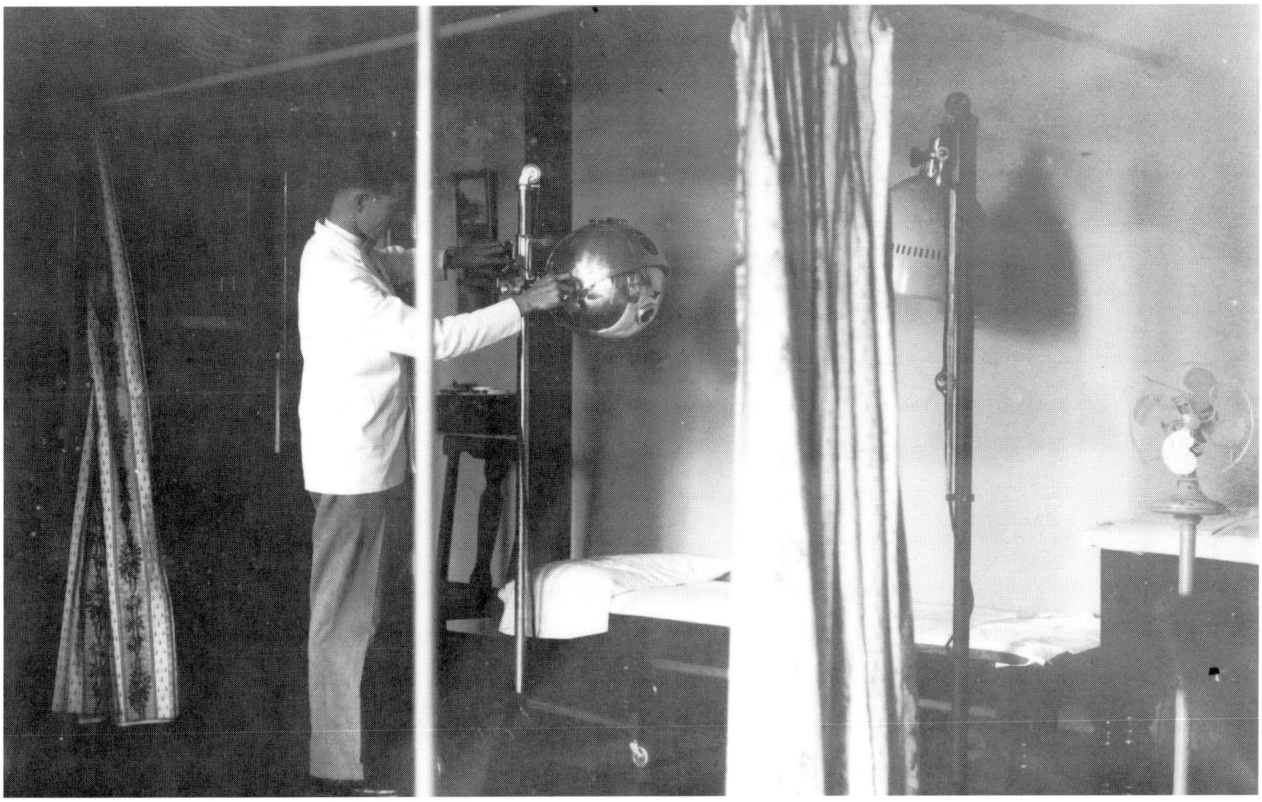

This is F. Gano Chance, circa 1927, in the new Chance Company Department of Physiotherapy and one of the apparatuses installed there. (Jack Chance Collection)

Incidentally, the three pieces of equipment included an ultraviolet ray apparatus with a mercury vapor arc lamp. This machine was a more powerful replacement for sunlight, which likely was a popular instrument during the winter months when sunlight was at a premium. In that day it was thought certain respiratory ailments such as tuberculosis, which would impact a later Chance family member, were related to such needs. Another apparatus was dubbed the deep therapy machine and was another light-type electrotherapy kind of machine. The third machine was a diathermic apparatus that produced a high-frequency electrical current. It's interesting that within this department there was also other equipment including the x-ray machine that Gano had been in charge of. It was recorded by the *Guard* that:

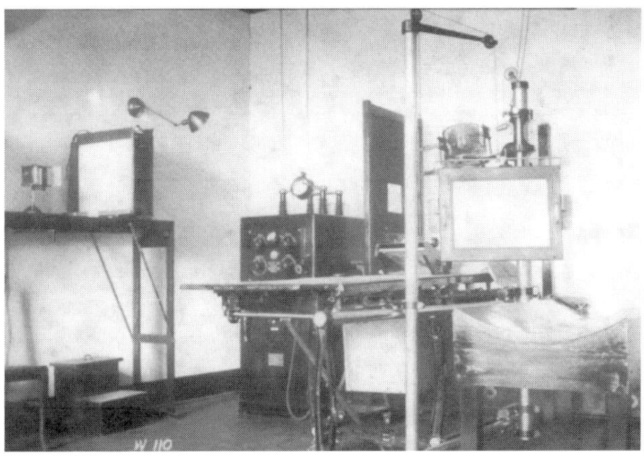

The Chance Company X-Ray Laboratory provided medical services for employees. Local doctors and patients could also utilize this advanced technology. (Centralia Historical Society)

> *the x-ray apparatus and other apparatus used by the company in relieving injuries or ailments of its employes* [sic]. *He tells us that it is claimed that 80 to 90 per cent of industrial injuries can be alleviated and cures effected so that an employe* [sic] *can be back at work in 40 per cent less time after an illness or injury than he could if he did not have access to this apparatus.*

Gano also said the company equipment was available to local physicians as had always been the case. The idea of helping out the people in the community was a longtime Chance practice and policy.

This information suggests the company was concerned about dealing with injuries in the Chance plant. With such an emphasis on effecting the curing of and the attention given to healing employee injuries, it undoubtedly also included dealing with safety issues and preventing accidents within the plant. Such information on safety programs within the Chance factory, however, is the kind of internal programs that can only be speculated about with no hard evidence yet to support it. In reference to the first-aid area or room at the Chance plant, with the just described equipment, it is interesting that Gano may have been one of the first company employees with nurse-like duties. Because of Gano's absences while attending college, it would seem logical that there was another person who doubled as a company nurse, along with their other duties. Mr. Chance's personal friend, Dr. Mayes may have been on call as were Dr. C. C. Parmer and later Dr. LaChance and Dr. Bradley for more serious cases. The need for local medical facilities was driven by the inability to reach distant hospitals quickly. Although there were several trains a day, medical emergencies requiring immediate attention couldn't wait for the next train. Roads were poor and many were of dirt making them virtually impassable during wet weather. So this isolated position drove A. Bishop's desire to be able to handle such needs locally. This early experience undoubtedly impacted F. Gano, who would in a few short years be directing his father's company. One continuing program he would support was allowing the loaning of company equipment to employees and assisting the community with special needs. He also continued a family concern about medical issues, and safety became important to him within the company during his long reign as its head.

As 1928 dawned there was an emphasis being placed on speeding up the production process. The January 13 *Guard* ran a news piece about what was occurring:

> *There was a meeting of the Chance Company and its employees Monday night to discuss the matter of inaugurating a system by which the various departments could be made to speed up production. It was decided to take several of the employes* [sic] *of the plant to other places where there are factories turning out products of iron, steel and other material of similar nature as that of the local plant, to let them form an idea of how this work is carried on, that they might be better able to help formulate the proposed system. It is said an out of town concern wants to figure on furnishing much or all of the products now made here and such a contingency would cut down the greater part of the labor now being done by home men.*

Was production performance below par in Centralia in comparison to other plants? Some will charge that management is never satisfied and always wants higher production levels and speeding up factory processes leads to more injuries. Employees tend to become complacent at times but around A. Bishop this didn't appear to be of a lengthy duration. One of the continuing challenges of local employees was to perform at or above the level of those brought in from the outside. Both A. Bishop and F. Gano would frequently bring in outside specialists to try and improve Chance Company. This always brought concerns from locals of being shown up or private jokes that another so-called expert was on the premises, an expert described as someone a distance from home. A concern for improvements by the founder may have given birth to what some would suggest was a fascination with nonlocal personnel, while management would suggest it was just good common sense to try and procure talented employees who could drive improved processes and procedures within the company. Of course the threat of outside hires didn't hurt in management's eyes of quickening the pace of the workforce. The same week as the "speed up" meeting was reported, the *Guard* carried another news item that is an early example of this. "A. B. Chance and Bryan Fenton left Wednesday morning for Chicago where Mr. Chance will likely buy new machinery, supplies and employ several expert tool makers, for his plant." This news item suggests that the kinds of investments being contemplated indicate a continued concentration on in-house production. These two contrasting news items appear legitimate, but suggest that A. Bishop may have become frustrated and this situation reflects his sometimes state of volatility. That same paper reported two Chance employees leaving employment apparently because of these changes or their status. Luke Horack, an employee for some time, left and went back to his former home in St. Louis. Joe Ulrich, a two-year veteran, resigned his position and also went to St. Louis where he took a position in a manufacturing plant. So there was fallout and changes resulting from this new program.

It is not certain who was the first of a multitude of people from out of town that came to Centralia seeking Chance employment. Not all these people were invited in, but some were opportunists who saw possibilities for a job at Chance Company. The most famous of that early group would appear to be the longtime Sales/Marketing head Jack Isbell, who came to work for Chance in 1928. Another expert with impressive credentials came on the scene in early 1926. The February 5, 1926 *Guard* had almost a full-column news piece on F. H. Dukesmith's employment at Chance. He had lived in Centralia some three months and had the charge of organizing a national sales force for piston rings. He got A. Bishop's attention by publishing the first city directory. The glowing article about him was likely spurred on because the *Guard* got the job of printing the directory, which didn't sell very well. A humorous story originated when this teacher of salesmanship tried to sell the city on purchasing several copies. The tight-purse-strings policy of longtime city head Mayor Bruton appears to have come into play as the city refused to buy any copies. So although the city directory may have gotten A. B. Chance's attention of one F. H. Dukesmith, its cost apparently had to be charged off to employment expenses.

Was the threat real of outsourcing some of the functions being performed in the Chance plants in Centralia? This idea of outsourcing is a common theme today and for the last several decades of the wholesale shifting of industrial work and jobs out of town and even to foreign countries. It is clear A. Bishop was becoming a seasoned veteran of competitive wars and was utilizing innovative and resourceful tactics, even in the personnel area. In fact, he could be considered a master in methods to pull his employees together into a company family while at the same time establishing the posture of pushing his employees for greater output. An early threat in a non-Centralia location of the Chance plant popped up again in the 1923 piston ring reorganization, although more subtle than this latest more confrontational posture. Many moves by corporations are blamed on the union, but that isn't mentioned here. Could the employees, or some of them, have become so stubborn and set in their ways to generate some drastic actions by A. Bishop? Also, what is the *Guard's* position? It is interesting how things have changed from hardly any company news to coverage of apparently every company meeting. Had A. Bishop become wiser in using the paper to promote his point of view? Had he become influential to the point that even the paper was promoting his viewpoint or were they displaying balance in their Chance Company coverage? The reporting of accidents or employee turnover wouldn't appear to be positive revelations for Mr. Chance. How was the press becoming privy to what was occurring at the plant? Clearly the success and growth of the Chance Company had made the activities

Baseball was a popular local competitive sport in the first half of the 1900s in Central Missouri. In this late 1920s, early 1930s photo, F. Gano Chance is shown with the "Chance Co." team. It was common for area teams to reach out to professionally skilled players to enhance their rosters. This picture taken near Allen Street with the Galvanizing Building, Chance Plant, and Wabash rail cars and water tower in the background consisted of at least two such talented stars, according to Jack Chance. The extreme left back row individual later became a St. Louis Cardinal pitcher. (Jack Chance Collection)

occurring there very newsworthy. In the search for conclusions it may simply benefit us to be aware that there was a volcanic-like surge of activities in Centralia and at the Chance plant. Brought with these changes are conditions and problems that are commonly associated with success and growth. There always seemed to be two contingencies of Chance employees, those loyal to there last breath to the company and those who tended to question and suggest that they didn't buy the down-home atmosphere and operating procedures at Chance Company.

Extracurricular, family kind of after-hours activities at Chance Company, were now reaching full stride. Many veteran employees, if their voices could be heard today, would indicate this was a very special part of being an employee at the plant. This did have many positive results such as forming the employees into a team, a cohesive work force, and in generating an employee society that those on the outside wanted to be a part of. Just the idea of the company head breaking bread with all employees was an original idea of leaving no employee behind! This in many ways was one of the most impressive aspects of the genius of A. Bishop in shaping and molding his personnel into a hardworking, loyal, and productive group. Thus the most significant revelation in unraveling the mysteries surrounding the founder is to show what an impressive, multitalented businessman he was. Even in his absence (world tour), Chance Company employees arranged for a Christmas tree to be erected in the Chance Gymnasium for the factory children. This was in the Anthony Building where many Chance Company Christmas parties were held. They also sent a cablegram, recorded in the December 24, 1926 *Guard*, "expressing their loyalty and greeting at Christmas time." This is an early example of the elaborate Chance Company Christmas celebrations to follow. In fact, the founder's presence at his last one in 1948, shortly before his death and prior to he and Mrs. Chance's departure for Florida, suggests the importance of this event.

One of the early sport teams organized among company employees was baseball. A reference was made earlier to a team that apparently wasn't very competitive. The May 1927 *Guard* told of various departments within the company playing each other. Besides revealing after-hours activities with company connections, there were four Chance Company departments identified: Tool Shop, Punch Press, Foundry, and the Piston Ring Department. A week later in the May 27 paper reports of one of America's greatest heroes, the Lone Eagle Charles Lindbergh had successfully completed the Atlantic crossing. Not to be outdone, Centralia had its own world travelers, Mr. and Mrs. A. B. Chance and Gano, who had returned home earlier in the month.

The July 22, 1927 *Guard* told of the Chance employee picnic held the previous Sunday in the woods at Poland. The exact location of Poland isn't known, but it could be on the farm west of town that eventually became Buford Park. The picnic was largely attended by members of the "force" and their families, said to number 150. They met at the plant at 11 a.m. and traveled by cars to the picnic location. A large dinner was followed by various games and contests. This was followed by a baseball game between the Whitewash Department and the Third Floor Green Caps. The employees capped off the day by presenting a gold chain to the boss. A. Bishop loved to be in the spotlight and responded joyfully to those around him who showed their appreciation toward him. This type of picnic and company gatherings would be continued and supported by A. Bishop, for over the next two decades. Employees also must have enjoyed it as veterans would suggest what the company used to be like and how impersonal a place it now was in comparison to the camaraderie of earlier times. Even though not everything always operated smoothly at the Chance plants, there was a fondness displayed by both the founder and his employees toward each other.

One veteran employee who began work during the mid-1930s suggested A. Bishop had a small inner-circle group of people he was close to. Apparently this young individual wasn't a part of this group. Most of us have those we are closer to and Miss Ida Lou, his multi-talented assistant, was part of another special employee working group that he depended on to keep the day-to-day activities at the plant fine-tuned. As other activities have shown, this did not appear to stop at the business end, but could extend over into the social area. The *Guard* captured such an event in its November 18, 1927 issue: "Mr. and Mrs. A. B. Chance, Miss Jennie Green, Miss Ida Traughber and Miss Maurine Chamberlain were in Mexico at the *Ben Hur* show. This would appear to be a movie of what is now considered a classic. A. Bishop enjoyed having people around him. A few months later on April 20, 1928, the paper covered an A. B. Chance business trip to Texas. While there he visited at the home of Mr. and Mrs. J. P. McKenzie. Mrs. McKenzie was a former employee at the Chance plant better known as Miss Maurine Chamberlain! A little different arrangement is suggested in the September 14, 1928 *Guard*. "Mr. and Mrs. A. B. Chance, Miss Ida Traughber, Miss Jennie Green and Mr. Taylor all motored to Jefferson City Sunday and visited the State Capitol and also visited in Wellsville."

Socializing with friends and family wasn't something only associated with A. Bishop as he also influenced his two sons to do likewise. Another picnic was covered in the September 9, 1927 paper, held at the clubhouse at Lake Deutcher. Supper was served and tennis was played. Centralia was gaining in sophistication with a game seemingly confined to large cities or special clubs of the elite. Even more interesting are the names associated with this gathering, especially Miss Anna Lee Toalson, the future daughter-in-law of Mr. and Mrs. A. B. Chance. Guests were Mr. and Mrs. A. B. Chance Jr., Miss Anna Lee Toalson, Gano Chance, Miss Anna Belle Fountain, Eddie Howard, Mr. and Mrs. Billie Atchinson, Abe Curtright, Miss Oneita Kleinsorge, Miss Hope Flemming, and Harry Jennings.

In mid-September 1927 the *Guard* covered an event at the Chance plant in a resourceful manner. Thirty Chance employees were temporarily laid off "to see the street fair." This may have been a benefit of their free time, but

Miss Anna Lee Toalson in a portrait used in the "Queens" section of the 1927 University of Missouri yearbook, the *Savitar.* She was chosen as one of the Queens representing the college that year. (Jack Chance Collection)

likely not something they anticipated. This piece was ended with the cause of this temporary unemployment, a failure of raw materials to arrive at the plant. This may have been an early example of "just-in-time" inventory control that faltered. In late November there was a business meeting at Chance Company held on a Monday night. The paper reported new rules were adopted and some contemplated changes worked out. There were a multitude of changes taking place within the plant that required a great deal of planning and coordination. In spite of 1927 being a bad year for the agricultural sector, it was a good year for the Chance Manufacturing Company, with 1928 anticipated to be even better. As is sometimes the case, one man's loss is another man's gain. One of the reasons for the success of Chance Company was its predominately farmer background workforce. This was true in the office as it surely was in the factory. For example, the drought year of 1954 brought several farmers to work at Chance and they never looked back. The same was likely true in 1927 as continuous company growth had established an employment of some eighty employees and hardworking new recruits were eagerly sought. As the year ended, a company Christmas party was likely the setting for the seventy Chance employees to receive a wristwatch from their employer on Christmas Eve. The *Guard* jokingly suggested this wasn't done to get the workers to the plant on time as the company had a time punch clock. Even news items such as this are valuable in revealing important information about how the company was developing. Closing the year, four participants in a previous fall picnic got together on New Year's Eve at the Globe Hotel for dinner. Gano Chance and Eddie Howard entertained their dates, Miss Ann Lee Toalson and Miss Anna Belle Fountain.

Introducing 1928, the famous evangelist Billy Sunday came to St. Louis in early January for an eight-week engagement. Although the previously covered Speed Up meeting at the Chance plant in the January 13, 1928 *Guard* suggested that the focus had returned away from social back to strictly business, the same paper covered an A. B. Chance–sponsored ball! Held the previous Saturday night, this masquerade ball at the Chance Hall for employees and invited friends awarded prizes for best costumes. Some were said to represent Chance products. Music was furnished by an orchestra from Mexico. Cakes with white icing and a large "C" in the middle were apparently provided by employees and it was suggested they were in compliment to the boss. This kind of atmosphere seemed to hold things together during a time that increased business, expansion and changes brought pressure to management and employees. Both A. Bishop and employee groups appeared to be promoting an organized employee entertainment agenda. These social activities were anxiously anticipated by hardworking employees, and like the later profit-sharing program, this was an additional form of compensation for employment at Chance Company. This was also an additional incentive in attracting those desiring employment. So it appears to be a win/win situation during a time when people had to be more aggressive in pursuit of entertainment. A week after the ball, a meeting at the Chance plant on a Wednesday night was covered in the paper. A club was formed among employees for social purposes and amusements. L. D. Coulter and Miss Fontelle Brooks headed the men and women of this group. All planned activities were to take place in Chance Hall, which was described as "the best equipped and arranged of any hall for the purpose in this section."

Speeding up production was being transformed from talk to action. It's unclear where A. Bishop was getting his ideas, out-sourced or from some internal expert. The *Guard* on January 20, 1928, carried news of another meeting at Chance Company involving the foremen of the various departments on a Monday night. The goal was to teach employees "better efficiency and the meaning of the term *Follow Workmen*." Another problem addressed was an age-old problem of more consistent relations between departments, although it was described "to secure better harmony in the handling of the departments." This sounds like what the boss would be suggesting when adding "make the work of the employes [sic] more pleasant and lighter if possible." The paper suggested it was sort of a "get-together meeting." Sometimes it is good to have competition between departments, when it can produce productive results. But this instead seems to be addressing the methods of some of the foremen and the disparity between departments in how they were being operated. Management was asking for a more sophisticated approach in teaching and handling employees. This is reflective of what could be expected from A. Bishop, as his higher production requests had likely generated some heavy-handed tactics by his foremen. He isn't necessarily backing off of desiring greater productivity, but he wanted his people to work smarter with a higher level of management skills being reflected. The article suggests there were limits to which he would go when more efficient techniques could produce the same or greater results with less stress on and cost to the employee. He and his management appear to be trying to grow his people at a faster pace.

The industrial sensation of the first quarter of the twentieth century, the plants of the huge Ford Motor Company complex, had been silenced for several months. On January 20, 1928, the paper announced the new Model A Ford car had finally arrived. The Centralia dealer, Donalson, took twenty-two orders in a week. They were promoting greater comfort and horsepower that would propel it down the road at sixty-five miles per hour! Although these kinds of speeds could not safely be attained on the dirt roads from Columbia to Highway 22, college students such as Miss Anna Lee Toalson and F. Gano Chance made frequent week-end trips home. The train was always a good choice, especially when it rained. As noted in the early employee interviews, C. C. Hough, a Columbia accountant, began working part time at the Chance plant. There were continuing reports of his presences at the plant, including a week in January, but usually only for a day such as on a Friday in late February 1928. He spent Tuesday and Wednesday of the week of August 13 attending to business at the Chance plant. This was the beginning of a long work-life association with Chance Company. This reflects that the efficiency and accuracy of accounting practices within the company had become more demanding with the plant's growth. The interviews confirmed some problems and revealed manager personalities on the premises during those times, some of whom had to be reigned in on occasion.

Amid all these goings on, Mayor Bruton was quietly reelected to another term as the head of the Boone County Bar. This is beginning to sound like his recordbreaking reign as mayor (eleventh election in 1925), and it would eventually lead to his departure from Centralia. In mid-November it was announced that newly elected Judge Bruton was taking office as probate judge of Boone County. He had been nominated to that position on the Democratic ticket in early September, according to the *Guard* with an accompanying picture. He resigned as mayor after seventeen years and it could be said the only way he could get away from the responsibilities of that position was to move out of town. It was noted in the mid-December, 1928 *Guard* that he was moving to Columbia and vacating his Centralia law office that he had established in 1902. So another chapter in the history of Centralia is played out by one who was deeply dedicated to public service and to the cause as he saw it. But the departure of the old is always sprinkled with reports of the new. A longtime Centralia landmark, Squires Café was being painted in late February 1928. It was an unsophisticated, hometown sort of restaurant that served good food that survived a decade or so prior to the century's end. It was always crowded at noonhour when plant workers were off for lunch prior to the fast-food invasion of Centralia.

In August there was the loss of a limb, but it was a Sturgeon farm accident. The irony of this is at the time of this writing another Sturgeon farmer by the same name (Pollard) lost his life in a similar silo accident! Backing up to the February 10, 1928 issue of the *Guard*, there was a serious accident reported at the Chance plant. On a previous Monday night a Marlow Wright of Columbia lost two toes when he stuck his foot through the safety bars of the freight elevator, operated by Night Watchman John Botkin. The attending surgeon was apparently able to reattach them. In spite of the many advancements that would occur in 1928, it was a year that would be remembered and forever marked with accidents and injury. It was a little more serious than a case of mumps in early February by future employee Paul Sturgeon. The March 30 issue of the *Guard* recorded another injury at the Chance plant of a production worker Clarence Bryson. A wiping rag got caught in his drill press and jerked his hand into the machine, which cut and crushed that member badly. Dr. C. C. Parmer was called in to give first-aid treatment. Spring orders were mushrooming production needs to the point a night shift was added in such departments as Die-Making, Punch Press, and Galvanizing, and the Foundry was run some nights and Sundays. This increased volume of activity undoubtedly contributed to work-related accidents, as did new employees. The May 11 paper reported that Mac McCardy almost lost his right eye when a chuck of metal flew off a grinding machine he was operating in the Chance plant. Although there was a surge in employment at the same time there were layoffs of some twenty-two in molding and the piston ring area, which placed pressure of another sort on workers. But nonrelated accidents were also on the rise, especially of a more serious nature from auto mishaps. For example, a Mr. Steelman lost sight in one eye while motoring to Mexico when a passing car threw a piece of gravel. These were more dangerous times with poor road conditions, machines, and autos that didn't have the safety features they do today.

Over the years the Centralia-based fleet of Chance Company vehicles were occasionally involved in accidents. The company policy of maintaining fair and positive relations with the public in such incidents involving company and public vehicles was reflected in the expeditious manner in which they were handled. Of course there were the incidents when a company vehicle would make contact with some static object or be damaged by the material carried. When these vehicles such as a maintenance truck were removed from service, this usually meant it was worn out and wouldn't run. It was amazing to see what kind of damage could result from daily workweek activities of a decade or more. There was silence as to how or when each occurred. Then there were the situations where two company vehicles would collide with each other, maybe the first of these or the first reported on April 20, 1928. The *Guard* version: "Two Ford trucks driven by Chance employees, collided at the south end of the plant at about 4 o'clock on Monday afternoon. No one was hurt besides a bad shaking up, but both of the trucks were badly stove up." A week later there was news of the widening and extension of the concrete driveway on the south side of the Chance factory. Some news coverage was flattering while others were not. Many personal concrete projects were partially derived from encountering mud roads in childhood. A. Bishop for most of his life contended with Missouri mud (remember highways out of Centralia to Mexico, Columbia, etc., were still dirt in 1928) and a passion of his was to hard surface streets, sidewalks, and vehicle passageways at his company and in his town to get out of the mud, especially in the winter and springtime. This also carried a functional advantage as the movement of materials and production at his plant was boosted when concrete floors and roadways were introduced. This particular driveway project also had a safety connotation as it suggested safer operation would result, undoubtedly referring to the recent truck accident. This is reflective of a company that was concerned about safety issues, especially during a time when injury accidents were occurring at the plant more frequently than was anticipated or would be tolerated. This is important in many respects, especially when the events of 1928 are taken into account.

As early June is entered, another serious injury is reported at the Chance factory in the Drop Forge Department. Paul Barnes lost his left hand while operating a trip hammer that developed a twenty-thousand-pound force. After being initially attended by Dr. C. C. Parmer, he was taken to the Coil Hospital in Mexico where the hand was amputated. The news article stated that this accident caused widespread sorrow. There was heavy activity at the Chance plant in filling orders in June, even some Sunday work reported. In spite of all the plant activity, A. Bishop was still making many business trips. The big Chicago trip in January was followed by an early March trip to St. Louis, an April business trip to Texas, and a week in New York in early June where he appears to have been when Barnes was injured. His travels seemed to have stopped at this point, as this is reflective of what would become a tragic latter-half of 1928 for him. Of course there was always enough occurring in Centralia to keep him occupied. In early July there was a report of the new Chance Never-Creep® Farm Post Anchor being displayed on the north side of the factory building. It was lighted up at night in an installed-like position. This is representative of the promotional side of A. Bishop.

There was no escaping the seemingly accelerated level of auto accidents. Auto performance was progressing faster than local road improvements or focuses on safety issues and operating discipline. But, accidents at the Chance plant seemed to be following the same trend. The July 27, 1928 *Guard* told of Chance employee Arnold Stewart becoming ruptured in lifting a heavy die block. Although there was progress in crane installations in certain raw material areas, many of the factory operations relied on brute human strength. The same paper spoke of Arthur Douglas being severely burned in the Chance Foundry the previous Saturday while pouring molten iron. He was said to be improving nicely. The factory was a hard and sometimes dangerous place to work and that is why Mrs. Fancy Chance didn't like her student son Gano working in the Foundry during summer break. But in spite of all these challenges there were the lighter moments of anticipating the annual Chance Company Picnic. There is no solid confirmation that it was an "annual" event, although one was held in 1927. The same late-July paper told of the cancellation the previous Sunday of this event because of a heavy rain on Saturday, but it was rescheduled for the upcoming Sunday.

The Chance plant attracted a lot of attention in Centralia with the continuous whirlwind of

No. 23–Operation 7 factory scene: The Chance factory scene pictures four identical electrical motored machines that were rough turning (machining) the outside surface of piston rings. The four operators were from left to right: Jim Miye, Charles Ancel, Don Sexton, and Bud Hicks. This is likely a 1930s scene on the third floor of the main factory. A special feature of this photo is the safety sign, "Wear Your Goggles," just below the clock where employees were sure to observe, from time to time. (Centralia Historical Society)

The Chance plant consisted of eighty-three employees in late September 1928 and the larger number of workers and plant activities held the potential for more accidents. The *Guard* of September 28 told of two more injuries, the first to Drop Forge's Arthur Douglas. His luck or work habits were no better in the Drop Forge than they were in the Foundry, where he had been burned in July. A piece of metal came off a hammer striking and fracturing his kneecap. He was said to be getting around on crutches. Another Drop Forge Department injury to Joe Drown, a punch press operator, occurred when a piece of metal struck his right arm. It was bruised and generated a great deal of pain, but he was treated by the First Aid Department in the plant. Such occurrences apparently didn't deter people from seeking employment there. Miss Irene Oldham, a veteran telephone employee (Missouri Telephone Company and former chief operator for Home Telephone) took a Mailing Department position in the Piston Ring plant, per the October 12 *Guard*. Her position at Missouri Telephone Company wasn't to be filled. This is an example of a continuous attempt by A. Bishop to bring quality and proven employees into his firm and in taking care of his own.

activities occurring there and the constant flow of trains through town. It had transformed the area into a big-city kind of atmosphere. This was clearly evident as a flood of employees entered and left the Chance premises at the beginning and ending of shifts. The noise and smell of factory operations extended even into the night and weekends. The general public becomes fascinated with such overnight sensations, although we now know of the long years of struggles to attain success. In response to the loose security and the problems caused by unauthorized people in Chance plants or on Chance property, a no-trespassing policy was established in September. Notices were put up all over the plant warning against trespassing during working hours. The reason cited was people coming into various departments at will, meeting with accidents, causing damage and costing production time. Tours were apparently allowed with proper supervision, much like it was in later years.

As October 1928 introduced fall once again, Chance employees continued to do things together as Miss Ida Traughber, Miss Nina Rowland, and Beverly Mayes and son traveled to Columbia on a Friday night to see the movie *Fightin' Fool*. As November was coming onto the scene A. Bishop was testing the economics of municipal-provided electrical power for his plant instead of his power generating unit. Another person often connected with economics, although he was a renowned engineer, Herbert

Hoover had been selected as the Republican presidential candidate. But November 1928 will for the Centralia people of that time always relate back to accidents and injury beyond that suffered by Chance employee Irl R. Hicks with an infected thumb. Miss Ida Traughber's sister in California suffered the loss of her husband Charles Bryson. Miss Ida Lou traveled west to be with her sister and while there visited with her brother Dr. W. F. Traughber and family. To support Chance Company concern about safety issues, what could be labeled as a safety meeting was recorded in the November 9 paper and was held at the Chance plant for all employees the previous Friday, November 2. The company was requesting that employees wear their goggles while working at the emery and grinding machines as supported by the factory picture on page 260. This confirms such safety eye protection was available and was being provided to employees. The company also told the gathered group to use all precautions for their own safety when at work, in and out of the factory and especially at the machines. Many employee injuries do occur outside the plant during off hours. Management and public concerns over the frequent injuries at the plant, especially during 1928, undoubtedly prompted this meeting. It is not an attractive image to reflect that one is calloused and unconcerned about such conditions, and it appears Mr. Chance was not happy with the plant's recent safety record. As the Thanksgiving holiday was approaching, this posture set a positive tone and one can sense that safety was becoming an important issue in the plant. This is a blessing in a way, before the tragic news of Buford Chance's death on November 24, 1928, especially since it related to an accident out of the plant, suggesting that the concerns voiced at the company safety meeting were prophetic in a sense.

There was seemingly nothing that could halt plant activity, as even the Christian Chance founder had resorted to Sunday work at his factory. This couldn't have been a very popular move especially in an area that had many churches and people who faithfully attended them. Working on Sunday was considered a taboo, but the local press was mute on this subject. Some might suggest A. Bishop had strayed from the beaten path and had become a success-driven man, but that seems unfair. The death of Buford was another test brought by the rigors and difficulties life can bring our way. The November 29 *Guard* revealed that the Chance plant had been closed since Monday, November 25, on account of Buford's death and Thanksgiving and would reopen Friday morning.

Suggesting that the momentum of problems at the Chance plant was continuing to raise its ugly head, a serious elevator accident was narrowly averted on December 12. When the meeting of several employees ended they crowded into the elevator and as the elevator descended, the steel safety gate slid out of place catching the steps on the second floor, buckling it in the center and thrusting it out upon the riders. Luckily a portion of the bent gate hit the safety stop, which prevented potentially more serious injuries. That same *Guard* issue told of a Christmas program being planned by Chance employees, possibly by the people in the elevator incident. This was likely being done with an emphasis on giving comfort to their Boss who was dealing with the loss of his son. Without A. Bishop's blessings, this apparently didn't occur as the Chance family instead desired solitude in their time of grieving. The Chance plant, except the Molding Department, was closed down during Christmas week. The local paper suggested this was for the employees to enjoy the festive season and because of a bad outbreak of flu. Chance Plant Superintendent Bryan Fenton, residing on Railroad Street, Daniel Sennett, and others were reported as being confined to their beds. So instead of the plant usually closing only on Christmas Day, the 1928 holiday season at the Chance plant was handled in a way that reflected the troubling circumstances that had engulfed the Chance family.

Chance Plant Transformation

THREE STORIES, FOUNDRIES, AND MULE BARNS!

Noting the progress of the Chance Manufacturing Company in the last half of the 1920s, one conclusion that can be reached is A. Bishop Chance's dreams had finally come true. This taste of success had many far-reaching, positive influences on the founder and his employees. An adrenalin rush can be felt as this exciting period is more closely examined. In this section will be revealed the structures being added, the products being built, and the activities taking place within the walls of the constantly enlarging and increasingly more sophisticated Chance complex.

The maddening expansion pace of activities orchestrated at Chance Company would appear to have reached its zenith in 1925–26. The closing of Rollins across the tracks and the opening of Collier, with all the business relocations and new Chance chemical and paint plants, and the transformation of the original Chance factory complex with a third-story plus a power plant addition, and the large Foundry Building added to its rear, transformed the whole between the tracks area. These changes in size, quantity, and quality of structures signaled that Centralia's Industrial Park was maturing in grand style.

With the water and sewer bonds paid off, the city was exercising its newfound economic clout in street paving, in purchasing the electrical distribution system, and the completion of the new City Hall. Electricity was a hot issue in the spring of 1925, brought in part by the increasing consumption by the Chance factory operations. At a special meeting with power and city officials, A. Bishop suggested the possibility of going back on line with the power company. He was asking if they could provide current without a break. A five-minute interruption could cost $5,000 in his Foundry operation by ruining "the cupola full of cold metal and leave him a mess of ruined iron and the framework of the furnace." He preferred using current from the transmission line if assured of continuous service and at a lower cost. Cost was very important as he said "he was in the manufacturing business with a line of goods that had strong competition with goods manufactured in cities where electric current was purchased at from 1½ to 2 cents, and he had to pay the city here 5 cents for what he used." The city was providing current at near cost to encourage his operation in town. He agreed to go to Chicago to the power company's headquarters to meet and consult with them about his requirements. A May 8 news item carried the results of A. Bishop's meeting with Missouri Power and Light officials, which hadn't been worked out to his approval. The construction of a new or larger generating plant was anticipated within Chance Company, but they would still use city power for regular and emergency needs.

The May 8, 1925 *Guard* carried stories about the new Chance buildings. From unofficial sources, it was suggested that the original two-story main Chance factory building at 210 North Allen would be extended to three stories. What was referred to as the power plant addition to its immediate rear would also extend to three stories. In addition, a new east extension to the Honey garage building was taking place. In the midst of this building frenzy an extracurricular activity performed by a Chance employee stole the spotlight because of its unusual nature. It was an example of the Chance community-minded service policy, possibly a result of company funds being firmly "locked up!" The safe at the Farmers and Merchants Bank had a faulty time lock and couldn't be opened. A called in "expert" of the $25 per day variety had failed to work the combination. As the seriousness of the dilemma intensified, A. Bishop must have

Plants: This collage of photos represent the Centralia Allen Street factory complex of A. B. Chance Company. It is the post 1929 period as the new Punch Press building, in the lower left, had already replaced the Mule Barn. The center building is the Main factory/office. Note the front and southwest main entrances which had some type of awning to protect the front office from the hot summer sun. The cupola on the rear top was not a permanent structure. The Galvanizing building is to the right in front of the Wabash water tower. The lower right structure is the Anthony building at Railroad and Rollins. The upper right plant is the Foundry that was attached to the rear of the Main building. The smoke stacks, etc. in the right center may pinpoint the location of the furnaces used to melt the metal for casting production. The center water tower is a city of Centralia structure that was the result of Mayor A. B. Chance's administration of 1909-1911. It was located to the rear of the Foundry building. The north side view in the upper left features the Chemical building. Note the street in front of it across the tracks is Collier. Those tracks are of the Alton Railroad yard that was taken out in 2006. The building across the street is likely a paint shop, storage, and possibly the Drop Forge building. These separate buildings were to the north of the Foundry that eventually became the modern Drop Forge that attached to the Foundry and extended north and east beyond the Chemical building. (Jack Chance Collection)

sensed the need to try another approach and sent one of his most talented employees, Irl R. Hicks, over to the bank with an acetylene torch. One can imagine the local chatter this generated and the crowd that gathered outside of the bank. After eight hours of being hit with a shower of heat and metal sparks from the huge door, Irl cut through so the door could be manually opened. Irl's only remark in a hero kind of atmosphere was he didn't see a great deal of promise in burglarizing a safe in that manner!

Centralians were supporters of the several Columbia institutions of higher learning. The convenience of its short distance allowed Centralia-area students the convenience of returning home on weekends. Local college interests were also due to the positive and sophisticated atmosphere in and around Centralia. Miss Anna Lee Toalson was a student at Stephens College, which suggests it wasn't uncommon for women to be attending college. For example, the graduating class of the Centralia High School in May 1925 reflected an astounding three to four ratio of ladies to

The "Cupola in Operation" as this picture is described is a vertical furnace for melting iron. This is a post-1925 picture within the new Foundry Building behind the main factory/office. The background is of the raised area at the east end of this plant. (Centralia Historical Society)

young gentlemen in that year's graduating class! A high school diploma was quite an accomplishment in those days as many young people entered the work force after graduating from grade school. The more affluent and those thirsty for more worldly academic adventures even went beyond routine college enrollment by taking special tours. A surprising development was the later Chance world cruise was first anticipated in June of 1925, according to the local paper. Students would receive a full nine-months credit for this "University Around the World Cruise." A. Bishop anticipated going with his son F. Gano if business affairs could be worked out. Departure was scheduled for September 25, 1925, but the *Guard* issue of that date stated the cruise had been postponed until October of 1926. Area students also approached their preparations differently for entering the job market. A future Chance employee, Harold Cox, graduated from Quincy Business School in June with accounting credentials and associated himself with his brother at LaCrosse Lumber.

There were many ways the Chance factory was impacting Centralia, including electricity. The area's consumption of about twenty thousand kilowatts per month in May of 1925 was represented by the Chance operation using about one-half of that total. The new Chance Foundry would run this up even more and the higher consumption brought more leverage in lowering rates. However, the city was concerned about agreeing to a new rate structure with higher consumption reductions. If they would lose the Chance volume it would actually increase local rates. A large plant in a small town brought challenges to local officials, although a cooperative spirit existed due in part to the overall positive influence of the Chance plant on the local economy. Apparently the city's two big boilers at the water plant, that earlier had generated city power, were available to Chance Company. The mayor, according to the June 19, 1925 paper, said it was possible they could be rented to A. B. Chance for steam heat for his plant.

The construction of a new warehouse on the Alton tracks at the rear of the present foundry, was to begin soon, according to the June 12, 1925 *Guard*. It would be two hundred feet long and forty feet wide. It was opposite the new Chemical Building, "and on the west side of the new section of Collier Street. It is to be a one-story brick structure and will house the anchors and other manufactured articles and supplies." This project may have been triggered by preparations for the new Wabash Station that required the moving of some structures.

The previously discussed plant shutdown and piston ring reorganization in December 1923 is reflective of inadequate factory space. A. Bishop wasn't afraid of change and his frequent reshuffling of the plant layouts to utilize every square inch of floor space, made for a pleasant place for those employees who were easily bored. On September 26, 1924, the *Guard* ran an article on a new Chance Foundry. A blast furnace and molding room were first fired up in early 1921 in a temporary building. This current building plan had actually been proposed four years earlier.

> *MAY EXTEND THE FOUNDRY. Report has it that the long contemplated malleable iron foundry at the Chance plant is to be built this fall or winter and that the plant is to be extended to the east as far as the city water plant will permit.*
>
> *It is said the Illinois oil station is to be moved to the eastward on Collier Street which is to be opened up to traffic across both railroads.*
>
> *We mentioned this project some months ago, but it was not under way at that time. The plant when completed, will take care of much of the work that is now being made for the Chance plant in other places, if, indeed, it does not take care of all of this work.*
>
> *Work of opening up Collier Street we are told has already begun.*

This confirms the location of the City Water Plant and the Illinois Oil Station in that Collier Street area. The Collier Street project may have begun, but it would be another year before being completed. However, this would tend to deter any criticism of the Rollins Street closure and would address Chance Company's growing needs back in that east area. Another aspect of the future Collier Street industrial area was it challenged Allen Street's importance. In fact, after all the mid-1920s upheaval, including the street's construction, it exceeded the Allen Street area in its impressive array of new buildings. There was another new Mule Barn, Crump and Dawson coal bin, Chance Chemical Building with west facing frontage, the City Utility/Water Tower Complex, Chance Foundry, and the new Chance Paint Shop buildings with an east frontage. This likely reached its zenith in 1948 with the removal of the Mule Barn and the completion of the Chance Switch and Tool Building. Some or all of these buildings had front facades that were designed to impress like the business building fronts used to attract customers. This surely rivaled or at least relieved Allen Street of some local traffic over to Railroad or to Highway No. 22. It is difficult to visualize this now, with the Foundry and Tool Building the lone reminders of a time when Collier became a hub of activity, especially with all the Wabash Driveway traffic and east end Railroad Express activity.

There was a significant change in the Illinois Oil Company location, as earlier stated in the September Foundry article. The November 28, 1924 *Guard* told of its relocation under the direction of Manager Hinton Gulick to the former Ira Smith property on North Hickman Street, north of Stone's poultry house. Stone's is the current location of a farmer's elevator on the corner of Hickman and Wilson. The 160-by-150-foot lot is where the new oil house and tanks, designed by local architect Ed Chamberlain, were to be located. Today this would appear to be the facility occupied by the Missouri Farmers Association jobber. This business's former location between the tracks on Rollins would become inaccessible with that street's closure (between tracks). The Alton was extending their switch track "on west to Barr Street as a team track, and this same spur will serve the Illinois station as a track for their oil cars." This track appears to have been just to the north edge of these businesses, located north of Wilson Street. This is another example of the improvements in function and appearance of Centralia firms. The movement of businesses out of the Rollins Street area would open the door to Chance Company's eventual absorption of the entire area, between Hickman and Jefferson.

To continue the marathon of proposed Chance plant construction, the June 19, 1925 *Guard* carried an article headlined, "The New Foundry." The Chance Malleable Iron Foundry, which had been spoken of for a long time, would begin as soon as the several other current projects were completed. Seven different firms were competing for supplying the steel structure for this large building. The Structural Steel Company of St. Louis was the low bidder for an estimated 225,000 pounds of steel to construct the 320-by-80-foot-wide, two-story structure. The upper area was to have a concentrated array of windows for summer ventilation and lighting. Veteran employee Paul Sturgeon suggested Moberly paving (street-type) brick was to be used in the walls. The castings produced in this large facility would be used in a wide variety of Chance products, but immediately concentrated in the piston ring and anchor lines. The article suggested this was a unique facility for the western area and could attract other factories with needs for malleable iron material. This didn't occur but it does show the boomtown potential of Centralia in the 1920s.

Bricks played an important role in the construction of buildings during the late 1800s and early 1900s. They are still preferred in certain applications for appearance and durability, but because of economics and time constraints most commercial buildings of today are sided with metal sheeting. Like most other popular products, the commercialization of bricks in these earlier times resulted in new applications such as sidewalks and city streets. So it isn't a surprise that when A Bishop selected a material for the walls of his new foundry building in Centralia, the material of choice was brick, and there was a Moberly connection. On Beuth Road is a large pit, now a lake and city park, where the clay was dug and transported by a short rail line up to the brick plant near the Katy/Wabash main line in south Moberly, where these brick were manufactured.

Roy Hansen and Bill Hollander gave an outline of the Chance products in the late 1920s and 1930s, some of which required a foundry to produce. Roy said, "So you had the product lines of piston rings, expanding anchors, Never-Creep® Anchors, and automotive parts." (1979) Bill responded:

> *That is right. The other anchor was the screw anchor which was cast iron. They made cast iron cone anchors and malleable iron Never-Creep® Anchors. The hardware items were the various eye rods. We did not make any headed bolts and such other than eye bolts. Hardware items were the thimbleye bolts, the anchor rods, twin eye anchor rods, eye nuts, the angle thimbleye that bolted down a guy, and the piston*

rings, The big part of the foundry tonnage was piston rings, cone anchors, and screw anchors. The tools they made at the time were the square wrench to install screw anchors and they forged what they called a lineman's wrench. I think there were open ends on it with a hammer on it. You could drive a pole step and it had a few open ends that fit 5/8 and 3/4" square nuts that were used so much on anchor and pole line hardware work at the time. They used to call it the 4 in 1 tool. It also had a hole that after you drove the pole step to start it, you could hook it through this hole and you could screw the pole step to the pole. (1979)

Roy responded that the "4 in 1 tool" was still in the catalog as late as 1950.

A. Bishop Chance was like a dynamo in respect to his seemingly endless generation of new ideas. Expansion requires balance, however, and one advancement requires a bevy of other changes to insure that the entire operation complimented each other, and the overall capabilities of each element reached an optimum efficiency level. The earlier offer by the city for rental of their heating plant to Chance was picking up steam, per another June 19 article on the front page of the *Guard* entitled, "A Heating Plant." The two boiler units hadn't been used by the city for two years, which earlier had been used to pump water out of the city well, now performed by electric pumps. According to an engineer called in to examine the possibilities, the plant could be converted to heating units, and being housed in the city building with a large chimney, they were ideally located in the area of the several building, Chance complex. It was suggested they would likely be converted from coal to oil fired. That same paper told of another new Chance building started in the past week that would soon be completed. All of the area brick masons and construction crews must have been in Centralia in 1925! This latest structure, the new Paint Shop, would be 40 by 120 feet and be located directly west of the new Chemical Building. A new "ware-room of the same size will be erected directly west of this building as soon as the former is completed." Notice the latest version was different than described in the earlier June 12 article. The start of the big foundry would begin when those buildings were completed. The paper suggested Mr. Chance would soon fill up the between the tracks area if this pace continued.

In early July 1925 the foundation for the new Foundry was being dug plus steel was arriving and a week later the erection phase was in progress. The addition to the rear of the main/original factory building was also in progress. The foundation that included a basement was allowed to harden because of the heavy electrical generating machinery and the three-story structure that would sit on it. The July 31, 1925 *Guard* ran an article on the new Chance Power plant. The brick masons were erecting the walls using the same mat surface brick as used on the J. Ford Honey garage (Anthony Building). The tall smokestack would allow the smoke from the electrical generating units to rise high into the air. These units would provide a large amount of power needed in the Chance plants. The article stated the Chance operation would continue to utilize some electrical power supplied by the city, but this in-house source would enable the new Foundry to have power in case of interruptions on the transmission line supplying current to Centralia. These proposed buildings were quickly becoming a reality as the between-the-tracks grounds were being transformed almost overnight. The task of financing all this construction would seem to be a full-time job for A. Bishop, besides planning and overseeing the daily operations of this growing concern. This was occurring as a relatively unknown by the name of J. E. (Edgar) Hoover became chief of the U.S. Secret Service.

In the employee interviews, Paul Sturgeon spoke about the rear addition to the original

Chance factory or main office building at 210 North Allen. Note that his remarks about the widening coincide with the eighty-foot width dimensions of the new Foundry Building, as covered a few paragraphs back. This area would eventually become office space on the second and third floors such as the mid-century conference room on the third floor and office services (mail room and office supplies) on the first. However earlier, part or all of the north side of Office Services and a small second-floor loft storage area was the machine shop and all or portions of the upper floors were factory areas. Because of the greater height first floor, there was a central ramp and short stairs on the second floor to reach the higher level of this new addition. The widening in the rear apparently lined up more closely with the traffic lanes as described by Paul Sturgeon:

> *In the office building, which is a building 60' wide by 100' long, the first two floors was built in 1917. I don't know whether it was 1924 or 1925, somewhere along in there, they added the third floor and the back section. The back section they widened to 80' width to go with the street. (1979)*

More importantly, this widening allowed for greater square footage in floor area. This was especially beneficial when the new Foundry Building was added to the rear, as for several decades this area became the hub of Chance factory activity. The additional floor space was vital during the later expansion periods as proven by later additions tacked onto both the north and south sides.

August in the heartland of America is often associated with fairs. The Centralia Fair of 1925 had an expanded opportunity of utilizing the new Chance Auditorium in the Anthony Building for some of the fair displays. The Chautauqua continued its late summer presence in Centralia with an August 22–28 schedule. However, the theft of a Chance Company car and the progress on the new Wabash Depot also shared the spotlight of local events. Entertaining was also a part of the summer scene at the A. B. Chance home with Mrs. Margaret Greenlee making one of her frequent visits in early September of 1925. As fall approached, the October 2, 1925 *Guard* captured details of "Adding a Third Story," an article about the third-story addition to the main factory building at 210 North Allen. The firewalls were taken down and the upper part of the building rebuilt. The new upper story was to be utilized in the manufacturing of the "various electrical and automotive products of the plant, and probably for storage purposes." The article traced the beginning, when Chance only utilized the first floor, and when the Foundry was put in, people thought he was "plunging," but even with all the other buildings today, he had run out of space. They predicted the construction of another building before spring 1926. This was the kind of boomtown atmosphere that permeated this rural, prairie community in 1925. Work on the new Foundry to the east had been slowed by the weather, but it would be rushed along as permitted. Amidst his company's success, A. Bishop Chance had become a local celebrity. Rumors and stories about him occupied the supper hour in area homes. One such rumor, which he told a reporter was false, was the possibility of him erecting a new three-story hotel in Centralia! He said all the major construction projects he was going to get involved in, were now in progress. He also stated that the rumor that he owned the corner lot on Railroad and Rollins Streets was incorrect as that property belonged to C. W. Settle Sr. It was an upbeat time, with a construction frenzy surely never again equaled in the company's history. To confirm faith in his town, A. Bishop addressed the Commercial Club in late September on the merits of new industry locating in Centralia. He told of the benefits of it being centrally located and the fine railroad facilities serving the Prairie Queen.

The fall of the year in Centralia was always anxiously anticipated, as it became a tradition of many area residents to attend University of Missouri football games. One individual who later influenced a multitude of people to head toward Columbia on Saturday mornings was Francis Gano Chance, who was listed in the paper as a University of Missouri attendee in late October of 1925. The company later developed a special program of inviting guests from out-of-state (Big 8 Conference areas) that were customers or associates, for "Football Weekends." The football season also ushered in thoughts about preparing for winter. There were concerns about streets and roads, especially one who wanted Railroad Street in front of his new and expanded Anthony garage complex paved. The Wabash opposed helping with this, awaiting completion of their new depot. Although the telephone business had fallen into the shadows of the news coverage generated by all the local construction activity, the Chances were still very much involved in Home Telephone. A new $40,000 "central" was being built at Sneed and Rollins, across from the Post Office or where Meador Funeral Home now sits. Progress on the new Wabash Depot was continuing in late November, as the roof was being completed after a month's delay, caused by the huge roof timbers being rejected.

As 1926 dawned, our nation's interstructure was being further developed which influenced future times. There was a movement afoot to provide greater electrical generating capacity, suggesting its growing utilization and importance. The primary justification for building dams on rivers was for generating electricity, but the resulting lakes became a boom for water sources and recreational purposes. Such a proposition received state approval for the construction of Bagnell Dam in south Missouri that would establish the Lake of Ozarks. In late January the local paper also covered an interesting foursome, not for golf but in divesting of the Howard Building property on Singleton Street that they had earlier purchased. They were A. B. Chance, H. E. Stone, O. B. Mayes, and Arthur Bruton. The new owner was J. H. Bagby. The previous mayoral candidates who at times represented adversarial camps in local city politics now seemed to have resolved themselves to a closer relationship. A case in point is a St. Louis trip in May by Mayor Bruton and A. B. Chance to discuss street paving with the Wabash.

By early March of 1926, a news item about the ugly coal sheds between the old station and the three-story Chance factory confirms that the third story was now on the main Chance building. The new and quality construction taking place had the immediate impact of making any poorly constructed or maintained structures look out of place. The Wabash was building a fence along Railroad Street in front of the new station to insure safety. They were also grading and sowing grass on certain areas around the new station, further adding to the beauty and well-maintained looks of the between the tracks grounds. There were also changes taking place inside the Chance plant as office space was established on the second floor of the expanded main building. The paper on March 5 reported that Mr. Chance would move his business office and force to the new suite, when completed. Along with gains come losses, as the April 9 paper covered the passing of M. S. Bush (1856–1926). This Baptist gentleman was mayor twice and one of the most prominent of early Centralia businessmen. He, of course, had an early influence and connection with A. Bishop Chance. His grandfather T. V. Bush invented and patented the first corn planter, which also has a Chance connection as will be seen in Volume II.

As mid-April and springtime once again came to central Missouri, negotiations were still taking place on the unresolved issue of the paving of Railroad Street. Since A. Bishop was not one who got stuck on something he couldn't immediately solve, he instead headed west across Allen Street in a significant expansion move. He had squeezed about all the plant between Allen and Collier that could be allowed at that time, and there were at least two

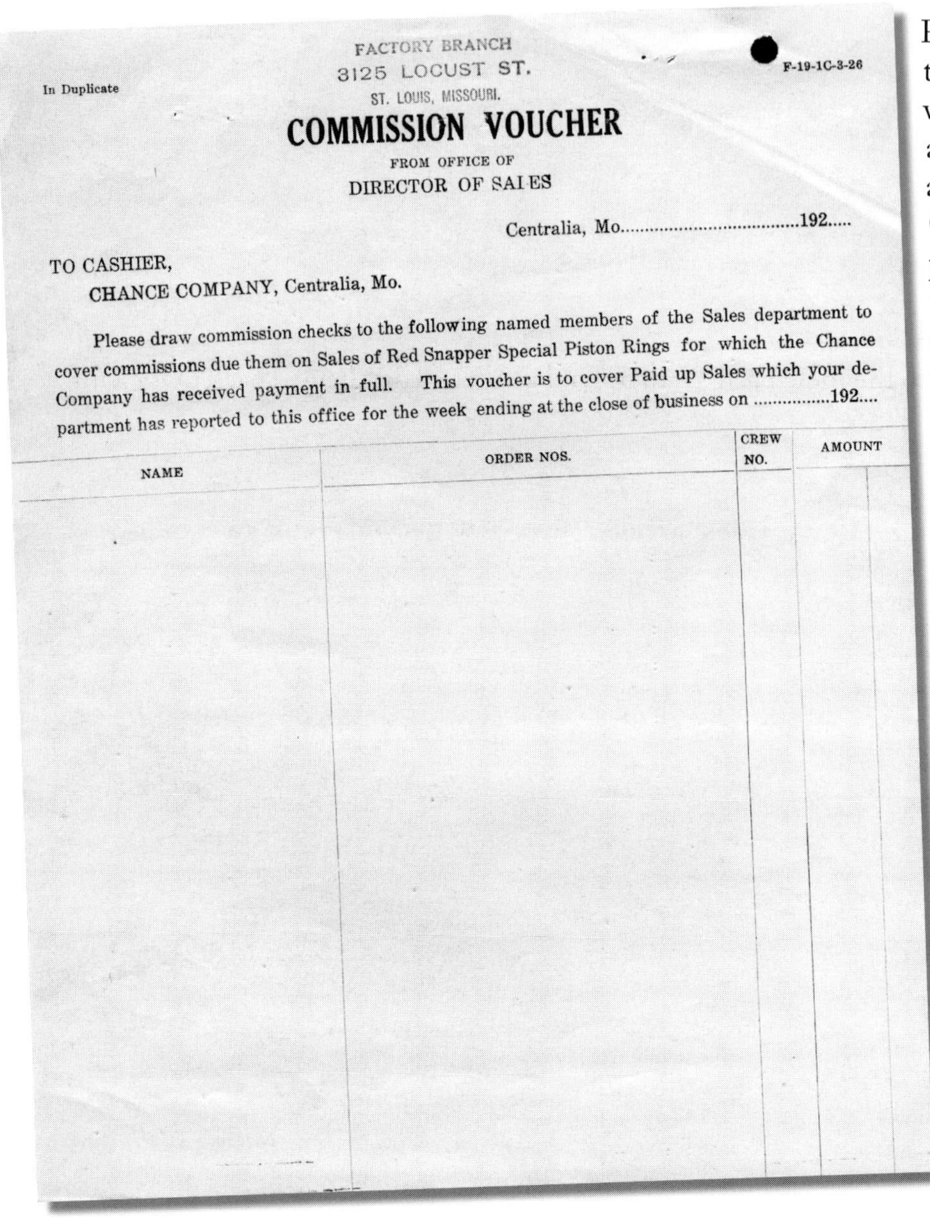

FACTORY BRANCH
3125 LOCUST ST.
ST. LOUIS, MISSOURI.

In Duplicate

F-19-1C-3-26

COMMISSION VOUCHER

FROM OFFICE OF
DIRECTOR OF SALES

Centralia, Mo.....................................192.....

TO CASHIER,
CHANCE COMPANY, Centralia, Mo.

Please draw commission checks to the following named members of the Sales department to cover commissions due them on Sales of Red Snapper Special Piston Rings for which the Chance Company has received payment in full. This voucher is to cover Paid up Sales which your department has reported to this office for the week ending at the close of business on ....192....

NAME	ORDER NOS.	CREW NO.	AMOUNT

Form-19-1C (Big C) dated March 1926, reflects a system to pay commissions to agents who were solely or partially responsible for the sales of Chance Red Snapper Piston Rings #1756534. (A. B. Chance Co.)

reasons why it was a good choice to proceed west. First, with the prevailing westerly winds through north central Missouri, the Green and Sons Mule Barn located directly across Allen Street was not an attractive situation either from the perspective of the eye or nose. The April 16, 1926 *Guard* covered this story they entitled: "CHANCE BUYS BIG BARN." The mule business was a declining proposition as mechanized farming progressed, and Mr. Green was no longer utilizing this very large building. Plans for a smaller Mule Barn to be located east of the city water plant (Collier Street area) was reason enough to accept an attractive offer from Chance Company. The close proximity and convenience of this building to the Chance complex across the street, just when they were needing more space for storage of increased production levels, sealed the deal. According to the article, Chance Company was going to remodel the interior and add a "spur switch track from the west down the center of the building to facilitate the loading and unloading of materials." The eventual Alton spur track was instead located along the south side of Wilson Street, or on the north side of the Mule Barn location. Pictures of the Mule Barn a few pages forward, show it to be located where the new Punch Press Building was erected in 1929. There is still today a large steel frame used to hoist loads in and out of rail cars on the mid-north side of the Punch Press Building, which defines the location of the 1929 Alton spur track serving the mule barn. However, the Wabash siding close to the south side of that building was also future and so it is unclear which rail line was being considered. A. Bishop did have additional leverage to persuade Wabash Railroad officials to help pave Railroad Street. A chance to charm them in late April at a city banquet held in the Chance Hall, to give thanks for the new Wabash Station, might do the trick!

The move of W. L. Green of Mule Barn fame was more specifically identified in mid-May as he would locate on Collier where the Davis Blacksmith Shop was situated. In mid-June further information of the between-the-tracks location told of his purchase of a sixty-five-by-one-hundred-foot area. Ironically, this would be near or the lot where the new 1948-era Chance Switch and Tool Building would be built. So in a couple of decades the A. B. Chance Company would again begin expanding to the east, a move that would bring closure to the mule barn era in Centralia.

Information about early Chance Company employees suggests they were a versatile group. Their experiences ran the whole gamut and qualified them as construction and transportation professionals. These men might laugh at the idea of being labeled professionals, artisans of a multitude of trades. They would more likely humbly consider themselves apprentices and simply laborers, but they were architects in the building of Chance Company just as A. Bishop was. They were the ones who put form to many of the founder's ideas and at the same time took care of the day-to-day business activities responsible for sustaining and supporting the expansion of the company. Although building craftsmen and contractors were brought in to handle some difficult jobs such as the assembling of a building's steel framework or the delicate skill of bricklaying and masonry, nevertheless Chance employees were involved in many construction details in the Allen Street factory complex. Going back to 1926, when Les Reynolds was employed, brought back remembrances of his involvement in the plants. "I went to work in the piston ring [department] up stairs [Main Office]. You might work up there two days, you got covered with water." (1979) The subject of the Mule Barn also brought back memories of Les's involvement.

> *When they cleared out that mule barn there was so much manure. When it was slow* [regular production] *we hauled all that dirt in those old Model "T" dump trucks right across the road. That concrete floor in the mule barn, I had to turn all of that over there and about the Main building clear on down, there was no concrete floor at that time. We poured all of that in there. (1979)*

Since the concentration of manure would not make a good base for a permanent floor, it had to be hauled out and dirt hauled back in for a fill. Centralia's water drainage problems also made it necessary to have the building sitting above the water level during rains or flooding. The concrete floor in the Mule Barn was completed in April of 1928. Les's reference of company employees pouring the plant's concrete floors brought up some additional details. When there was a need to pour more concrete after regular hours, Les said that the Boss would tell them:

> *Those it don't suit, you go home and the ones it suits can stay. Well there wasn't anybody that went home. There was no overtime, everything was just straight time. He would come back about the time it would take* [to complete the job]. *Nobody checked out, he would take the names of the ones there and check your card. He would give you no overtime but gave you extra hours. (1979)*

This information revealed the fairness and the appreciation A. Bishop showed for his dedicated employees. It's very gratifying to see a businessman who did over and above what was expected or required and shows he was ahead of his time as eventually time-and-a-half pay would be mandated. This kind of act and treatment is another reason why his employees were so loyal to him and the company.

In May 1926, a utilizer of Chance anchors Commander Byrd, passed over the North Pole. Locally a Miss Francis Chance, with a name similar to F. Gano, passed away at the age of seventy-seven and was buried in Bluffs, Illinois. She was likely the daughter of Garrison and

Martha Ann (Roberts) Chance, Sarah Ettie Francis Chance. The name Francis appears to have originated from Francis Anderson, Garrison's mother, who in 1818 married Thomas Chance in Delaware or New Jersey. Garrison and sister Margaret and family, according to the family Bible, moved to Boone County together in 1840. From Columbia they moved to the north part of the county in 1841.

Every period has its share of unique problems that could be labeled growing pains. By 1926 the telephone poles were carrying more wires and weight and locally cut timbers were rotting out. This was before treated poles were in use and machines with augers to dig deep pole holes were in service. Wet weather added to the problem by allowing the poles to shift in the soft ground and in some cases leaning over into the roadway. Such was the case on Highway No. 22, as vehicles had to be alert to dodge occasional poles and wires leaning over into the road! These maintenance problems and costs were another reason for some exchange owners, including the Chances, to exit the telephone business. In early August, the sale of Home Telephone was announced.

At Chance Company, many lifelong relationships were being established. In addition to the Chandler brothers, there was at least another Hallsville Chandler employed at the plant, Austin C. Chandler. There he met his future wife, Miss Jenness Forbis, who was part of the clerical force. The Chance influence also extended to the Chance Hall, which had become an important gathering place for community activities as the Centralia Masons held a banquet there in late May. The *Guard* office on the north side of the City Square was surely influenced by all the local construction as they added a second story to their back room addition in June 1926. Not to be outdone, especially by the Wabash, A. Bishop continued the railroad's east side of Allen Street sidewalk project north in front of his plant on to the Alton rail line. This was a time to perfect and complete unfinished details after the all-out building boom of 1925.

Continuing to complete earlier major expansion projects, the recently installed electrical generating units at the Chance plant were tested in late July 1926. Another test to see if they could handle Centralia peak loads was anticipated. The ability for city residents and businesses to continue their activities in case of a breakdown in the Missouri Power and Light source, was eagerly anticipated, especially by the *Guard* with its electrical powered presses. But the self-sufficiency of the Chance plant to insure its continuous operation was the big advantage, especially when electrical reliability was not up to today's standards. In the future A. B. Chance Company would commit itself fully to the electrical arena that positively impacted the dependability of electricity around the globe, just as A. Bishop envisioned and put into operation in his plant.

August ushered in the usual and the new. The Chautauqua was again scheduled for the twenty-first through the twenty-seventh and Chance employee Irl R. Hicks and wife had a new daughter. The world cruise was on again for Mr. and Mrs. A. B. Chance and son Gano, as early September found them in Columbia working on passport papers. Scheduled departure was September 11. The tour was made up of university students and their families and was of an educational as well as sightseeing nature. A week later the *Guard* on September 10, 1926, said they would leave on Saturday for New York and would return the following May. This was the "Ryndan University World Cruise—1926 and 1927," according to F. Gano Chance papers. The name *Ryndan* originated from the name of their host ship, the S.S. *Ryndan*. The next week's paper told of a large crowd of relatives, friends, and employees that gathered at the station to send them off. Progress of the world tour, as covered by the *Guard*, found the Chances at the Bixby Park, Long Beach, California gathering of some sixty-five former Centralians on October

Official company photo of the Galvanizing Department dipping operation in Warehouse No. 2 or the old building that sat between the Foundry and the 1925 Wabash Station. The overhead conveyor system may suggest this is a post–September 1927 photo, although the galvanizing operation came on line in late 1926. (A. B. Chance Co.)

10. Progress in Centralia continued as in mid-October, the Crump and Dawson coal bin was built near the Chance Chemical Building on Collier Street. In other news Chance client Honey Ford (Anthony Building) sold their dealership to Donelson Motor Company of Salisbury. Honey, a former St. Louis businessman, appeared to be a sharp promoter and may have reacted from the erosion of Ford sales due to competitors of the Model T Ford, which were getting an antiquated image.

In the founder's absence, there was expected to be a slowdown of new activities at the Chance plant, but they instead seemed to continue unabated. An early November 1926 *Guard* news item stated that the Chance Manufacturing Company had started a School of Instruction in electrical appliances, tool making, engineering, drawing and other lines so their employees would become more familiar with the work at the plant. This was to lead to greater efficiency and acquaint employees with the specialties they manufactured. In the 1960s, there was a resurgence of this program, but we now know it was a revival of an earlier originated program. As the Chances visited Hawaii in the Thanksgiving time frame, another article described the enlarging and further progress at the Chance plant. Two new departments, Stamping and Galvanizing were begun and when in operation more employees would be needed. This outside work would now be done in-house which is just the opposite trend in many of today's firms throughout the U.S.A. The tradition of no lengthy plant shutdown was mentioned in comparison to instability seen at other large plants. Helping this cause was a shipment of Chance anchors to Japan indicating export business was filling production voids. This same November 26, 1926 article also suggested plans were being made to erect a brick building on the Mule Barn site. It further added that no action was expected until spring, (apparently upon A. Bishop returning), but he had long anticipated improving that part of Allen Street. This is very important information, especially in view of the 1929 fire and immediate construction of this proposed building. We now know why it was possible to rebuild it so quickly, because it had been anticipated for several years! Today this is Building No. 7, commonly referred to as Punch Press at Allen and Wilson.

One conclusion reached is that A. Bishop was not only the early architect of Chance Company, but his influence penetrated deeply into the future of the company. In the previous paragraph are several subjects touched upon that would become company trademarks. F. Gano would loyally sustain and grow many programs that his father had earlier instituted.

In a way he couldn't go wrong as they had already proven their worth, although there is always some apprehension in applying old ideas to future challenges. However, certain things are worthy in any time, including the education of Chance employees, becoming one of the hallmarks of Chance Company. There was the hope that employees would continue to grow and the company supported such efforts in many different ways, as specifically will be focused on in Volume II. The idea of doing internally what had earlier been farmed out defined security and growth. Today the opposite seems to be true with employment erosion and future opportunities lost for young people. This loss of control and expertise is also a frightening aspect of this trend. Coming from a lifetime of steady employment and a time when the ability to harness and produce locally, what was earlier considered too difficult to do in-house, was a confidence, loyalty, and allegiance builder. Last but not least was this idea of steady employment, a concept that was always important to A. Bishop and F. Gano Chance in their operation of Chance Company. Often lost in the more publicly accepted idea of employee benefits was that constant plant production generated cash flow, maintained a solid structure, and promoted future planning that helped insure a company's success and longevity. This was a principle that the Chances learned well as forward looking and thinking business people.

The November anchor sales to Japan is noteworthy. Chance Company did have a good business relationship with Japan prior to Pearl Harbor on December 7, 1941. Piston ring sales to Japan in the late 1930s were significant and as the Chances sailed toward the Far East in December of 1926, fifteen years prior to war, one wonders if A. Bishop hadn't become anxious about what was occurring back in Centralia. Although most current goings-on were on the order of internal changes and the initiation or completion of current projects and plans that he had likely charted out, it is impressive that he had established an organization that was forging ahead with significant progress in his absence. Not only was he a hands-on person but he also possessed the abilities to hire, educate, and organize capable personnel into an efficient operating team.

In early December of 1926 there was another significant addition to the list of materials being used in the Chance plant. A rail car load of ingot steel was received from an Alton, Illinois plant. The news article stated it would be used in the anchors of various sizes where more strength was required. This is significant, especially in view of the new Foundry being built for casting Chance products. The article also noted the operations of the large steel stamping presses that were now running to form steel into anchor parts and other items of the Big C line would also be formed from this process. This change to steel and presses would become a lower cost, more reliable and quality method of producing anchors, especially the new Taylor Expanding Anchor discussed later in this chapter. In fact, most of the other anchors were still being cast or made of iron. But this new expanding anchor would influence other types such as the blades of screw anchors being made from pressed steel. According to Chance executive Nathan Toalson, the purchase of the new expanding anchor is why the substantial punch press capability was born at Chance, in early December of 1926. The new expanding anchor was basically made up of pressed steel components.

Another example of the progress occurring in A. Bishop's absence, was the bringing online of a second furnace in the Foundry. The molten metal produced, according to the December 17, 1926 *Guard* article, was used for casting anchors and piston rings. Additional people would be needed to operate the second furnace, which for the present was going to be dedicated to piston ring production. This points to growth in capacity and employment at Chance. The previous year it was noted that Chance had decided to put in their own Foundry to fulfill casting needs then being supplied by out-state vendors with resulting high freight costs. It was a questionable move in the eyes of some but "Factory mechanics have completed making a number of sets of molds on the pattern machines, for casting the piston rings by the hundreds of thousand, where they have been able to make only thousands with the old one furnace." Louis Edelman, a master mechanic, was directing the process of getting the new furnace operation going by year's end or early 1927. The article continued to recognize the specialness of the large operation in Centralia and "he never asked the city to donate a penny toward locating his enterprise in operation here." In spite of Mr. Chance's desire to remain in Centralia and attempts to get help to get his plant up and running in its beginning years, he didn't reap the full benefits of incentives that exist today or even in earlier times. The article continued about company growth that was exceeding Centralia's. It again touched upon the Mule Barn replacement with a new brick structure across the street, west of the Chance plant. They also could foresee other improvements in the Chemical Building, Paint Shop, and other buildings away from the main plant. It was also noted the receipt of a big shipment of cast steel from the east, "used in making the famous anchors."

As December of 1926 came to a close the Chances had reached China in their around-the-world trip. Back home it was business as usual, although one surprising revelation of the 1920s was numerous bank failures, something associated with the financial catastrophes that occurred in the 1930s. In spite of industrial growth under the surface there was economic fragility, but at Chance Company green were the lights of progress. The new Galvanizing Department came on line and was said to be operating at full force with seven employees. With the news of new processes at Chance Company came a focus on the choosing of veterans to manage it. Family members are often different in certain ways as were the Chandlers. Flynt remained a factory employee most, if not all of his long career at Chance, but junior in employment time Ollie entered the management area. Flynt appeared to be a valuable and versatile employee who liked factory work. Ollie became the manager of the new Galvanizing operation, but it was likely a foreman-type position like he would continue to occupy for several decades at Chance. Having worked in the Galvanizing Department, comments made about the new processes of dipping materials in six different chemicals before the zinc or galvanizing coating was applied, brings back memories of the smell, hard work, and dangers of that process. Because of the demands of the utility areas for reliable and long-lasting parts upon a pole line's harsh environment, this was a process that was destined to be required even more in the future. Chance Company was marching ahead with capabilities that would allow it to produce in-house more and more of its products, especially those directed toward the utility industry.

The year 1927 revealed that America was coming of age and the Chance Company, as the *Guard* was now referring to it without the manufacturing insertion, was also maturing with what appeared to be growth in both its auto and utility product lines. At the beginning of the New Year the Chance Foundry force was working on Sundays due to shortages. This could be construed to be a drastic measure in a traditionally religious community that was proud of its churches, especially with the company head being a longtime active member of the Christian Church denomination. But A. Bishop was on the world cruise, with late February finding the family in Greece. Usually when a firm has a shortage one automatically thinks in the opposite direction, of a shut down, but this was a "good shortage" of a lack of goods on the shelf to ship sold business. The paper suggested it was a shortage of goods to fill orders for the Big C products. So even in the founder's absence the company was on the

A photo of the Chances around-the-world tour group in Turkey. They are in center area of picture. The message side is reproduced as it was addressed to "Chance Co." who Mr. Chance addresses as "Dear Bunch!" (A. B. "Buff" Chance IV Collection)

move with at least one new employee, a new toolmaker H. L. Reed of St. Louis. He would work in the Machine Shop to make tooling for the new anchor, which may suggest the new Taylor Expanding Anchor. To further support the idea that business was good, Chance Company received a carload of steel and a carload of iron from a Marion, Indiana source, in mid-February. As April began to allow outside work, a new 150-foot driveway was completed on the north side of the plant. All the success of the Chances attracted some relatives as early April found another A. B. Chance from Illinois moving to a farm east of town. The commercial success of Centralia was continuing to also drive railroad improvements with the Alton completing a new siding by their depot up to the Negro school area.

Not only was May of 1927 ushering in the vegetative state to northeast central Missouri, but the A. B. Chances were due back Friday night, May 6 at 2:00 a.m. To commemorate this occasion there were several events planned, although the decision to put in a concrete walk from the Chance stucco house on east Railroad Street north across the railroad tracks to the Thomas yard, was likely not one of them. The city would put in its portion plus a four-inch concrete floor in the water plant. The Alton had agreed to do their part, but the Wabash was not responding to the request. The Wabash officials in St. Louis must have wondered at times: What do they want to do—concrete the whole town! Centralia wasn't bashful about requesting help to improve their town and playing the railroads against each other was a convenient leveraging opportunity that offered certain advantages. So what would happen when the person most responsible for generating a volcanic-like atmosphere in Centralia, return? The May 16 *Guard* suggested the first indication of this arrival was the blowing of the electric siren on the Chance plant and other "demonstrations were used to make visible and audible effort in welcoming the wanderers home." This was a big deal, especially since it occurred at 2 a.m.! So even if you chose not to participate or get caught up in the Chance roller-coaster-like events, about the only successful avenue available was to leave town! The celebration of the return of the Chances included a band that went to the Chance home and serenaded the party. The M.W.A. brass band played such favorites *Let Me Call You Sweetheart* and *Home Sweet Home*. One of the first events upon leaving the station, which wasn't identified, was a dinner at the Mr. and Mrs. J. A. Chance home on west Singleton. Some of the Chance office force were invited guests. Later the Mr. and Mrs. Buford Chances entertained with a six o'clock dinner. The itinerary of the trip included most of the principle ports in almost all of the major countries of Europe and the Orient.

Not all was well upon A. Bishop's return, as his new expanding anchor needed redesigning and there was a substantial recall problem, according to employee interviews. Although the 15 millionth Model T had rolled off the assembly line in mid-June, Henry Ford had a similar problem in delaying getting their new car out and Ford dealers across America were suffering, signaling a lengthy period with no car production. Donelson Motors in Centralia wasn't exempt as all they could do was to promise the new Ford was coming soon and to try and fill their shop with maintenance work. Without missing a heartbeat, work on completing Chance plant expansion projects continued. From interview information it is evident that employees were heavily involved in this work. A news item in the June 10 paper said the plant was putting down a concrete floor in the Foundry that was about a 60-by-420-foot slab. This would make it the largest single floor in the city. The same week found father A. R. Toalson visiting his daughter Miss Anna Lee, who was attending the University. A former employee Eddie Howard of Slater, an electrician, was back with Chance. As July came, change was the order of the day at the Chance plant with the Tool Making Department moving into a portion of the new Foundry. There were thirty-seven different machines including punch presses, drill presses, automatic saws, lathes, etc. according to the news item. The "superintendency" was placed with Donald Wood and Joe Uirich. Another July 15 article identified the first floor of the main factory building, as the location of this new department: "where tools of all descriptions used in the plant will be kept and made." The Galvanizing Department was also moved from near the city water plant to Warehouse No. 2 in front of the Foundry or between it and the Wabash Depot. The Wabash's request to remove a building between the Chance plant and the new Depot, during the new depot's construction and the spur tracks relocation, may relate to this or another building.

Baseball was continuing to be a popular local sports activity. There was also extra inning action in and around the Chance plant as adjustments continued being made to take advantage of the expansions earlier completed. A. B. Chance, according to the July 15, 1927 *Guard*, was before the city asking for a fifteen-foot strip along the south side of Wilson Street on the north side of his Warehouse No. 3 or the Mule Barn. He wanted to put in a spur track that would connect with the C&A (Alton) rail line (not completed until 1929). The city suggested such a move would destroy the hitching racks, but it was a troublesome area as cinders were constantly needed to keep it in usable shape. The Board discussed the possibility of removing the rack to the east side of the Thomas Lumber Yard. The approval of this proposition would allow the flexibility of Chance Company utilizing the Alton for incoming raw materials and the shipment of finished goods. It would also permit the removal of an unsightly and unsanitary situation at his front door, another step in A. Bishop's desire to shape up this area. He also was going to have a spur track on the south side of the plant run from the Wabash, but since this was on Wabash property no city action was necessary. Although it was earlier pointed out that the Wabash spur could have been the result of the proposed Wabash and Alton connection during World War I, no solid proof has surfaced that it was built. The spur today would appear to be the one discussed here in 1927, which would continue in the future across Allen to the east, serving the

Chance factory, a portion of which was built in 1925. The article further stated Warehouse No. 3 would house the finished anchor products of the plant and much of the raw materials. The new spur tracks would facilitate the shipping in and out of the company's products and materials that "Mr. Chance has been figuring on this improvement of shipping facilities for some time." There was a further suggestion that the production from the new drop forge would benefit from the Alton arrangement, which may relate to this spur eventually running across Allen to the east. One picture shows this spur crossing Allen and proceeding east between the Chance factory and the Alton Depot. This was a move that would not only increase the capacity for shipments into and out of the Chance Plant, but was the first substantial attempt to centralize and organize the expanding shipping operation into a designated westward area.

To remain competitive in the marketplace, A. Bishop had learned that he had to match the lower prices of his competitors, but still produce a quality product. To do this in his out-state located firm is an issue that provokes a focus on its advantages and disadvantages. As his firm grew, so did his competition. There is a sense during this time in the mid-1920s that he not only had his eye on how he could grow his company and reap the benefits of a larger operation, but he was also integrating more economical methods into the plant's operation. He appeared to have been driven to cut costs and increase productivity during this time that indicates he was feeling the pinch from big-city producers, especially in automotive piston rings. It seems that some of the inherent disadvantages of his isolated, rural out-state Missouri location such as increased freight cost, availability of a skilled work force, and higher utility costs intensified the problem of remaining competitive. Earlier discussions about electrical costs suggest that his big-city competitors had an unfair advantage of lower electrical rates and very likely had a considerably more reliable service than did remotely located Centralia. So the Chance operation wisely investigated the lowest cost power source available, with the built in advantage of controlled reliability. Of course, what resulted was the new Chance Power Plant with its oil-fired generator arrangement. A news article in mid-July 1927 about City Superintendent Gillam's investigation of a new source of power for pumping water from their deep well stated that Mr. Chance's power plant was being monitored. Its operating cost was projected to be one-half the present cost of electrical driven equipment.

Harris Davis, drop hammer operator, is forging steel cone anchors in this Drop Forge Department scene, a couple of decades after these capabilities came on line at Chance Company. (Cover, *Chance Tips* January/February 1950)

As July 1927 continued, traditions, such as the all-employee picnic were being established that carried on the social aspects of employment at Chance. The manufacturing progress being made would also propel the company into a self-sufficient posture for decades. This would allow Chance Company to become and maintain its position as the world's leading producer of earth anchors. Instead of certain parts of Chance utility products being produced elsewhere and shipped into Centralia, they would begin to be produced in-house with the installation of his new Drop Forge Department. One of its main purposes, per the July 15 *Guard* was in "making the round iron anchor rods used to clamp on the anchor plate when it is placed in the ground." These rods were surely steel, and the reference to round was the shaft of the rod. Per the article, the rods were previously made in Eastern foundries and with the new Drop Forge, Chance Company could be making the entire anchor (anchor + rod) in Centralia. The importance of this development holds much future significance as the substantialness of the machinery along with its maintenance and operation requirements wouldn't make it easy to move out of town, as would be done with certain other departments. Today the big hammers still forge cherry red steel into utility-destined products as they have for over three quarters of a century in Centralia, but of course that too could change.

The growth taking place at Chance Company and thousands of other firms like it across America, would provide a platform off of which the middle class would develop. The post–World War II G.I. Bill would provide for the educational needs of this growing group, which exploded from newfound war technology and industrial might. This foundation for our future progress and prosperity was being firmly laid by such aggressive entrepreneurs as A. Bishop Chance. The Chance Company influence and impact upon the surrounding Centralia area is immeasurable! Today this is in reverse and becoming a hollow shell of our former self, one wonders if our economy isn't living on the past. Such nations as China, in a similar growth mode as the United States during the early and mid-twentieth century, may become the Super Powers of tomorrow.

As July continued, *Guard* coverage suggests A. Bishop's return to work brought additions and progress in the plant at an alarming rate, many of which are still in place eighty years later. A new die-sinking machine was installed for making the dies required for manufacturing the patented Chance anchor rods (Thimbleye® and Twineye®). The importance of this uniquely designed product is clearly driving investment in it. A new "up-setting" machine that "hammers the hot metal into molds or forms" was to be applied to those products not adapted to castings. This big, low-profile, box-shaped machine was different than the big vertically positioned hammers, as its piston-like action mashed and squeezed hot metal into products such as pole top pins. A Drop Forge tour of duty included heating metal for the operator of an up-setter. Before all of this machinery and the new Drop Forge were in full swing, a carload of anchor plates and rods from an Illinois source were received in late July. This was thought to be the last of these shipments, as these products would soon all be made in Centralia.

The summer of 1927 was a defining period in Chance Company growth and development, especially in view of the future implications of the utility business and the intensified focus on those products. One mistake that could have proven disastrous if not fatal would have been a complete reliance on piston rings and inattention to Chance utility products. One can shudder at the thought of how easy it is to get on the wrong track, but this wasn't to be. In fact, the focus on utility products was made much earlier than the late 1930s as is normally thought. From the very beginning with the arresters, a focus was always on telephone and electrical utility business. After the 1912 sleet storm, A. Bishop appears to have never lost faith in his guying solution and its potential in the utility market, although it took years of concentrated efforts to get his anchors and anchor rods recognized and accepted. The issue instead involves the "full line" involvement although new items were continually being developed. A limitation at that time was that electrical utility development was still in a more primitive state and so later more fully developed product lines like hot line tools, had little or no market. The Chance penetration was instead deep in anchors and rods due in part to A. Bishop's unique inventive designs and promotional efforts. Sales volume was growing and it became a very financially lucrative proposition, which drove further attention. The importance of these products to the survival of Chance Company during the Great Depression can't be overstated. They also financially supported various expansion projects and investment back into the company, community, and employees throughout the Company's remaining life.

Closing out July 1927 with even more "Changes at Chances" as the *Guard* addressed it, brought confirmation of the just-mentioned points. Two rail carloads of new machinery were received, which were to be used in the new expanding anchor that was being put on the market, per the article. This is likely referring to the newly acquired Taylor Anchor. However, this raises the issue if it was previously marketed, as it would appear he hadn't had time to revise this acquired anchor. The "put on the market" suggestion and the next statement concerning inadequate production levels appear to be in conflict. As the article continued, the demand for it was "great" and the company had been "hampered" in putting them out fast enough. Although it was thought this new machinery would alleviate this problem, what later would be realized is these machines would also be applied to the manufacturing of other products. This was one of the keys to Chance Company success, the pool of in-house factory engineers that increased the capabilities and versatility of machinery, initially purchased for other specific purposes. One such drop forge machine, weighing forty thousand pounds, required a ten-foot foundation! It was said to strike a sixteen-hundred-pound blow at 105 repetitions per minute. This could be referring to the huge hammers, still a vital part of the Drop Forge Department at Chance, although that many strikes per minute may suggest another machine. In operation these machines do shake the ground like a passing freight train and when in operation they can be heard deep within the residential area of Centralia. What is remarkable is for nearly eighty years this has been as natural to Centralia as the surrounding prairie grass and the howl of a distance coyote once were. The equipment just received was said to be some of the most up-to-date in the country. Later stories, especially during the tight financial years of the 1930s, suggest worn-out machines were brought into the plant and even some machines were built from scratch utilizing extra parts and inoperable machines. But the paper's account here in 1927 suggests that new machines were being purchased when finances appeared to allow it. Little has been found of how A. Bishop was financially able to buy all this new machinery and construct all of these buildings, undoubtedly a substantial story in itself. This is information that the paper apparently wasn't "privy" to. What is revealed is a name being developed that we at Chance labeled Pole Line Hardware. This was the nut and bolt or the hardware line as described. The *Guard* article called it the pole and line hardware, and suggested it also included the anchor line. In later years anchors were split off into a separate line becoming the first product line, the 100 Series, followed by Pole Line Hardware, the 200 Series. One of the benefits and revelations of this story is to view the refinement in products and descriptions taking place as this industry was evolving.

Deeper within this extensive article were described the changes necessary to make room for these giant machines. The Paint Department would have to go, but since it would "be rebuilt larger," it is clear there was an organized plan in place. This would put Drop Forge across Collier to the west of the Chemical Building. The new building construction had been necessary to get a roof over all these new and enlarged production departments, which were pushing the storage of finished goods out. This function would move west across Allen and is a reason why the Mule Barn was acquired. Another significant revelation was that a S. A. Clouse had been brought in to organize the installation of this new machinery. He was "said to have been in the pole and line hardware manufacture since he was quite small." Yet another so called "expert" was on the premises, but it's not said if Mr. Clouse was temporarily brought in or remained as a full-time employee. In retrospect this was a wise policy to draw off of experienced personnel, which both A. Bishop and F. Gano continued throughout the history of the company. Suggesting that Mr. Clouse was "quite small" could have multiple interpretations. One take is that he like many Americans of that time left home and began working at an early age. It's unclear where this machinery came from but was likely purchased new from a machine manufacturer. Mr. Clouse could have been working for such a firm or came to Centralia in search of employment. It was a personal feeling that production (factory) personnel were not always fully appreciated after A. Bishop's departure. A move to a new assignment to become someone directing important transitions and being heavily relied upon, could be an attractive proposition. What is revealed is that there were a number of specialists coming to 210 North Allen very early to help launch Chance Company into a self-sufficient posture of manufacturing their own products.

Further details of the July 29 Chance news piece covered the installation of a second air compressor powered by a ten-horse electric motor. Setting on a seven-foot concrete foundation, it was to be utilized in the Foundry to blow out and to shake and compact the sand in the molds. There is quite a contrast in the older motors and new ones. The old ones were very large in relation to their horsepower output, so a ten-horse unit would have been a very large electric engine in size and weight. In the September 2 *Guard*, a news item explained expanded uses for the compressed air. In addition to operating the "vibrators" in the Mold Making Department, it was applied to the city's deep well to lift out water. This was a test in case the city's compressor broke down. The test was successful except when the Foundry vibrators were turned on at the same time it reduced its effectiveness to perform the city function. In addition to those activities, Warehouse No. 2 was being rebuilt to accommodate a more modern galvanizing plant, with a new fourteen-by-eighteen-foot pot. The "pot" was a large metal container with a heat source, which melted the zinc. Chemically treated and cleaned manufactured items were dipped into this molten pool of zinc to become galvanized. A system of overhead, motor-driven conveyors was also being installed to move castings and other materials, but it's unclear how extensive its coverage was. As fall approached, the Chance plant would be poised to handle new manufacturing processes at production volumes not before possible. All of these developments in the last half of the 1920s, suggest that Chance products, especially of the utility nature, were becoming firmly established and any comparison with the Chance operation began two decades earlier, is hardly recognizable. But there was commonality in that each day would bring new challenges and problems that would require further changes, resourcefulness, and hard work to overcome.

An early September news item told of the Chance Company purchase of several new molding machines for the Molding Department, used in producing Chance piston rings. The new Galvanizing Department was running with a night shift in addition to the day crew. This was said to be necessary to speed up the galvanizing of the new expanding anchor, Chance Pulley, and other Chance® products. This may be the first reference to the plating of Chance anchors. This raises the question if all or only a part of the expanding anchor production at that time were given the more expensive to produce galvanized coating. It was common to paint anchors instead of plate them, but to overcome early skepticism especially when the more fragile appearing steel plate anchor was introduced (versus castings), it may have been expedient to give them the long-life galvanized finish to satisfy customer concerns. But the expanding anchor wasn't the only anchor generating business as the September 9 paper told of a carload of Chance Never-Creep® Anchors being shipped to Chicago the previous week. However, one key employee, Miss Ida L. Traughber was heading in the opposite direction to vacation at Hot Springs, Arkansas, undoubtedly a well-deserved rest from the hustle and bustle taking place on a daily workday basis at 210 North Allen.

As September 1927 closed, the Drop Forge became operational with eleven male employees. The news article suggested their activity would be focused on manufacturing eye-bolts, anchor rods, and various parts of the new expanding anchor. These were hardware-type products produced for the telephone and electrical utility industry. In mid-October, the idea of laying underground cable by the telephone industry was now being applied to electricity. The former Chance paint shop on Collier was receiving a new 220-volt underground electric line to operate machinery newly installed in the plant, which suggests the new Drop Forge machines. This wasn't foreign to A. Bishop, as he had directed the laying of the underground telephone cable in town. To confirm further the involvement of Chance factory employees in performing many construction and refurbishment activities at the Chance plant, the *Guard* article reported: "The work is being done by the local force at the plant."

At the east end of the new Foundry Building was a raised area that had three stories. There was a basement and the raised portion had factory offices, restrooms, storage areas, break-time and lunch benches, Drop Forge check-in and Foreman and Clerk office on the northeast corner, Hardware Assembly on the south-central corner, and canteen machines for surrounding departments (Galvanizing, Drop Forge, Shipping, Hardware Assembly, Anchor Welding, etc.). This more recent posture, especially on the second-floor area, reflects changes from the 1920s time frame. It is a place one can associate with certain people: Glenn Lane, Jerry Cox, Jess Smith, Dave Green, Maurice Kinkead, Dee Daughtery, Bill Bradley, Jerry Robinson, Perry Roberts, and Bill Roth. The third story was sort of an attic with offices and meeting rooms. When the Main Office was being gutted and redone in the 1990 time frame, the offices had to be temporarily moved and the Production Planning group ended up in this upstairs area on the east end. It brought back factory memories with the metal grit that saturated everything like Oklahoma dust during the Dust Bowl Days of the 1930s. The rhythmic bang and vibration from the huge hammers and machinery in the nearby Drop Forge also reminded one what Chance Company was all about and how close this group was to being transferred back into the factory environment. Just to the rear of this building was a concrete oasis of sorts where Drop Forge and other productions were stored, flanked on either side by later additions of the Drop Forge to the north and the warehouse to the east. Earlier this area was where the city water tower and power plant had stood. Today there remains a low-walled shelter with a roof and also a similar structure just to the east across what would have been Collier. The nearest one close to the tower must have been what is described in the *Guard* of October 21, 1927, as a settling basin for the municipal water/power plant. Likely Maintenance professionals Bob Jones, Al Hajduk, or Don Smith could have quickly answered what these objects were years ago. This is one of the benefits of this work, answering some long-sought questions, but it is now more enduring and gratifying because it documents company history.

A news item in the early November paper told of the continuing progress at Chance Company to make room for more Drop Forge machinery. This appears to be additional expansion as the article said the Forging Department was "enlarging." These changes would force the Toolmaking Department to move from its first floor, main factory building location. What appeared to be an organized plan at times looked more like a session of musical chairs with rearrangements of department locations. In spite of the unstated new location of the tool group, two new "skilled" Toolmaking employees were added to its roster. The new Galvanizing Department was surely staying put, but the article stated that Warehouse No. 3 was being partitioned by a battery of carpenters so it could be heated and operations could be continued through the winter. Certain paints and chemicals had to remain warm to effectively perform their function. A carload of zinc pigs was just received and were being stored in Warehouse No. 2. They were said to appear like a bar of silver as they were clarified zinc used in the galvanizing process. To continue building on earlier changes, the main foundry was getting automatic overhead tracks and carriers installed, as previously announced. They would permit the transporting of heavy machinery and supplies from the railroad tracks to the proper departments. They were electrically operated.

As November rolled toward the holiday season, Chance Company's metal-casting production kept the railroad busy shipping raw materials in and finished goods out. According to employee interviews, the Foundry operation wasn't a perfected art and continued testing and experimentation brought various solutions to solve bad pours or flawed batches of castings. Mid–turkey month 1927 found Chance receiving a carload of limestone, usually destined for application on farmland, but instead going to the Chance Foundry to be used "for tempering processes." Within a block of advertisements in the *Guard*, the November 25 issue carried a Big C ad for "Shinever," (see page 210) an auto and furniture polish available in three sizes that Buford Chance undoubtedly used and endorsed. This was one of many auto-related products Chance Company offered including their popular piston rings. Garages in area towns routinely ordered a ring set that was to be shipped overnight by Railroad Express or R.P.O. (Railway Post Office) showing that railroad's efficiency in transporting mail and goods commonly exceeded our transportation capabilities of today.

The *Guard*, in its excellent coverage of Chance Company activities, ties employee remembrances and other information neatly together. As covered elsewhere, there was an employee comment of how the operation of the Chance spot welder would dim lights all over town. The November 25, 1927 issue covering Chance news told of the purchase of a new "Acme Spot Welder." It was electrically operated and first installed in the Punch Press Department. The use of Chance-generated power would have prevented the sapping of local current, but Chance was also hooked into the local power system. This is another aspect on the negative side of the affects of having the Chance plant in town. Other references in this book point to the tolerance of local citizens as they wisely never forgot the benefits of having the factory located there. This was a very special relationship grown by a large payroll and the frequent Chance contributions made directly to schools, other local institutions, and the community. Some of these new machines, like the spot welder, were being brought in for use in producing the new steel plate, expanding anchor.

The late November article continued with news of the receipt of another carload of anchors and a carload of steel from Iowa. This was likely giving the Wabash additional tonnage on their Moberly to Iowa north branch. The purchase of malleable anchors from an Iowa source is discussed later in this chapter. The steel was said to be utilized in the Chance Punch Press Department. The receipt of further new equipment was noted with the arrival of a Howell overhead hanging scales, attached to overhead runs, which allowed movement. It would weigh in anchor plates and other materials brought in and out of the factory. Only about two feet square, the scales had the capacity to weight up to two-thousand-plus pounds. These were used, in retrospect, to keep raw material vendors honest and records straight. The weight of outgoing goods was also necessary in the shipping function.

The introduction of *Guard* readers to December 1927 included another Shinever® ad, suggesting it would be a good Christmas gift. Names of products and the reasons behind it are interesting. One could only hope for a car wax that lasted forever and even today this is more of a positive, suggestive condition, than a reality. The Chance firm may have been operated by country folks, but they had learned their A. B. C.'s in sales promotion. The only thing seemingly winding down at Chance Company, per the December 9 *Guard* in their routine plant coverage, was 1927. Bigger things were hoped for in 1928. One of the aspects of the early company not routinely covered, was the sales and marketing group. Most local coverage involved the plant/machinery and the Centralia work force. The news item on the ninth departed as it told of the salesmen of the Chance Company being called in Wednesday night the seventh and given new ideas on selling methods for the newer products that would soon be put on the market. The article continued in stating what a fine organization: "from foundry to office and selling force and there is no doubt about the value of the products the plant is putting out." One of the new products that Jack Isbell and the sales group would have in the Chance catalog for 1928 would be the revised two- and four-way Expanding Anchors.

The local paper's December 9, 1927 Chance factory report covered some of the final activities of 1927, such as placing into operation new machinery and systems recently purchased, so the company was poised for entering 1928 in which production could hopefully be concentrated on without interruption. Increased factory activity was noted by the adding of "two forces on the night shift in the Die Sinking and Toolmaking Departments." This further points to increasing night work in Centralia, the small rural community where the lights burned all night. The new Assembling Room, the Spot Welding Department, and the new overhead carrier rail system (Warehouse No. 2) were completed and ready for operation. The Assembling Room, if associated with the spot welder, may be referring to the assembling of the new expanding anchor. A familiar part of anchor production was the final operation of painting or coating the anchor with a heavy paint solution, especially related to the cast anchors. Since the cosmetic quality of the finish (lack of runs, etc.) wasn't an issue, the quickest and least expensive method was to dip the anchors and let the excess paint drip back into the paint vat. This was first done manually,

Pictures of anchor dipping (*Chanco News* November 1946)

confirmed by a story by a young Gano Chance, but in a quest for a more efficient, automated method, the Chance Plant instituted what was called the Dipper. The article suggested this electricity driven contrivance would carry anchors from the rail car over to the dipping area where the preserving mixture coated the anchors. The conveyor system then carried the anchors to the storage areas where they were dropped into their designated bins. One would assume that the anchors were still wet when stored and actually dried while in storage. Pictures show expanding anchors stacked to the roof in the later Punch Press Building (No. 7) or in the 1927 Warehouse No. 3 (Mule Barn). However, since the rail car is mentioned here in 1927, these anchors being processed by the Dipper would have been the malleable or cast anchors that were then out-sourced from Iowa and or other firms. So what may be considered a much later developed automatic system of handling anchors in volume, actually occurred in late 1927, fathered by A. Bishop and crew.

One surprising conclusion in summarizing the progress of the Chance operation, is how advanced it had become by the close of 1927, or twenty years after its official birth. No other year likely holds the significance as do the sevens, 1907, etc. This is a year within the seven decades of the company's existence that is reflective of what the company had become and the direction it had charted. The two longtime leaders of the company have frequently been compared and as is often the case the more recent usually has the advantage, for obvious reasons. However, this coverage is bringing an abundance of fresh information and remembrances to the aid of the Founder. Although A. Bishop displayed leadership in all facets of the company, his hands-on involvement in the factory area was somewhat unique and likely not always welcome. In 1927 he had others assisting him, but this kind of personal factory relationship would end with his reign. Certainly as the company grew larger, designated engineer types would chart out and design new operations and F. Gano naturally became less involved than his father had been in factory operations. They had certain similarities yet were different, as any two people might be. Backgrounds play an important role in how a job or position is handled and the progress and continuing sophistication of the Chance operation is a special tribute to A. Bishop, a man who never stopped growing and changing just as did his company.

Closure of the Roaring Twenties

Business Rush, Mishaps, and a Fire

In spite of only one year separating 1927 and 1928, they seem light years apart. Like Lindbergh's crossing of the Atlantic or the Model A replacement of the Model T Ford, the big things had been done, now was the time in 1928 to get on with the business of America and Chance Company, more business. Early January began in an unglamorous way with several racks installed for the purpose of holding steel stock for the Punch Press Department. After the last few years, such minor activity was welcomed, but like keeping children occupied to deter mischief, the boss considered the importance and advantages of keeping his employees busy. One current focus was what earlier was termed the "speed-up." Soon after that policy was instituted, he and his plant superintendent, Bryan Fenton, traveled to Chicago on a week-long business trip. A week later, the January 20 paper told of what occurred on that trip. Three new drill presses and a "powerful" automatic trip-hammer for the Tool Making Department had been purchased and immediate delivery promised by the manufacturer. An early February article covered the opening of the new Forge Plant, located back of the main plant in its newly remodeled building. The machinery was now installed and ready for operation. Previously spoken of in 1927, it had taken several months to get it into full operation.

The machinery wasn't the only thing getting more sophisticated at 210 North Allen as a professional accountant, C. C. Hough, who would later become a longtime Chance financial manager, began spending time in Centralia. A. Bishop continued to be on the road, traveling to St. Louis in late February on business. An early March news piece told of the busyness of the operation, such as the employing of two blacksmiths, one for the day shift and for the night in the Foundry. They were Tom Shepard, late of Tulip, and M. R. Roberts of Mexico. Activity was increasing as an influx of new spring orders from Southwestern Bell Telephone and other large pole line companies arrived and were booked. The plant was running day and night to try and catch up and one reason was the new expanding anchor that had become a hit and was in great demand. This was the improved version in which A. Bishop had made substantial adjustments to correct problems. The March 9 *Guard* article told of the Die Making and Punch Press Departments working day and night.

> *There has been a slight change made in one part of the expanding anchor and the drop forge department is being prepared for the change and will begin a 24-hour operation about the latter part of this week.*

The Chandler brothers had become a fixture at the Chance plant and typify what America, the land of opportunity, was all about. It was no little thing to dream about a steady job and in working hard one could anticipate being promoted to a more skilled and higher paying job. This was part of the draw of job seekers to Centralia, during the 1920s and throughout the Company's history. Flynt, according to the lengthy March 9 article, had moved from the Tool Shop, a position held for a number of years, to the Tool Making Department. This was said to be: "a step upward." His brother Ollie, the paper described as "another old wheel horse" was promoted to the Tool Department. "These boys have practically grown up with the big plant." This kind of scenario would play out a multitude of times over the life of Chance Company and throughout factories across America.

A rule of a manufacturing concern is most of what comes in, hopefully will go back out, but in a processed and refined form. Late March saw a carload of steel received and on the same Friday, a carload of the new expanding anchors left town for some unknown destination. It's mind boggling to attempt an estimation of how many tons of raw materials and finished goods have entered and left Centralia. By the mid-1920s and certainly in the spring of 1928, this revolving conveyor belt of activity was beginning to accumulate some significant numbers. The *Guard* of April 6 pointed to Saturday work in addition to night shifts in an attempt to fill increased customer orders. The previous Saturday, two carloads of anchor plates and one of rods from Illinois were received. Maintaining raw materials for this plant could be likened to the consumption of a smorgasbord enthusiast. There were three carloads of steel currently on the road destined for Centralia: "to be used on the spring rush for anchors." It seems that outside construction was more seasonal in the past. This was true in the utility business as in forecasting Chance Company business the springtime activity would jump, as it apparently did here in 1928, with an influx of anchor orders. Anchor orders often reflected new construction activity, which opened up in the mid and north regions of America in the spring. The Drop Forge Department was operating full time, as the news article continued, "pushing out a large order from an Iowa concern." This could have been from one of Gary's telephone groups or for agricultural purposes in fencing, as Chance anchors were promoted in such farm publications as the *Wallace Farmer*. The Galvanizing Department had slowed as weather and lack of supplies had caused part-time activity, but now they were back on track running day and night. After all the hard work over several decades of promoting and building an organization and plant to satisfy current demand, Chance Company represented an American success story.

Amid the rush of manufacturing activity in the spring of 1928, there was time taken to add to the quality of life around Centralia. Entertainment took on many forms with a modern version, the radio, having a growing impact on American society. An amateur broadcasting station at the Chance factory was licensed as 9-CUQ and had developed over the winter with W. E. Jones and D. C. Wood as engineers. They apparently were Chance Company employees, but that and what was being broadcast wasn't mentioned. The April 13 *Guard* told of improvements at the Chance factory. A substantial project of putting in a concrete floor in Warehouse No. 3 (Mule Barn) so anchor rods could be stored was described as extending to Barr Street. This may have instead meant Hickman as the Stone's Poultry House was thought to be across Hickman to the west of the Mule Barn. What this as well as other information points to is the No. 3 west area had become the storage and shipping point for Chance anchors and rods. These were the volume utility products and owing to their bulky nature a large storage area was required for incoming receipts, warehousing of finished goods, and shipments.

The extensive April article continued by describing an improved and redesigned, new style drop forge pulley block in which samples were currently being sent to dealers. It replaced the old-style cast-iron block, formerly made by the Big C plant. This is interesting from at least two angles. First, the drop forge capabilities of the plant were now making it possible to redesign and construct stronger products. The article continued by stating the new block was made of entirely malleable material by drop forge machines and "it is not possible in ordinary practice to break any part of it." This is crucial to the safety of electrical and telephone linemen in having reliable tools in performing maintenance and construction work high up on a line. Many utility products were also under extreme stress during storms and Chance forged products were well known for their strength and reliability. This is an important development with substantial future implications that helped further the Chance image as the producer of quality products. This is remarkable from the standpoint that there were no known professional engineers on the premises, but that was changed a year later by Bill Hollander and his classmate F. Gano Chance. Although employees weren't as "expert" as needed at times, they and A. Bishop Chance were generating some impressive results!

The second significant point of the change in the Chance pulley was it was formerly "made by the Big C plant." As earlier discussed, the cast piston rings of the automotive line took a lot of the Chance Foundry capacity. So in respect to this apparent division between the two lines, auto and utility, the forging operation became a part of the Chance Company Utility operations. Although the Foundry operations would continue beyond the life of the Big C line in producing castings for utility products, it was currently considered a separated entity of the auto area. From employee interviews, there were problems in the Foundry, of course some of this was the nature of the beast. But any chance of failure in the field for equipment like a pulley block could be a very dangerous situation and to construct it from very reliable forgings was a move in the right direction. Keep this in mind when the hammer is discussed during the Depression.

The Friday, April 13, 1928 article continued with more good news, not bad. The previous

Monday, a carload of anchor plates was received from Illinois and was being dipped and stored in the warehouse. The outgoing included a carload of Giant Never-Creep® Anchors to San Francisco, California, and another load to Cinroe, Texas. The reference of "Giant" Never-Creep® Anchors confirms there were more than one size being produced and sold. It's not said if a business trip to Texas by A. B. Chance at that time was related to that shipment or other business opportunities. But the founder wasn't the only Chance in the news as his younger son Gano, a student at the University of Missouri, frequently came home over weekends as he did in late April. In retrospect, Buford seems somewhat out of the spotlight. Gano seems to be a very responsible youth and his active involvement in many advancements of the day was reflective of his father, which garnished attention. It's unclear what the relationship between father and sons were during that time, but the father was very busy, as he previously had been, suggesting he was not available at times as the two sons grew up and entered adulthood. How this impacted the two sons is unclear, although one could speculate that Buford could have been affected more by his absences. This particular weekend Gano spent remodeling and improving the company radio station. The paper said Gano had made radio and branches of electrical equipment a special study and his pioneer radio work in Centralia was now followed by a continuing interest while in college.

Hidden from our view on most occasions, A. Bishop's inventive activities were alive and well. A "Pole Guy Fixtures" patent application was submitted on April 26, 1928, and became a seventeen-year Patent No. 1,804,202 on May 5, 1931. It was a guy-securing fixture that spread the pole loading so it could sustain greater stresses or normal guying loads without damaging the pole. It also allowed messenger or back guy fixtures of various types on the opposite side of the pole. It consisted of a downward loop designed two-ended bolt (threaded legs) that fit through a saddle to spread the load. It then allowed other attachments on the opposite side to be attached to the two bolts protruding through the pole. A similar filing on July 11, 1930, resulted in another seventeen-year Patent No. 1,813,482 issued on July 7, 1931. It applied the Thimbleye® contoured eye to this Pole Guy Fixture with what appeared to be a forged yoke for holding the guy cable. It was a two-bolt (standard straight type) with a pole plate on the opposite side to spread the guying load. Like the previous patent, it protected the poles, but it also protected the cable. The emphasis seemed to be on maintaining the integrity of the guying cable as was earlier cited with the advantages of the Thimbleye® eye design. This is another example of A. Bishop's blanket coverage of product designs so no window of opportunity presented itself to competitors. What resulted were multiple patent filings on the same general items with a different approach or emphasis and enough design change to insure another patent. An interesting recipient of the issued patent was Hubbard and the Rural Electrification Administration (Note: The Rural Electrification Administration, R.E.A. was formed on May 11, 1935 by a President Roosevelt executive order). Hubbard, a large pole line hardware manufacturer, was on notice of the patents legal coverage, and if they wanted to build it this would require compensation for that right. The Rural Electrification Administration's notification could allow their recognition and hopeful acceptance, which could also result in eventual compensation for Chance patent efforts and expenditure.

An April 27, 1928 news article told of a Monday night meeting of Chance foremen. They were given a handout of working instructions at the plant and other matters were discussed. This meeting could have had safety or speed-up implications and/or was oriented toward maintaining organization and standard operating procedures. The level of activity within the Chance plant required disciplined control and direction to maintain a smooth operation. This increased activity brought on many new problems that had to be addressed such as "specials" or MTO's (made to order). The Foundry ran all night the following night after the foremen's meeting. Both furnaces were run at "full heat" to supply some very hard steel for special anchor plates. It's possible certain experienced and knowledgeable foremen and laborers worked through the night. One of the continuing requirements of working at Chance was the expected all-out effort by the workforce to meet special or storm order business. The article closed the very busy week at Chance by recording the receipt of another carload of anchor plates from the Illinois source. Following what was becoming routine was their unloading, dipping, and storage in Warehouse No. 3.

In spite of Chance Company taking on the airs of a world-class manufacturing firm, there were conditions around Centralia that reminded one that the frontier landscape had not totally disappeared. Wolves, for example, were still being killed as efforts to exterminate them were in its final stages. Today it could be cited that this upset nature's balance and overpopulated deer and turkey have become a menace. But the Prairie Queen was in a little world of its own, with its many modern conveniences. There were even service clubs such as the Rotary, which Mayor Bruton and former mayor A. Bishop Chance were members of, and other reminders that Centralia had advanced much quicker than other area, rural communities. It must have been a treat for rural people to come to town even though dirt roads would have to be traveled. Important visitors to Chance Company, whom management were trying to impress, were likely never advised to come by car. This was one reason A. Bishop pushed for modern conveniences as in his frequent travels he saw firsthand, especially in larger cities, the advancements of the day and it wouldn't sit well for Centralia visitors to suggest that Chance Company was located "out in the sticks!" Although an unlikely setting for a substantial manufacturing firm, early May of 1928 continued to reflect the steady hum of industrial machinery as Chance Company shipped out three carloads of anchors to Chicago on May 4.

The level of activity at the Chance Plant not only instigated night work but weekend work. A news item in early May told of Miss Ida Traughber being on crutches as the result of a badly sprained ankle she received while returning from work the previous Saturday. The six-day workweek at that time was a Chance Company standard, as it likely was at other going concerns around the country. One major reason for America's success was people who put in long hours of hard work. During that time a sixty-hour workweek existed as remarks abounded at Chance about ten-hour workdays being the norm and not the exception. The other significance of this news item is Miss Ida Lou was the person in the office who opened the incoming correspondence and customer orders and then arranged the shipping papers so orders could be quickly sent on their way. She was a key figure who was depended upon to insure that operations within the office ran smoothly. It would be a good bet that the "Boss" made sure Miss Ida Lou got to and from work in her current less than mobile condition.

Local news coverage in late May caught an unexpected layoff in the piston-ring area and also news of both Chance brothers, Buford and Gano. Buford and wife had left for Des Moines to transact company business and Gano was home the same Sunday visiting and preparing to open his laboratory in the Chance factory after the University of Missouri closed for the summer. Like his father, Gano always seemed to have something in orbit and the world cruise surely made Gano and his parents closer. His frequent trips back home while in school suggest they had a good relationship and he had strong family ties. His parents must have been proud of Gano as he was entertaining to be around and was successfully fulfilling his college requirements. Two weeks later the June 8 *Guard* told of Gano's election as Worthy Master of Alpha Tau Omega fraternity in Columbia for the coming year. He had finished his winter term at the University and would again join the family firm through the summer. This early example of the student worker program at Chances was a long-standing tradition.

The supercharged atmosphere around 210 North Allen was little changed as a late May 1928 surge saw three cars of anchors shipped out. The new expanding anchor "seems to be making a hit with everyone who has use for ground anchors of any kind." A. B. Chance per the June 8 *Guard,* was in New York on business, suggesting he wasn't about to confidently settle into his rocking chair believing success was his and the pump no longer needed priming. Success had come too hard and he knew it could just as quickly evaporate. When A. Bishop was in town he exchanged his suitcase for a toolbox and rolled up his sleeves and entered into the factory environment. This was a dangerous proposition for employees who had to make changes from his abundant supply of ideas. The huge volume of materials coming into and out of Warehouse No. 3 ate up hours of manual manpower even requiring local part-time workers to speed up what was becoming a bottleneck. The ability to move these materials was plugged into the success of the company as sales volume growth could only result if the plant had the capacity to handle increased activity. What was labeled a Swinging Track was as much a common-sense solution as a brilliant idea to speed up the unloading of rail cars. If you were one of many who were performing the backbreaking job of unloading a carload of one-hundred-thousand-plus pounds of steel, a swinging overhead track was the kind of automation that could readily be understood. The solution from an unnamed source proved there was a better way. They had some smart, part-time college kids doing some of that work, and such toiling gave Jack Curtright and others renewed incentives to study harder and stay in school.

Rail cars once ran on the north side of the Punch Press Building along Wilson Street

For DEFENSE and the FUTURE

A car load per day and still going strong • • •

More than a carload, 40,000 pounds, of anchors has averaged leaving the A. B. Chance Company warehouse every day this year. That is nearly 4000 anchors per day. Many of these have gone to army posts to hold communication and electrical distribution lines, or for temporary use in connection with secret maneuvers. Many have been shipped to regular users, who have undoubtedly been pleased, in most cases at least, with the promptness with which their orders have been filled.

It takes fourteen distinctly separate operations to complete one Chance Steel 4-Way Expanding Anchor, 33,670 operations per working day to meet the demands for this, the most widely used of all expanding anchors. So far, this demand has been met. Four hundred eleven thousand Chance Steel Anchors have been drafted this year. Eighty-six thousand four hundred and two Never Creeps have gone out to make pole lines more secure. Screw Anchors, Cone Anchors, Rock Anchors, etc., have all been conscripted heavily and have gone promptly. The A. B. Chance Company and its employees are proud of our service record. We plan to keep up the good work, but priorities and defense demands are continually getting tougher. We, therefore, urge you to anticipate your anchor needs, obtain priority ratings, and send complete information, date needed and priority ratings with your orders. This will help us to help you by having the anchors on the job when you need them.

Follow your trade papers for the latest information on how to obtain priorities.

Pictured is the conveyor system at Wharehouse No. 3 either loading or unloading anchors. Since this is on the south side, this is the Wabash spur track. (*Chance Tips*, March 1941)

and the land was requested from the city in July 1927 by Mr. Chance to build the Alton spur. There now is a drainage ditch, like all over town, plus a small electrical substation in front of a substantial metal framework in the wall near the center of this building. It was assumed that this frame was for loading and unloading of heavy materials or machines as described by a June 1, 1928 news item about the new swinging track in Warehouse No. 3. But since the Alton spur along Wilson Street wasn't completed until 1929, this new system described in the article, appears to have serviced the Wabash spur track on the south into Warehouse No. 3, although later as noted, there was such a capability installed on the Alton side. The swinging description suggests it could be swung or maneuvered over the load or car body and the track suggests the carrying box or mechanism could roll along the overhead rail to the interior portions of the building. The track braces were said to be made out of very heavy steel capable of carrying very heavy loads. The article continued by stating this was another of Mr. Chance's plans to make his factory up-to-date and efficient in quickly handling supplies. After employees installed the overhead track, a carload of anchor plates was received on Saturday May 25, 1928, and this new system was apparently utilized to unload them into the wareroom of Warehouse No. 3 where finished goods were stored. The in-service date of May 1928 would refer to the Mule Barn before the fire in early 1929. Two cars of sheet steel were also received for the Punch Press Department.

Chance Company news items in the local paper covered a variety of topics, suggesting approval and the information collected by the *Guard* surely had company assistance. Was this an official or one or more employees reporting on their own? The swinging track article closed by identifying more early employees. This practice will identify relatives that worked at Chance, possibly unknown to later generations. Jesse Roberts was promoted from a molder to the foremanship in the Casting Department. This is an early example of what was recognized as "promoting from within" considered by most employees as a positive and long-standing policy of Chance Company. Another veteran employee Clifton Hulen was given the same kind of position in the Ring Department, having charge of all the ring machines and workmen there. As is customary, it is not mentioned what happened to the former foremen. Two new employees were also added in late May 1928, Garland Donelson and Chlore Curtright. It is possible both were temporary summer student employees as Curtright had just returned from Kirksville (Missouri) where he attended school the past winter.

Saturdays around the Chance plant seemed to be one of the busiest days. A couple of news shorts in the June 8 paper told of the receipt of a shipment of tool steel used for the making of dies for the Punch Press Department. The company had also received a large order for expanding anchors, which resulted in a decision to have tool room workmen continue into "Sunday to complete the dies to fill the order in time." It is not clear why new dies were needed, possibly for a special feature, mark, or size anchor, but at the rate they were producing expanding anchors, they may have simply worn the old dies out. Bill Hollander (veteran engineer) said as the dies wore with use, the end product would begin to look less and less like the first one. Sunday work at Chance Company on those occasions continued to be significant, although it was no longer unexpected. Mid-June brought the receipt of another carload of anchor plates from Illinois on a "Monday-Monday." During the week, two carloads of new expanding anchors and heavy duty, Never-Creep® Post Anchors were shipped to Birmingham, Alabama, and to Illinois points. According to this there appeared to have been at least two types of Never-Creep®, the H.D. and regular type. This could relate to the "Post" designation, which was a fence (farm) type. The various Illinois points suggests that a carload could have multiple customer orders and destinations. Some of the locations mentioned could mean they were being shipped to company warehouses, but such an organization at this point in time is uncertain.

During the summer of 1928, there were some complaints registered about high electricity

The Chance Chemical Laboratory was established in 1928. Since Bill Hollander is pictured on the left, this would be a 1929 or later picture. (Centralia Historical Society)

costs. Most users hadn't accumulated the variety of appliances that eat up a lot of kilowatts, but those heavy users like Chance Company felt the sting of the higher-cost, lower-volume rate structure. Showing support for his youngest son, A. Bishop seems to be trying to accommodate his new student employee. The July 24 *Guard* told of the new Chemical Laboratory that was taking over the former drafting room on the northwest corner of the second floor of the main Chance plant. One reason for its selection was its many windows that provided excellent lighting for laboratory work. There was also something else interesting about this location and lab now "under the management of Gano Chance who has been taking chemistry and other work at the Missouri University to equip him for this branch of factory work." This would later be the location of the executive offices of the longtime president and C.E.O, F. Gano Chance! It was frequently repeated "about the man in the corner office" or "you had better not be called up to the corner office." The corner office was prime real estate at A. B. Chance Company in later years as apparently it was in 1928.

The lab was also referred to as an experiment room. Gano later talked about his future role in the company prior to his brother's death, as quite different than what occurred. He was preparing for a technical position instead of one in management. Recalling Gano's early childhood years, this is the kind of lab that father and son were constructing or planning at the time their Snow Avenue home burned. Although Buford showed no such inclinations, his son, little A. B. Chance III, reflected impressive skills and interests in the chemical lab area. It takes little imagination to visualize the presence of the founder visiting and working in this new "experiment" room. There was an eight-by-nine-foot case on the north wall where the chemicals were kept and a large table with other related items in the center of the room. On the south wall were various pieces of testing and other equipment. This facility was an early example of the focus at Chance Company on R/D (research and development), fostered and continually practiced by the founder and supported by the son, which would culminate in the construction of the F. Gano Chance Research Center in 1961.

Local agricultural-driven business conditions weren't all that good in 1928 and one potential casualty was the Chautauqua that had come to town for a long time. It was an "iffy" situation, but it appeared to be on for now. In spite of Chance Company's growing size it was only a part of the local economy, as agriculture was a key element then and today. The big auto and utility products plant between the tracks appeared to be on pace for record sales, if that plateau hadn't already been achieved. A record-size piston ring order was received in late July to support this goal. It's not clear how much profit was being made, but if expenditures for new plant and equipment were any indicator it was adequate or generated the necessary cash flow. The competitive climate frequently spoken of suggests the profit margins weren't excessive, which seemed to be the normal conditions over most of the company's life. As mid-August was reached, it signaled the summer was ending and students would soon return to school. At Chance Company, a *Guard* article stated they were "rushing work." A portion of Chance anchor sales was being generated by farm demand. Besides a new post anchor being introduced earlier in the summer of 1928, the "Rushing" article wasn't begun by telling of some large customer order, but instead revealed some important visitors to 210 North Allen, headquarters for A. Bishop Chance's vaulting empire. Agricultural professors from the state colleges of Iowa, Kansas, Minnesota, and Wisconsin visited on Wednesday, August 15. Although they were mainly interested in the fencing demonstration at the plant, they were inspecting all Chance farm hardware items. This confirms there were other products being produced specifically for, or general-purpose items being utilized in agricultural applications. Because the University of Missouri was an early agricultural originating college, it was noteworthy that they weren't a part of this group. They may have already made one or more pilgrimages to Centralia on prior occasions, especially in view of the short distance, and may have even influenced their out-state colleagues to visit central Missouri. So what is the significance of college interest in the Chance anchoring system? The idea that these people were often focused on the cutting edge of new technologies, doesn't seem to hold water in this case as there was an early effort by A. Bishop to promote his anchors for farm use as he did for utilities. It is a gray area as to what proportion of Chance sales promotion efforts were directed to various markets and neither is it known how early anchor sales were divided. Agricultural applications were a sizeable market that was likely never deeply penetrated, becoming a distant second to utility sales volume. Farmers didn't have deep financial pockets and were resourceful in using naturally available, zero-cost materials. The new Post Anchor was stirring up some interest in the agricultural sector as Chance advertisement efforts apparently caught the attention of these college officials and their interest would appear to be a positive situation, laced with a level of approval. Their expected positive influences in their respective states would give Chance Company a dose of prestige and recognition, plus hopefully boost sales.

The August 17, 1928 "Rushing" article did reveal the fury of activities occurring at the time of the professors visit and on into Saturday and Sunday after they were safely back home. Chance Galvanizing Department employees didn't have the liberty of going shopping, picnicking, or attending church on Sunday as they worked through the weekend galvanizing parts for the new Chance Pulley. The Foundry furnaces had to be run several days at a time to complete the demands of this job, but what drove such a concentrated effort to complete these production orders? This does reflect the nature of A. Bishop in his relentless struggle for success and a concern to get it while it was hot. This may not have been special in Chicago or Pittsburgh, but it was big in Centralia. Even when this employee came to Chance in 1961, there was an intense concern for new business to sustain plant activity. This atmosphere was still new and exciting in 1928 and it points to why Chance Company grew and succeeded in a tough business climate by immediately taking care of current demand. It suggested the only promise for tomorrow was to take care of your customers today. Customer Service was a watchword at Chance and it is clear the founder instilled this into the hearts and souls of Chance employees, present and into the future.

Capturing the public atmosphere around Centralia is important as it directly impacted Chance Company. The prevailing cooperative spirit in this time frame provided a willing workforce and a supportive public attitude toward the company. There must have been continued excitement over what the Chances were up to and rumors prevailed over what they were and were not doing. The Chance firm was a big duck in a small pond and the contrast in its development with the rural surrounding area made it even more of a sensation. It was like

Barnum and Bailey's "Big Top" coming to town in the excitement it generated and A. Bishop was the ringmaster. The nation was still young and industrial development was applauded and approved. Instead of firms taking a photo of their plants, they often hired an artist to make their Hardee's-size complex look like Willow Run (huge World War II Ford plant constructed to build B-24 bombers)! They even had heavy black smoke shooting from their large smokestacks, a sight that in a few short decades would bring strong opposition. This smoke was an action-filled sign of industrial might and self-sufficiency, utilized to further boost their images. It was comforting for people to see such progress when only a few decades earlier the area was an untamed frontier. It was always uplifting, especially in remote areas, to see railroad tracks as it was a sign of man's progress and also an operation that ran twenty-four, seven. People find comfort when stores are open and a radio announcer is on duty around the clock, as long as it's not them working. This kind of never-closing situation was beginning to be associated with Chance Company, as various locals passing up and down Allen and Collier Streets during all hours of the night, could usually see some kind of activity occurring within the Chance works.

The mid-August "Rushing" article continued with the announcement of the installation of a new Heat Treating Department located in the Foundry that dealt with the tempering of the cutting and pressing dies. It consisted of oil baths and preheating equipment with new oil and cooling filters. On the third floor of the main factory where the Ring Department was located, a new ventilating system was arranged. And what was becoming routine, a carload of rough anchor plates came in from the Illinois plant on Monday and were being dipped and stored in Wareroom No. 3. Although certain wording in the paper may have changed to simply not appear repetitious, words such as "rough" may in fact be more descriptive of the natural state of the material being received. Within the article company chemist F. Gano Chance had returned the previous week from Minnesota "where he has been on the selling field for the Chance anchor line." The "rush of work" at the Chance plant saw the startup of the Drop Forge and the large Foundry furnace, which would require seventy-two hours to bring up to temperature.

Within the August 17, 1928 article was additional news on the Jiffy Rim Tool. There were five hundred built and shipped to Sedalia to be exhibited at the Missouri State Fair. This would appear to be taken as a sales opportunity, as why would you need that many on display? This news piece did identify a Mr. Van Orsdel for whom Chance was making the tool. The first of the week also saw the receipt of three carloads of steel from Gary, Indiana, to be used in the Punch Press Department. The "Jiffy" was again mentioned, among other articles utilizing this steel. The August 31 *Guard* also had a lengthy piece on: "The Jiffy Rim Tool Co., of this city." This would suggest Mr. Van Orsdel was located in Centralia and may represent what was long hoped for, the locating of firms near the Chance plant, which would draw upon its multiple manufacturing capabilities. A fall 2004 article in the Moberly paper about the 1907-built former Brown Shoe Factory (eventually eleven hundred employees) being torn down in Moberly, suggested that larger towns around the area got the bigger plants while smaller towns like Centralia were passed over. But Mr. Van Orsdel had a problem, one of name, as another "inferior article" used the same name. Although not mentioned, this would also raise registered trademark issues. The new name for Jiffy was the "By-Jo" which may have been selected because it was unusual enough no one else would be using it. The news item continued with information that reflects marketing involvement. The sales pitch suggested: "The 'By-Jo' is a first class tool and will become as standard in car and rim repairing as the Stillson wrench is in plumbing." The nature of auto travel in 1928 is also revealed as this tool could also be used "in camp" for various purposes. The auto camp was a primitive forerunner to motor courts and later motels that were commonly established along main highways and in many cities and towns. Crime apparently was also a problem in those days as one general purpose of the tool was "being a good weapon to fight bandits and car thieves with at a pinch." This sounds more like the Old West than the heartland of America in attempting to become more civilized and sophisticated. This rim tool may be another auto-related product that was added to the Big C line. Early September ushered in a new school year and students traded their work clothes and fishing poles for pencils and Big Chief tablets, including F. Gano Chance. His summer company activities would be difficult to top by the multitude of future student employees.

The legacy of Irl R. Hicks is enhanced further as he again makes Chance Company news in the September 14 paper. His inventive talents must have been appreciated as again he came up with a clip "which holds the cable and rod on the Chance farm post anchor." It was said to be less bulky and would reduce the stress on the anchor. Until an advertisement is found picturing the post anchor with all this mechanism, will a greater understanding of how the cable was involved, be known. That same week the Chance Drop Forge was busy on the new Chance Pulley. In spite of all the new machinery in place, a carload of anchor plates continued to be received from their Illinois source.

A news article introducing the new Ventilating Ring in the Big C line also revealed internal details of the Chance operation. Chance Company now had eighty-three regular employees and very importantly their distribution in the various plant departments were identified. There were six in Drop Forge, eleven in Tool and Machine Shop, seventeen in the Foundry, ten in Punch Press, eighteen in Piston Ring, four in Assembling, seven in Wareroom No. 3, four in Wareroom No. 2 and Galvanizing, one night watchman, and five other employees. This number of employees at quitting time generated a momentary traffic jam of people moving in all directions from the Allen Street area. But unlike future times when caravans of vehicles headed toward Mexico, Madison, Paris, Higbee, Clark, Sturgeon, and Columbia, most employees at that time lived in Centralia. In fact the article stated that the Chance Plant helped provide homes for 408 Centralia citizens. This positive impact on the local and Boone County economy was beginning to be felt in a substantial way, although that was only a fraction of what would occur in the future in a larger multicounty area.

For A. Bishop Chance this may have been his finest hour and a period he could fondly recall. There was a concern future events would impact his spirit, as one of the consequences of aging is not only the physical affects, but also being increasingly exposed to the suffering and losses of those around you. Showing concern for worker health, local designer of all types of structures, Ed Chamberlain tested his hand in designing a twelve-by-twelve-foot tower that was twenty-four feet tall and located at the east end of the Chance factory. It would house the new blower and replace the old one that cleared the machine rooms of steel dust. As mid-October came, more employees at the plant were trying to complete the many fall orders for the new expanding anchor and piston rings. In fact, the factory had a backlog of over a hundred thousand rings. A Mr. Taylor who had been with Chance for the past three years was being credited for some of this success plus the selling force of the company. This may indicate that the Big C Line had changes in its marketing personnel. Rings were said to be only one of several automotive products of the company. The ring factory (machinery/boxing) was located on the third floor of the main factory building and had a workforce of both men and women.

The 1920s weren't laced with constant economic stability, although growth of modern conveniences such as the automobile, telephone networks, and electrical systems had favored business activities for the Chance Manufacturing Company. Plant activity could be erratic in response to or lack of incoming business and shortages in raw materials. The diverse but limited product offerings lent itself to such an on and off cycle. The Chances would state this was an important reason to establish a full range of products, especially those not so seasonally impacted so the plant operations could be more constant. The firm was displaying it was adjusting to growing business opportunities, as in the fall of 1928 it was a race to fill customer orders and reports of considerable night work was becoming the standard operating procedure. The October 19 *Guard* "Going Strong" article continued with news of a new automatic nut-testing machine in Wareroom No. 3. It threaded nuts on anchor rods and placed them in stacks. It ran the nuts up the threads, testing them, and then ran the nut back to the normal operating position. Although automation in certain areas would benefit production, it also suggests A. Bishop's lifelong fascination with new advancements. When he took visitors for a plant tour, it is almost a certainty he stopped at this new machine and proudly pointed out it magically installing these threaded nuts. Many people born before the dawning of more modern times could never quite make the transition from horses to the auto, but this wasn't one of A. Bishop's weaknesses, as in fact he had more trouble with horses than the auto!

The October article continued with news of a carload of steel from Illinois received on Monday and two cars of Never-Creep® Anchors shipped to neighboring states. Production, shipping, or receiving schedules seemed to favor Mondays as a popular day for incoming shipments. Instead of progress concentrating only on initial changes and additions, the old ways were now giving way to improved second-generation methods and materials. The carbide gas plant used to heat the oil in the ring-tempering bath was replaced with a new bottle-gas system. This may have been propane such as used in rural areas in mid-century times, to supply fuel to kitchen ranges and furnaces. The Drop Forge furnace was also replaced with a "late design" one. It was placed between the two trip hammers "to supply both of these ponderous machines with the red-hot steel used in the various products of the plant." A heavy user of this process was the forming of the eyes on anchor rods. The Assembling Room was also being equipped with the latest tables and stones. Taking up his former position in the Pattern Room after a several-year bout in Mexia, Texas, Roy C. Chamberlain returned to work with veteran pattern maker Irl R. Hicks. This was an important function demanding accuracy in the thousandths of an inch. A less precise procedure was the bringing in of metal to be melted down for foundry operations. Local metal such as the *Guard's* old Columbian web press was also a quality source for the hungry Chance furnaces, as proudly acknowledged by the Rodemyre Brothers.

Late October brought the recurring subject of electrical power for the Chance plant to the forefront. In reexamining the costs between the Chance electric generating plant and city current, a test of the new city rate would be undertaken. Not only were there cost and reliability issues, but there was also a concern over the capacity of the public source to supply the large Chance operation during peak-usage periods. There was a Chance family trait of constant vigilance and not being bashful about trying new methods and bringing in outside help. This process was alive and well under the founder's watch and frequently visiting the power question was due to its financial impact on Chance Company's bottom line. Electrical power was a costly portion of the plant's overhead, as Chance factory pictures of that period indicate many machines powered by electrical motors. It was also a costly proposition to maintain and operate its own generating plant, but it was beneficial in applying leverage to reduce public electrical rates and in providing reliable power.

In November 1928 there were concerns over safety issues, and goggles were being promoted and used in the Chance plant. It was a time of departures of two prominent Centralians, Mayor Bruton whose county judgeship would end his seventeen-year reign as mayor and the death of Buford Chance, expected leadership heir of the Chance firm. The Chance plant continued in a free-fall of activities already initiated, but the Chance family was shattered and plant activities halted as the family went into mourning the rest of the holiday season. As early December came, a carload of oil was stored in the Chance storage tank for furnace use and a carload of steel from Illinois was stored in the Punch Press Department. The Chance Company electrician had been kept busy connecting up all the new machines such as a monster one-hunded-horsepower electrical unit in Drop Forge, receiving five hundred feet of "double aught" cable run from the Chance switchboard. How better to close out the year, in view of the atmosphere in Centralia, than to be positive and try to be uplifting. The *Guard* thought this might be a good time to reflect on the positive aspects of Centralia, clearly influenced by the Chance family. The article told of how clean a town Centralia was, especially in comparison to other communities in a fifty-mile radius. Tumble-down shacks had been replaced with homes that reflected the pride of their owners. Stores were inviting and clean and another plug was made to support local enterprises. So the year ended differently than it had begun twelve months earlier in Centralia, but there were all those things that would carry over and one could be thankful for. However, inescapable is the thought that a chapter had played out and there was a finality to a glorious period in the history of Chance Company, the local community, and even of America.

One way to cope with the loss of a family member, after a period of mourning, is to stay busy. This may have been more difficult for Mrs. Fancy Chance than for A. Bishop because demands at the Chance Plant had taken no vacation. To confirm that the whirlwind of activity at the Chance Plant was impacting the world, there was no better publicity or prestige builder for a small firm than to have a prominent person or organization utilizing and promoting your product. Commander Richard E. Byrd on his two-year exploration of the South Pole was keeping in touch with the civilized world by sending daily radio news bulletins out. The Chance anchor was "the mainstay brace used by Commander Byrd to brace his camp in the Antarctic." This news item in the January 4, 1929 *Guard* is an uplifting way to begin "29," a year not remembered as friendly to our nation's economic well-being. In addition to the Byrd story, that initial issue also revived the coverage of events within the Chance Plant. Certain industrialists were adamant about keeping their factories clean and well organized. Various examples of A. Bishop's concern for a quality appearance in the community have been covered and this concern was also applied to his factory. At Chance, it was common practice for an ending shift to clean up around their machines and like the dress code in the office, there was an expected office and desk appearance as well. There was also a full-time company painter in the later years and employees called him "Dauber," a humorous play on his image of daubing paint here and everywhere. It may be suggested this was long after A. Bishop's time, but if one focuses on all the things that carried over and that he initiated, you get the picture. A January 4 Chance news article was entitled "CHANCE IS OVERHAULING." The Chance Plant was receiving fresh paint on the interiors of several departments. More substantial changes were also exercised in this clean-up and fix-up project by the painting and overhauling of machinery in the factory. This was a slower time that was utilized to prepare for the spring rush of activity, but it also reflects a concern about appearance and its positive impact upon the work environment of employees.

As the "overhauling" article continued, gadgets of various sorts were a lifelong fascination of A. Bishop. It is uncertain if needs drove purchases or a procurement of an item drove a search for an application for its use in the Chance Plant. It undoubtedly would be fascinating to hear some of the conversations of employees—what is that thing? Oh, it's another one of "Boss's gadgets" he purchased on his last trip to Kansas City! The latest was a Ground Detector that was mounted on the switchboard at the plant that would detect and locate any "ground" on the service wires at the plant, on the power lines, or in the city distribution system. This device would not only detect, but also identify where the fault was located. This kind of sophisticated, complicated for the times system, surely made A. Bishop's day. On the practical side, the reliability of the power system was vital to the Chance Foundry, especially if the company was utilizing public power instead of self-generating it. Being able to quickly locate an outage was important. So even though it wasn't the type of device less-talented factory owners and managers had the skill to install and operate in their plants, the electrical background, experience, and genius of the Centralia Wizard made it routine in Centralia. Those around him never grew bored in what the next day might bring. Incidentally, this repeated use of "switchboard" isn't referring to a telephone nerve center, but to a centralized control panel of electrical switches and assorted instruments that controlled the source and monitored and distributed the electrical current within the Chance Plant.

Early January also found the large air compressor shut down to repack the main shaft. The need for a large volume of compressed air was another costly element of a factory operation. Certain processes generated metal shavings that could be blown off critical surfaces of the machines, plus the several Foundry requirements. The maintenance crew was kept busy repairing and performing routine maintenance and lubrication procedures on the many machines and systems now in the Chance Plant. When a new machine was purchased, it had to be set up with all the various requirements for its operation. The new punch press received in early January 1929 made a total of eleven such machines on the premises.

A January 4, 1929 article covered additional details of the new Alton switch track. As might be expected, A. Bishop was a central figure and even financial contributor in its development, although it would service several businesses, first identified as Stone Poultry, Producers Grain, and Wareroom No. 3 at Chance. A February news item also included Illinois Oil Company and a public platform off which the general public could utilize for the receipt of personal items. For example if a stove was ordered from a Sears and Roebuck catalog, it could be shipped and received on the public platform. To allow for right-of-way in the Barr Street area off the main line, A. Bishop purchased Mrs. Zurelda's residence/property (Barr and Wilson). The track then proceeded east by Stone's/Illinois Oil and on the north side of the Mule Barn (Wareroom No. 3), or on the south side of Wilson, narrowing the street slightly. It then proceeded across Allen and hooked back into the Alton line to the east. Construction of the siding was begun around the first of February 1929, but because the new Chance (Alton) overhead conveyor didn't arrive until after mid-March, the new spur track apparently wasn't used to ship the first outgoing until a few days later. The Wilson Street Alton switch track, first discussed in 1926, took three years before it was completed. The Mule Barn rail shipments up to that time, apparently, were being handled exclusively by the Wabash spur on the south side with its already-installed overhead conveyor system. The long periods between initial visions for future changes and actual construction seems out of sync for an A. Bishop Chance, but the third story on the Main Building, the Foundry, the Punch Press (Wareroom No. 3), the paving of Allen, and the Alton switch track are prominent examples of proposed projects that took several years before being realized. One conclusion is that he was more calculating, firm, and patient than many stories portray him to be. This also supported the idea that he was constantly thinking of ways to advance. His preplanning unquestionably resulted in a more organized arrangement of the Chance factory complex and in Chance Company's growth and development, although it required continuous adjustments and at time appeared disjointed.

One important raw material in man's march into modern times was steel. Chance Company, in early January 1929, received another rail car load of steel from Illinois on a Monday. Another vital component in the industrial process is manpower and attention to this area is due in part to the variations in workmen and the potential it held in increasing performance. Although the feeling of loss was likely close to the surface, the Chance family carried on and A. Bishop seems to be actively leading his firm in a way consistent with the past. Sometimes our lives are on autopilot and it is comforting to just be able to continue on with that we are familiar with. Monday, January 7, included an after-hours foremen's meeting that wasn't the kind of get-together that allowed for absences as all foremen were reported to be present. It was an advantage in certain ways of living close to the plant, but a disadvantage if your closeness was considered too convenient to be called after hours. Management and exempt (off-the-clock) employees residing in the Centralia district would become a company requirement. It probably wasn't necessary to spell this out in 1929, when area roads were still dirt and local travel was not always convenient. Like the fix-up work taking place in the plant and on the equipment, additional preparations were being made for the anticipated spring surge of business that could overwhelm the operations if they weren't properly prepared to handle it. Being able to promptly fill customer orders was the goal as it continued to be throughout the life of the company. As noted earlier in such meetings, there was an effort to perfect the workers' relationships between departments. Personal recollections suggest that simple things such as the organized movement of in-process materials from one department to another, was important. Many search and rescue missions were assigned to find lost materials, especially as the plant became more dispersed and spread out. And then there was that portion of the meeting dedicated to promoting a fraternal feeling, a Chance trademark. This feeling of family was considered important by the founder and this idea of "work to a better understanding among the employes [*sic*] generally." An employee who feels they are part of a team are considered more valuable as a positive attitude and good feeling among the people in the plant would in the end bring more favorable results. A. Bishop also felt, as most people do, that it is more personally gratifying to have people under your leadership or influence, happily going about fulfilling your instructions. One of the first impressions of being employed a little over a decade after A. Bishop's departure, was the friendliness of the people, a reflection of the type of atmosphere desired and enjoyed by the founder.

Producer's Grain Company on the north side of Wilson across the street from the main Chance factory, was identified as connected with M.F.A. (Missouri Farmers Association) based out of Columbia. A. B. Chance Company later would buy this property from M.F.A. to use as a storage area and parking lot while utilizing the west area with the odd-shaped red block building. This allowed the cooperative to move east off Jefferson where they could expand further, but in 1929 they were doing well and had transacted $400,000 in business the previous year. Allen Street was a busy area between the tracks and the street continually deteriorated as a result of the heavy traffic. A. Bishop decided he wanted it paved and planned to first have concrete curbs put in but this fell through. After several months of frustration, in early November he decided to instead have it graveled to get through the harsh winter and spring weather to come.

The variety of Chance products was a special aspect of the company that is even reflected in mid-January 1929. A shipment of eleven thousand pounds of chemicals and supplies went to S. A. Shs. Supply in St. Louis, which would appear to be automotive items. The Chemical Department at Chance was said to be working right along with a small force, which likely had previously been managed by Buford Chance. Another shakeup in the foremen registry suggests the stability of a core group of foremen in the 1940s, 1950s, and 1960s can more readily be appreciated. Maybe this testing and attempting to find the right combination in the 1920s was setting the stage of requirements and confirming the importance of stability in the factory floor managers. The entire Punch Press Department including Drill Press and Assembly were turned over to Daniel Sennett, a one-year veteran. Ralph Tucker, formerly in Assembling and the Drill Press Department, was made foreman over Galvanizing. Elsewhere the Molding Department was found running at partial capacity, but by the first of March it was expected to be operating at full force.

Amidst the arrest by local authorities of a lady bootlegger of moonshine liquor, Chance Company was also becoming busy in late January 1929. A load of Illinois steel went to the Punch Press Department as they now were rushed with orders coming in for various anchors. One carload of anchors went out on the previous Wednesday the twenty-third. Centralia's Home Oil Company officials and its three directors, one of which was A. B. Chance, were anticipating more sales of home and business heating oil as the winter month of February loomed and the unpredictable weather of March approached. Vacationing Chance employee Miss Ida Traughber, while in St. Louis visiting the Leslie B. Traughber family, may have attended a new talking picture (movie), which the larger cities were now enjoying. As January ended, the Chance factory was running several departments overtime to catch up on what the *Guard* labeled as "spring orders."

In early February 1929 the Chance factories were plugged back into the city's service power line, as the result of a breakdown in the large diesel engine powering the air compressor. The engine on the electrical generator was apparently identical as a major component was scavenged off it to put the compressor back in operation. A six-days-a-week work schedule was again in vogue for employees or for some at least as a carload of "pig iron" was unloaded on Saturday the ninth to be used in the Foundry. Another carload of steel was received and stored in the Punch Press Department, to be used for Chance expanding anchors. That business was in "full swing," per the *Guard*, turning out hundreds of these anchors daily. The Drop Forge was also busy and more "hands" would be required when activity became even more intense, as spring orders hit with a vengeance. Continuing work in Wareroom No. 3 would see more overhead tracks installed to service the Wabash and Alton sidings and to connect these together so a quick unloading could be handled from either line. Away from the hustle and bustle around 210 North Allen, quieter moments brought thoughts of family and Gano made another of his frequent trips home, as he approached his graduation in the spring.

Railroad shipments into and out of Centralia for 1928 were important because area roads were poor and the truck not well developed. Material was primarily moved by train, but the *Guard* in its February 22, 1929 article said truck shipments were growing. There were 1,248 inbound and even more (broken number) outgoing rail car shipments. Not all this activity was Chance Company related such as 14 incoming carloads of automobiles and the many loads of livestock (310), grain, and produce going out. Fortunately, this is further broken down such as 287 carloads of metal coming in, a Chance staple, and 287 (identical) loads of Chance patent anchors being shipped. Other Chance shipments included 10 cars of piston rings, 6 of pulleys and blocks, and 5 of various other products. The plant, for heating purposes, had used 30 cars of oil! Shipments included 1928 activity through February 20, 1929. It is virtually impossible to convert this data into pounds or dollars with such items as piston rings being boxed. This is the reason there were more loads of products going out as the bulkiness of packaging and the irregular shapes of manufactured goods increased more volume, but lighter carloads.

Will Rogers was making a strong emergence on the public scene in 1929, as were the violent activities of Chicago gangsters with Al Capone and Elliott Ness of the "Untouchables" coming to mind. After a four-month shutdown, the Piston Ring Department started up in late February and the March 1 *Guard* said it was "during the dull period for this article on the markets." Remarks about the inability to be competitive and the discussion of moving piston rings out of town appears to have been impacted by a lack of orders to keep it in constant operation. Although some veteran employees said they were not laid off during their forty-year careers at Chance Company, it appears that this wasn't true for all employees. It wouldn't seem logical that they could have been furloughed to other departments during the slower winter months for such a sizable group. Taking advantage of such lulls, a Kansas City firm was waterproofing the basement at the main plant in late February. A business trip to Kansas City in February by A. Bishop may have been connected. The Galvanizing Department was working steadily to keep up with orders, but the explosion of their oil furnace on Tuesday, February 26, temporarily stopped operation, as did a blower failure for four days. Employee Ralph Tucker received several burns but no serious injuries from the explosion. There was always a chance of getting burned in Galvanizing, even during so-called normal operations. The Chance article closed with a personal interest story of Sol Sexton who had worked in St. Louis for the past thirty-five years and had returned home to begin working at Chance. His position was in the Foundry and this connection may have initiated the contact

Anna Lee (Toalson) Chance (A. B. "Buff" Chance IV Collection)

with a St. Louis foundry operation discussed later in March.

Two main departments of the plant, Galvanizing and the Foundry, ran on Sunday in late February because of factory part shortages and an influx of customer orders. Galvanizing would continue day and night activity for that week. Chance Company news had become so common the *Guard* established a weekly column headed "At the Chance Factory." Two carloads of anchor plates from Fairfield, Iowa (change from Illinois source?), and one of steel for Punch Press were received on Monday, March 4. Wednesday, two carloads of anchors headed to Eastern states as overhead track work continued on the Alton side. Another gadget, akin to the fault indicator recently discussed, a registering meter had been installed. It indicated the amount of current being used in the plant and kept a chart showing peak loadings, etc. The factory was still using city current, but the Chance generating unit was ready for operation if needed. Spring fever had surely hit campuses as the last weekend of February found Gano, Miss Anna Lee and Nathan Toalson, Eddie Howard, Miss Anna Belle Fountain, Ralph Denton, Morris Cox, and Frank Edwards home over the weekend. On the national scene an indicator of a bleak future revealed itself on March 4 when President Herbert Hoover took the Presidential Oath of Office in the rain. But for the time being, life seemed to continue as it had with University students Jack Curtright, Gano Chance, and Miss Anna Lee Toalson visiting parents in early March 1929.

Noting the importance of photography and advertising as an in-house operation, R. L. Angell and Son were plastering the "Camera Room" at the Chance Plant in mid-March. Punch Press and Drop Forge received two cars of steel and the Galvanizing Department was frequently running day and night to plate many of the products being manufactured. The reporter suggested that some of the work turned out looked almost like silver-plating, it was so smooth and fine. Chance factory activity and the local agricultural economy were not only driving local railroad activity, but a sixteen-car train was put on the Columbia branch to serve college students during Easter vacation. Students not only utilized the train but also automobiles and buses out of Columbia. However, the Chance factory continued to work with a carload of pulleys heading to Kansas City on a Monday, followed by three cars of anchors to Eastern states. Such volume also required incoming raw materials such as three carloads of steel from Illinois for the Punch Press, which suggested a lot of volume was being generated by the new expanding anchors. Overhead conveyor materials were being received and installed on the west side of Allen Street. With the completion of the Alton siding, the first Chance shipment went out on March 22. Amid the spring rush, several employees of the Foundry were temporarily laid off in late March as the Tool Making Department was busy making special tooling to get the Foundry back into full operation. These layoffs were noteworthy, but they could be considered relief from the nonstop kind of work activity at the plant. The only way to get a day off it seemed was when a machine broke, delays in materials arrivals, or some glitch in scheduling. This wasn't very susceptible for planning a family reunion or taking a trip, but it does reflect the all-out effort by the company and employees to satisfy business activity at a level that had earlier only been dreamed of.

An avenue traveled by Chance Company was the acquisition route. The volume of business in the Centralia Plant in the spring of 1929 was apparently bursting its capacities at the seams. In a move to expand productivity, employee Clifton Hulen (May 1928 Piston Ring Department foreman) traveled to St. Louis to inspect a foundry. The thought was, if acquired, it might be left there to add to Chance Company's casting capacity. The idea of a plant in a major city was an attractive proposition to A. Bishop, but after he began the construction of his plant

Mule Barn Fire and Punch Press Scenes (*Chanco News* June 1957)

complex there isn't the feeling he would have made a decision to move the entire operation out of town, although he occasionally publicly discussed the problems and disadvantages of his plant's remote location. Although Kansas City was the city where relocation had earlier become a serious possibility, St. Louis was getting some attention in 1929–31. It is interesting that competition with a St. Louis based firm was the hypothetical basis of his complaints to the city about high water and electrical rates in late 1931, as the grip of the Depression was being felt. A. Bishop suggested the difficulty of being competitive with shipping costs to and from Centralia when a firm in St. Louis had neither. This may reflect the fact that one Chance steel source was identified in nearby Alton, Illinois. So it was not surprising that he was interested in city-based factories in 1929 for both capacity reasons and to be able to compare its costs against his Centralia operation. This was just common sense in his view and such opportunities were wise to pursue. The future would continue to hold such possibilities, but Matthews (St. Louis based) would, for example, be moved into Centralia after their purchase several decades later.

The latest machine to be added in April was a L. W. Pound metal-planing machine placed in the Machine Shop. Employees continued to be transferred to fill immediate needs and foreman positions more perfectly. Ollie Chandler became one of the most traveled of this foreman pool and must have developed abilities that management needed, as fast-changing plant areas required it. He was sent to be in charge of Wareroom No. 3, which had become a center of activity and significant automated changes were recently added to enhance its ability to handle these increases. His departure from the Tool Room was filled by Austin C. Chandler. Greeting Ollie on a Monday were two cars of steel from Illinois and two outgoing cars of anchors. But what confronted Ollie and Chance Company a week later on Tuesday, April 9, 1929, was to bring more problems to Wareroom No. 3 than the talents of its new foreman could overcome. At about 9:45, Tuesday morning a fire broke out in the dipping room (where anchors were dipped in paint), and the combustible materials and the openness of the large wood structure allowed the fire to spread quickly. Only about three minutes after the first explosion, the fire was already coming through the roof. Employees and the fire department could do little to stop it, but did save the scales and part of the sheds at the west end.

Prior to 1929, the so-called Punch Press Building, across Allen Street to the southwest of the Main Office, was formerly the location of what was labeled the "Mule Barn." A local story suggests that area was a stable for many decades, as earlier information from the local paper confirms. Green's Mule Barn, per the

April 12, 1929 *Guard* fire article, was built around 1905 and purchased by Mr. Chance in 1926. It is thought there was a stable located there prior to 1905 and it was the place where a post–Civil War story originated. After the Civil War, a stranger rode up one day inquiring about a certain Centralian with Confederate persuasions. This stranger was believed to have been the most famous of all American outlaws, Missouri's own Robin Hood of sorts, Jesse James, returning to the scene of the Centralia Massacre. There is debate as to whether Jesse was part of Captain Bill Anderson's party and participated in these Centralia skirmishes, but it appears that both James brothers were present.

The fire that destroyed the Mule Barn in 1929 was remembered by four employees with different perspectives. The first by then student Paul Sturgeon:

> *Well, I remember it was early one morning and I was still in high school. During recess this billow of black smoke showed up in the western sky, creating quite a bit of excitement. What it was, was a metal barn that had been used in World War I as a mule barn and livery stable. Mr. Chance had acquired it for warehousing and cleaning anchors, particularly Never-Creep® Anchors. The process that was used was a dip vat with a trough on each side of it and they would pick up these anchors with a hay hook, drag it through the asphalt paint and throw it on the drain rack on this side. When they got that filled then they would drag it back through on the other direction so that all of the anchors were double dipped. (1979)*

Although combustible fumes from cleaning solutions or the anchor paint only required a spark to ignite, the fire's source is unconfirmed. As Paul alluded to, this was the anchor building and Flynt Chandler explained what kind of work he did there by saying, "Painted the anchors, tied them together, and shipped them."

P. D. "Paul" Sturgeon (*Chanco News* March 1946)

(1979) In remembering what happened to the Mule Barn he said:

> *I was working in the machine shop at that time and Mr. Chance came running out of the office next door and said grab a fire extinguisher we have a fire over at the mule barn. Shoot, by the time we got there, there was no need of taking a fire extinguisher, it was gone. (1979)*

Although employees can suffer loss of work time or even their jobs from a fire, Les Reynolds had an unexpected loss:

> *We were working with Uncle Ed Chandler in that house where Hough lived and there was Ward McDaniels, myself and I don't know 2 or 3 more down there. Ollie came down there and said, Uncle Ed, they want you to clean out and build some bins in the basement. He said, Les, can you get your car and take Uncle Ed up there and just drive it over in the mule barn. We got up and went over there and Uncle Ed said Ward,*

you and Les go over to the back of the mule barn over there in the lot (they had all kinds of lumber) and he said, I want about 8 nice 2 x 4's about 10' long. We went over there and got them and came back. Now he says, I need about 3 or 4 2 x 6's. Ward had started up the steps. He says, good gosh Les, look over yonder that mule barn is a fire. I thought he was joking with me because my car was in it. I said, yea you ain't fooling me. Man she was done gone! (1979)

Nathan Toalson recalled the Mule Barn and the fire that consumed it:

I remember when it burned in 1929. It was originally a mule barn and I guess Chance Company bought it and used it as a warehouse. Actually, when it burned I was attending school. I remember the pictures of how the anchor rods were stacked like—or tepees. They were stacked in a big circle. (1979)

The April 12 *Guard* article confirmed and added to the recollections of these four veteran employees. Paul's recollections of the dipping process was knowledge he likely acquired later as a factory employee and eventually as the vice president of manufacturing. The paint was described as "largely tar" and there were seventy-some barrels of paint and solvent stored there before the fire. The paint or dipping tank was about a four-by-eight-by-six-foot open container. W.T. Fountain who gave the alarm was twenty to forty feet from where the fires started, or where the explosion took place as he heard a noise behind him. It's thought an accumulation of paint fumes were ignited by an unknown source. Moments before Flynt Chandler was told to get a fire extinguisher by Mr. Chance and to follow him, Mr. Chance, from an upper story of the main building across the street, saw what was occurring. The fire article also told of the vehicles in danger including the loss of a car by Les Reynolds. Two cars were pushed from the fire area, one of which was Mr. Chance's car parked in a driveway. Three carloads of anchors and other products were pushed out of the way and saved. And what could be considered more appropriate at the Mule Barn fire than for Burl Hisle to save a team of mules hitched to a hitching rack near the fire!

Fires were a constant threat in earlier times and generally were more destructive because of the lack of proper equipment and an adequate water supply. The first running water system in Centralia, according to the November 8, 1890 issue of the *Guard*, was a large water tank being built near the railroad depot. It was for the iron horses of the Wabash Railroad. The steam locomotive required a lot of water to operate and to avoid boiler explosions you didn't dare run out. An engine pumped the water from Lake Deutcher about a mile east of Centralia into the tank. *Guard* editor A. Rodemyre suggested the possibility of laying lines in downtown and using this water source to fight the kind of devastating fires Centralia, like other towns, had experienced over and over. Problems must have arisen as discussion about it ceased. It's interesting that the Alton Railroad, per the November 7, 1913 *Guard*, made a request to use city water for their locomotives and this occurred several times over the years by both railroads, especially during droughts. After being defeated in November of 1906, the waterworks was first approved by Centralia citizens according to the *Guard* of March 1, 1907:

This has been an eventful week for Centralia, and an anniversary that may be celebrated in after years. We have bonded the town for waterworks and municipal lights, launched a canning factory upon the new wave of prosperity and organized a brass band to head the procession marching out to lead the New Centralia to her future greatness.

F. Gano Chance said when his father became mayor:

> *At that time we didn't have city water. As mayor he instigated putting in water lines and digging a well. He put a light on top of the city water tower. Some people thought it was crazy idea but as I got into the teen years and was riding around with a girl at night, I could always find Centralia because the county was flat and you could see it for 7 or 8 miles. (1981)*

The August 12, 1910 *Guard* suggested that the city planned to put an electric light on top of the water tank, or what is called a water tower today. But it wasn't until September 1914 that discussions about it again arose. The needed $50 to cover costs was donated by local businessmen. The *Guard* of November 27, 1914, confirmed the light was turned on and it appeared like a "diamond in the sky."

One of the advantages of a city water system was of course, fighting fires. The original Centralia water tower was located near the Chemical Building and behind the Foundry. The Chances direct involvement and the Chance Company policy of actively supporting the local community was a win/win situation for both Centralia and Chance Company. But in spite of a modern source of water in close proximity to the Chance plants in 1929, it didn't save the Mule Barn.

A. Bishop seemed to be his old self in the way he handled the fire disaster, in spite of losing a son a short time before. As is said, "when it rains, it pours," and in spite of setbacks things could have been worse. At the time of the fire, it was raining with a mixture of snow and the wind was blowing from the northeast, which allowed the fire's airborne debris to harmlessly be carried to the southwest, leaving Illinois Oil, Producers Grain, Stone Poultry, and the Chance factory to the east safe. What are the odds that these favorable weather conditions would exist at the exact moment the fire occurred on an April 9? Winds in these parts don't normally come from the northeast and although it does snow in April, this is not all that common. What makes it even harder to understand the seriousness of the loss is the manner in which the Boss took charge and didn't spend three weeks wringing his hands and feeling sorry for himself. This is a special time in examining the optimism and positive attitude of A. Bishop. He had never allowed setbacks to dampen his enthusiasm and his strong spirit again is reflected in his optimism toward the future. The Bible speaks of good coming from crisis in our lives and what had been envisioned for several years would quickly be erected where hot coals and ashes now lay smoldering.

The Mule Barn was not a fireproof building and with all the combustible materials and the many activities taking place there, an accident was just waiting to happen. The *Guard* fire article revealed about a third of the northeast side was used as the dipping room and joining this was the drying room. On the south side was a storage room (Wareroom No. 3) for the various sizes of anchors "including the new expanding anchors and a lot of new galvanized anchor rods." Two floors of storage area also included other products such as wooden maul handles, cement, lime, building materials, and other supplies. This area included the newly installed overhead loading devices. January inventory figures showed a value of equipment and stock at around $200,000. "The building was insured at a fair valuation, but on account of the high rate, the material, paint, and stock was never adequately covered by insurance." The adjuster came in a few hours after the fire and began work. Mr. Chance said that as soon as the loss was adjusted and the lot cleared up, they would begin to rebuild. The new warehouse would be of brick and steel or tile block and steel. He was reportedly concerned about the safety of his workmen immediately at the time of the fire, which included eleven regular employees besides those operating between the plants.

The fire affected the entire Chance operation, but the April 12 fire article stated production would continue on. One of the big losses was storage room but this wouldn't be allowed to slow activity, "as these can be shipped out to other points if necessary." Because of the way A. B. Chance Company later organized its so-called "ServiCenters," which normally consisted of a sales office and warehouse, it was suggested earlier that A. Bishop may have first developed this when growth of the company had skyrocketed. The "other points" as mentioned in April of 1929 confirm either Chance warehouses in various sections of America and/or some type of distributor arrangement. This additional warehouse network is confirmed further in stating that orders for Chance products could be handled, "as a large supply is stored in the various cities where the company maintains offices and storage plants." Not all the stock in the Mule Barn was lost as fire items would be inspected and that found salvageable would be repainted, etc.

The corner lot at Allen and Wilson, where the charred rumble from the Mule Barn laid, was not destined to be ignored real estate. Like a whirlwind, A. Bishop swept into this project and grabbed the opportunity in making yet another modern expansion in his mushrooming enterprise. This wasn't a new idea, however, as after his purchase of the Green and Sons Mule Barn in the spring of 1926, a *Guard* news article in November 1926 told of Chance plans to erect a brick building on the mule barn site! Although there were losses in materials, finished goods and in production stoppage, the fire wasn't without benefits as it cleared the way for the new facility. This was a period of optimism and growth for industrial firms across the country, but soon like a candle without a wick, these times would flicker and disappear. Not until the world got caught up in war would profitable factory activity again reach such heights. The expansions of the 1920s would thus be a blessing in disguise during the early 1940s in helping the Allies win World War II.

Paul Sturgeon, with the aid of his photographic-like memory, reveals what was about to be erected along with some thoughts developed from being around Mr. Chance.

> *After the warehouse burned down I suppose, Mr. Chance was such an optimist that before the fire had cooled and everything he was on the phone ordering a new building. You probably don't realize it but that was one of the earliest types of Pre-Fab buildings in that it was a Tucson building. The structural steel was expanded and the building was composed of 2, 40' low bays on the side and 2, 60' high bays in the middle and the building was 100' wide.* [Note, recent measurements found these dimensions across the front or rear were 20' x 30' x 30' x 20' =100' which agrees with Paul's 40 and 60 foot dimensions that were split on either side.] *He bought the steel framework and the contract was handled by Hathman Construction Company of Columbia. You notice the design of the building, in order to get more clearance they built the pier up. I forget how high up, about a couple or 3'.* [Note: These are raised 2 feet, adding credibility to Paul's remembrances.] *They set the steel on top of that so that they could take a standard building and get higher clearance. At this particular time, when they built the new building, they put in the more or less automatic dipping of anchors in the NW corner. (1979)*

The *Guard* news article stated the name of the building firm was "TrusCon Steel Co." The representative of the TrusCon firm was already working with Mr. Chance on the plans for the new building a few days after the fire. Its larger area would require changes with the Wabash and they were cooperating. Mr. Davidson of the Wabash was there the next day, Wednesday the tenth, to work out a time lease on the additional grounds and to plan changes on the sidings to better serve the new facility. The new building would be one story in height, built in a modern fireproof manner, with a double driveway (truck) down the center for loading in the dry.

During disasters there is the opportunity for people to reveal their inner strength and true character. Heroes are born in such instances and there are also those who are supportive, helpful and compassionate during such times. The employees of Chance Company were generally a loyal group who were supportive of their boss. One of the realities of Centralia was how the lower living costs related to lower wages as Chance employees were often torn between leaving for a higher salary or staying put. This was particularly true for skilled workmen and management personnel such as veteran employee Irl R. Hicks, who had been offered a larger salary from a firm in Oklahoma. There usually were other reasons for leaving and the Chance operation like any other employee environment, could at times breed dissatisfaction. The fire generated even greater anxiety, but such a situation also ran emotions high in regard to A. Bishop who had yet another mountain to climb. As Irl took the situation into account, even while the fire blazed away, he fired off a telegram to the "OK" state rejecting the job offer. He later said in a *Guard* interview that Mr. Chance needed him during this crisis and he felt it his duty to stay on. He had grown up with the factory and he considered his loyalty was worth more than dollars and cents. This kind of loyalty to duty is more often cited in law enforcement circles, with firemen, and especially in our military forces that too often sacrifice all in serving their community and flag.

One telltale sign of using brick on additions can be its variations in appearance from the original. Unless the same color, size, and texture of brick is utilized, it becomes obvious what has been changed. In addition, it's virtually impossible to close windows, doors, and other openings with brick without it standing out, unmatched like a sore thumb. The third-floor addition onto the Main Office Building is a case in point as Roy Hansen pointed out, "You can look at that building and see that the third floor is a different construction." (1979) Paul Sturgeon also said, in reference to the use of brick on the Punch Press Building: "It's funny about that building, when it was first built, the face of the building is the same brick that Mr. Chance used when he put on the third floor and the back part of the office building." (1979) Further into the discussion, Paul recalled an incident that identified the source of these brick:

> *Back to the basic brick in the warehouse in place of the one that burned down. We had occasion in the later 40's or early 50's to do some modification on that building and Bill Decker was the Plant Engineer at that time. They were going to close up the two garages that were in the SE corner where Mr. A. B. and Miss Ida stored their automobiles and make an office out of it. So we retrieved the brick out of the wall, to see if we could match it, and we discovered the brick was made by a brick company in Streator, Illinois. We got on the phone and called and told them who we were. Didn't even have to send him a brick. He said, 'Oh I remember Mr. Chance, he was in here and picked out those bricks!' He said, 'I will send you the brick to match,' and this was like 20 or 30 years later! Imagine getting the same brick to match that building. (1979)*

What is remarkable about this story is not necessarily brick, but the people involved. The memory of, what appears to be the owner or manager of the brick company in Streator, was very impressive. But maybe he had some help as Mr. A. Bishop was in many ways an unforgettable character. Recorded stories and situations such as this point to his ability to make lasting impressions upon those he came into contact with. One would assume that in 1929 the brick for the new Punch Press Building from Streator, Illinois, would be shipped direct to Centralia by rail. There seems to be some confusion as to whether it was the Punch Press Building or the factory on the east end of the Main Office, as Flynt Chandler recalled, "I helped haul all those bricks that are in that building over there, from Columbia, paving brick to build that building." (1979) Before discussing this further, an additional comment made by Paul Sturgeon on bricks used in the Chance factory construction might clarify this:

> *Then they built the building right straight back, which was another Tucson* [TrusCon] *building, the building behind the Main Office where the foundry and the press shop and everything was. That building, curiously enough, is made of paving brick that was manufactured at Moberly, Missouri, and used for paving lots of roads in this area. For instance, from Moberly to Cairo and then the Moberly to Huntsville road. (1979)*

It was confirmed that the east Foundry Building was in fact walled with the large paving brick. The Main Office expansion and the Punch Press appears to have utilized the same brick (Streator) as Paul eluded to here and in previous comments. It's possible Flynt meant Moberly instead of Columbia or a distributor arrangement allowed the Moberly brick to be obtained in Columbia. The Chandler brothers could have been involved in hauling both the Streator and Moberly brick but likely only the Moberly brick.

It wasn't until this plant material was examined more closely, that there developed the realization that the parking garage was located in the Punch Press Building and not the east extension on to the Main Office. This is logical as Mr. A. Bishop and Miss Ida Lou were just across the street from their offices. The earlier days found the Main Office entrance situated to the south side instead of the front or north where the GM&O (Alton) Rail Station and Producers (M.F.A.) elevator were located. These areas would be absorbed into the Chance complex several decades before the final swept to the south which allowed finished and unfinished Chance goods to reside in spaces formerly occupied by Stephens College coeds and Proctor and Gamble salesmen on their way to Columbia via Wabash rails.

Brick also played another role in Flynt Chandler's life. Referring to the "building across the street" he related how he had fallen off a high section of that building (Anthony?) into a pile of brick. As a result, he experienced a hip injury, which is covered earlier in 1925. This indicates continuing involvement of Chance employees in the construction of Chance plants. The hauling of the brick required vehicles and apparently in the 1920s–early 1930s, the company had, according to Flynt: "Yea, we had the Apex, two old Oldsmobiles, and a solid tan Maxwell." (1979) A Model T Dump was also on the premises in the 1929 time frame, at the time the Mule Barn burned. These vehicles also had a dual role in transporting factory raw materials and finished products.

The fire may have dominated the news at Chance Company the week of April 8, 1929, but there were other significant activities occurring. There had been many surprises of advanced developments taking place at the plant, such as the latest of castings being made from aluminum! This was made possible with their new aluminum furnace in operation, which according to Bill Hollander interview material was used more for in-house needs than on production items destined to be sold. Also on this Monday, they had received two carloads of anchor plates of various sizes and because they weren't unloaded before the fire on Tuesday morning and stored in Wareroom No. 3, they were instead later stored in Wareroom No.2 in the Galvanizing Department. But the April 12 *Guard* article generated a big question of what it meant by: "These are anchor plates made in one of the other plants of the company and brought here to be finished." It is unclear where this other plant was except perhaps Wej-

Lock of Winterset, Iowa, which could suggest an earlier than expected working relationship between these two firms. More likely it was referring to Iowa Malleable who regularly supplied anchor castings to Chance.

A more normal occurrence, although one made more difficult by the fire was a shipment of three carloads of anchors to the east on April 12 and 13. This Friday and Saturday activity suggests that anchor production had again been continued at some level to satisfy either customer orders or to relocate material that was overflowing Wareroom No. 2. An unwelcome event, another injury accident was reported on Monday the ninth when Bill Kersey suffered an injured finger at the plant when it was caught in a vice. Another person getting a lot of attention was the surviving son of Mr. and Mrs. A. B. Chance. It must have been difficult for F. Gano to not only deal with spring fever on the M.U. campus, but to be absent from all the Chance plant activity. Events such as the fire must have given him the urge to come home instead of concentrating on finals and studies, a disadvantage in that respect.

Two weeks later the April 26 *Guard* again brought local news items of interest. April showers were temporarily dampening building activities on the new Allen, Wilson, and Hickman Street Chance plant, to be labeled the Punch Press Building. Remarkable was the news that all the material for this new building was on the ground! This apparently was made possible with inventories that could be substantial because TrusCon had developed the "pre-fab" type of construction, per Paul Sturgeon, that allowed commonality in materials. However, a Chance news article two weeks later on May 10 stated warehouse work had been held up because the structural steel was not on site, but had been shipped. Even that level of performance is still noteworthy. Bricklayers were already laying the brick walls between showers. Other items to be installed in the new wareroom were received on the previous Sunday the twenty-first. Two tanks weighing five thousand pounds would be used to store a preservative utilized in the pickling of some manufactured articles. Continuing with a regular activity around the plant, was the painting of the tool shop and tool room and "other departments will be brightened up as fast as they can get to them." In spite of advancements of the day, some situations put progress in reverse. Long sought hard-surfaced roads were bringing higher speeds and more serious injuries and deaths on American roadways. Discussions about cracking down on reckless driving and establishing highway policemen or patrols would eventually bear fruit.

A significant *Guard* news item of changes at Chance on May 10, 1929, made more impact in view of the upheavals generated by the recent fire. Changes were likely needed to keep abreast of business with limited plant capabilities. In response to the spring rush winding down, and with the new Punch Press facility coming on line, maybe this was an appropriate time to update the Chance layout. This was described as many new changes in the departments and reorganizing the personnel to place them in the best position for work. Like the earlier "speed up" situation, it appears to indicate management dissatisfaction with plant performance. The article said the goal was "to obtain better working conditions and to get the plant on an efficiency basis." It would be beneficial to know who was calling these shots, but such widespread and substantial reorganization would likely be above the plant manager's authority and point to A. Bishop's involvement. The new machine plans (machine movements from one department to another or within a department), resulted in sixteen men being laid off. To employees, this hot and cold kind of environment wouldn't seem to be a very attractive situation, but as discussed, this may have been the only way some workers got a mini-vacation. It is very difficult to imagine the "Boss" laying-off workmen if they had business enough to generate production orders. The machines weren't the only thing being shuffled as employees "will hereafter be placed in the department best suited to their ability to turn out the best work." This perceived problem, alluded to in the past, seemed to be an ongoing challenge. There are always jobs filled by individuals that could be performed better by someone else, so there is hardly ever perfection attained. The machines as mentioned were "being arranged to insure an orderly flow of the product thru the plant." It seems like these exercises were frequently occurring, but the continued purchase of new machinery and processes made this an ongoing process. For example, this news story concluded with word of new grinding machines being placed on the third floor of the Ring Department. New machinery and supplies for the new Punch Press Building was also arriving and it was calculated that in two weeks, when the roof is completed, this new machinery could be installed. Not to be ignored are new products such as the expanding anchor's success that required a different group of machinery and operations. One area not regrouping was Galvanizing which was "in full blast!"

Alpha Tau Omega

Mrs. Margaret Greenlee
Chaperon

Fifth row—J. Barnes, Drake, Coggins, Lancaster, Kearney, McKee, Jennings
Fourth row—McClanahan, Edgington, Robinson, Evereman, Scothorn, Ausmus, Lloyd
Third row—Waite, Padgett, Bodendieck, Allen, Chance, Morris, W. Barnes
Second row—Noll, Rector, Minor, Karsch, Early, Hughes, Miller
Bottom row—Landrum, Porter, Viles, Null, Ambrose, Richerson, Swink

Alpha Tau Omega Fraternity was founded at Virginia Military Institute September 11, 1865
Gamma Rho Chapter was established April 21, 1906

Colors—Old Gold and Sky Blue
Flower—White Tea Rose

Page 314

Pictured is the A.T.O. Fraternity featured in the 1926 University of Missouri *Savitar*. The chaperon, Mrs. Margaret Greenlee, and F. Gano Chance are already familiar to the reader. In addition Mr. Ausmus will begin a Centralia law practice that will eventually lead to a long association in representing Chance Company. The name Kearney will also become familiar to the reader in Chapter Six. This is Jim Junior. The name Bodendieck in addition became a familiar name in Volume II, in his association with Tips Tool. (Note: According to Gano this A.T.O. Bodendieck from Kansas City was not the Illinois-based Tips Tool Bodendieck.) The Chance, Kearney, and Bodendieck families influenced and controlled a substantial portion of the Live Line Maintenance Tool product area over several decades. (University of Missouri Yearbook)

Lighter subjects such as Babe Ruth's marriage were a topic of conversation in mid-1929. The Centralia paper also covered a local celebrity, A. B. Chance who was on the Board of Directors of the Rotary Club. With at least four grocery stores and a new J. R. Allen Plymouth dealership, Centralia was a lively place. C. C. Hough of Columbia was in Centralia over a Tuesday, May 7, on business. Internal accounting challenges in the Chance Plant, according to employee interviews, required his occasional expertise. Part of the problems at the plant may have been due to the absence of J. B. (Bryan) Fenton who was the plant superintendent from 1926 to November of 1928 when ill health forced his departure. There is a sense that he was a

A. R. Toalson Home. Corner of Hickman and Sneed Streets. (Centralia Historical Society)

stabilizing force at Chance and a man A. Bishop leaned and depended upon. He is likely one of the reasons why the world cruise could be taken with the assurance the plant would continue to run smoothly. Illness was, however, a constant companion as Mr. Fenton was "gassed" during World War I service and his immune system was very weak. Mr. Fenton passed away on May 10, 1929, in a Colorado military hospital and showing the appreciation for his former employee, A. Bishop traveled to Kansas City to meet the funeral party bringing the body back to Centralia. According to Tom Fenton, whose family operated the local funeral home, they weren't relatives. There were also other unwelcome circumstances at the Chance Plant when Don Ingrum's forearm was injured by a piece of steel. But on the brighter side the May 31 *Guard* carried an ad for: "The Old Reliable Standard Fence, Never-Creep® Anchor" at a special price of just $1.50! See page 123.

As June 1929 unfolded, news of Francis Gano Chance's accomplishments at the University of Missouri was reported. He was a member of the Alpha Tau Omega fraternity and had just graduated with a B.A. degree in arts and science and a B.S. in chemistry and engineering. This is quite an accomplishment, especially in those times. This would bring Gano into Chance Company full time and he also introduced a fellow classmate Bill Hollander to his father. Being a self-made person, A. Bishop wondered how he was going to use two educated professional engineers in his firm, but time would prove that Gano and Bill represented two of the most influential employees that worked there. Gano was into medical issues at the plant and the recent Ingrum injury, representing the entrance of metal into a worker's body, may have driven the Chance purchase of an electro magnet which was powerful enough to draw metals from injuries. It was installed in the office of Dr. G. Ben Kesler. As mid-June came, A. R. Toalson (1859–1929) became ill and passed away several days later. Born in Boone County, he had been a city resident since 1867. He was a Baptist and had a son and daughter in his second marriage and had lived in the same home for sixty-two years! His daughter Anna Lee, a senior at the University of Missouri and a student in the Rosenthal School of Commerce in Columbia, had in early June received a statewide award for a speed record on a typewriter. She would be a candidate for a B.S. degree in social science in August. She was a member of Phi Mu and Theta Tau Epsilon Sororities. Mr. Toalson was able to enjoy at least a part of his daughter's scholastic accomplishments. Miss Anna Lee began a Commercial School, a branch of the Rosenthal College in Centralia in late September at the Toalson residence. This would indicate she was out of school and back home in Centralia, as was F. Gano Chance.

Work on the new Wilson Street complex was progressing with corrugated steel roofing being installed. Installation of the overhead tracks would also begin soon. A week later, the June 14 *Guard* told of the assignment of a reliable figure to take charge of problems associated with

Portraits of F. Gano Chance and Anna Lee Toalson: The photo on the left is a late 1920s reflection of Gano. On the right Miss Anna Lee, in this late 1920s portrait, was described as a light blue-eyed, blond-haired, young lady with a Centralia address c/o A. R. Toalson, 117 Hickman Street. She was an attractive, popular, and talented young lady who would marry Gano on June 28, 1930. (Jack Chance Collection)

Warehouse No. 3 and to get it completed in a timely fashion. Ed Chamberlain had late material issues and "other vexations" to contend with. He had to redraw the plans to correct some disputed windows and doors and then employees and local carpenters could proceed in finishing the building. A week later there were eight brick masons at work. At the same time, Gano had not yet gotten out of college and made a routine visit back home. Gano's returns weren't just coffee and tea social events as his many past activities and involvements in the community indicated. Not unlike his father, he could often be found orchestrating some movement or event and the latest was a team effort with his father in taking the initial steps in getting Allen Street between the new Chance plants, paved. The June 21 *Guard* article suggested the city had spent enough money in the past forty years on this heavily trafficked strip to have it paved twice over. Mr. Chance had also worked hard to keep this area serviceable. Interview material from Gano about his activities and relationship with his father a half-century later reflect a limited plant partnership. But what is being revealed is Gano was becoming immediately involved with his father at the plant. This is further reinforced as previous and future activities are viewed. A person like A. Bishop may not have been the easiest person to work with, but this isn't new for father and son relationships. Being the founder, he was no stranger to making decisions and sharing this now with a son, especially with the educational and professional credentials of Gano, must have occasionally developed some friction. What is clear in mid-June of 1929 is the combined forces of father and son attacked the Allen Street paving issue with the son's engineering capabilities being put to the test in surveying and having the curbing installed. Once this was done, paving could easily be laid to conform to the grade of the curbing. This section was 350 feet in length and was in front of the Chance plants on both sides, Producer's Elevator, and the Alton and Wabash Stations, making it one of the most heavily traveled sections of roadway in the area.

A few months after a crisis in one's life, the stress from such an event can impact new or already existing health conditions. The Sanitorium at

Mayo's in Rochester, Minnesota, was visited by people with medical problems from all over the world. The July 5 *Guard* told of the departure of Mr. and Mrs. A. B. Chance for Mayo's, but it didn't specify which of them was ill. This was a holiday week with the Fourth of July falling on Thursday, which allowed most of the Chance Plant to enjoy a fabled four-day weekend! The following week the Chances returned home just in time to see the final preparations for completion of the Chance warehouse. Although it was a storage area especially for expanding anchors before the completion of the new warehouses across the street east in the 1950s and 1960s, this building was later considered primarily a manufacturing facility. Its warehouse function was the primary focus in 1929 because a portion of Chance anchors were out-sourced. After painting (dipping), the anchors could conveniently be piled to the ceiling awaiting customer shipments, as anchors produced in that plant were for several decades. Regardless of this activity, Mrs. Chance and her mother Mrs. Gano left a week and half later for Hot Springs, Arkansas, which suggests she was having medical problems. The paper a week later said she was vacationing in Colorado, which is surprising but the mountain climate may have fit her medical needs. A quote suggested A. Bishop didn't accompany his wife to Colorado but: " Mr. Chance is at home and on the job again," as the rush of activities at the plant apparently dictated.

The last phase of completing Warehouse No. 3 was the job of using a special type of putty in installing over four thousand window panes into steel sashes. The contractor estimated that this was a four-day job. Amid this activity, a week later, the Chautauqua had been approved for another run beginning August 23. The Wabash had also completed the new siding on the south side of the new Chance warehouse and all was in readiness of celebrating the completion of A. Bishop's last major construction project for his company. It was closing the chapter on a building boom over a four-year period that would not be equaled again. This was a time that defined the height of the A. Bishop Chance era, although it isn't conceivable that he had exhausted all his dreams, visions, and ideas. There were still adventures to come in continual patent work, confronting a Depression, terminating unproductive relationships, reorganizing his company, establishing the Wej-Lock association, community service work, farming projects, and constructing rock gardens!

The completion of the new Chance structure across the street to the southwest from the main factory at 210 North Allen Street, was an imposing addition per coverage by the July 19, 1929 *Guard*. It had come at the hands of a "big loss to Mr. Chance" but the results had boosted the image of Centralia and Chance Company. The march of the modern mechanical age in man's quest to increase the quality of life through the use of machines included a Centralian who was a forward-thinking person who brought reality to these visions. It is special to see someone who worked through the whole cycle, from a rough and tumble frontier prairie town to a maturing and sophisticated home for a significant and innovative contributor to the established telephone and automotive industry, agricultural area, and the growing electrical utility business. As the *Guard* and A. Bishop had done in the past, they again partnered in a needed and unfinished project, the paving of Allen between the tracks. Modern buildings were now facing a thoroughfare that reflected more primitive times. So the celebration was abbreviated in respect to unfinished business and they wouldn't have it any other way but to complete in a professional manner, the work that had begun years earlier on the between-the-tracks grounds.

More recently the Punch Press Building has been given a more modern numerical designation of No. 7 for fire identification purposes, etc. The number seven, according to Scripture is a perfect, complete number and in many ways this building was perfectly suited to house anchors, the oldest existing and financially most successful Chance product line. It could be a more attractive building even today if the white masonry stone, tastefully positioned over refurbished windows/doors and the brick walls were cleaned and blasted back to their original sheen. In some ways it represents a time warp in which the multitude of windows reflect an earlier time when natural lighting and summer ventilation was desirable and necessary. Even today the summer ventilation offered by windows that would open, along the sides of the high bay areas signal that not all has changed in the factory environment. But time has resulted in worn concrete slabs, which now reveal pretty river rock stones richly populating the surface of concrete poured by employees of an earlier time.

What was the impact of this large building in 1929? It must have been an imposing structure to locals. Those traveling through Centralia on the many daily passenger trains must have also been impressed. The spread of the Chance Plants now signaled this was a substantial operation and had already quelled any public notion that Mr. Chance was simply a dreamer or tinkerer. This building became a billboard of sorts with the company name, A. B. Chance Company in large block letters across its high front and rear. Also heralded on its front for a long period was a sign proudly signifying that this was a "Profit Sharing" Company. But there was something else, another symbol that flew in front and above the building that is very important to Americans, especially the Chance family. Strategically centered is the flagpole for the United States Flag, which can clearly be seen by those traveling past or those entering the Chance Company headquarters' doors on Allen Street. So along with this building's several names or designations could also be added "Flag Building," although each major separated plant of A. B. Chance Company was provided its own flagpole by the company so the Stars and Stripes could proudly be displayed to employees and visitors alike. This undoubtedly was a result of the Chance family's patriotic views, especially those of F. Gano Chance. Out of respect for our many blessings in America and to those who have sacrificed in various ways in service, the flag became a focal point most mornings. It also became a gauge of the wind's direction and power and cold windy days might generate multiple focuses on the flag. These glances constantly strike one with the visual beauty of the flag and the soundness of the principles for which it stands and flies. The flag this day (October 19, 2000), however, was at half mast and instead of one, there were two reasons: the U.S. Naval personnel lost on the destroyer *Cole* and the tragic death of Missouri's Governor Mel Carnahan, his son, and advisor. This reminds one that there are sacrifices even during peacetime for those who unselfishly serve us in our military and in the political arena which we sometimes view in a condescending manner. Maybe we should try and understand both groups better and be more appreciative of the service and vital services they provide to all citizens in our continual quest for maintaining our freedom and way of life.

The August 2, 1929 *Guard* brought news of a possible local expansion project with no indications that the national economy was going to falter. It concerned a relocation of a going concern on the West Coast, if local capital could be raised. It was not easy to raise capital in Centralia because of its small size and since investors had been burned when earlier firms went sour. This new firm would use castings made by Chance Company and so A. Bishop was at the Board of Aldermen meeting in support of this new business. What he had to say is interesting and although it was a time to point out the positive aspects of Centralia, he also mentioned some of the labor challenges. Skilled labor was difficult to be found in the area and at times also hard to keep. However, he said he had always been able to secure that type of man and at present he had two of the best men in their line. They had been with him for seven years or more and one of them was surely Irl R. Hicks. He continued saying these men had adjusted to local conditions of a small town and their salaries had an advantage in such a setting over large cities. In choosing Centralia, one reason cited by the owners of this automotive accessories firm was the two railroads. A. Bishop's next point concurred with an explanation of the advantages of local railroad service. He pointed to a recent instance of an emergency shipment that began at 7:00 a.m. with a rail car request and by 1:09 p.m. it was picked up by an express train and was on its way to a distant customer. He suggested such service in the city may have cost several days of delay. The reporter concluded that Mr. Chance had faith in the future of his hometown.

By late August 1929 the just-ended Chautauqua had a deficit of over $400, which meant the fifty-two guarantors had to cough up $8 a piece. This was a lot of money in those days and may have been an early sign that the local economy wasn't all that strong. However, this program, after an over two-decade run apparently wasn't as popular. Reflective of those times, business owners and managers were trying to gain more production and efficiency because of increased cost and the unsteady economy. The Alton had new engines with a second engine (booster) that allowed greater starting power, which meant longer trains could be handled from fifty cars to an optimistic seventy-five to one hundred. Another sign of the future was the impatience displayed by Americans in trying to do more things quicker. One result was the airplane, which was beginning to make its mark. The roof of the new Warehouse No. 3 got a huge sign (20 by 150 feet) painted on it so flyovers could in the daylight tell they had entered Centralia air space, not the kind of thing that generated confidence for early airline passengers.

The September 6, 1929 *Guard* announced the move of the H. T. Hasted family into Centralia from Cleveland, Ohio. Mr. Hasted, formerly with Hubbard and Company, was the new vice president and general manager of the Chance Manufacturing Company. This is the earliest vice president position found at Chance, even preceding F. Gano Chance's vice president designation on December 31, 1932, but Gano was the second in command versus the Hasted position of being subordinate to Factory Superintendent Carpenter. There is a problem with the *Guard* news item, however, as the name of the new Chance general manager appears to be N. C. Husted and not H. T. Hasted. This is based on the one letter difference in last names, the only record of H. T. Hasted found in the *Guard* and in company material, and the constant use of Husted from May 1930 into the mid-1930s.

The same fall 1929 news article explained that A. Bishop was providing housing for his new plant manager. One of the reasons why he got involved in local real estate was to provide homes for high-ranking company officials as he had previously done for the Fentons and now the Husteds. The February 15 paper stated the two-story house on Railroad Street vacated by the Fentons was being remodeled. This would appear to be the home on the corner of Railroad and Collier. The Husteds were going to move into a Chance home on Jefferson temporarily until the house on Collier Street had been remodeled by Mr. Chance. Similar accommodations were provided in the Chance Apartments, the Garden House, and later the Guest House as there always seemed to be housing and motel shortages in Centralia and few apartments to assist those needs.

Even prior to the Depression there were frequent bank failures appearing in newspapers, the source for the public's local, state, national, and world news. A bank at Tibbetts failed because of shortages per the October 4, 1929 *Guard*. Closer government scrutiny and the F.D.I.C. would help curb future instability in banking institutions. Other signs of a bleak economical climate were emerging such as Producers' Grain changing their payment policy to a cash-only basis. Although out of school, Gano

Chance still was drawn to Columbia as he visited friends there in late September, unaware at the time he would inherit an abundance of problems and responsibilities in the not to distant future. It was still a time to enjoy the advancements and building boom of the 1920s and there was a sense that the *Guard*, surely like other newspapers, was taking a positive approach to troubling signs that were making more frequent appearances. Keeping the faith, they mixed negative hard news with those established things that gave people confidence and could be depended on. This was easier for Centralians as a stroll or drive down Allen or Collier Streets brought one to the Chance Company factory complex and its size and activity instantly suggested all was well and at once one was renewed!

The *Guard* of October 4, 1929, again went to the well and pumped up a bucket full of company news. Chance Company was described as the:

> *largest factory of its kind in the world. . . . pole line hardware, automotive products, paints and polishes, together with other lines such as friction tape, tube flaps, tube patches, value grinding compound, oil and waterproof cement and other articles including the Chance pulley.*

This detailed product summary is clearly more oriented to the Big C automotive area and identifies products that hadn't been covered elsewhere. Another concise listing of distinct operations within Chance is also informative. The plant consisted of:

> *a cast iron foundry, a malleable iron foundry, an aluminum foundry, galvanizing foundry, a factory in which the Chance products are made and finished, a paint factory, and three large warehouses which with the extensive grounds, make a plant that is as modern and up-to-date as any in larger places.*

This listing also appears to be weighted toward foundry and not to the additional extensive machinery operations such as Punch Press or Drop Forge. The Centralia plant had placed Chance products on the markets of the world, according to this piece. An interesting question arises of how many of these automotive products were made in-house versus those bought in bulk and packaged at Chance. Both seem to be true. Paint was covered in both above listings and the "paint factory" suggests a portion of the Chemical Building operation was dedicated to the mixing and packaging of paint, which may have been or included the tar kind of compound used to seal canvas auto roofs.

This elaborate news piece continued speaking of A. B. Chance being the original inventor of items that "have no peer on the market today." They were so unique that there was little competition on the world market. They included ground anchors and lightning protection devices especially used by the telephone industry and in other construction projects that would include the agricultural and electrical power industries. It isn't mentioned that Chance brought out competitors and products to expand their lines and insure maximum volume by eliminating competition, revealing the business prowess side of the "Centralia Wizard." This industrial activity required about ninety full-time employees, not including salesmen and at rush times other part-time people were utilized. Previously it was estimated there were four grocery stores in Centralia, but this large number of Chance workers was the reason the Prairie Queen could support nine grocery stores! So north Boone County, Missouri, had much to be thankful for in late 1929.

The October 4 Chance article also included additional details about the newest plant on Wilson Street, Warehouse No. 3. It was a 120 [*sic*] by 350-foot modern fireproof structure that was the center of incoming and outgoing shipments, although special overhead loading devices also served the Foundry, Warehouse No.

2 and other areas. The roof of No. 3 included the twenty-foot-high by ten-foot-wide lettering for airplanes, which pilots had reported was the largest and most easily read of any they had seen. A picture of the entire Chance complex as seen below, shows several smaller buildings that were later torn down or replaced with larger factory buildings. The building to the north of the Wabash Station, was the Galvanizing (Warehouse No. 2) as discussed in an October 11 news article. The many public-spirited aspects of A. Bishop was mentioned plus his establishment of a department in the factory for the recreational purposes of employees.

It is not over until it is over, aptly applies to A. Bishop Chance. Comments about the leaking roof and metal sides, deteriorated by the strong galvanizing chemicals or its age, dominated discussions about the building where the galvanizing operation was located. Such references may have been referring to the building where the galvanizing operation was earlier moved from the city water tower area. The October 4 Chance article and illustration clearly showed Galvanizing was now located between the Wabash Station and the main Chance factory. This area is now covered by storage and/or a large warehouse built in the 1950s and then expanded in the 1960s when the tunnel under Allen and the Allen Street truck loading docks were added. The "old steel-clad building to the east of the factory" as the October 11, 1929 *Guard* article described it didn't live up to the modern look of this area and A. Bishop was planning another new building. There is a rule that when improving one portion of a multi-pieced project, that after its completion, the least desirable remaining object will come into focus. Images were very important to Mr. Chance and today when an eyesore is seen it most often reflects that someone has a reduced respect for the quality of their surroundings. Sitting right behind the new station in open view of arrivals and hundreds of rail passengers passing by each day was a structure that no longer fit this new setting. Amongst all the clamoring for a new station and moving the stockyards out of town came renewed responsibility to complete the cycle. One reason why so many brick buildings were built at the turn of the century (1900) was for fire protection. If a structure next door caught fire, the brick firewall usually protected it and even whole blocks of surrounding buildings. The Chance plan was to replace the Galvanizing Building with a new fireproof brick and steel structure and relocate it to the northeast corner near where the Drop Forge, Paint, and Chemical Building were

This south view of the Chance factory complex from Railroad Street appears to be an early 1930s photo. It is representative of the basically unchanged appearance from 1929 (Punch Press in left corner) to 1938 when the ramshackle coal shed beside the main building was purchased and removed. The other building in the right front, housing Galvanizing, was also an appearance problem, but a 1947 photo shows it somewhat refurbished. (A. B. Chance Co.)

located. This would add to the centralization and efficiency of the manufacturing operation and remove the negative elements of the galvanizing operation away to a more nonpublic area. It was to continue its warehouse function and was "part of the building program both for the utility and beautification of the plant and grounds and to cut down the fire hazard." In spite of delays, it is interesting that this proposed location was near where a new plant was constructed over two decades later to house the galvanizing operation.

The design of many buildings in the late 1800s and early 1900s took on a similar appearance. The shape was like three buildings in one, with either side attaching to a high bay center section with a loft on either side. Farm barns and stables undoubtedly promoted this design that allowed a driveway down the middle with livestock hay stored in the upper loft. This design also allowed the upper midsection to have windows for light and ventilation possibilities when natural conditions ruled a building's environment. The new Foundry and Warehouse No. 3 surprisingly repeated this design concept, which was earlier displayed in the Mule Barn and the Galvanizing Building. The October 11 Chance article also described what originally occupied the between-the-tracks area in a most vivid and nonpositive manner. In spite of it now being one of the most "modern looking in the city," in earlier times it was sunken and weed-grown with old frame sheds and piles of lumber, poles and other debris scattered over it. Viewed from passenger car windows from both main line railroads "gave the impression that the city was in a run-down condition and located on a dumping ground." Chance Company had a vested interest in the impression reflected to visitors in attracting business and in relocation possibilities, in addition to personal public concerns of A. Bishop in improving his community where he lived and worked. This is an argument for local ownership that no amount of public relations expenditures by corporate holding firms or chains can ever convincingly suggest their priorities represent local interests.

The Saturday entourage from Centralia to Columbia for University of Missouri football games was now a traditional autumn activity with F. Gano Chance as an attendee on November 2. Although not appearing the least bit seasonal, continual construction at the Chance plant proceeded forward. T. J. Windsor from Columbia took charge of the eight hundred feet of overhead steel railings installation for anchors and other products according to the November 1, 1929 *Guard*. The source of materials driving the overhead installation of conveyors was said to be from the Wabash, which is logical for Iowa-sourced anchors. They were then dipped, weighed, and checked. The product could then be transported for shipment on either the Alton or Wabash, through thirteen switches and sidings, not unlike a railroad. Chance employee departures were becoming more common as the plant became larger and established. L. D. Coulter, employed at Chance for several years, was leaving for Eldon to be near his new job involved in the construction of Bagnell Dam. Additional local news told of Chance employee Ben Brumley being injured by a fall while installing an overhead pulley block. Dr. C. C. Parmer was called to attend the injured man who was reported to be improving from the fifteen-foot fall to the concrete floor.

A. Bishop continued to bring the latest advancements of the day to Centralia. His concern for safety issues as the first anniversary of his son's untimely death approached, were continued. A new sprinkling system for the Anthony Building, still an auto dealership and garage, was the newest entry. The Viking Sprinkling Company was in charge of the installation costing around $5,000. Having the advantage of knowledge about future economic conditions such expenditures make it clear A. Bishop and other business leaders didn't yet realize the extent of the impending disaster. Painfully, many people would look back and ponder earlier decisions, but it is the spirit of

Americas to have faith and to forge ahead with the limited view one can muster at any given time. The ill-fated Houchins Overalls Factory would come to Centralia and it was first planned to open in December with employment of sixty to seventy women. These activities speak of an upbeat time as indicated by the completion of the Chrysler Building in New York City, which became the world's tallest building. It of course would have a relatively short reign when the Empire State Building was completed in the early years of the Great Depression. There were other signs of change that Gano Chance later indicated he was right in the middle of during a trip to New York City. The November 15 *Guard* stated Gano was in that city on business that week!

Another sign of change was the slowly growing dependence on the automobile for transportation needs, instead of the train. This independence would eventually spell doom for non-Amtrak rail passenger service, which had difficulty in running in the black even during its heyday. The new all-weather-type highways being constructed lured people to the open road. Chance employee, Miss Ida Traughber, along with her sister Miss Lizzie, their father F. M., Mrs. Mary Waller, and Mrs. Emma Garrard motored to Moberly on a Sunday. Although Centralians seemed to have frequented Moberly on occasion as they did Mexico and Columbia, one Moberly resident wasn't one of the *Guard's* favorite sons. The late December issue told of Temple Stephens having a chain of eighteen area grocery stores, including Centralia. January and February 1930 editorials about the menace of chain stores was an early sign of people trying to sort out the causes of the encroachment of tightening economic conditions. As the new year of 1930 was ushered in, the paper began to acknowledge that "business was lagging." Farmland had no stated value, it was charged, and farmers and bankers suffered failures and losses. Life was continuing on, however, as Mrs. Chance continued to entertain locally which reflects that the A. B. Chances were trying to maintain normalcy in the face of new adversities. Miss Ann Lee Toalson and Gano Chance attended an A.T.O. (Alpha Tau Omega) dance in Columbia on a Saturday evening in early February 1930, continuing that theme. Later information points to the Chance Company sustaining momentum well into the early stages of the Depression, but local businesses were failing because of the poor local farm economy intensified by recent crop failures. These poor farm conditions and lower valued investment opportunities were likely one reason that brought A. Bishop into the farming venture.

The economy hadn't stopped Chance Company business travels although during tough times there is the thinking that a greater sales effort and more concentration on advertising is beneficial. C. B. Taylor, the Big C sales manager, per the January 24, 1930 *Guard*, returned from Chicago and other points where he had promoted Chance piston rings. Taylor had been with Chance since around 1925, the article stated, but it's not clear what his relationship with F. H. Dukesmith was. Dukesmith was hired in February 1926 as director of sales of piston rings. The two positions could be considered separate jobs, but the smallness of the Chance operation would suggest that Mr. Dukesmith had already departed. A meeting of Chance salesmen in late January at Centralia indicates the utility sales effort was handled by different personnel. In early February A. Bishop was in Kansas City for several days on business, which may have been associated with Chance salesman J. H. Montague of Kansas City, who was ill. Centralia Rotarians were also active as Gano Chance and a group visited the Boonville Rotary Club. A side part of this trip was Gano's and H. Jones' interest in a used airplane they were considering purchasing.

Business transactions and moves continued, some of which could prove disastrous. E. O. Turner, a local resident since July 1926 who managed the Gary controlled Missouri Telephone Company, was leaving that blanket of protection to venture out on his own by acquiring the Frederickstown, Missouri telephone exchange. This February 28, 1930 *Guard* article cleared up several issues as covered in detail in the telephone section in Chapter 2. There were other activities that held future significance as Mrs. A. B. Chance, friends, and Miss Anna Lee Toalson visited in Columbia. As March brought uncertain weather, the same kind of uncertainty was getting a grip on America. Bank officials were leaving positions routinely suggesting it was for better jobs when it may have been a move to escape impending disasters in their current banks. The movement of people was noteworthy and appeared to be on the increase. Job losses brought people back to their parents' home or the homes of relatives and friends. It was a slap at a person's pride and confidence, with suicides on the increase. Although this usually focused on business-suit-clad New Yorkers jumping out of skyscraper windows, it also routinely struck all over America and even the frequently portrayed strong and independent farmer wasn't an uncommon victim.

The spring of 1930 would bring the world's problems to Centralia's doorsteps, as was occurring all across America. Factories were closing, especially those in small communities that may have mustered up bonuses to bring them in. The March 21 *Guard* carried an article with the following remark: "Since the market flurry and the general business depression of the past winter many of these small concerns were hit in the general slump and some of them have had to close down." There is always disparity it seems in how certain businesses and institutions weather economic downturns. The contrast couldn't be more evident than in Centralia. Banks were failing, as did the Fulton Bank one week and a week later in late March,

Department of Commerce letter: An example of the U.S. Patent Office and the patent attorneys representing Albert B. Chance, working out problems on previous filings. This particular one was for Patent No. 1748109 involving anchors. (A. B. Chance Co.)

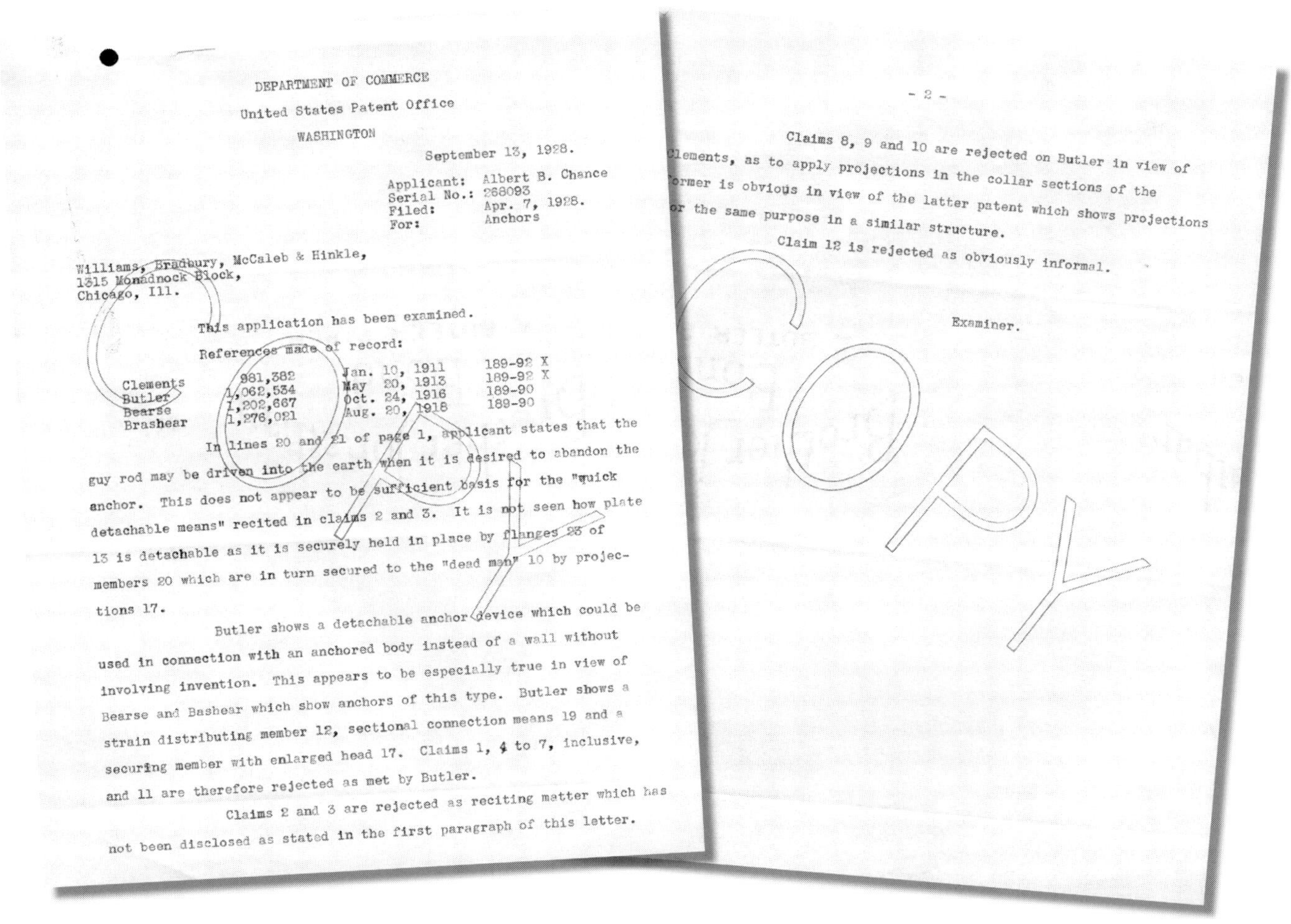

DEPARTMENT OF COMMERCE
United States Patent Office
WASHINGTON

September 13, 1928.

Applicant: Albert B. Chance
Serial No.: 268093
Filed: Apr. 7, 1928.
For: Anchors

Williams, Bradbury, McCaleb & Hinkle,
1315 Monadnock Block,
Chicago, Ill.

This application has been examined.

References made of record:

Clements	981,382	Jan. 10, 1911	189-92 X
Butler	1,062,534	May 20, 1913	189-92 X
Bearse	1,202,667	Oct. 24, 1916	189-90
Brashear	1,276,021	Aug. 20, 1918	189-90

In lines 20 and 21 of page 1, applicant states that the guy rod may be driven into the earth when it is desired to abandon the anchor. This does not appear to be sufficient basis for the "quick detachable means" recited in claims 2 and 3. It is not seen how plate 13 is detachable as it is securely held in place by flanges 23 of members 20 which are in turn secured to the "dead man" 10 by projections 17.

Butler shows a detachable anchor device which could be used in connection with an anchored body instead of a wall without involving invention. This appears to be especially true in view of Bearse and Bashear which show anchors of this type. Butler shows a strain distributing member 12, sectional connection means 19 and a securing member with enlarged head 17. Claims 1, 4 to 7, inclusive, and 11 are therefore rejected as met by Butler.

Claims 2 and 3 are rejected as reciting matter which has not been disclosed as stated in the first paragraph of this letter.

COPY

- 2 -

Claims 8, 9 and 10 are rejected on Butler in view of Clements, as to apply projections in the collar sections of the former is obvious in view of the latter patent which shows projections for the same purpose in a similar structure.

Claim 12 is rejected as obviously informal.

Examiner.

COPY

even closer to home, the Bank of Centralia failed. The other two Centralia banks were okay for now it was thought. Charges were made that false rumors had initiated a "run" resulting in the Centralia's bank failure. When a bank went under, it carried with it the savings of people and businesses which had a snowballing effect on bringing hard times to those who one day were solvent and the next without funds. For the time being, Chance Company seemed to be holding its own and one explanation by F. Gano Chance was that utility purchases were committed so far in advance that it took some time before it wound down. Fortunate were certain Chance employees such as Ida Lou Traughber, Maurine Grubbs, Verna Tucker, and Mr. and Mrs. Beverly Mayes captured by the March 14 *Guard* as being business callers in Mexico the previous Monday. But why were they not working on a weekday? Another revelation besides the possible shorter workweek was the *Guard's* earliest use of adding Miss Traughber's middle name as employees commonly referred to her simply as Miss Ida Lou. An event that could suggest all was well at 210 North Allen was the next evening employees entertained their families with an oyster supper in the Chance Club Room. The same March 14 paper included another news item that was likely connected to the supper as representing a going away event for the "Boss." The A. B. Chances and Mrs. Gertrude Gano were leaving on Saturday the 15 for a month's stay in California where they would visit relatives and Mr. Chance would concern himself with business interests of Chance Manufacturing Company. They returned per the April 18, 1930 paper. The Chances seemed to be doing well, but it could also suggest spring activity was less rushed than normal and the boss's absence less crucial with a full-time son and a capable group of managers taking care of business. However, a letter to Chicago patent attorneys on March 28, 1930, by "I. L. Traughber, Secretary," suggested the impending piston ring patent filing problems would have to wait for Mr. Chance's return!

The Depression years were a time that those born in its later stages are akin to in a strange way. This may result because our families spoke of it so much and the stark, vivid pictures painted of those hard times imprinted our psyche. It can even be a romantic period in relation to Franklin Delano Roosevelt's leadership and his positive influences on common people plus the New Deal's revolutionary programs. Being reexposed to this period in a more personal way, bank failures are no longer shrugged off as affecting only financial institutions, but their hardship carried forth to real people and businesses trying to maintain employment and solvency. Casualties continued to mount and some in a not so obvious manner to future prominent Chance employees. Nathan Toalson and friends visited Columbia on a Sunday, meaning he was apparently now at home in Centralia. Of course the Depression plus his father's death were hard blows and this March 21 news article may identify roughly Nathan's employment at Chance, which other sources suggest was 1930. A mid-April news item indicates that Granville Ryman from California may have lost his job and his visit to his mother's home may have been more permanent than reported. A week later news of his obtaining a job in Joplin as a state highway inspector, likely confirms this scenario. But some were sheltered to an extent by their youth, as was Sam Sneed who was on the Eighth Grade Honor Roll in late April of 1930, as another school year was coming to a close. All wasn't short term with no lasting foundation, however, as O. M. Squires bought the Model Cafe. So traditions continued to be carried forth in spite of the challenging times as the best and worst of people tend to shine forth during such times. Robberies and burglaries seemed to be on the increase and auto thefts and fires became frequent. It's unclear if vehicular fires were just carelessness, poor auto designs, or acts performed to get an insurance settlement. In early May, the Chance trait of helping their friends took on a new twist when the son F. Gano accompanied Stanley Garrett to Mayo's. They returned home with a prognosis for Mr. Garrett of bad tonsils that needed to be removed. The removal of tonsils, like "bleeding" in early times, was a cure all for several decades.

Often spoken of traditions by early veteran employees were being established at Chance Company in a more secure and permanent way. Repetition often makes this true as N. C. Husted packed his suitcase, caught the train and headed toward the southern states to promote Chance products, per the May 9, 1930 *Guard*. Travel in and out of Centralia to and from all over America, plus the world, would continue to be routine throughout the life of A. B. Chance Company. Husted no doubt hooked up with Chance salesman A. C. Wilson, working out of Atlanta. Centralians were proud of Albert Bishop Chance and his accomplishments and a banquet held at Chance Hall on the evening of May 14 celebrated that fact. The date had a special significance as it was the twentieth anniversary of the birth of the Chance anchor and also close to Mr. Chance's fifty-fifth birthday, according to the May 16 *Guard*. To show that the "hands-on" spirit of this now successful entrepreneur still burned deep within his soul, H. S. Howland went to the Chance home and told A. Bishop "that his new sprinkling system at the hall was kerbluey, and wanted him to fix it as the Methodists were holding a meeting at the place." This got Mr. Chance into the surprise gathering before he knew what was occurring. F. C. Harris was toastmaster and Judge and former Mayor Arthur Bruton spoke of Mr. Chance's life. Mr. Husted, representing Chance Company, spoke of how all arrangements had been handled. Elder C. Allen Burrus in behalf of everyone, presented Mr. and Mrs. A. B. Chance with a set of silverware. Judge H. A. Collier was called on to also speak. After becoming a controversial local figure as mayor some two decades earlier, he had come full circle by obtaining star status in this small mid-western community. However, for the most part it would remain there, except within the limited circle of the industry Chance Company served, as we focus on his life some seventy plus years later.

The 1920s building boom at Chance Company had become a closed chapter in the firm's history. The balance of this section and the last half of 1930 will instead wind down with some significant social events. At the plant, carloads of incoming and outgoing materials were apparently occurring at reduced levels, although it wasn't in vogue to vaunt success as business failures and increased unemployment was settling in. Some prominent people preferred to be seen in a Chevrolet instead of a Cadillac. It was a time that the economy progressed rapidly from bad to worse. Tough times would eventually hit Chance Company with a vengeance that would test the mettle of a father and his son in their struggle to maintain survival in the face of impending disaster.

One of the special aspects of management types becoming employed at Chance Company

F. Gano Chance and Anna Lee Toalson at the Chance Bungalow around 1927. Pictures shot that day appear to have included the Buford Chances who were living in that home at that time. (A. B. "Buff" Chance IV Collection)

The June 28, 1930 wedding picture of Miss Anna Lee Toalson and Mr. F. Gano Chance. Attendants were Mrs. Bertha Chance and Mrs. Barbara Ausmus on the left, with Harry Jennings and Nathan Toalson on the right. (Jack Chance Collection)

was the smallness of the town and the outgoing nature of the "Boss." This at times led to little or no division between social and business relationships, as they often became one that shared a possibility for negative or positive consequences. The Chances, both A. Bishop and later F. Gano, enjoyed the friendship of employees, especially those from far away places that brought varying backgrounds and fresh approaches to local social life. In mid-1930, the Husteds and the A. B. Chances were becoming socially involved, and a May 16 news item covered a trip to St. Louis on a Wednesday by the A. B. Chances, Gano, and the Husteds. As June dawned, two banks in nearby Mexico were closed and in a couple of weeks the Wabash Railroad reported another sharp cut in employment. Another sign of things to come was orchestrated by Mrs. Buford Chance in a bridge party held for Miss Anna Lee and Gano. The purpose was clear a week later when the *Guard* of July 4 announced the marriage of Miss Anna Lee Toalson and Francis Gano Chance, which took place Saturday noon, June 28, 1930, at the home of the bride, at 117 South Hickman. The choosing of this location was undoubtedly in tribute to her father, who had passed away a year earlier and had lived in that home sixty-two years! A college classmate of Gano and local attorney R. E. Ausmus played a couple of songs on the piano after the ceremony. A honeymoon trip to Winnepeg, Canada, by car followed. On their return, they would be at home on East Singleton Street.

As of July 1930, Chance Company had a vacation policy or employees could take a couple of weeks off. This had much earlier been confirmed by Miss Ida Lou, but now a Miss Verna Tucker, who had a position with Chance Manufacturing Company, was spending her two-week vacation with her mother near Sturgeon, per the July 4 *Guard*. This may be Mrs. Verna Palmer who worked in the office and became the longtime secretary to Jack Isbell in Marketing/Sales. A two-week absence from Centralia seems to be about right as Mr. and Mrs. F. Gano Chance returned home from their two-week honeymoon. There was also an attraction of former residents to return to Centralia to visit friends as the J. F. Honey family (former Ford dealer) annual visit and Judge Arthur Bruton and wife frequent returns attested. Centralia continued to progress and grow as the need for more water dictated the drilling of a second well only some thirteen feet from the original. Doubt about its success was removed when early December saw it come in like an Oklahoma gusher. Back at Chance, N. C. Husted left in late July for Chicago on company business. Later Bill Hollander spoke of "Nate" Husted who he said was a plant manager. Bill didn't portray Husted in the best light and suggested A. Bishop also grew less fond of that association, but Bill didn't mention the closer social relationship that earlier existed. As spoken of previously, modern communication to and from the outside world was vital to Chance Company's success. A. Bishop had helped his own cause by beginning and advancing telephone sophistication in the Centralia area. But an early competitor, the telegraph, was again back in the spotlight in an August 1 *Guard* article about Chance Company getting a new double unit Simplex telegraph printer, receiver, and transmitting machine. This may be what later became known as a "Telex." This was the latest, as they were said to be, installed currently across the country. It was an electronic method of receiving orders and transmitting messages to distant points. It printed a tape like a typewriter and its transmitting/receiving generated separate records serving as a means to save errors in orders. Chance Company continued to be involved in the latest innovations of the day and this desire to take advantage of such developments continued to be standard operating procedure at 210 North Allen. Any thought that the rural, remote setting of the Chance firm made it backwards isn't confirmed by its utilization of the most modern methods. This Chance trait continually demonstrated by the founder was also supported and repeated by his son F. Gano Chance during his long reign.

It was a roller-coaster time with the emphasis on the lows and few highs to celebrate. The B/O (Baltimore and Ohio Railroad) took control of the Alton, which had fallen on hard times into receivership. In late August the business on the northwest corner of the Square at Allen and Singleton, Farmers and Merchants Bank of Centralia, failed. The death knoll also sounded for the short-lived overalls factory a week later, although officially, it was already temporarily shut down. It was wise to maintain contact with the home folks, although that may not have been the Husted's purpose when they visited relatives in Niles, Ohio, and other eastern points in early September. The crime of stealing appeared on the increase as some people became desperate and those charged with the white-collar variety in banks were facing jail and prison time. But travel continued as the R. E. Ausmuses traveled to Sikeston to visit Mrs. (Beck) Ausmus's parents. Mr. and Mrs. A. B. Chance also in late November, near the second anniversary of their eldest son's fatal accident, left for Battle Creek, Michigan, for treatment of a stomach disorder that was plaguing Mrs. Chance. She was improved after two weeks there. Being the wife of a businessman such as her husband surely was not always an easy proposition. It was also a difficult economical time when ulcers and other stress-related health issues were impacting the public. The December 12, 1930 *Guard* gave another good example of why, when a few miles to the west in Sturgeon, the bank closed. It earlier had been consolidated from two banks. On a more positive note, Highway No. 22 from Mexico to No. 63 (twenty-eight or twenty-nine miles) would soon be graveled, although that was premature optimism, as it didn't occur until June 1931.

In spite of the difficult times, those connected with Chance Company seemed to be doing fine and there were pockets of enthusiasm, especially for young adults like F. Gano Chance. He and O. G. Wilson contemplated opening a miniature golf course next spring, but Gano would likely have to continue traveling to Columbia to play golf, as the gathering economic storm clouds were surely more visible in the spring of 1931. A reality now becoming evident was the constancy of the expansion and growth period of the 1920s was now past and those abundant accumulations would now be thoroughly tested during a new age dominated by decline.

The Early Sales Force

Having original unmolested material is valuable in many ways such as capturing the language commonly used in former times and places. The following assortment of randomly generated material about the early Sales Department is such an example. It yields some fascinating perspectives and revelations about one of the more prominent Chance marketing persons, Jack Isbell. The story of Isbell's Chance Company experience makes for some interesting reading and the hiring of Isbell is proof that A. Bishop Chance recognized talent. Both he and Gano's efforts to keep him on board is confirmation of Isbell's value to Chance Company. His guidance and direction was sought within the company including a heavy reliance upon his marketing expertise. During his long tenure he assisted in the formation of the modern Field Sales Group and the company's expanding involvement in export/international marketing.

A Centralia native and longtime Chance Marketing Department employee, Granville "Granny" Ryman, recalled some of the early sales personnel:

> *There were such fellows as Jack Milligan, Al Wilson, Mr. Mahieu, Charley Clark, and Jack Isbell in later years. Such guys as Charley Clark and Jack Milligan used to call Jack Isbell* [a newcomer] *a young whippersnapper. Jack Isbell was probably a pretty fair whippersnapper because he became the boss. Charley Clark would call up and they would have an argument over the phone. Jack would tell Charley that he was fired! The next morning Charley would come in, get a better job and more territory. Territories were important* [size + makeup of customers] *because you got paid on a commission and they got pretty fair commissions. (1979)*

In the late 1920s, when Verna Palmer began working in the Marketing area, she became familiar with the small Chance Company sales force. "The first salesman I remember working for the company was Jack Milligan in Chicago, Charley Clark in Pittsburgh, Mr. Montaque in Kansas City, and Mr. J. H. Lyons on the west coast [thirteen western states]." They were salaried with a commission for extra sales, according to Verna. Al Wilson was also an old-time salesman in Atlanta and Dally/Meckler were in New York as agents. Jack Rice was hired by Isbell and moved to New York and later Rice hired Jay Blanton who worked out of Richmond. "Then Dave Garroway Sr. was an agent in the New England states, I believe working out of Boston." (1979) Garroway's son became a TV star in the early years hosting the *Today Show*. Paul Sturgeon remembered, "Dave Garroway's son worked at Chance Company, I believe, selling piston rings, as I remember one time on the *Today Show* he made mention of the fact." (1979)

Details about Jack Isbell, whose sales career at Chance Company would span four decades, was provided by F. Gano Chance:

> *Jack Isbell came to work for us in 1926* [sic]. *He was promoting the fence anchor; he and my brother. After a couple of years in that operation, Jack came into our Utility Sales, as a salesman. He had the west coast and he quit when we got together with Kearney. Mr. Kearney never did hire Jack and he had agreed to hire all the* [Chance] *salesmen.* [Note: the Kearney situation will be covered later in Chapter 6] *He then went to work for the Wej-Lock Company in Iowa and when the Kearney deal was thrown aside, he came back to Centralia. The Wej-Lock Company became a division of the Chance Company.*

There is a question of when Isbell first came to Chance Company, but most sources suggest 1928. In the early 1930s Jack Isbell came back to the company as sales manager, as Verna Palmer remembered:

> *He hired the guys for the territories and saw that they did their work properly. I worked with Jack for many years. Jack was a very knowledgeable person and he knew people from coast to coast, but he liked to live on the west coast. (1979)*

In Memoriam

AN old friend's work has ended. John J. Milligan, who devoted the last twenty-five years of his life's work to the electrical industry and the A. B. Chance Company, has closed his books on a job well done and gone on to rest. He was the second to join the sales department of this organization and has contributed much to its progress since his first day in 1916.

We know "Jack's" many friends join us in this tribute to one who has served the industry so well and so long. "Well done, good and faithful servant." You will linger long in our memory.

From Chance Tips December 1941. Memoriam for J. J. Milligan.

Jack's return was also remembered by Paul Sturgeon, but from the perspective of the type of automobile he was driving. This kind of thing didn't go unnoticed in a small town. "When Jack came back to Centralia, I guess about 1932, he was driving a brown colored Studebaker automobile, two door, with great big wide doors." (1979)

Some stories are soon forgotten or lost while others appropriately survive and linger on like a galvanized anchor rod or house payments. Bill Hollander, in his very colorful fashion, described his version of how Jack Isbell came to be an employee of Chance Company. Recalling a comment Bill made about "outsiders being considered experts," this situation no doubt was one example that helped generate this opinion:

> *We were right in the middle of the Depression* [Jack was first hired in 1928] *and one of the darnedest things happened. Jack Isbell was selling razor blades and whatnots* [salesman for Ivory Soap] *through drug stores and he was on his way over to Columbia, traveling on the Wabash Railroad. He had an hour or so to wait for the Columbia branch train and he looked over and here is the Chance Company, the sign said. It didn't mean anything to him and so to kill time he went over and introduced himself. "I'm Jack Isbell, just saw your factory here and would like to go through it." So A. B. gave him the "cook's tour" showing him around the plant, what they made, what it was, and how it was used. Jack said, "Well I'll tell you Mr. Chance, you are losing money!" A. B. said, "Is that so? How do you figure that?" "Well," he said, "I think I could sell these outfits and I can make this company grow." So they sat down and worked out a deal and Jack went to work for Chance Company as the Sales Manager at a higher salary than any of our other salesmen. There were only about four other salesmen*

at Chance Company at the time. There was Charley Clark, Jack Milligan, Montaque back east, and another in Atlanta. For a period of time they were pretty jealous of him but in a short period of time Jack was outselling all of them. He was a salesman! A. B. did things like that, very impulsive. When he got an idea, he got right on it. (1979)

Some might suggest A. Bishop was an easy mark and that every drummer who came through town would sell him some of their wares. One has to consider that he was an accomplished salesman in his own right and his success in selling was important to the growth and success of Chance Company. This experience no doubt revealed to him the attributes of a good salesman, which he correctly recognized in Jack Isbell.

The early sales force and the marketing arrangements were briefly discussed by Nathan Toalson in answer to a question of how Chance sold their products around 1930:

Well they had salesmen. Primarily sales were through distributors; Hubbard sold to Graybar—Joslyn to independent distributors. As I recall, Chance Company had about five salesmen in the 1930s. We were really calling on the utility companies to get them to put our anchors on the approved list. We actually sold through Joslyn and Hubbard and then distributed to the utilities. (1979)

The products that Nathan remembered when he came to work in 1930 were: "Well of course, they had anchors and rods and also the expanding anchor which was the old sliding type. Practically all of the Thimbleye products that we have now. Piston rings and all that of course." (1979)

A January 31, 1930 *Guard* article about a Centralia sales conference told of "traveling salesmen for the Chance Manufacturing Co., selling the anchors and pole line hardware." The members of the group were identified as: Proctor Montague of Independence, Missouri; J. J. Milligan of Chicago; J. H. Lyon of Oakland; C. L. Clark of Pittsburgh; Alfred C. Wilson of Atlanta; Fred O'Brien of Boston; and J. T. Isbell of Des Moines. It's interesting that Jack went to work for a firm near Des Moines when he quit because of the Kearney situation in 1932. Jack did return and would finish his long career out at Chance over the next three decades. A couple of weeks after the sales meeting, the February 21 paper carried a news item about a J. H. Montague, a former Chance salesman, who had just undergone an operation. A. Bishop traveled to Kansas City to be with the family during that time. This is not the same person previously described as Proctor Montague of Independence, as Proctor attended the January 1931 sales meeting. According to personal papers of Mr. Chance, Proctor was the son of J. H., who was employed in 1928 when his father became ill. Later, reported in the *Guard* of June 6, 1930, J. H. Montague passed away. Mr. and Mrs. A. B. Chance, Gano, and Miss Ida Lou Traughber traveled to Kansas City to attend the funeral. This is another instance of the closeness among Chance employees, and the dignity of caring for others in their time of need. This was also significant because Mr. Montague, which company interview material spelled as *Monague*, was the first Chance salesman, if you dismiss A. B. Chance's role as being the first in most every position within the company.

Another tradition being established in the 1930 period was the annual sales meeting. During January 13–16, 1931, "the six territorial representatives of the Electrical Division of the Chance Company of this city met here last week for a four-day sales conference," per the *Guard*. Note the designation of "territorial representative" replacing salesman. A lot of effort was expended to make the sales conference a significant and beneficial gathering. A portion of one was attended at Tan-Tar-A at the Lake of the Ozarks in the early 1970s. There was a lot of work in investigating a proper location with adequate facilities to handle a hundred or so people. A. H. Knepper, longtime vice president of sales, was overheard speaking about the Elms in Excelsior Springs, Missouri, as a potential site for a sales meeting and voicing his concern that the hotel rooms were very small. These were glorious events because of the personalities, as anything could and usually did occur when these extroverted, aggressive, self-driven-type individuals were assembled together. It was a pleasure to work closely with the Field Sales personnel over five decades. There were difficulties but the challenge and satisfaction of working with a group of dedicated, quality individuals was rewarding and a privilege. Several of those gentlemen are now gone such as Don Dennhardt, a regional engineer out of the Chicago area. Don always came by to visit with those he had contact with, while in Centralia, as did most others if time allowed. This was out of courtesy and a good business habit of maintaining contact with the support staff that they depended upon back in Centralia. Sales personnel would visit Centralia occasionally when they brought customers in to confer with headquarters personnel or to visit the facilities. Product problems or conferences could also generate a visit.

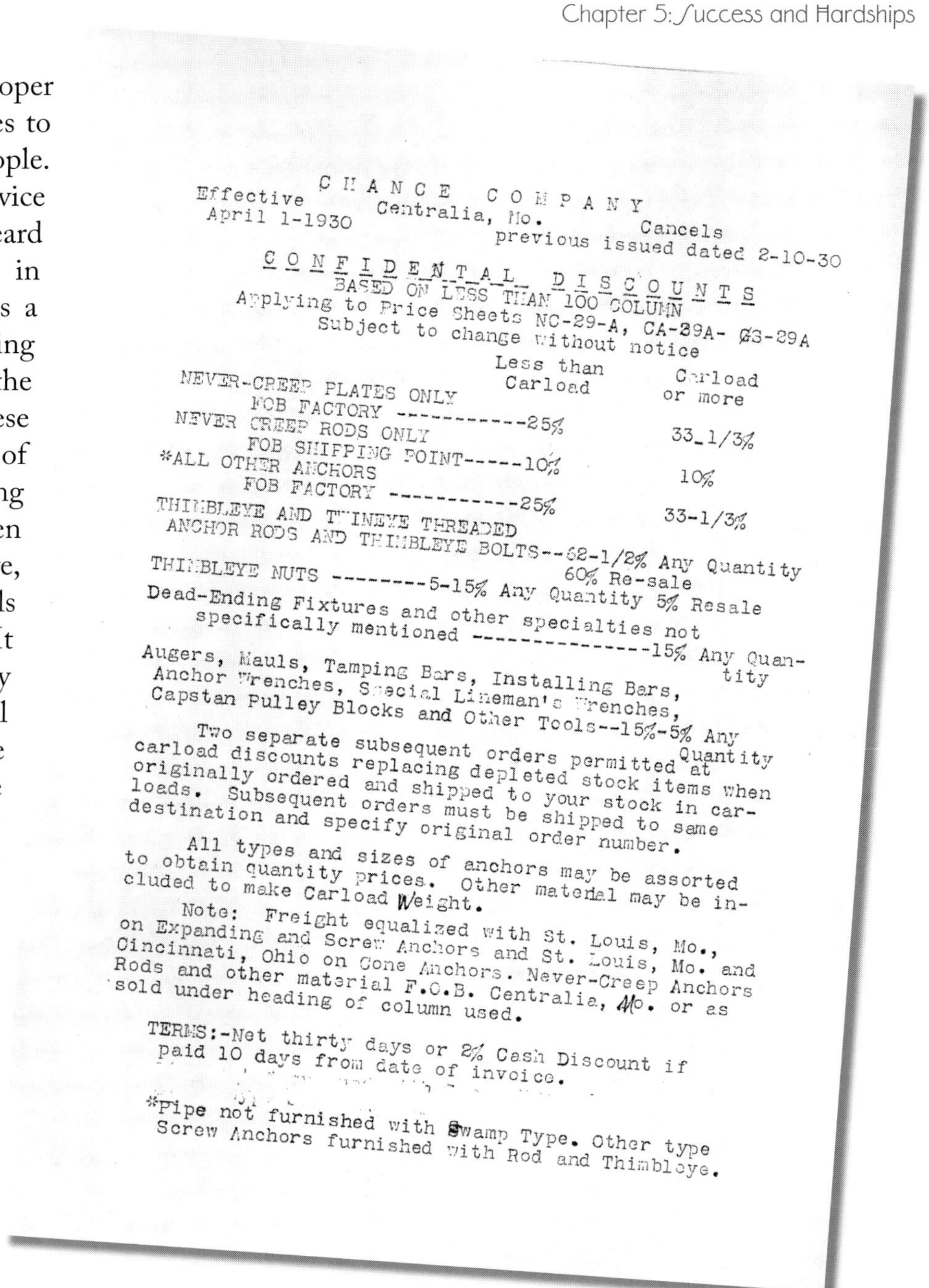
C H A N C E   C O M P A N Y
Effective April 1-1930
Centralia, Mo.
Cancels previous issued dated 2-10-30

C O N F I D E N T A L   D I S C O U N T S
BASED ON LESS THAN 100 COLUMN
Applying to Price Sheets NC-29-A, CA-29A- GS-29A
Subject to change without notice

	Less than Carload	Carload or more
NEVER-CREEP PLATES ONLY FOB FACTORY	25%	33-1/3%
NEVER CREEP RODS ONLY FOB SHIPPING POINT	10%	10%
*ALL OTHER ANCHORS FOB FACTORY	25%	33-1/3%

THIMBLEYE AND TWINEYE THREADED ANCHOR RODS AND THIMBLEYE BOLTS--62-1/2% Any Quantity 60% Re-sale
THIMBLEYE NUTS --------5-15% Any Quantity 5% Resale
Dead-Ending Fixtures and other specialties not specifically mentioned ---------------15% Any Quantity
Augers, Mauls, Tamping Bars, Installing Bars, Anchor Wrenches, Special Lineman's Wrenches, Capstan Pulley Blocks and Other Tools--15%-5% Any Quantity

Two separate subsequent orders permitted at carload discounts replacing depleted stock items when originally ordered and shipped to your stock in carloads. Subsequent orders must be shipped to same destination and specify original order number.

All types and sizes of anchors may be assorted to obtain quantity prices. Other material may be included to make Carload Weight.

Note: Freight equalized with St. Louis, Mo., on Expanding and Screw Anchors and St. Louis, Mo. and Cincinnati, Ohio on Cone Anchors. Never-Creep Anchors Rods and other material F.O.B. Centralia, Mo. or as sold under heading of column used.

TERMS:-Net thirty days or 2% Cash Discount if paid 10 days from date of invoice.

*Pipe not furnished with Swamp Type. Other type Screw Anchors furnished with Rod and Thimbleye.

A 1930-era discount sheet was apparently sent to selected customers and/or for internal use by Chance sales personnel. (A. B. Chance Co.)

Additional details of the 1931 sales event included those in attendance, which was the same as those listed in the January 1930 sales meeting, except that J. T. Isbell was now headquartered in San Francisco instead of Des Moines and had replaced J. H. Lyon of Oakland. Paul Sturgeon said Jack really liked the West Coast. H. J. Trotter had been added to the Omaha territory and Fred O'Brien of Boston was removed. This was a net loss of one although the shuffling could have been the reason. The central section of America was well represented with less coverage in the East and West. There was always some turnover in the sales area, due in part to the opportunities they routinely encountered. During times of cutbacks, it seemed that it was also easier to make changes in the field involving people that were distant and thus a less personal decision than say someone located in Centralia. Hardly any sales information is available for the early years, but the article did state that 1930 had been Chance Company's largest year in sales volume to date and the optimistic attitude of the salesmen thought 1931 would be even better. There must have been some doubt by the writer: "If this is true it will mean employment for many Centralians throughout the year." But the company was still reaping the benefits of good business and had not felt the real impact of a major slowdown that progressed steadily during 1931.

The subject of the 1931 sales meeting was directed toward discussing matters pertaining to the problems of distribution and expansion. It's ironic that both these concerns would bring failures in attempting to fulfill these eagerly sought goals. Expansion surely involved a larger sales force to generate more business, but with the Depression already knocking at the door, this would remain a hope not soon realized. The distribution issue will be addressed in the next chapter when a relationship developed between Chance and Kearney failed to address the needs of Chance Company. Although this is 1931 material, it was based on 1930 history and the success that had been achieved, but with an expected eye toward growth instead of entrenchment to face difficult future times. Discussions during the meeting concerned territorial promotion work with the salesmen giving talks about the business situation of 1930 and the outlook for 1931. This was followed by covering distributor, jobber, consumer, and their organizations. Reviewing sales policy, price to consumers, profit to distributors, materials, designs and selling suggestions were also subjects of discussion. C. L. Clark received a special prize "for the best treatise on Chance products."

The home office employees involved in the sales meeting included N. C. Husted, Chance Company vice president, who was chairman of the meeting and had charge of arrangements. As expected, A. Bishop Chance gave a welcome address to the gathering. Husted then discussed the purpose of the meeting and "I. L. Traughber gave a discussion on correspondence." F. G. Chance gave a featured talk on advertising, which he was a part of at that time. The second day, N. C. Husted discussed the very important contract on Bell business. Concerns for that business had generated several trips to New York over the years by Chance officials. The Bell Telephone business was likely Chance Company's largest and one of the oldest customers. When sales to them and other customers dropped off, it had immediate and devastating results as will be seen in late 1931 and early 1932.

Earth Anchors

Earth anchor use on power and communication lines has become a firmly established and enduring staple. Utilization of the anchor resists disturbing natural forces, reacting upon man-made structures that appeal for assistance in maintaining their up-right integrity. However, it wasn't an overnight sensation and just as line construction grew across the country, so did the efforts of one A. Bishop Chance after his primitive anchor application in 1912. His dedication in developing the technical aspects of his invention, maintaining a watchful eye on anchor activity and his negotiating prowess in corralling other inventions and competitors could be likened to the scrutiny and loyalty of a lighthouse keeper. This was vital in protecting

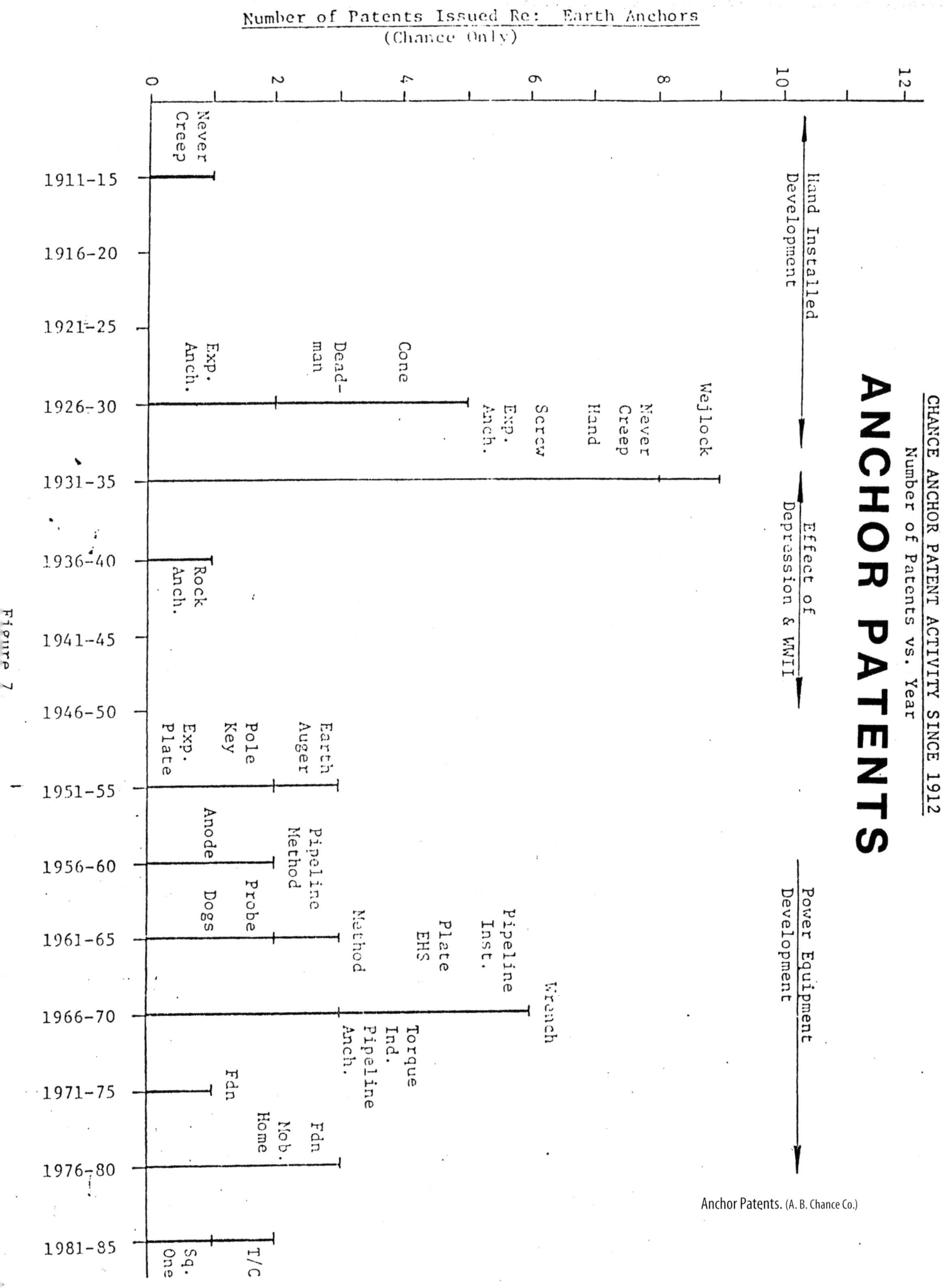

Anchor Patents. (A. B. Chance Co.)

Chance interests and establishing leadership in the anchor product line.

Chance Company's prominent role in the commercialization of earth anchors generated an interest in the history of anchors. An engineering source authored by Samuel P. Clemence, provided details in a textbook manner about how anchoring developed. One of the first uses found was an application on tents, an early structure that was "subjected to uplift and overturning forces." The anchor used in this case was commonly referred to as a stake. The Bible, in Isaiah 54:2 refers to their use. These stakes were of a temporary nature and were removed and reused as warranted by these people's nomadic movements.

As man-made structures grew, the "Dead Man" type anchor was developed to hold greater loads of five hundred to ten thousand pounds. These were usually logs or rocks that were buried in the ground up to ten feet. The early screw anchor developed so holes could be dug in soft soil. The anchor could be screwed in and removed, leaving the disturbed soil which then could be easily removed, generating a hole. Concrete-like material could then be placed in the hole to generate an anchoring point or in establishing a foundation. The anchor could be left in the ground and later this developed a limited system similar to the modern Chance Foundation Anchors. The cost of the installing tool was recovered by a German patent that used a retractable pipe that could be replaced by rods or cable attached to the in ground positioned screw blade or what Chance labeled the "Helix."

Although it is thought to be modern-day developments in anchor technology, Alexander Mitchell, a blind English brickmaker in 1830, developed powered installed foundations. By 1833 Mitchell began constructing lighthouses in the tidal basin of England using his newly developed "screw pile." They were said to be successful, yet preconceived notions about foundations needing to be substantial in bulk and weight limited his invention's advancement. Story after story evolved a century and half later in trying to overcome the construction trade belief that a quickly installed, compact, steel Chance Foundation Anchor couldn't perform as could a slab or pour of concrete.

Various pile and screw anchor foundations continued into the late nineteenth century. At that time large suspension bridges were being built which generated tremendous tension loads to the foundation. Cables were anchored in rock to help support them. Mining applications also began to use wood anchors in holes drilled in rock. A West Virginia coal miner in the late 1800s is credited with the forerunner to the mechanical rock bolting system, which led to the rock anchors of today. As the need for anchoring systems increased in the early 1900s with the development of radio transmission towers, electrical transmission lines and pipelines, the hand-installed anchor was the dominant type. After World War II, when mechanized construction equipment became the rule, the utilities needed an automated installation system to cut manual labor overhead and develop more uniform and greater anchor performance. This led to the development in the 1950s of the power-installed anchor that revolutionized the anchor industry. Using high-capacity torque devices, the anchor could be installed in all types of soils to tremendous depths. Larger diameter, multi-helix anchors were also developed to extend the range of anchoring capacity needs considerably beyond that possible with previous manual systems. Briefly, this has presented a condensed overview of earth anchor development.

A power line pole out in a central Kansas wheat field is like a fan's head at a St. Louis Cardinal baseball game, if you pop it up, there are all kinds of things working to get it back down. Natural things like the wind, ice, and sleet or a pole set into earth too soft to maintain its vertical integrity, all work to lay it over. Add to that, loadings from untold pounds of metal lines twisting and pulling on the line pole, especially at corners where the line has to change directions, and you get big-time stresses quickly followed by a distressful failure situation if the line isn't properly guyed with earth anchors. Thus the anchor becomes the unsung hero, only seen by the installation crew, but pulling duty around the clock, day after day, helping maintain the dependable flow of electrical power just as vitally as an insulator or electrical switch. But that is an expected performance, just as surely as taxes and the availability of a Coke. However, it wasn't always that way. The Chance earth anchor had to be invented, developed, tested, demonstrated, and approved by the customer all of which becomes a very important part of the Chance Company story.

After the Never-Creep® Anchor design was set in metal, A. Bishop then had a multitude of challenges to overcome. His manufacturing facility off the City Square was small and directed toward telephone and lightning arrester needs. Reports indicate this is the period when the Kansas City connection threatened the future for Centralia as the company base. So it was a time of adjustment, searching, and refocus in which F. Gano Chance spoke of the outsourcing of the early Chance anchors:

> *He* [Mr. A. Bishop] *had been buying anchors from a St. Louis concern—cast iron about 1913 or 14. That was only a small anchor 5 x 10 (inches). In a year or so he needed some bigger ones so he went up to Iowa and worked out a deal with Iowa Malleable Company, Fairfield, Iowa, and they started making it out of malleable but still made the 5 x 10, the little one, out of cast iron for many years. But the 615, 6" x 15" long was the next size and they started to make that out of malleable. About that time we had occasion to use them for fence anchors. The first fence anchors were made of cast iron. (1978)*

As Gano indicated, the early part number of Chance anchors revealed its size, as the 615 was six inches wide and fifteen inches long. This was a small anchor as the size of an anchor was basically its surface area, which determined its holding power. A larger surface would hold more than smaller surfaces with identical installation procedures in the same soil. Since the casting procedures to make these anchors required tooling, it was common to build the part number, company name, etc. into these molds. The item being formed would thus reflect these numbers and other features engineered into the mold or die. An interesting situation resulted in 2000 when an individual from Estes Park, Colorado, called into Centralia and the call was transferred to this employee. While the city was digging for a water main, they found a strange cast-iron plate. Cleaning it up, they noticed raised numbers and letters "615" with the name "Chance" on its surface. They wisely felt this must be valuable and worthy to keep in a local museum. It was explained to them that this number was both an early Chance anchor with the part number additionally indicating its size. This was a testimony to the long life of cast iron or malleable, and the philosophy at Chance Company to build quality products. This seventy- to eighty-year-old anchor, even contending with the harshness of underground elements, had faired well.

There was also another fascinating aspect to this story owing to Estes Park personally being a favorite area and also becoming a retreat for the A. B. Chance family. Coming up the beautiful valley along side the Big Thompson River into Estes Park in the 1950s on a family vacation trip, made a lifelong impression of the beauty, power, and pristineness of the cold, crystal-clear waters of a mountain stream. But in relation to the anchor, one of the most prominent Chance engineers, Bill Hollander, who was the anchor man over portions of six decades, had a vacation home in Estes Park!

Bill Hollander (*Chanco News* June 1950)

Trying to identify an anchor in the ground is a difficult proposition, unless it's dug up. Changes above ground on the line could raise issues of the capabilities of the installed anchor in meeting those needs. The late 1939 issue of *Chance Tips* magazine revealed a new Chance anchor tag, which would help remove this problem. It was made out of brass to resist rust and revealed the catalog number of the anchor, its area (size), anchor rod length, and the year produced. It could be attached to the guy line or nailed to the pole after the proper information was punched on the tag. Railroads used a similar system in dating railroad ties with date nails, which were also used by some utilities to date poles. This employee recalls scrapping surplus Chance past-year date nails, which had become obsolete. *Tips* of June 1941 told of the decision to halt the brass-tag program because of defense requirements.

A story by an early, veteran Chance employee, Flynt Chandler, describes another anchor size and part number, the "620":

> *We got a big order, one time, for a bunch of "620" anchors. They* [customer] *wanted them shipped by express and they wanted them right now. So he* [Mr. A. Bishop] *came over . . . 4 or 5 of us, maybe you* [Les Reynolds] *got in on it. Parked the little Maxwell truck over by the railroad crossing and took the other trucks and hauled these anchors over there. Piled them up on it until the train came, #12 passenger train had a couple of extra baggage cars on it. We went in there that night and loaded them anchors for him. Old man Diggs was the station agent and of course he run the express* [Railroad Express]. *Of course he was going to get a heck of a drag* [benefit from business]. *So Mr. Chance called him and had him give each of us fellows $2.00 a piece. (1979)*

Les Reynolds suggested it was, "That was about 9:30 or 10:00 o'clock." (1979) Which was a very long day's work!

Cone anchors, above all else, were simply different. They weren't the most popular anchor type and seemed to have an on-off status, but the cone was a longtime member of the Chance anchor family. On June 21, 1927, Albert B. Chance petitioned for a patent to cover an improved cone anchor. The proposed anchor would have less material with operating advantages over previous designs, especially in providing for a guying rod that could be threaded onto a nut on the inside top of the anchor. The anchor had two distinct surfaces, a steep upper and a flatter lower area. The upper had a multiple layered pyramid shape which under guying tension, per Mr. Chance, would generate a compacting ring of fill material which would generate more holding power for the flatter lower surface. As the patent process slowly evolved, there became a problem with a Mr. Bierce of Cincinnati having a similar filing. The patent examiner then requested detailed origin dates, and related information from Mr. Chance. A reply letter reproduced on page

343 provides development and other valuable details of this improved cone plus the pattern of activity from the initial invention to the ultimate goal, the customer sale. As the Bierce situation unfolded, it was found that he had filed on April 21, 1928 (began selling on April 5, 1927), which gave the legal advantage to the Chance filing as it was submitted in June 1927. To counter the slightly earlier Bierce sales date, Chance Company had likewise begun building and selling the Chance version in late 1926. It could be charged that the Chance anchor was copied, but that legal proceedings wouldn't be necessary. If the patent hadn't been applied for by Chance, then it's clear how someone can gain control of an idea or invention, possibly not their own. A seventeen-year Patent No. 1,756,534 was issued to Chance on April 29, 1930, and a Canadian Patent No. 219,482 in June 1929. As further details were revealed, it gets more complicated during the examiner's investigation, when it was revealed that Bierce Patent No. 1,070,603 issued in August 1913, was about to expire. Another Chance Patent No. 1,134,820 (Land Anchor) was also cited and included in those existing patents having possible interference with the improved cone. A competitor, Specialty Device Company of Cincinnati, was the holder of the Bierce patent. Incidentally, A. B. Chance Company purchased that firm in 1953! So it's unclear who copied whom. It could be Mr. Bierce wasn't aggressive enough in getting another improved patent on his original 1913 patent. Gano mentioned his father's practice of continuing trying to obtain new patents and extend legal coverage by improving an existing patented invention. Our privileged insight into this chain of events in the waning years of the 1920s, reflects A. Bishop's continual attention to detail and aggressive manner in developing and maintaining business and legal control of his growing anchor family.

This was a very busy and productive period in A. Bishop's business life, contrary to the image commonly associated with his winding down post 1930 snapshot. In addition to the cone there was other earth anchor patent activity. On March 14, 1928, Albert B. Chance petitioned for an earth anchor patent that was issued on April 22, 1930, as No. 1,755,811. On January 23, 1932, he applied for a reissue and was rewarded with No. Re. 19,177 on May 22, 1934. From a complex overview of what this patent covered comes the initial conclusion that it simply related to all earth anchors having guy rods that were fastened to the anchor by a nut or other means. As eluded to earlier in the Thimbleye® section, the anchor and anchor rod were a team, which are no better portrayed than in this system being patented in the 1930 period. The first-generation anchors, except a few screw types, required that a hole be dug to install the anchor. The anchor rod became an installing tool in a way as it was used to lower and position the anchor in the hole. One weakness was the anchor didn't have a retaining or positioning area or device to hold the proper alignment and keep the anchor from sliding up the rod. Neither did it provide a lock for the nut so the rod could be removed or adjusted once installed. So the area where the rod entered the anchor would now have lugs, clips, or other features designed into the anchor that would lock the rod nut in a position that made a stable, workable link between the two. After a period of time, the nut would rust and bind to the rod. In those conditions where a line was moved and the guy rod, protruding above ground level became an obvious hazard, a means other than the expensive and time-consuming job of digging up the anchor was needed. These lugs or clips were designed with enough strength to handle normal installation conditions, but yet purposely fragile enough so a sharp sledge-hammer blow would break or bend them so the rod could harmlessly be driven through the anchor into the ground with the above ground portion now below surface. The application for reissue of the subject design featured an expanding and simple cone anchor design (as opposed to improved type discussed in previous paragraph). Concerns were being recognized and addressed for hazards generated by guying requirements. Since the guy line to be effective had to be angled out from the pole, the pole cable and rod head would also spark the development of the guy guard so it visually could more readily be seen than a thin cable, again reflective of safety concerns.

Another patent application of February 4, 1929, was related to the socket idea in the top of an anchor for engaging the rod or tool for installation, apparently first applied for on October 24, 1927, Serial No. 228,463. This early 1929 filing resulted in a seventeen-year Patent No. 1,800,504 of April 14, 1931, described as "Earth-Anchor Tools." This was comprised of an installing tool for placing screw-type anchors in the ground. One of its features was the handle area that was adjustable up and down the shaft with recesses so it could be retained at different heights to best suit the operator. The cross shaft-handle could also be slid to extreme side positions to help the operator generate more installing torque. The head was also adaptable to other fittings that would allow its use with various anchor designs. The head contacted the top of the anchor hub, making for a more stable installation, which also reduced component breakage. This case was also apparently accompanied by another anchor filing which appears to involve the pilot-point idea. The following such June 23, 1930 filing, covered in the next paragraph, continues in that vein. However, it would seem to follow one filed over a year earlier.

Earth Anchors that became the "Pilot Screw" project, according to a scribbled remark, was filed for a patent on June 23, 1930, Serial No. 462,954. Patent No. 1,940,938 was issued on December 26, 1933. Requiring three years for completion of this process appeared to be the norm and not the exception. The ideas and new concepts surrounding this invention, was oriented toward the screw-type anchor. The focus was on establishing a pilot hole by driving the guy rod in the direction the anchor was to be installed. This allowed screw anchor installation into hard soil not before possible or very difficult at best. This soil disturbance or pilot hole allowed the anchor to more quickly and easily be installed with less physical effort and man-hours expended. There were several ingenious features of this proposition, in fact two varying designs that provided similar results. Both would allow the guy rod to be driven below ground surface, if they were abandoned. There were a couple of features that were similar to the more modern P.I.S.A.'s (Power Installed Screw Anchors). One design featured a square-shaft guy rod with two helixes that apparently were castings. This "hub" and "screw flanges" (blades) now carries the more-modern "helix" designation. The screw flanges coiled around the hub like the outer surface of a drill bit. Both designs appear to have required different components with one exception being the rod eye that was threaded onto the top of the guy rod and served as a contact point for driving the rod into the ground and an eye for guying attachments. Both hubs had an open center for passage of the guy rod, a locking lug on the top for the installing tool version and a square hole for the square guy-rod design that performed both installation and guying functions. The sharp point that fit on the bottom of the hub and allowed the soil to be spread so the large hub could more easily pass down the pilot-hole, could either be of the screw-on type or forged into the rod end. As Albert B. Chance stated in his petition for a patent, "I have invented an improved method of installation of earth anchors of the screw type, which may be practiced by several different forms of apparatus." The last part of this statement is key as the Centralia Wizard was advancing anchor technology, he was also locking up certain features that Chance Company would have future control over in the manufacturing process, vital in maintaining their leadership role.

Gano explained the patent system and his father's approach in protecting Chance interests:

Our sales were running along about $250,000 and we were making it all in profit in the one anchor that my father had a patent on which was about to expire. A basic patent is 17 years and a design patent is only 14 years. If the idea is merely looks and doesn't have any unusual basic principle why then you have to get a design patent. Of course you get additional patents, you don't renew a patent. You change the design a little bit and you get a patent on the improvement. (1981)

Gano was suggesting the importance of the anchor to the financial health of Chance Company. Other income sources such as the royalties on anchor rods were sometimes grouped under the anchor, consequently becoming anonymous in the process.

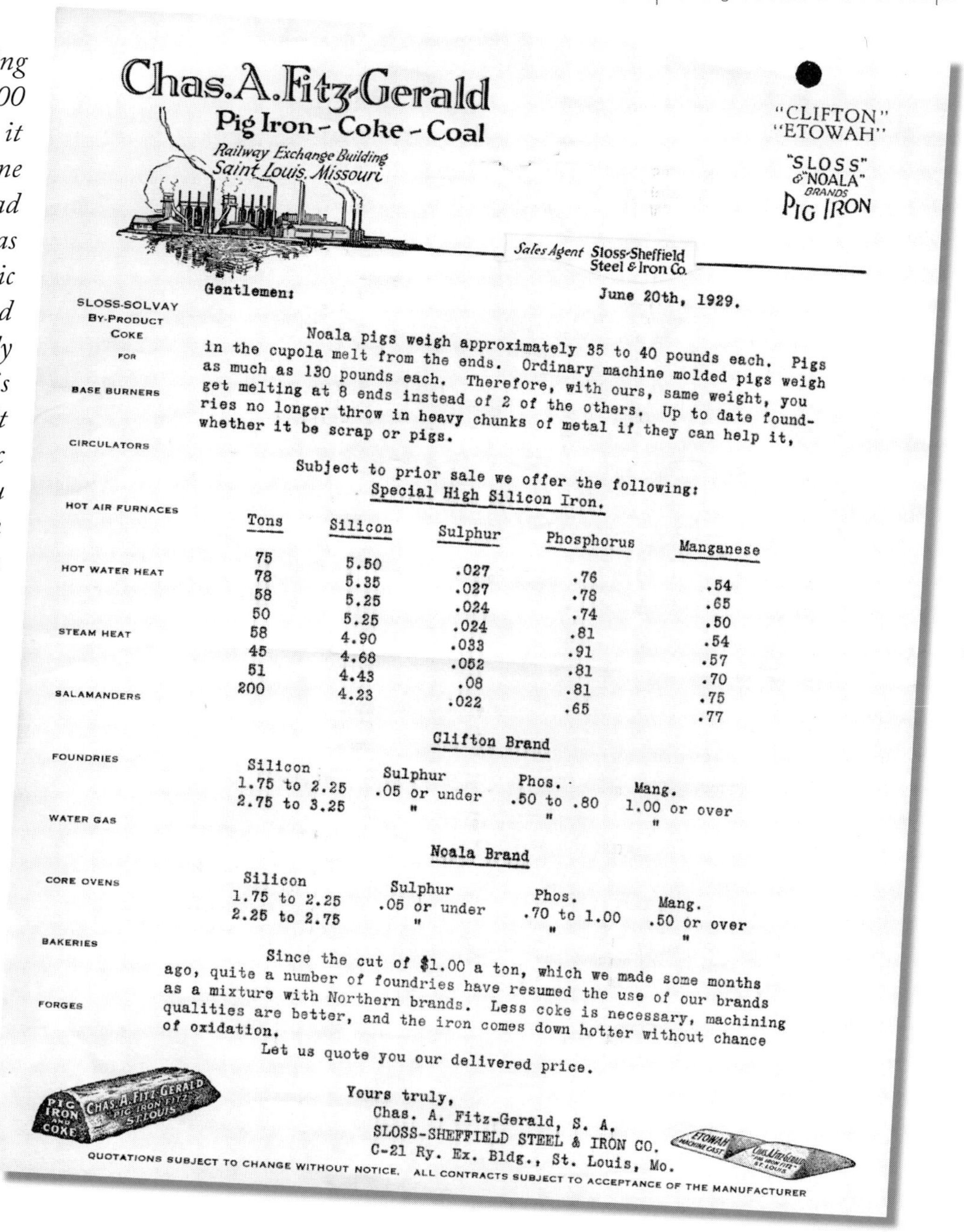

Chas.A.Fitz-Gerald
Pig Iron - Coke - Coal
Railway Exchange Building
Saint Louis, Missouri

"CLIFTON"
"ETOWAH"
"SLOSS" & "NOALA" BRANDS
PIG IRON

Sales Agent Sloss-Sheffield Steel & Iron Co.

SLOSS-SOLVAY BY-PRODUCT COKE FOR BASE BURNERS CIRCULATORS HOT AIR FURNACES HOT WATER HEAT STEAM HEAT SALAMANDERS FOUNDRIES WATER GAS CORE OVENS BAKERIES FORGES

Gentlemen: June 20th, 1929.

Noala pigs weigh approximately 35 to 40 pounds each. Pigs in the cupola melt from the ends. Ordinary machine molded pigs weigh as much as 130 pounds each. Therefore, with ours, same weight, you get melting at 8 ends instead of 2 of the others. Up to date foundries no longer throw in heavy chunks of metal if they can help it, whether it be scrap or pigs.

Subject to prior sale we offer the following:
Special High Silicon Iron.

Tons	Silicon	Sulphur	Phosphorus	Manganese
75	5.50	.027	.76	.54
78	5.35	.027	.78	.65
58	5.25	.024	.74	.50
50	5.25	.024	.81	.54
58	4.90	.033	.91	.57
45	4.68	.052	.81	.70
51	4.43	.08	.81	.75
200	4.23	.022	.65	.77

Clifton Brand

Silicon	Sulphur	Phos.	Mang.
1.75 to 2.25	.05 or under	.50 to .80	1.00 or over
2.75 to 3.25	"	"	"

Noala Brand

Silicon	Sulphur	Phos.	Mang.
1.75 to 2.25	.05 or under	.70 to 1.00	.50 or over
2.25 to 2.75	"	"	"

Since the cut of $1.00 a ton, which we made some months ago, quite a number of foundries have resumed the use of our brands as a mixture with Northern brands. Less coke is necessary, machining qualities are better, and the iron comes down hotter without chance of oxidation.

Let us quote you our delivered price.

Yours truly,
Chas. A. Fitz-Gerald, S. A.
SLOSS-SHEFFIELD STEEL & IRON CO.
C-21 Ry. Ex. Bldg., St. Louis, Mo.

PIG IRON AND COKE — CHAS.A.FITZ-GERALD PIG IRON FITZ ST. LOUIS

ETOWAH MACHINE CAST

QUOTATIONS SUBJECT TO CHANGE WITHOUT NOTICE. ALL CONTRACTS SUBJECT TO ACCEPTANCE OF THE MANUFACTURER

A St. Louis source, Chas. A. Fitz-Gerald, was prepared to supply the Chance Foundry with pig iron and fuel. #1756534 (A. B. Chance Co.)

In a discussion about industrial processes that were in their infancy and generating serious quality and reliability problems, Engineer Bill Hollander explained these conditions in the 1920s–30s time frame:

Pot welding for example, you just put some current across it and sometimes it stuck and sometimes you burned a hole through it. Malleable iron was a case. You took some castings and annealed them, sometimes they were malleable and sometimes they came out like an old crock. Just break up if you looked at them. In the gray iron foundry sometimes there was a good grade of gray iron, sometimes it was full of chill spots, and other times it was hard all the way through and brittle. No one really had an idea why but it was just the way things were at that time. This was in the years right after the World War. (1978)

Flynt Chandler shared some additional information including the outside sourcing of the early anchors as Gano had earlier confirmed. "They were shipped in after we went from the gray iron to the malleable. They were made up here in Fairfield, Iowa. They shipped them down in carloads and we would unload them, paint them, pile them up, and stack them." (1979) Malleable was cast iron that was reheated to remove the carbon and/or a heating (annealing) process that relieved stress and toughened or tempered the casting. Malleable could also mean it was iron that was reshaped by forging, rollers, hammering, and other equipment.

Paul Sturgeon also had some remembrances about early anchors and the manufacturing process at Chance:

> *They made cast iron anchors when we put that foundry in. . . . The foundry was built 1924 or 25 or 26, along in there and somebody had an idea we could anneal gray iron and malleabilize it by some scientific medium. They played with it, for, I don't know how long. It never did work, of course and then as I said we went to malleable anchors. Would you believe it, not many years later we went back to the cast iron anchor as an addition to the malleable line. (1978)*

Paul said that was done for economical reasons to lower the cost of the anchor. Paul further recalled:

> *We had a very elaborate test system and they would at that time test for soundness of the metal castings. They had an iron bar and would hit the anchor and if it had a good ring to it, it was a sound casting. Well, I guess I should go back and say originally that was the process all right. But originally it was cast iron but along about that time* [around 1930], *the anchor had been switched over to malleable iron and they did not work so well under a sharp blow. The anchors would be shipped in here in carloads and I believe about that time they were buying them from Iowa Malleable. . . . It was unloaded on the south side to begin with but later when they rebuilt the warehouse, they put a railroad spur right down the side of the warehouse and they came in on the north side.* [Note: This would refer to the Alton spur track on the north side of Warehouse No. 3 on Wilson.] *Then later on, of course, they changed suppliers and bought them from National Malleable. (1979)*

Even F. Gano Chance got into the act of working with anchors. As later covered, Buford and Gano were likely the first of many hundreds of students that worked at Chance Company. Gano was discussing the new sixty-by-one-hundred-foot-long, two-story building that his father had erected between the tracks (Main Office Building) in 1917. They had a machine shop on the first floor and:

> *He also stored these anchors in part of the first floor and used to dip the anchors by putting a hook in the hole in the anchor and dipping it down in asphalt paint. I used to do that and some anchors were 40" long. I was a little bit short to pick them up, let them down into this vat, pull them out, and hang it on a rack so it could drain. I remember one time a hook came loose and a 10' x 40", steel malleable plate anchor went into the vat of paint and splashed paint all over me. (1978)*

According to comments about the hook, it appeared to be a tool like a large fishhook with a hand handle. It was used by farmers to get a firm grip on bulky hay bales. Ice and meat packing plants utilized similar instruments.

In 1929, new employee Bill Hollander came upon the scene. Upon being interviewed, Bill didn't come away with the idea that Mr. Chance had discovered him anyway close to being perfectly suited for employment at his Centralia firm. As Bill recalled:

> *Mr. Chance, after he interviewed me, told me that coming out of school I really didn't know anything that was of value to a manufacturer. With my school training, he believed that I would learn where I could be of value. So I went to work at a salary of $15.00 a week and very happy to get it because in 1929 the market crashed. Most of the fellows in my class were selling apples or insurance, begging, in soup-lines, or anything; trying to get by. I did work. And when I came to work for Chance, there wasn't a single drawing in the plant. (1978)*

These comments attract and deserve further analysis and discussion. Perhaps A. Bishop was explaining that a position was something you worked into at Chance's in an apprenticeship fashion and that formal education in itself didn't give one the necessary credentials for such a post. Most people of that time had earned a different kind of diploma, given by life's hard knocks. They were sometimes leery of "those college kids" who had yet to prove themselves in actual service and recent college graduates could be a bit cocky at times on their way to conquering the world. It was an early classical example of the old and new worlds colliding with each other. The world that A. Bishop came from was quite different from that which an educated, specialized, professional like Bill Hollander represented. There is a beautiful verse in Scripture that seems appropriate in explaining this more completely. "To every thing there is a season and a time to every purpose under the heaven." (Ecclesiastes 3:1, KJV) Gano's and Bill's arrivals at 210 North Allen in 1929 signaled a transition, although it would begin gradually but explode in the postwar period with the wholesale entrance of engineers into A. B. Chance Company.

Further analyzing the first impressions the founder made on Bill, implies that you contribute by applying educational knowledge to actual factory operations. This was a practicing philosophy of A. Bishop. The student work program and new hire policy of the placement of new employees in the factory environment were examples of this tradition at Chance Company. It is easy to lean heavily on the first sentence but the second one may be more significant. Bill did learn where he could be of value as he relates, "You just worked from a model as there were no drawings. There were no records of what it had been (tolerances-dimensions). So when I came here, one of the first things I had to do was to start making drawings of all of our products." It would be a fair assumption that A. Bishop wasn't opposed to progress being made in the engineering area by new hire Bill Hollander, the one man Chance Engineering staff. This appears to have occurred more by accident than purpose, but having professional engineers on the property in 1929 would benefit the company in immeasurable ways, although as patent information confirms there were some early engineer drawings of Chance anchors and rods.

Steady and efficient performers like Bill, although not always purposely ignored, can allow attention to be diverted elsewhere. At times it becomes an attractive alternative for employees to contemplate performance failures to reveal the importance of their punctuality of performance. The view of Chance Company from the outside for most or all of its life, would be of a heavily oriented manufacturing company, long after the focus had also shifted to marketing. This proves the old adage that reputations die slowly. This employee can relate to the emphasis on marketing from the balanced perspective of having worked in both areas. Eventually the engineering people felt ignored as they had no substantial or advanced facilities, but F. Gano Chance would spearhead the construction of the Research Center in 1961, after his and Bill Hollander's service period had reached thirty-two years!

Tractor snow scene: The March 25, 1944 photo of an apparent anchor test would appear to have taken place on the Chance Farm. The stream would suggest it was near the Buford Memorial Park area through which a small stream (branch) flowed. The tractor, a McCormick Deering on steel wheels (common during war), was likely one of Mr. Chance's four farm tractors. This is a mechanized example of an anchor pull test that Bill Hollander witnessed in May 1929, described on the next page. This picture was part of Art Houston's successful attempt in accumulating historical material for a Chance Company Museum, as directed by F. Gano Chance in 1973. (Chance Foundation Files)

Another intriguing aspect of A. Bishop's attitude toward Bill was how he viewed his son who was also in some ways in the educational doghouse. Bill was a college classmate and valuable discovery of F. Gano Chance, who had also become an "educated engineer." On several occasions, it was repeated that Bill was a top-notch scholar who tutored some of his University of Missouri classmates, including Gano. Bill Hollander would become, however, the Chance anchor engineer, whose contributions to anchoring development was immense. His knowledge and experience spanned a time from somewhat primitive anchoring to the sophisticated and innovative anchoring systems of today, many of which he was involved in. Bill had come a long way from that spring day in May of 1929 when he made an exploratory trip to the anchor capital of the world and found out that this small Midwestern firm wasn't in the business of supplying anchors to shipyards. Maybe, A. Bishop's initial apprehension was justified, as he came to the realization that he not only had to deal with a son full time who had all of these bright ideas, but also his sidekick Bill Hollander just as they were about to enter the Depression!

Some fifty years later, Bill recalled his trip to Centralia as if it was only yesterday:

> *I got to know Gano when I was in school over in Columbia. He invited me to come over and meet his Dad and see about going to work in Centralia and we planned to come up on a Saturday morning. Gano, at that time, had a Model T Ford Coupe and it rained the night before and there wasn't any chance to drive the 25 miles from Columbia to Centralia after a rain. So we came up on the Fog and Humor train. Fog and Humor is what they used to call the old Columbia Branch* [Wabash Railroad]. *They hauled everything and*

even picked up milk cans and everything else. (1978)

It is considered important in protecting the method of speech utilized by Chance people from earlier times. A result, however, is the changing meaning of words, especially of the slang variety. A compliment one day might become an insult the next. In the same manner, younger people can't imagine a world where rain made it too difficult to travel by automobile and instead reverted to one of the more reliable modes of travel of the day, a train.

Bill continued to explain his visit to Centralia in May 1929:

When we got there, Gano took me on the cook's tour of the plant and for the first time I knew an anchor was for something other than to hold a ship. After the tour he inquired of Ida Lou Traughber, where Dad was. She said, "Oh he is down in the lot testing anchors." So we went down and at that time the lot was next to the old mule barn where the old Tips Tool building (1948) stood. They were testing screw anchors. Ollie Chandler was one of the installation gang. I don't recall who the other strong ones were. They were twisting them down, they had a saw buck built up using a structure to pull the anchors and had a fence tractor providing the muscle power to pull them and a dynometer to hang on to, to see what they would do. This particular test, they had made cast iron screw anchors with a variety of pitches. The first screw anchor Chance made, why I think he took a Matthews anchor and kinda polished it smooth and used it as a pattern. They were very, very narrow pitch and A. B. had decided that perhaps a greater pitch or less pitch might make a better anchor. Bud Hicks at that time did model work and was the electrician and kinda Jack of all trades. He had made a series of screw anchors with a variety of heel spits and Ollie and this other boy would get on and they would say OK start, and they would go around like horses in a treadmill. They would screw them down as deeply as they could, get them until they just ran out of "oomph." They would measure the depth they were installed and then they would pull them with this old fence tractor and check the depth they had got the different pitch into the ground and what they held. [Note this is similar to the procedure pictured on the previous page, except a tree was used to pull off of instead of a frame constructed from large timbers, etc.] *One thing about installing the screw anchor that way was when you get one down and get a good bite into the ground, it got pretty hard going and it was much easier if you rocked it a bit. When you broke the thing loose, you could turn it a lot easier. So the result was, whatever depth you got your anchor to, the earth above it was all loosened and they did not hold too much. But that was the first anchor I ever saw pulled and that was in about May 1929. (1978)*

An interesting sidelight to Bill's description of anchor testing was the application of modern techniques of the nonphysical types to advance Chance anchoring science. In reference to the world cruise Gano mentioned:

My father bought a movie camera—about 1920. He had many tests made of the anchor, putting it out, the machine, etc. to see how much it would hold. So he decided to take movies of these tests. The first one I remember I went with him and operated the camera. You had to turn the crank to operate it. I had to practice to get an even turn. It was at New Jersey Western Union—was testing. . . . So he had two 35 mm cameras that you turned the crank with a great big, heavy tripod . . . a Pathe Camera. (1981)

These movies were surely utilized in the Chance sales promotion efforts and show the early time frame of their use in a smaller firm. The confirmation of such a film is revealed at a showing in February 1925, covered earlier in this chapter. Two of the largest firms of that day with far greater resources, International Harvester and Ford, were using such techniques. A. Bishop noted what was occurring across the country and applied certain selected advancements to his operation. Some of this advanced technology even came to Centralia such as a Honey Ford Demonstration Day event in February 1923 in which movies and talks were given in promoting the Ford Motor Company products. An example of what such period cameras looked like and how they were operated, is shown in the aborted marriage scene in the Academy Award winning 1934 film, *It Happened One Night*, starring Clark Gable and Claudette Colbert.

Confirmation of the Chance movie camera was noted in a September 26, 1924 *Guard* news piece within the city's minutes of their normal monthly board meeting. Mayor Bruton informed city officials that A. B. Chance had taken movies of certain Centralia roadways and would show them to the city board. However, the machine refused to work and had to be sent off to the factory. The pictures would be shown later. This may have been the situation Gano describes elsewhere, about running the camera inside of an open car, while the car drove down the rough city streets.

Reminiscing about his first trip to the anchor capital of the world, Bill Hollander was getting out of school soon and was looking for a job. A collection of remembrances from different periods produced the following thoughts:

> *I was finishing up early in June, but you know oddly about twenty years later, I drove about 1500 miles south of Centralia to test anchors in a Central American swamp, the very southernmost part of Mexico, and it was screw anchors again. I will say that there wasn't very much difference in the anchors but there was a heck of an improvement in the roads and the cars. So this particular test* [1929] *did not show too much because none of the anchors held very much. When they run out of oomph, why none of them were installed too deeply in that light clay that was down in that lot. But it more or less reflected what the whole industry thought of a screw anchor at the time. A screw anchor was all right for a light temporary load, but it did not hold very much and it took an awful lot of muscle to install it. The screw anchors that we made at that time were cast iron and used a square wrench to install it. We made them here and at that time they had a gray iron foundry and also nonferrous. We made some aluminum castings and some brass casting, but not very much brass. A number of aluminum castings were more for in-plant use than for product use and were for making match plates and the like. At that time there were expanding anchors and they were made of steel. There were two or four blades opposed to each other, formed to give a lot to nest together and you had expanding links in the top plate that drove until you got them expanded and then you backfilled. (1978)*

An expanding anchor was purchased from Taylor, although it was modified by A. Bishop and is covered in the following section. There were many different types of earth anchors developed over the years, some of which had as strange a name as their appearance. Andy Anderson (veteran Chance demonstrator of Hot Line Tools, known worldwide) had saved one such example called the "Harpoon" Anchor. There is a danger in applying another sea and ship name to an anchor product that is confusing enough to the unwary, but maybe it just fits. Andy said it had four folding blades that were driven in the ground. As pulling force was exerted on the anchor the blades would shoot

October 26, 1929 Letter and #1,756,534 Cone: The October 26, 1929 "Cone Anchor" letter is in response to the patent examiner's questions concerning the time frame origin of this improved Cone Anchor. The overlay copy displays the Chance Improved Cone Anchor engineering drawing, enclosed with the patent filing of June 21, 1927. (A. B. Chance Co.)

Oct. 26th, 29

Mr. Robt. H. Wendt.,
Williams Bradbury, McCaleb and Henkle.,
1315 Monadnock Blk.,
Chicago Ill.

Dear sir:

In looking up the cone anchor we find the following:

REPLY TO A: We have been unable to get any exact dates just when we first conceived the idea of the operative structure of the CONE but it was in the early spring, sometime in May or June of 1926.

REPLY TO B: Our first drawings were made about June 1926.

C: Our first written description of the invention was about August 1, 1926.

D: It was about the first of June or July when this anchor was first discussed with different members of our Company.

E: It was during August that castings were made and tests were conducted on the anchor.

F: We startedthe sale of the CONE A`GHOR late in the year 1926. From that time to present time we have added several little improvements, lastly, the nut retaining feature, and have sold these anchors all over the U S.

No one, to our opinion or belief has ever used a set anchor of the cone or pyramid shape having camming surfaces of different angularity and having a guy rod of any kind. Further the flare of the anchor at the bottom is also original with us, at least we have never seen or heard of anyone else using the same

Trusting the above furnishes the information you desire, we are

Very truly yours,

A Bishop Chance, Pres.

ABC::IT::::

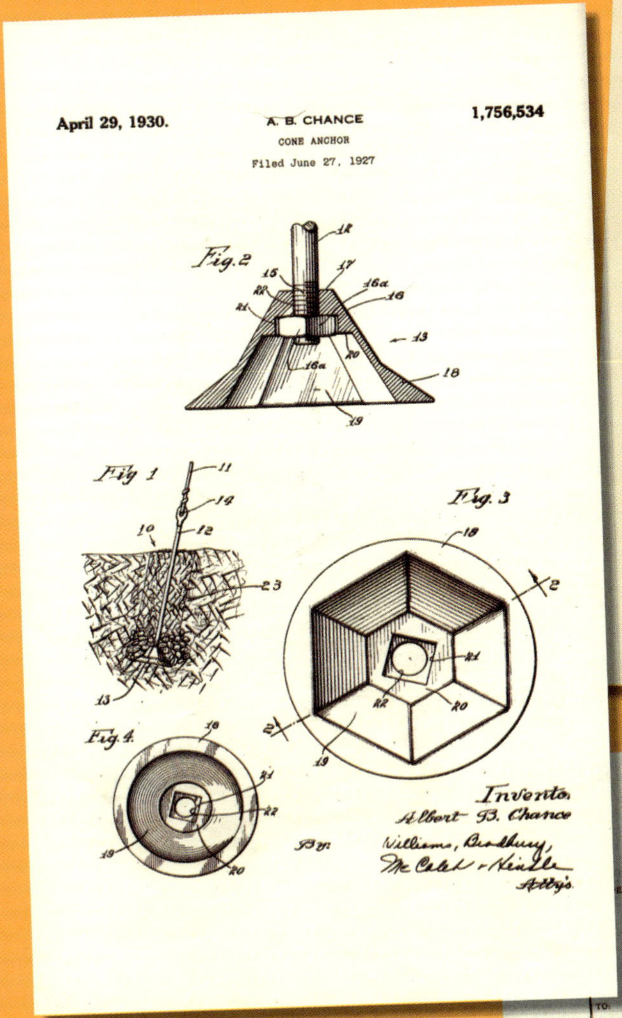

GRAYBAR ELECTRIC CO.

Chance Company

CENTRALIA, MO., U. S. A.

TERMS

2% CASH 10 DAYS, 30 DAYS NET. Subject to sight draft when due. Account to bear interest after due.

YOUR STOCK 275 SPRING ST,
NEW YORK CITY N Y

TO: INTERNATIONAL STANDARD ELEC CORP
25 BROAD STREET, EXPORT DEPT,
NEW YORK CITY N Y

DATE SHIPPED: 12-14-27
VIA: FREIGHT
YOUR ORDER NO.: 64721-0-7
DATE: 12-12-27
INVOICE NUMBER: 7 - 3678-BK8
REQ. NO.

130 6" 2-WAY CHANCE EXPANDING ANCHORS PTD 95 123 50

DISC 25% 92 63

B/L ENCLOSED

Sales Order 130: Chance sales order #64721-0-7, recovered as a surplus copy that's backside was re-used in later patent correspondence. It likely represents a customer order for the Taylor Expanding Anchor, which by December 14, 1927, has surely undergone a certain unknown level of modifications by the "Boss," truly becoming a Chance 2-Way® Expanding Anchor. (A. B. Chance Co.)

or harpoon out. There was a suggestion that it may have been manufactured in St. Louis by an unknown source.

The 1929 edition of *History of Manufactures in the United States*, Volume III for the 1893–1928 period, made no mention of Chance Company. Any such effort to cover all the small firms and cottage industries springing up in the United States during that period may have been like trying to cover a football field with a blanket. Industrial products were mentioned, including the anchor made from iron and forged at Camden, Maine. This obviously was the ship-type anchor and not the land-based type.

The earth anchor had all sorts of uses, even intruding upon the better know ship anchor's sacred domain. Gano suggested such a uniqueness of earth types superiority in tying down small boats in 1941. A four-inch screw anchor was sold to fishermen to lock their boats on the banks. The anchor was screwed into the bank and the boat locked to the anchor with a chain and padlock. The anchor could be removed and reused when the boat was taken from the lake or river. One of the earliest applications for the Never-Creep® Anchor, besides telephone lines, was their usage on farm fences. All through the years, Chance Company was continually trying to find new uses, in promoting the earth anchor's versatility in various applications. Agriculture was a neighbor in and around Centralia owing to the rich prairie surrounding the town. So agricultural applications and products were a natural for the company to branch out into as seen in certain new Chance farm products introduced in the 1930s. Gano again spoke of this area, "See story of Never-Creep®. I think you will find this in an ad run in the *Wallace Farmer* on Fence Anchors, about 1925." (1941) *Wallace Farmer* was an Iowa-based farm magazine that in 1940, for example, had a circulation of 260,000. Mr. Wallace at that time was Secretary of Agriculture in the Roosevelt administration.

Jack Curtright, a longtime Chance Marketing employee, recalled the progress of anchors and their installation. "There were patent anchors. They used a lot of different things. Expanding anchors were not the first anchor by far. The Never-Creep® was way ahead of it but it was a little harder to put in. Mr. Cole had an expanding anchor made of malleable iron." (1979) Jack considered the Never-Creep® Anchor as A. Bishop Chance's most important invention, and per *Chance Tips* in 1939 about 2,500,000 had been produced and sold:

> *It was a profitable thing and for a long time it was the major anchor. Then Expanding Anchors came into the picture and the malleable anchor and then eventually Chance Company made the steel expanding anchors, the old 8-Way. As we progressed, we found out that the screw anchor was made by Matthews in St. Louis. We made a No-Wrench Anchor with a heavy rod and used the rod to put it down in the ground. (1979)*

Andy Anderson remarked that, "We used a triple eye rod large enough to get a bar through it with manpower to put it down. (1979) Product information in 1939 revealed the current Chance anchor offering included the Never-Creep®, two-, three-, and four-way Expanding, Screw, Cone, and Rock anchors. In 1940 the Chance Pole Key Anchor was introduced per *Chance Tips* of January 1941. Jack continued:

> *We used up a lot of steel, I will say that, but it was a good anchor. We used to say, "There is an anchor for every type of soil." I remember there was a fellow that worked for Chance Company named Kunkle* [Dick]. *We were putting on a test down at St. Louis and he used to say, "well he could tell the texture of the soil by putting it in his mouth, pulling it between his teeth." You know anybody can do that, if it is gritty, then it's sandy soil. But he didn't know where he was, but I knew where*

he was. He was in an old sewer bed! We were putting that anchor down for Union Electric, to see how much it would hold in that type of soil. (1979)

Several more decades from the 1930 time frame would have to evolve before the design and installation of earth anchors would fall prey to automation. This doesn't mean that R/D at Chance Company was not reaching out into the future as longtime Shipping Foreman Roy West witnessed:

I'll tell you what, though, back in 1940 I saw Hollander walking around the plant packing that multi-hely (helix-blades) rod, which they didn't think it was any good. Now it's one of their best sellers. So Bill was about 30 years ahead of them! (1979)

The Taylor Expanding Anchor

THE DESTRUCTIBLE "INDESTRUCTIBLE"

A statement that has generated a lot of personal thought is the Biblical verse in St. Mark 10:31, "But many that are first shall be last; and the last first." (KJV) There is a special distinction of being first, but this in itself doesn't guarantee the maintenance of a position of recognized importance. Even those with a knowledge of Chance history will generally point to Tips Tool (1937) as the acquisition of first choice while the Taylor (1926) and Wej-Lock (1932) associations are ignored. So not only does obscurity surround both of these subjects, but also a shared commonality that both firms (Taylor/Wej-Lock) were involved with anchors, but Taylor came first. One of the reasons for lack of recognition, besides Taylor being a minor purchase in comparison to Tips, was the problems it brought. Even A. Bishop ignored it later in listing the highlights of his life.

In 1926 a sales agreement had been reached by Mr. Chance and a Canadian by the name of Mr. Taylor. According to Bill Hollander, this purchase was consummated in New York City just before the Chance's began their Ryndam world tour. This was the source and beginning of Chance Company's involvement in an expanding anchor. With this purchase, the family of Chance anchors would be broadened to include what would eventually become the most popular anchor type for a time. It became so important to the bottom line of A. B. Chance Company in the mid-century period, that to know the economical health of the Company was to simply know how many No. 88135, 8-Way® Expanding Anchors were being sold and shipped to customers. Ironically, it would be the founder of the other company, Wej-Lock, who would later reinvent a new 8-Way® Expanding Anchor that revolutionized this anchor type and boosted its success in the late 1940s.

Further comments about the Taylor Anchor giving Chance Company an expanding anchor was made by Nathan Toalson:

Now that was the old sliding type expanding anchor.... I guess that was what put the Chance Company in the punch press business, because these were all steel stampings and was one of the items which gave us the punch press which was put in this new building that had expanded from the old office in the back. [Note: Machine Shop/Tool Room and later for several decades the location of Office Services] *I remember I worked in the punch press when it was in that location. That was acquired, as I recall, from a Canadian Company who had been approved by Bell Lab for this particular anchor as an expanding anchor. The Everstik Anchor*

From Chance *Tips* November 1939. Build Security.

was basically the same principle, which was malleable castings instead of steel stamping, which was the old sliding blade type of anchor. . . . But the Everstik was the competition back in the early days, as far as the expanding anchor was concerned. Then the Wej-Lock anchor came along. . . . I think the anchor was really bought originally to be used with guard rails and then they started selling it to the electrical trade. (1979)

Bill Hollander continued to explain, in his colorful manner, what he had learned about Mr. Taylor and his anchor:

> *Taylor had developed the anchor and he sold them with some success but he had a lot of problems with them. They were approved by Western Electric and by Bell Properties but he was having health problems and his doctor told him he was going to have to get out of that business as he was worrying himself into the grave.*

So he contacted A. B. and they dickered for a period of time to buy the anchor named "The Indestructible." Actually, it was anything but! They met Mr. Taylor in New York and they presented him with a certified check for a down payment on the Indestructible Anchor. They had their lawyer present in New York and after they had closed the deal, he said, "We [Taylor] *have had a lot of problems with it but all of the problems are worked out. We will ship all of the dies to your plant down in Centralia and all the back orders." So A. B. said on his world cruise that he was a very happy man. . . . So they sent everything, all of the stock, raw materials, and dies to Centralia. While A. B. was on his trip around the world, Buford* [his oldest son] *got the job of setting up to start making anchors. They got quite a stock of anchors from Canada and they produced more of them here. They started shipping them* [to customers] *and by the time A. B. got back there was a great heap of anchors that had been returned. They were put together by spot welding and about the third time you hit the anchor with the tamping bar, why they would go all to pieces, and they would come back. There were a number of industrial processes at that time that were in their infancy for sure. (1978)*

All of the shortcomings of the modern processes with their unpredictable results and the difficulty of finding the correct design generated a great deal of worry, as Mr. Taylor had found. It was not mentioned if A. Bishop began to get telegrams from back home of the impending disaster that the Destructible-Indestructible Anchors were generating. Neither has it been found what his reaction was on his return to Centralia when he first saw the pile of returned expanding anchors. It's doubtful that U.S./Canadian relations were on the upswing in Centralia, Missouri. Before A. Bishop's return, Buford had apparently headed a troubleshooting group, as according to Bill:

They had started work trying to determine why there were good and bad. At the time they were mostly bad. So one of the first things A. B. did, at that time, was to eliminate the spot welding that held the links to the blades and the tamping plate on top. Also, he added the base blade to the bottom as they used to have a loose washer on the bottom of the anchor. He added a base plate to the old expanding anchor and it began to move a lot better. (1978)

Bill came to Chances right after that, but apparently the problems encountered were the talk around the Company for years to come. Paul Sturgeon recalled the original Taylor expanding anchor that A. Bishop had brought back to Centralia. He also described its weakness. "It did not have a base plate on it. Just the sliding plates and a retainer. It did not have a base plate or a nut retained." (1979) Solving this problem reveals A. Bishop's practical and back to the basics type of approach in replacing both a too fragile design and modern industrial techniques that were not yet perfected.

It is gratifying to gain further insight into early company acquisitions and history, especially when it is generated by official documents. A key to revealing more about the Taylor acquisition and anchor came via patent material generated by no other than A. Bishop Chance. A most valuable discovery is finding that Herbert M. Taylor's July 14, 1925 patent for an "expansible" type anchor No. 1,546,327, was shared with fellow inventor Frank Sutcliffe. The word *expansible* suggests this was the description Taylor/Sutcliffe gave their patented product although expanding would become the norm, or that is how Chance Company described it. Chance Company had acquired this patent, but found problems with the anchor. What wasn't revealed was Mr. Chance's extensive reworking of the Taylor/Sutcliffe Anchor that led to at least three more Chance patents! His discussed improvements to the 1925 patent was possible because the Taylor patent had been re-assigned to Mr. Chance and it could be used as a basis on new claims for his new versions. The quick attention he gave his new anchor, on his return from the world tour in May 1927, culminated in the submission for a new patent on May 26, 1928, resulting in the issuance of No. 1,714,766 on May 28, 1929. This is where it gets sticky with Patent No. 1,786,228 of December 23, 1930, also filed on May 26, 1928. An application for its reissue was filed on January 23, 1932, and Patent No. 18,697 issued on December 20, 1932. Patent No. 1,786,228 was surrendered when No. 18,697 was issued and all three appear to be a continuation of the two original patent filings of May 26, 1928, labeled as Case 9 (two-way) and Case 10 (four-way).

A patent filing is assigned to an "Examiner" in the U.S. Patent Office, who determines if there are any previous patents in the files that are in conflict or covers claims made by the new filing. In this case, there were six other patents that were cited, the most significant of the group appears to be the Blackburn Patent No. 1,041,233 of October 15, 1922. Incidentally, this filing and process was orchestrated and monitored by patent and trademark law attorneys, Williams, Bradbury, McCaleb, and Hinkle of Chicago. Correspondence between them and Mr. Chance on the problems and conflicts on one occasion, included copies of the drawings of the other patents in question (Quinn No. 1,504,899 of August 12, 1924; Stewart No. 864,391 of August 27, 1907; Justus No. 756,374 of April 5, 1904; Blackburn No. 725,398 of April 14, 1903; plus the 1922 Blackburn and the 1925 Taylor patents) to which Mr. Chance replied that they already had those patents on file. This suggests, as has and will be shown, that anchor activity was closely monitored by the founder, which proved beneficial in allowing his firm to continue its long leadership position in anchor production and sales. Notice that the Blackburn patent consisted of two patents of 1903 and 1922, which was apparently the one (1922 version) that was causing the most concern. Another revelation of this material is the inventions of the same item apparently weren't uncommon, but the way they were designed was an issue. Being first doesn't appear to be as important as covering unique and improved designs of an item in patent claims. In respect to A. Bishop, he was much more than an inventor, as revealed by the challenges in developing a viable product out of the Taylor Expansible Anchor. This was nothing new, as this process can be seen on his very early arrester, even the Never-Creep® Anchor and rod and on a variety of products. He had the skill to reduce parts while simplifying manufacturing and assembling operations of the Taylor anchor to generate a competitive and profitable product. At the same time, by modifying the anchor's operating characteristics, it became a stronger, more reliable and a practical product that the customer could rely on in performing successful guying installations.

Several years have expired since the initial two-page story of the Destructible-Indestructible Anchor was written. That was before patent materials unlocked secrets of the past and illuminated details of its Chance transformation. The Taylor Anchor is a perfect example of the difficulty of covering an important subject without hard evidence, plus no reference to it has been found in company material, except the patent information. Even then, it was used as a reference to support the request for a Chance patent. Without the Hansen/Chance interview material, the Taylor story of what appears to be the first Chance expanding anchor, may not have seen the light of day.

Company material in 1928–29 points to the lively sales activity and quickly gained popularity of the new Chance expanding anchors. They generated a sizable business for two decades, until the 8-Way® was introduced in 1948. There are reports that it supported a good portion of anchor production, especially after it was formally introduced in 1927. Correspondence of October 4, 1927, confirmed the "Chance Two-Way Expanding Anchor" of the six-inch size, cost around ninety-five cents and weighed six pounds. It becomes more difficult after the Wej-Lock acquisition, to always determine the anchor type being discussed. Although it isn't known what the original Taylor Anchor exactly was, when the Two- or Four-Way Expanding Anchor is mentioned during 1927–31, it is a Taylor derivative, and it included one or both of these multiple-blade configurations. A December 1927 customer order reproduced on page 343 for a Two-Way Expanding Anchor, adds credence to the Taylor being at least a Two-Way as the new Chance versions were not filed for a patent until May 1928.

The attention and changes A. Bishop applied to the Taylor anchor and the patents obtained, may have led him to think of its as "his anchor." It had become more of an in-house generated product, but it was transformed from an acquired product. After turn-of-the-century patent information revealed the first Chance Company acquisition goes back to the lightning arrester era, the proverbial argument is raised of what came first, the chicken or the egg. Out-sourced products may have ignited the birth of Chance Manufacturing Company instead of the later Sky-Rocket®! Like the early expanding anchor, the Sky-Rocket® was a revised and improved product that grew out of existing, cruder, early non-Chance inventions. The anchor also had patents pending prior to 1912 and Gano even said that his father had to buy some of those people out to get a good patent. What was different about the Never-Creep® Anchor was an emergency forced a crude, quickly made solution that turned into a later, much-improved version that was patented before or after other anchor patents were acquired. Doesn't this sound familiar? Although on the surface it's more sensational to be viewed as the sole inventor of an item, in reality A. Bishop's ability to refine an item hinged to his aggressive business abilities, including his constant vigilance, foresight, and gifted negotiating successes, are even more impressive!

Although the original Taylor/Sutcliffe patent file wasn't found, drawing copies of the two revised versions that Mr. Chance patented are reproduced on the next two pages. It is a complicated appearing apparatus with Figures 1 through 3 picturing its uninstalled posture and Figure 8 as it appears in the expanded position. This first series (on the first page) is the Two-Way, represented by Patent No. 18,697. The second series is the Four-Way (four extending blades in installed/expanded position), with Illustrations 1 and 2 representing the uninstalled configuration and 3 and 4 representing its expanded profile. It is the version patented under No. 1,714,766. It's not entirely clear if the allowed features are represented by these drawings or they are the ones initially issued when the patents were submitted in 1928. What it does reveal is Chance Manufacturing Company had a workable Two- and Four-Way Expanding Anchor prior to the time the Great Depression hit and became a most serious proposition!

From accumulated information about the early Chance Expanding Anchors, the following conclusions can be reached. Before the Depression came, the revised Chance/Taylor Anchor had become a success. Since it utilized several lower-cost, pressed-steel components, it may have been produced cheaper and sold at a lower price than the Wej-Lock Malleable Iron Anchor. Of course, another factor impacting pricing was fewer components. The irony of suggesting this is a new and radically revised version of the original Cole Wej-Lock Anchor converted to pressed-steel components (8-Way® of 1948), could be made and sold at a lower price than other existing expanding anchors. This is supported by the 1939 Chance ad on page 346 stating the Chance Expanding Anchor had become all steel. This suggests the bulk of Chance Expanding Anchor sales were the Taylor derivative and a lesser volume done by the original Wej-Lock designed for guard-rail use. This 1939 ad, when compared to the patent drawings of the revised Taylor Anchor of pre-1930, indicates it had been further refined and could be produced even cheaper than the earlier version with what appeared to be cast

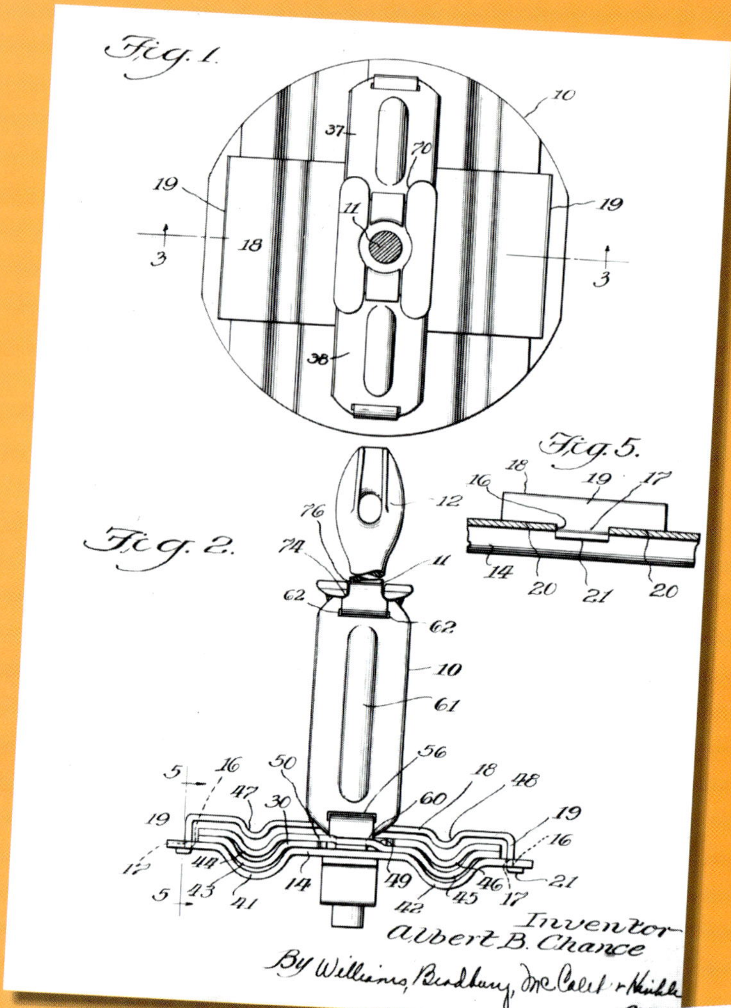

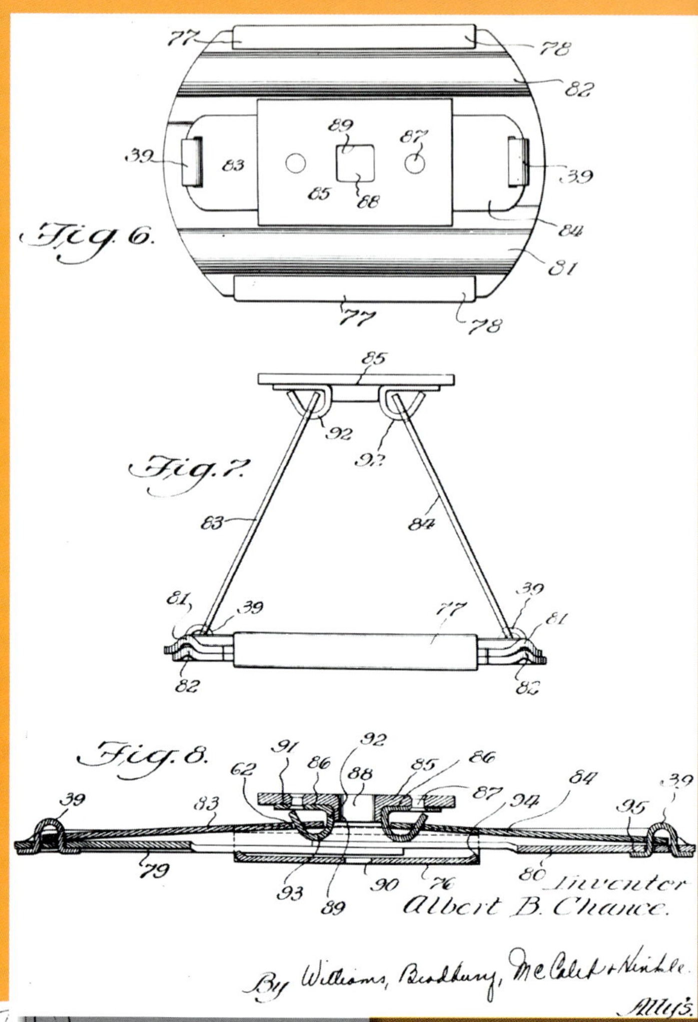

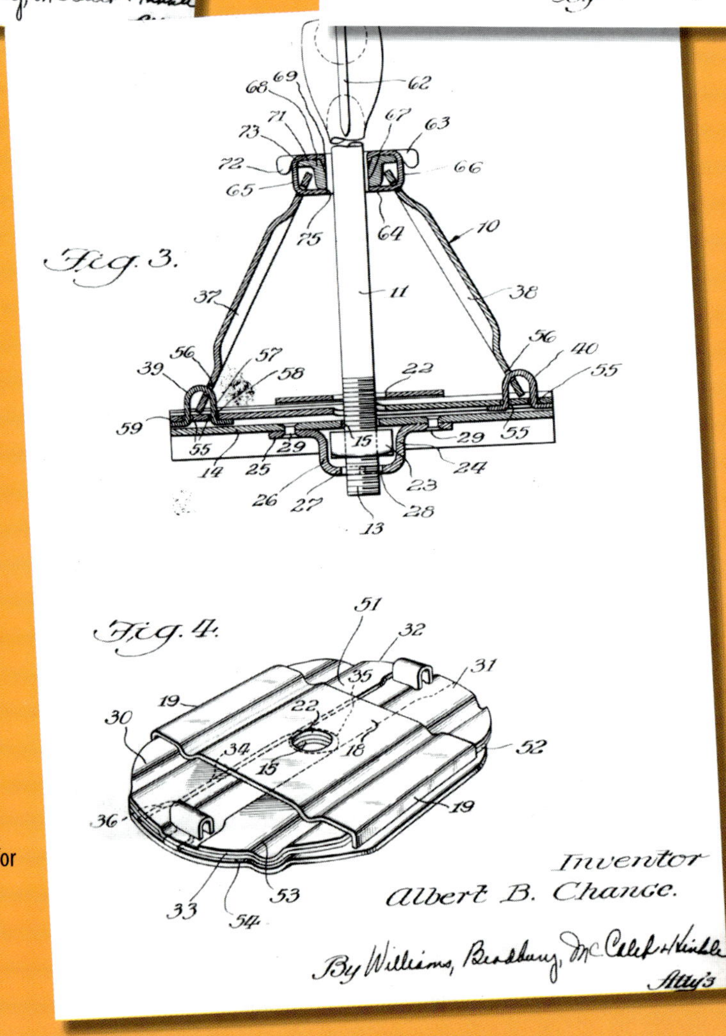

Six pages of patent technical drawings for the Expanding Anchor. (A. B. Chance Co.)

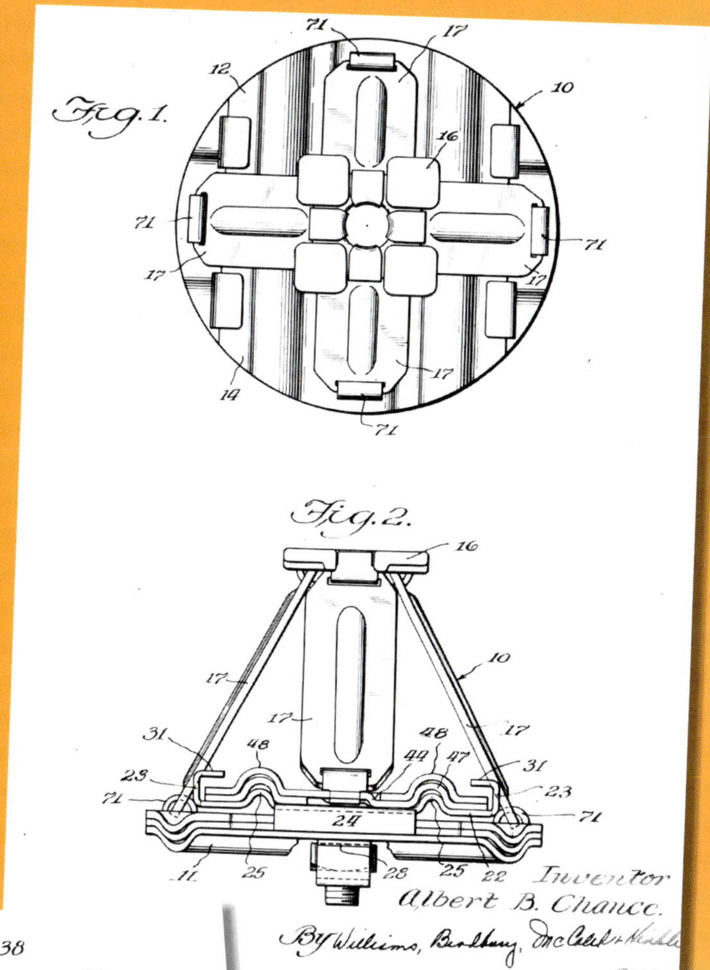
Fig.1.
Fig.2.
Inventor
Albert B. Chance.
By Williams, Bradbury, McCaleb & Hinkle

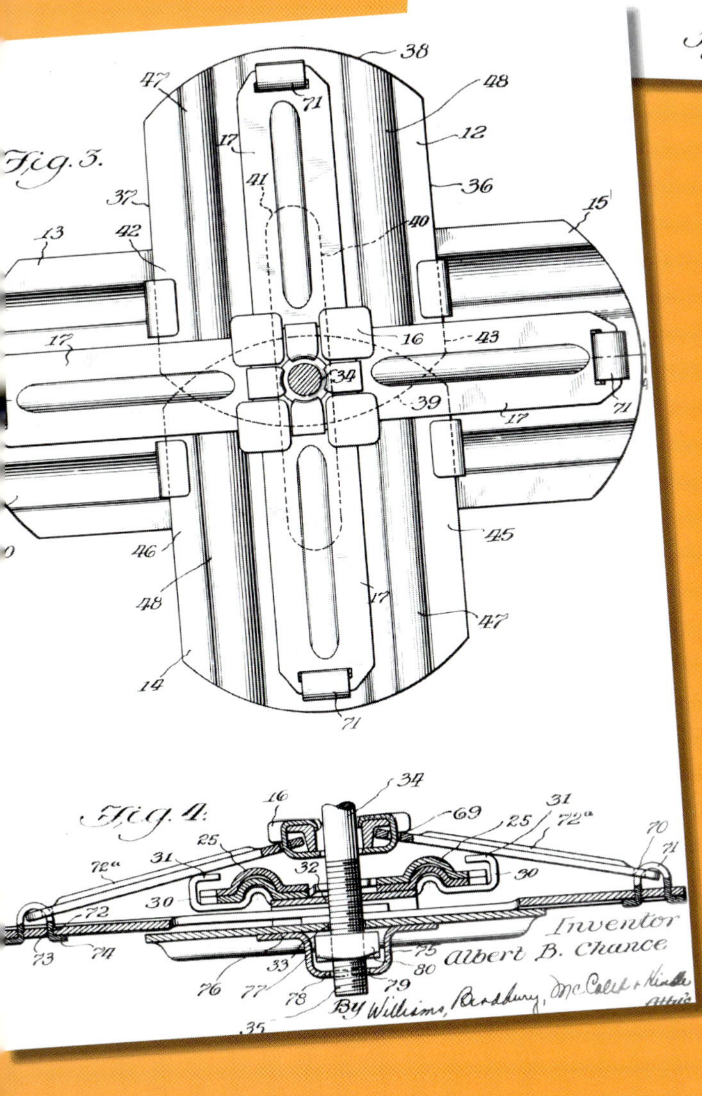
Fig.3.
Fig.4.
Inventor
Albert B. Chance
By Williams, Bradbury, McCaleb & Hinkle
Attys.

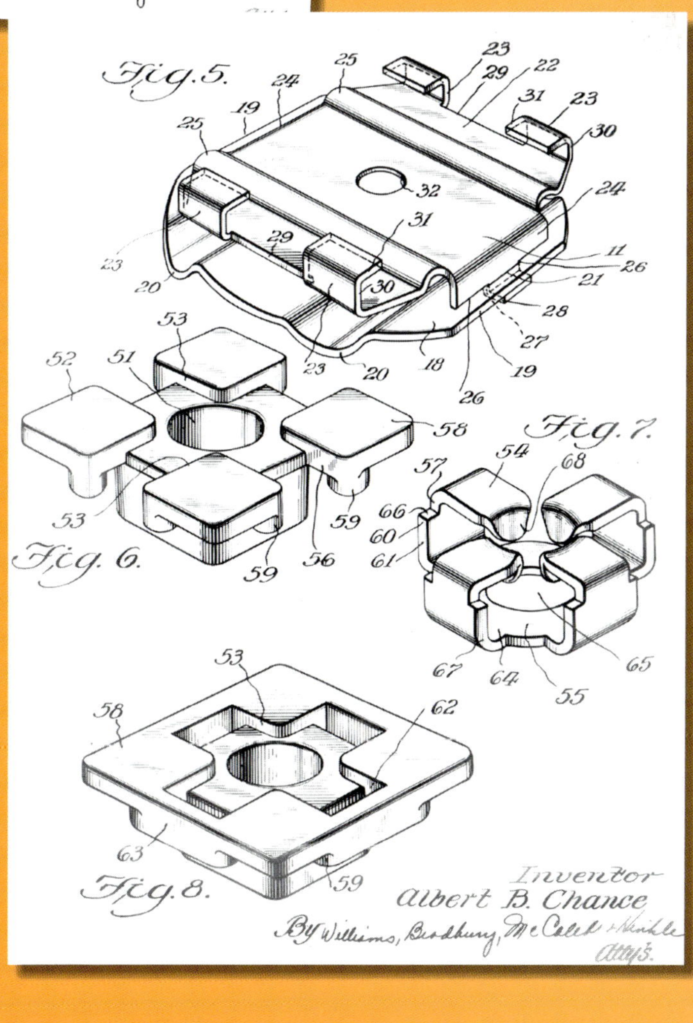
Fig.5.
Fig.6.
Fig.7.
Fig.8.
Inventor
Albert B. Chance
By Williams, Bradbury, McCaleb & Hinkle
Attys.

base plates. An interesting question thus arises, was it A. Bishop or Mr. Cole and his sidekick Bill Hollander, who had perfected this anchor even further by 1939? Of course, it is possible the 1939 Expanding Anchor pictured could be the Wej-Lock version, but the blades and other features appear to be the Taylor derivative. The benefit of analyzing this is the refinement of the Chance Expanding Anchors points to their importance in more appropriately sharing the spotlight with the sensational 8-Way®. Two decades of sales success and refinements preceded what could be considered chapter five in the Chance expanding anchor story, played out in 1948: Taylor/Sutcliffe; Chance revised Taylor/Sutcliffe; Wej-Lock malleable anchor; Refined Chance Expanding Anchor of 1939 or before; and the Cole 8-Way® Expanding Anchor of 1948.

Albert Buford Chance

1901–1928

This is Buford Chance, eldest son of Mr. and Mrs. A. B. Chance, in a light moment. The information on the backside of this photo suggests he was twenty-three or twenty-four years old, which translates to the early/mid-1920s. (Jack Chance Collection)

There is a constant struggle taking place in our lives between external and internal stimuli vying for attention and control. From a Chance family perspective, after twenty-seven-year-old Buford's death at around 11 p.m. on November 24, 1928, he became a silent, internal influence on his father, mother, younger brother, and of course his wife and young son. It did impact their perspectives and influence decisions that were made concerning Chance Company.

Comments that Gano made concerning his older brother's death speaks about his thoughts on family responsibilities. Gano's remembrances and understanding of what occurred is as follows:

> *I was in the university and he* [Buford] *was driving home from Mexico* [Mo.] *by himself in an open car and fell asleep. It was a straight road and I think it was about 7 o'clock Sunday evening and he had not been drinking. It had gotten dark and it was in the wintertime. They sent someone from Centralia over to get me and bring me home when my brother was killed. I got home and found, of course, my mother in bed crying. It was about 9 o'clock by that time. I remember making a dedication. She felt she had lost a son and half of her children and I promised to do the work of two. (1981)*

This gave Gano a singleness of purpose, in certain respects and he became a man who wasn't afraid to shoulder responsibility. The seriousness he developed could also be reflective of the influence of his brother's death, but it appears Gano already had a strong dosage of this trait. Gano said: "It stuck with me that I would make up for him." (1981)

A *Guard* front-page news item, "BOY HURT," on March 13, 1914, had an interesting relationship with our subject:

> *Buford Chance, the eleven-year-old son of A. B. Chance, met with a most painful accident last Sunday that came near being fatal. It seems that the little fellow was playing with some other boys of the neighborhood and they were jumping from the back porch when young Buford jumped so far that the clothesline caught him under the nose and threw him violently to the ground. For a while it was thought that his neck had been broken. He is getting along nicely at present, we are pleased to state.*

This could be shrugged off as just another childhood kind of accident in 1914, but knowing what occurred fourteen years later gains greater significance for at least two reasons. First it could show a very aggressive, not afraid of taking chances, kind of individual, much like his father. This early accident could have a medical connection in potential damage to Buford's neck/spine with a lingering effect that more easily resulted in or impacted his death in 1928.

Additional news items over the years were carried in the *Guard* about Buford. The winter of 1918–19 was a flu winter as outbreaks in Colorado were even locally reported. On January 17, 1919, there were reports of flu striking in Centralia including the Chance home. "Buford Chance and Joe Morris are out of the high school with a slight attack of flu." The condition of the two lads comes into question when the reporter used the descriptive word, "slight." Maybe this isn't surprising as Buford could be pictured as a typical teenager who was more interested in cars than in going to school. The same issue also carried another news item: "Mrs. A. B. Chance gave a dinner Sunday to the teachers of the high school. Music and games were also enjoyed by the assembled guests." This was a very nice gesture of Mrs. Chance in entertaining the teachers. Apparently during those times generous acts directed toward under-compensated teachers weren't uncommon, and being the nice person Mrs. Chance was, there may have been no motives other than to give support toward sometimes unappreciated teachers. That fall in the November 14 paper, Buford was a weekend guest of University of Missouri student Harold Shelledy. They attended the Nebraska-Missouri football game, which the Big Red won. A couple of weeks later on December 5, the A. B. Chance's entertained with a dinner that appeared more suspicious than the earlier one. The guests were Elder and Mrs. E. S. Priest and daughter and Elder and Mrs. A. L. Wills and daughter. Elder Wills was the Christian Church pastor

This is a young F. Gano Chance, second son of Mr. and Mrs. A. B. Chance. It appears to be a college picture from the 1926–28 time frame (Jack Chance Collection)

where the Chance family attended, but Elder Priest's position wasn't found. How could you have a more balanced engagement than having two young ladies as guests, although Gano and Buford's reactions are a mystery.

Gano spoke of what he knew about his brother in an open manner:

> *My brother was in the University and I was still in high school. At that time our little Company looked like it was going to be a good, prosperous, little Company with about $100,000 sales. I think that my brother was going to college at that stage, when business was easy for everybody, including our little business. He thought he was in college to have fun and in about 1½ years he had flunked a few courses and my father brought him home. I think a lot of things happened as I am sure he was as smart as I or smarter. But under the circumstances in the 20s, we had this little company with this patent on this anchor that was going some place. Everything was easy and easy for him. He was a likeable, personable guy so when he went to college he'd had a good time. When I got to college in the last few years things were pretty tough and I naturally learned to be more conservative and learned to study a little harder than he did. Perhaps that is the way things were and I actually believe my brother would have been an excellent partner, an excellent man in the business because he was well liked, more like my father. He had a lot of personality, was a good salesman and had a lot of enthusiasm. (1981)*

Buford was in college in 1921 per a news item in the February 25, 1921 *Guard*: "Mesdames A. B. Chance and O. B. Mayes visited their sons at Columbia Friday, and were chaperones at an informal given by the A.T.O. Fraternity." His father's close friend O. B. Mayes, apparently had a son who was at M.U. at the same time. Another revelation of that news item suggests Buford was involved with the same fraternity as Gano later would be. Another news item, a couple of years later on December 15, 1922, reported Buford Chance was in Slater on business the previous Monday. This would suggest he was then out of school in view of what Gano said in the previous paragraph about his college departure. Buford's obituary confirms he attended the University of Missouri in the 1920–22 time frame.

On April 20, 1923, the *Guard* carried a somewhat unusual piece about Buford and his mother visiting in St. Louis for several days. The purpose for this trip and how she was able to corral her twenty-two-year-old son will likely remain unknown. There is a sense that Mrs. Fancy Chance was a caring and responsible mother who took advantage of opportunities given her to instruct and influence her sons. A. Bishop may have still been absent on a West Coast trip and Mrs. Chance was dealing with the Buford developments alone. This wouldn't hold as much significance if the same paper hadn't covered an event on the previous Sunday that impacted the A. B. Chance family. The St. Louis visit as the *Guard* described it, "several days the last of the week" would indicate that it followed an important Sunday event, although it could also refer to the previous week:

> *CHANCE-PEMBERTON. Miss Bertha Bernard Pemberton and Alfred* [sic] *Buford Chance, Jr., were married Sunday night, April 15, 1923, at the home of and by Rev. F. W. Emerson in Mexico.*
>
> *The young couple went to Mexico on a visit and while there they made up their minds to get married and did so at the earliest moment the license could be secured and the parson visited.*
>
> *The bride is the daughter of Hon. Morton H. Pemberton and wife of this city and is a*

Pictured around 1925 is Anna Lee Toalson and Dorothy Rowland. Dorothy was also a friend of Bertha Pemberton and would later marry Ed Farrell who was involved in the Games Division at Chance. (Jack Chance Collection)

pretty and talented young lady, charming and possessed of good sense and the ability to fill her place in the home. The groom is a son of A. B. Chance and wife of this city and is a rising young business man, at present holding down the position as general manager of the Chance Manufacturing Co., in this city.

The newlyweds will continue to make their home in this city.

The Guard joins in congratulations and best wishes for the future of this excellent young couple.

This article, although family related, does reveal Buford's general manager position at Chance Company.

As May 1925 approached, the mid-1920s building explosion in Centralia was in full progress. Portions of Rollins Street were being paved and the new Wabash Station was a reality, at least in plans and the Chance Anthony Building was being completed. Chance Company was also planning to add a third story to the main, two-story factory building plus adding a huge foundry and warehouse east of it to Collier. On the east side of Collier the roof was being put on the new Chance Chemical Building, according to the May 29, 1925 *Guard*. At that time the "Chem" Building was set out by itself, separated by the proposed and difficult completion of Collier across the tracks in front of it. As the name suggests, this building had strange materials to deal with in comparison to the rest of the Chance operation. This is in keeping with the future in which diverse operations and materials would separate many Chance Company factory operations. This Collier Street Plant was the site of automotive-related items that were to accompany the piston rings into making Chance Company a substantial supplier of automotive-related products. The success of these products, however, is much less known than piston rings, which had become a substantial part of Chance factory operations. The full-line approach at Chance is showing its early roots. This clearly was an area that Buford was interested in because of his apparent fascination with automobiles. At least one employee suggested Buford was manager of that operation. Buford represents what could be considered the first American generation that grew up fully exposed to the automobile, and a strong interest was generated by many in his age group.

The employment of F. H. Dukesmith in heading automotive sales at Chance, immediately placed him on the road to auto shows, such as one in February 1926 in nearby Kansas City. He was promoting the "Chance Big C piston rings" line and developing a national sales force. As was common, a contingent of representatives from Chance Company usually attended major trade shows. A. Bishop had begun this practice much earlier with telephone items and he also attended a St. Louis auto show in 1924. An initial reaction of his presence at St. Louis was to see all the new vehicles as a spectator, but it likely was in promotion of the Chance auto line. Further confirming that, the company was getting too big for a one-man show, his eldest son also attended the February 1926 Kansas City auto show. Since Dukesmith's position is only mentioned with piston rings, one could assume that the other products of the Big C auto line were Buford's, but his position may have been more of a plant manager or in the overall management of the Big C line. The entourage that he headed in traveling to Kansas City included his wife and a Miss Dorothy Rowland who was a friend of Mrs. Buford Chance. Three years into his marriage, he was involving his wife in his business travels.

Although it is not clear who was in overall control of Chance Manufacturing Company while A. Bishop was on the world tour, three names that instantly pop up would be Miss Ida Lou Traughber, J. B. Fenton the plant superintendent, and Buford Chance. It was expected that when the boss was away cruising on foreign waters, the plant would sort of operate by the numbers with few if any new developments, but that wasn't the case at all. It could be that the founder was working hard generating new projects prior to his departure in mid-September 1926 and it took employees several months to just catch up. In early November, a school of instruction was instituted within the company for employees. New departments were added in late November for work formerly out-sourced and a second furnace was being set-up in mid-December to exclusively make piston rings. Without modern communication, it's doubtful if A. Bishop could be orchestrating all these activities, but neither was any information found that Buford was in complete charge. There appears to have been key employees who were completing and activating that which had been put into play. He must have been confident that those he had trained and left in charge would see things through, even in his absence for many months.

The Buford Chances were getting coverage in the local paper, of their business and social activities. In early May 1927, Mr. and Mrs. A. B. Chance and Gano returned to Centralia from their world cruise. Buford Chance and wife were involved in the welcoming home celebration by entertaining with a six o'clock dinner. The July 1, 1927 *Guard* told of "Mrs. Buford Chance returned home Friday after an extended visit in the East." There were no further details of this trip. The practice of referring to Buford as A. B. Junior was used officially as early as October 28, 1927, per a *Guard* news item, but at times this was referring to his son: "Mr. and Mrs. A. B. Chance, Jr. were in Mexico, Saturday visiting friends." As the year of 1927 came to a close the December 23 paper carried an interesting news item in regards to the makeup of traveling companions: "Mrs. Buford Chance and son and Miss Anna Lee Toalson were in Columbia the last of the week on business." This is an early mention of a three-year-old Little A. B. traveling with a parent and another single lady. In this case, Miss Anna Lee would become Mrs. Buford Chance's sister-in-law, but after Buford's death. As heavy activity in the Chance Plant continued and spring approached, the April 27, 1928 *Guard* found Buford and wife visiting in Columbia the previous Sunday. Another Columbia visit to friends was recorded in the May 11 paper by Mr. and Mrs. Buford Chance and son Little A. B. on a Saturday night.

Further confirmation of Buford's involvement in Chance Company business, is evident as 1928 progresses. A news item of May 25 told of a Des Moines, Iowa trip to transact company business. His wife accompanied him, which shows a pattern that had developed. This was becoming their practice, and as young people they apparently enjoyed all or a part of such benefits of traveling, shopping, or the after-hours entertainment advantages as a couple. It's unclear if they functioned as a team from the business point of view. Their Sunday departure, was likely on the Wabash as it ran to Des Moines, suggesting a common business practice of being on location the first day of the workweek. A huge piston ring order enhanced the importance of the automotive area in which Buford was involved, which must have kept him in town as no further trips are reported through summer and fall. Although his younger brother Gano was still in college, he had become very active in plant activities during his summer vacation period. It is not clear if one or the other were the favored son, although publicly there seems to have been a great deal of recognition and support for Gano, who appears to have been successfully fulfilling his college requirements with honor. One could speculate that Buford was becoming more responsible and attentive to company duties than he had given to his earlier college experience, but in the almost weekly articles about Chance Company growth and activities nothing further is mentioned of Buford until his death in late November. This doesn't read like an heir to his father's leadership position as company head, as there is not a public image of his father aggressively pushing him for an eventual takeover. Neither did Buford appear to be routinely getting public recognition or inserting himself but appears to be going along with the flow. The father at this point didn't appear to feel a need for preparing to turn the company operations over to either son. Besides providing the normal insertions of the sons into the business, in the fall of 1928 there was no apparent immediate situation that warranted this kind of serious preparation.

Two forty-five-year employees made some observations and keen insights into the three Chances: A. Bishop, Buford, and Gano. Flynt Chandler said Buford was like his father in that, "He was a chip off of the old block. Gano is a good guy, I liked Gano. He just didn't get around and mix with the workers like Buford and Mr. Chance did. He took a hold when things were in pretty bad shape." Les Reynolds first speaking of Buford said:

> *He was a whole lot more like Mr. Chance, I believe, than Gano. He would come out there and talk to you. He was kinda like his dad. He got out to see some of the outside work. If you had to work overtime, he would just leave your cards out there. It always seemed to me like Gano took everything more seriously. It didn't seem like he had much time to do much talking and laughing because he was thinking about something else. (1979)*

Another veteran employee, who came to Chance after Buford's death, Bill Hollander, recalled many things that he had heard:

> *Buford worked at the time A. B., Mrs. Chance, and Gano went on their world cruise. Buford was more or less managing the plant when Taylor sent all of the old Indestructible Anchors down and got it into production but I don't know how many years he was active in the Company. I don't recall ever hearing that. I heard so many*

The Buford Chance family. (A. B. "Buff" Chance IV Collection)

tales about the early expanding anchors and the automotive supplies and what not. Buford, from what I have heard, was very much like his Dad. He was like A. B. He was a hale fellow. He knew everyone, he talked to everyone. He must have been a very outgoing person. He and young A. B. must have been very much alike and both of them were like A. B. Chance [Albert Bishop]. *They were very outgoing, hale fellows, well met and would try anything. But I did not know Buford. (1979)*

According to Les Reynolds:

Buford worked in the office when I went over there [1926]. *There was just Mr. Chance, Miss Ida, and Buford there for awhile. Maureen Chambers* [Note: This is Maurine Chamberlain and future references to her will so reflect] *was next and then Verna* [Palmer] *came. I think Maurine came in the next spring after I went to work. I remember one time he lived down here in this little brick house on the corner* [Chance Bungalow?]. *The furnace blew up over there and it was a mess. Buford come up there and he said: "I want somebody to go down and clean that up." He said: "Les, who do you want to go with you?" I said, "Well, old Slim will be all right." Old Slim and I worked together. "All right, you all go down there and whatever you have to have, come back up here and get it." All the walls had to be washed and everything. We were down there, I guess, about 8 o'clock and we got everything started about 9:30. He came down there and he said, "You boys wash your hands and come up for breakfast." I said that we had already had breakfast! We went up there and she* [Bertie, Buford's wife] *had bacon and eggs and toast. We got through eating and he said, "Well, I'm going back to the factory. You all rest a while before you go back to work, if you want to. Don't come over and check out before you go home. I will check your cards." About the middle of the afternoon, he came over and said, "come up here," and she had pie and sandwiches. That is just the way he was. (1979)*

Another incident involving the eldest son of Mr. and Mrs. A. Bishop Chance was remembered by Flynt Chandler:

Buford was quite a guy. Maurine was late coming to work one morning and they lived the second house up there where the employment office is now. Buford came out there and said, "Where is Maurine?" I said, "I haven't seen her this morning!" I was up there getting anchor orders ready to ship. Out that door he went and went over there. (1979)

Maurine had overslept. Apparently, Buford was prepared to wake her up but her mother thought she should do that.

Flynt also recalled some of Buford's other activities and interests:

Well, he was kinda overseer. He didn't take too much interest in it like Gano would but he was generally there every day. But he was usually messing with an automobile or something like that. One time he bought an old Chalmer roadster and I bet I spent a month working on it, painting it and polishing it and getting it ready for him to drive. (1979)

The comment that Flynt made concerning Buford's more-relaxed approach to the business is revealing although it's difficult to determine if this was reflective of his more laid-back style and nature or he was just less interested in the Company than his four-year-younger brother. Of course, the Depression had a way of making people more serious, but any such affects would appear to have been potentially greater for Buford than an already serious Gano. Good times also have a way of spoiling us and this was a time (mid/late 1920s) when the Chance ship had docked, and from a financial point of view the family was doing well.

Like many longtime Chance employees, there was a natural inclination in knowing more about the Chance family. After hearing various stories and remarks about family members such as Buford, an image would develop of that person. One image of Buford over the years was that of a glamorous Hollywood-leading-man type who drove fast-and-flashy sport cars. Some of the things learned about Buford support that, while other information suggests that he was also made in the mold of his fascinating father.

Being described as one with similar traits and a likeness to his father, the idea of Buford carrying forth his father's first name (Albert) and initials (A. B. C.) is appropriate, especially since he was the eldest son. One veteran employee, David Green, suggested that the "B" in Albert Bishop's name and in his first son's Albert Buford name, weren't the same. Since Buford and his father had the same first names, in such circumstances it is common to refer to that individual by their middle name, as was Buford. It also became obvious that Frances and Francis connected mother and son, although Francis was also a Chance family name used in at least two other instances. The use of Gano as a middle name was also derived and sourced from his mother's (Mrs. Frances I. [Gano] Chance) maiden name. Isn't it interesting that Francis Gano Chance was also known and addressed as Gano, his middle name, thus each son representing in name, both their mother and father's families. Occasionally the suggestion is made that one son or daughter is the father's child and another the mother's. What one could conclude with a reasonable level of certainty is that the mother of Buford and Gano Chance was a lady of character who felt strongly about family roots as in noting some of her son Gano's comments about his famous ancestors, confirms she did pass on her family history to her children.

When talk turned to these earlier times, one repeated comment was that the elder son Buford was destined to be the heir to his father in directing the Company. This may have resulted in a different Chance Company. The death of Buford not only affected Gano, but also his father, especially in his relationship

Four generations of the Chance family are represented in this November 1925 photo. John is in the center holding "Little" A. B., Buford's son. A. Bishop is on the left and Buford on the right. This picture was taken at John's home, west of downtown near the Columbia Branch railroad track. Notice that Buford appears to be about A. Bishop's height of six-foot two-inches. (Jack Chance Collection)

with Gano and how he viewed the Chance Company and the involvement of Gano in it. From Gano's perspective, his brother's death did affect the destiny of Chance Company and how his father would handle the transfer of it over to him. In fact, Gano in talking about his brother Buford said:

> *He had a good personality and was very much like my father, but he never got an opportunity to run the company or to do very much with the company. That may be the reason why my father turned everything over to me. . . . He relied first upon his secretary and both of them had been working hard together and when my brother got started as a side party, he wasn't able or did not get into the the company. . . . My brother had been with the company about 3 or 4 years and he left the University of Missouri and went to work for the company. But my father never did really turn too much over to him. I think it was also with respect to parties, my father's secretary who thought that nobody could run the company but she and Bishop Chance. (1978)*

Buford was involved in the auto area, in the Taylor anchor, other anchor promotions, and in the overall company, but Gano didn't give much credence to that. There are strong feelings pouring out of a seemingly resentful nature when Gano discusses this subject. With the same strength Gano points to the transfer of the company to him when it appears to have been a more gradual process with some of the same problems pointed out that Buford faced. These strong feelings could be related to the earlier workings within Chance Company, but it could be a response to the seemingly unfairness of life and the disappointment of losing a brother. Although we somehow continue to get on with our lives, there are some losses that hover not far from the surface for the rest of our days.

Although Buford's image is somewhat tarnished by what has been revealed, he didn't have the opportunity to be fully tested and to develop beyond his early age. And too, we don't know that Buford didn't do a commendable job in the company during the cruise around the world his parents and brother took. Neither do we have much information on Buford's contribution to the Chance Company during the eight or so years he worked. It's obvious that Buford was the forerunner who made it easier for Gano to later take over the company. Possibly the situation, in which one stayed behind and one went on the cruise, displays the differences in these two young men with a high degree of clarity. Of course, it could simply be explained as a four-year difference in age and the station in life each had reached, especially in a matrimonial way. This does reflect Gano's thirst and interests beyond the normal day-to-day routine and a certain closeness with his parents.

Understanding that Buford's death was related to an auto accident, it became a goal to develop an early history of auto accidents and road conditions in the Centralia area, especially on Highway No. 22. Another purpose was to examine how road conditions around Centralia impacted Chance Company. Numerous accidents, some of which have already been recorded from 1910 forward, when the automobile became more commonplace, indicates that autos, like runaway horses, were not without serious safety concerns. There were no padded dashes, seat belts, or culvert and bridge guard railings, and inexperienced operators who still said "whoa" instead of stepping on their one-axle mechanical brakes. Auto registration was exploding according to a news item in the September 1913 *Guard*, as there were now 35,987 licensed vehicles in Missouri. As roads and vehicles improved, speed increased and turnovers and more serious accidents occurred. In July 1915 a turnover was recorded and then another in September. On November 19, 1915, the *Guard* captured another turnover in which two people were seriously hurt. A year later in September 1916 another vehicle turned over. A May 1919 turnover near Clark left Miss Ginger of Clark dead. In September 1919 two of Centralia's young men were killed south of town when their necks were broken in an auto accident. These early vehicles were very "SUV" in handling characteristics, being tall and narrow. Centralia, previously faced with livestock and chickens turned loose upon local residents, now faced a problem with young people being turned loose with the keys to the family auto. Speeding autos had become a problem in the city and additional complaints even came from local farmers, according to the *Guard* of August 20, 1920. Head-on vehicle collisions also became more serious as roads and vehicles improved and speeds increased. In September 1922 a Mrs. Glenn Hulen died from an auto accident. In 1925 there was one episode involving a smaller Model T Ford and a bigger more powerful auto, reported to the local editor by the Ford operator. As the big auto bore down upon them at high speed on the narrow road, the other driver appeared to be more involved with the lady close beside him than in avoiding an accident. Miraculously, they avoided contact in passing, but the people in the Ford were stunned which brought a strong response from the *Guard* in criticizing such calloused behavior. The auto and its operators were still infants in development of safe and responsible operating procedures and had many growing pains to overcome. However, this problem hasn't been entirely conquered eighty-plus years later!

Just as injuries from Chance Plant accidents grew, auto-related injuries also increased in

Buford Chance in his big roadster in the 1928 period. (A. B. "Buff" Chance IV Collection)

frequency and seriousness. The North Highway or No. 22 as it became known, was little improved into 1928. Lack of attention even involved Theodore Gary who had ignored its importance, per a December 12, 1926 *Guard* article. The need for a highway to connect up with Louisiana (Illinois connection following the Alton Railroad route) and across eastern Missouri connecting No. 1 (apparently No. 61) and No. 7 (No. 63), which would give direct connections with the across the state U.S. 40, was pointed out to Mr. Gary by his close friend E. O. Turner. This situation became even more important when considering the initiation of construction on a new Mississippi River bridge at Louisiana, on Highway No. 22. Mr. Turner had a vested interest in Centralia because he had purchased the Centralia Home Telephone system and credit goes to him in pointing out the gaps between Highway's 15 from Paris and No. 7 from Columbia to Moberly, leaving Centralia miles from hard-surfaced roads. Mr. Gary agreed and would consider it to be hard-surfaced at the earliest date. The decision to concrete No. 22 from Mexico to connect up with the Illinois system, had already been sanctioned. The Illinois stretch up to U.S. 36 was paved apparently at or before the time when the new Mississippi River bridge was completed. Local road conditions were slowly improving. Highway No. 7 (No. 63) from Columbia to Clark was being paved in July of 1926. This generally meant one side only was paved, for cutting costs and expenditures of monies that the state didn't have. Mr. Gary suggested that when people could see what a difference this made, they would favor state taxes to support a modern highway system. In early 1927, Mayor Bruton commented about the roads to Sedalia, comments the *Guard* said were unprintable. So it was an adventure to travel the area roads at the time when the A. B. Chances were taking a world cruise. This shows the advent of the automobile and the means of utilizing it was moving more slowly than one might think in relation to established methods of travel that were becoming much more sophisticated. The production of the 15 millionth Model T Ford in 1927, a practical and uncomplicated vehicle, suggests that its success and popularity were due to its ability to operate well on unimproved, dirt roads. Buford Chance apparently was driving a larger, more powerful type of automobile that in some ways was out of its element on the narrow, rough roads in north-central Missouri, but it could be more comfortably operated and attain higher speeds than say the Tin Lizzie. This roadster, sports-car-type vehicle suited the taste of young men like Buford, especially since he could afford its luxury.

Highway No. 22 began to get attention that suggests Mr. Turner's concerns voiced to Mr. Gary in late 1926, were beginning to bear fruit. Crews in mid-July 1927 were now surveying west of Mexico to east of Centralia, around the Boone/Audrain County line. They were designing a loop north around Thompson to clear buildings near the Alton tracks. While back in Centralia even city streets such as Allen were unimproved. An August news item told of A. Bishop bringing in three carloads of cinders for Allen Street in front of his factory buildings and warehouses, hoping to get them compacted before the heavy rains of fall began. Back out on the open road cars turning over continued to be reported. Neither were employees at Chance Company immune from such accidents outside the plant as Donald Wood, one of the Chance Plant superintendents, demolished his car two miles east of Mexico in mid-August. A week later, the Leonard Palmer's car overturned. One reason for the many accidents was rut-sewn, narrow roads with no shoulders. A mid-September report in the paper told of No. 22 being widened from Centralia to the Clark junction, and shortened between Clark and Sturgeon. The holiday season initiated by Thanksgiving and the increased travels generated more auto accidents such as a report on November 25 of a man being killed near Sturgeon on No. 63 as the result of a car turnover. As December 1927 introduced itself, another item revealed its much celebrated and anxiously awaited arrival, the new Ford car. Initial production and extraordinary demand was such that each Ford dealer would only get one. The momentum of the automobile and its influence on America was becoming even more entrenched, and road improvements were becoming a critical and necessary need.

As the latter stages of spring 1928 is entered, accidents continued for Chance employees both outside the plant and on Highway No. 22. A faulty steering gear was blamed by Clarence Douglas and his brother for their car turning over near Thompson. Although injured, they were able to quickly return to work at the Chance plant. In that same May 4 paper, it was revealed that No. 22 was still dirt from Centralia west. A week later, No. 22 was being graded and ditches cut in. Its condition was described as dusty and fair, east of Centralia. Americans have seemingly always been fascinated by speed. One of the causes for the *Titantic* disaster was to have the bragging rights, plus the commercial implications, of speed. It would be a while before America would challenge British ocean liner speed records, but railroad locomotives with recorded speeds of over 100 miles per hour established pride in the U.S. engineering capabilities. The British in early May 1928 also held the world's automotive speed record, but that changed when Ray Keech set a new record of 207 miles per hour at Daytona Beach, Florida. The coverage of this in newspapers across America inspired an even greater attraction toward speed.

By mid-May, highway officials were attempting to get additional land along No. 22 so it could be widened west of Mexico into Centralia. The only significant curve would be at Thompson to clear structures along the Alton track. Improvements apparently hinged upon this as it was suggested Centralia was in mud seven months out of the year. Highway No. 22 was said to be a graded dirt road for twenty-eight to thirty miles between Mexico and Highway No. 63. Would improvements come soon enough to help avert more accidents and would its users use caution in driving it? Another report of a car turnover resulting from a locked steering gear graced a news column in the May 18 paper. Even concreting didn't solve all the problems as another turnover on Highway No. 40 resulted when a front wheel dropped off the pavement. This kind of problem would be solved when a side curbing was included in the concrete slabs of many highways during this period and into the 1930s. However, it was found to generate another problem of shooting the vehicle back across the road into the path of approaching vehicles. Meanwhile, the May 25, 1928 newspaper told of the new Champ Clark Mississippi River bridge at Louisiana, Missouri (Highway No. 22), unofficially being opened to traffic for a couple of weeks. As a traveler of this route crossed this bridge and headed west, they were in for some unexpected, less sophisticated progress as they entered the heart of North Central Missouri.

As mid-July of 1928 came to the farms of rural Missouri, the row crops were all in and haying season was in full swing. Dryer weather during this time may have also been a reason for talk concerning the oiling of Highway No. 22. Late that month, another No. 22 accident occurs on one of its narrow bridges. Another overturned car killed a young lady from Oklahoma traveling to Columbia. In June on Missouri's hard roads, sixty-three people were killed and 820 accidents reported! By mid-September the oiling of No. 22 for thirteen miles over to Highway No. 63 was completed. The September 21 paper reported the bid for bridging and grading of the eleven-plus miles from Mexico to the Boone County line was let, going to the Gidinsky concern of Kansas City. The work began immediately west of Mexico and those fills would likely take most of the fall and early winter months before it could proceed west. The completion of the project would end with the big bridge over Young's Creek east of Centralia being anticipated next year. The improvements on west wouldn't occur, it was thought, until

1933. The contract let had nothing to do with the graveling or hard surfacing of No. 22 from Mexico to the Boone County line but was preparing for such future eventualities. So as the fall of 1928 is entered, Highway No. 22 is much as it has been for decades, as to its dirt surface and narrow roadway. It would also appear that the bypass or "S" curve around Thompson's business district near the railroad tracks had not yet been completed. This unimproved situation is definitely an inconvenience in trying to travel during wet weather or at higher speeds. The safety aspect of such roads is a debatable issue. Even today it is suggested that current roads are inadequate and must be improved but in comparison to 1928, there is very little in common. There was a personal responsibility for drivers to adjust to conditions even in that day.

Noting other accidents in the Thompson area, in mid-September 1928, a Robert Rippeto fell off the rear seat of an open car a mile west of Thompson. He suffered head and other injuries. In a way this is a preview of future events as the Thanksgiving holiday of 1928 approached. One obvious problem with a weekly newspaper is much of the coverage in its Friday's edition is old news. The *Guard* news items of November 29, 1928, were no different, especially an article of a fatality auto accident occurring about a half-mile west of Thompson, the previous Sunday night. Reports over the airwaves or on the printed page, however, do bring finality to such occurrences. It can be suggested that the death of a young person from disease, accident, or war are little different, regardless of whom they are, as they forever remain young. There is a sorrow in that alone. For reasons that one would assume were influenced by his father A. B. Chance, Buford Chance was also known as A. B. Chance Jr. This explains why Buford's son, Little A. B., would also sometimes be known as A. B. Chance III and his only son Albert Buford "Buff" Chance would also be know as A. B. Chance IV. The obituary of Buford in the November 29 paper did use his full name, Albert Buford Chance Jr. There were different stories of why Buford was coming back alone late on a Sunday night from Mexico.

Even the coroner, S. C. Adams of Mexico, came to Centralia the next day, Monday, and held an inquest to determine the cause of Buford's death per the *Guard* article. A deposition by Ralph Ferguson stated he and his wife and baby were driving back from Centralia to their home in Thompson in their Ford truck when he saw a car with bright lights coming toward them at high speed, he estimated at fifty miles per hour. The bright lights may have been the result of Buford having his running board spotlight on. The disposition continued as Ferguson gave road until his truck was almost in the ditch, but the westbound car gave no road and struck his truck knocking the rear end into the ditch. Buford's car continued about 130 feet and overturned. The Ferguson's had numerous bruises, but their baby wasn't injured. The death was attributed to the running together of the two vehicles. The body of Mr. Chance was found by R. H. Ferguson and Clinton Gibler and killed where found. The body was moved to Centralia from instructions of the father of the deceased. Mr. Ferguson wasn't held responsible for the accident. The surprising information about the accident was contrary to one local version of the story that surrounded it, although one tends to fabricate a story from bits and pieces heard over the years. The coroner's report destroys the myth of it being a one-vehicle accident, while crossing the railroad tracks. More details of Buford's accident were included in the news item of the auto crash in the November 20 *Guard*. The events of the accident are primarily from Mr. Ferguson's point of view and of course not Buford's. The sideswiping of the two vehicles tore a wheel off of Ferguson's "small Ford truck" and the "Chance car veered back into the roadway, ran across to the south where it struck the rise of dirt in the Alton right-of-way and turned completely over with the car upside down and the front end turned back toward the east." If

Buford's car "veered back into the roadway," this must indicate that he was giving road and was in the north side of the road partially off the roadway in the ditch? This is in contrast to the official report and it's not clear which is correct, possibly a degree of both.

The details of this 1928 accident depict a time much different than today's No. 911 emergency call convenience and immediate response situation. Buford's car came to rest upon the Alton embankment along the track and this appears to be the source of the railroad connection. A turnover in an open car is a dangerous proposition. Buford was pinned underneath his car by the spotlight and his injuries were of a fatal nature. The Fergusons went to the home of the Clinton Gibler family just west of Thompson and Mr. Gibler and Mr. Ferguson returned to the wreck scene to see if they could help Buford. Unable to turn the car over, Mr. Ferguson slumped in a shocked condition and they returned to the Gibler home and telephoned Centralia. The three Bonney brothers drove up to the accident shortly after the crash and found Buford under the car. They couldn't lift the car off Buford either, but knowing him, they telephoned his father, A. Bishop. Mr. A. B. Chance, Joe E. Green, Dr. O. B. Mayes, Mrs. A. B. Chance, and Mrs. A. B. Chance Jr. came to the scene of the accident and found Buford had passed away. They wrapped a blanket around the body and returned to Centralia with it. Other witnesses said Buford was driving fast that night, but this was normal for him. He began driving at an early age. One wheel on Buford's roadster was destroyed, likely from the initial collision that led to Buford losing control of his vehicle.

Albert Buford Chance was born on July 31, 1901. Graduating from Centralia High School, he entered M.U. for two years (1920–1922) leaving to take up a position in his father's factory. Married to Miss Bertha Pemberton on April 15, 1923, one child A. B. III was born of that union. His funeral was held at the First Christian Church in Centralia. The pallbearers were Harold Mayes, Jack Isbell, W. B. Atchinson, H. Jones, Joe E. Green, and Harry Jennings. The honorary pallbearers were Harry S. Rowland, David P. Hulen, Roy C. Chamberlain, Flynt Chandler, Ollie Chandler, Clifton Hulen, Ralph Tucker, and Ralph Elkin. These were obviously representative of family, friends, and fellow Chance employees.

It was life as usual, one might conclude, as there were no warnings as to what would occur although as pointed out, danger was present, but then no one is free from occurrences of a injurious nature. Buford's future looked bright as he had worked himself into the Chance Company, although it is not entirely clear what that totally encompassed. It is relatively safe to suggest that the pre-Buford A. Bishop Chance was not as inclined to share the company as the post-Buford father did with the remaining son, F. Gano. The dual blows of Buford's death and the arrival of the Great Depression heavily impacted A. Bishop Chance, but accumulated information about health issues appear to also be a key factor in his decision to become less involved in his company.

After his father's death, Mr. and Mrs. A. Bishop Chance provided a permanent or part-time home for their grandson Little A. B. or A. B. Chance III, as he was publicly known. Because of questions about his childhood and his mother, news items about Buford's son and widow were noted. It does suggest that the mother of Little A. B. (a name that apparently was given him as a child, but stuck into adulthood) appears to have been absent most of the time, but routinely visited in the years after Buford's death. There was a touching story that Flynt Chandler told when he lived behind the big Chance house before the rock garden was begun in the mid-1930s. Although there are no dates recorded with this story, it would have been in the 1929–35 time frame when Little A. B. was four or more years old. "My wife used to take quite a bit of care of little A. B. He would come over

Mrs. "Fancy" Chance with grandson, "Little" A. B. III. (Jack Chance Collection)

there and eat with us and drink his milk and she (Mrs. Chance) couldn't understand why he would drink his milk over there when he would not at home. He was just at home over there with us." (1979)

One positive way to deal with the loss of a loved one is to associate with friends and family, as Mrs. Buford Chance did in early January 1929. Mrs. E. P. McBride of Springfield, Illinois, and Miss Ross Rosenthal of Columbia were Sunday guests of Mrs. Chance who was apparently still residing in her Centralia home. The February 1, 1929 *Guard* noted she was now residing in Columbia and she was attending the "school in the Athens," which was an unfamiliar institution. This may have been referring to the Greek columns of the University of Missouri, which agrees with later confirmation she attended the University as well as Christian College where her father had taught. Following her February visit, Mrs. Buford Chance visited her parents, the M. H. Pembertons and the A. B. Chances on Sunday in early March. A couple of weeks later her situation was explained in the paper as attending school in Columbia, and also holding a position in the offices of the Chance Manufacturing plant. It's interesting she was addressed as Mrs. A. B. Chance Jr., a name possibly influenced by her Chance Company association. The May 17 *Guard* suggests Mrs. A. B. Chance Jr. was still residing in Columbia and was Sunday guests of the Mr. and Mrs. A. B. Chance. A November 15, 1929 news item told of Mrs. Bertha Chance, Bernard Pemberton, and Laura Mae Brown, all of Columbia, spending a Sunday with home folks in Centralia. It is not known if all of her visits were recorded or found but in mid-January 1930 Mrs. Bertha Chance of Columbia made a visit to Centralia over the weekend. The next visit found was in late May, to her parents and the A. B. Chances. In late June 1930, Mrs. Bertha Chance held a bridge party in the A. B. Chance home in preparation for Gano and Anna Lee's marriage.

In a more dramatic move in early August 1930, Mrs. Bertha Chance left on a Saturday for Chicago where she had accepted a position with the Title Trust Company of Chicago. This Chicago move ushers in a new chapter in her life. A month later in September, she was home on a Monday and her and Mrs. Gano Chance and others motored to Columbia. The *Guard* of February 27, 1931, recorded another visit of Mrs. Bertha Chance of Chicago over a Sunday, with her parents and the A. B. Chances. This would suggest she still was single and was maintaining relations with her in-laws and parents. In early May another visit was noticed, like one in late February 1932, when Mrs. Bertha Chance of Chicago visited her parents and the A. B. Chances. So in spite of the progressing Depression, she apparently was still employed. It wasn't until December 1932 that anything was found of her son, A. B. Chance III, who was a dinner guest of his great-grandparents the J. A. Chances. In the May 5, 1933 *Guard*, Mrs. Bertha Chance of Chicago, accompanied by Miss Betty Merritt, made a weekend visit to her parents and the A. B. Chances. This visit may have included a discussion about future plans. The official announcement of her marriage to Allan Price Born took place a month later on June 3, 1933, per the June 9, 1933 *Guard*. Her brother, Morton Pemberton Jr. was one of few in attendance, said to include only intimate friends. Mrs. Born was a graduate of Centralia High School and also of Christian College and the University of Missouri of Columbia. It would appear she attended Christian College before her marriage to Buford, and the University of Missouri in the 1929–30 period. This helps explain earlier questions raised about her educational activities shortly after Buford's death although it's not clear if this means she had degrees from both colleges. Since August 1930 she had held a position with the Chicago Title and Trust Company in Chicago and Mr. Born was in the real estate business in Chicago. This was the city to be in as two days prior on June 1, 1933, the Century of Progress Exposition opened its two-year run in the Windy City.

The Chicago Title and Trust Company was busy during this time taking care of their financial interests and investments. The turmoil in economical issues likely resulted in their becoming the receiver of the Illinois Life Insurance Company. To oversee their real estate and loans more personnel were required such as Centralian Harry R. Jennings who accepted a position with the firm. Mr. Born's real-estate interests likely also brought him and the former Mrs. Bertha (Pemberton) Chance together. After the marriage, Mrs. Born and her son became closer beginning with a July 1933 visit by A. B. Chance III to Chicago. Where he was living between then and December of 1934 is unclear, but a Christmas dinner covered by the G*uard* of December 28, 1934, with guests Mr. and Mrs. Allan Born and son A. B. Chance III of Chicago suggests Little A.B. was residing with his mother and step-father in Chicago. In mid-July 1935 A. B. Chance III came to Centralia to stay the summer with both his grandparents, indicating he still had been residing in Chicago. His Uncle Bernard Pemberton, after two years at the University of Missouri, had also gotten a position with the Chicago Title and Trust Company in Chicago, likely with the help of his sister Mrs. Born. But he had some serious health problems and a fear of insanity that his illness might cause. As a result, without the medical advancements and society's negative views of certain illnesses in that day, suicides were an all too frequent solution. Per the September 6, 1935 *Guard*, Bernard Pemberton, twenty-three (1912–1935), was buried in Centralia. Pallbearers were Gano Chance, J. Boulware, Nathan Toalson, Art Houston, H. E. Stone Jr., and Jack Curtright. This wasn't a good example for his young nephew A. B. Chance III, but life isn't always kind. The M. H. Pembertons and Mrs. Born also had a second son and brother, M. H. Jr. In late December 1935, the Allan Borns and A. B. Chance III came to Centralia for Christmas holiday and visited with both the Pembertons and A. B. Chances.

It is interesting that Gano spoke of his mother's response to Buford's death, but not his father's. This could be the result of the idea that big men don't cry and show emotion. Private papers of A. Bishop Chance did reveal how Little A. B. was to be taken care of by inheriting his father's share of the Chance Company. This was not only thoughtful, fair, and showed compassion toward Buford, but it also shows the strong sense of family that A. Bishop held. In his private papers of late 1930 he wrote:

> *So far as I know, I am in good health—except, recently my heart is "skipping" again but the doctor says it should give me no anxiety. I must not forget, however, regardless of my feeling of my past activity, that over a few years will step me up to 60 years and I am further mindful that we are some times snapped away, as was Buford, and it is only good judgment for one to get*

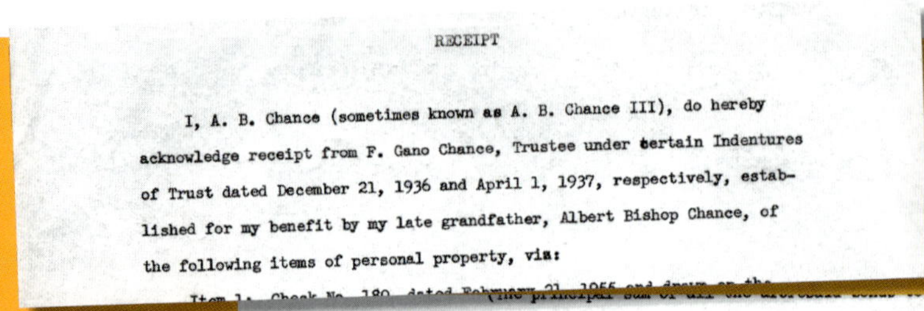
RECEIPT

I, A. B. Chance (sometimes known as A. B. Chance III), do hereby acknowledge receipt from F. Gano Chance, Trustee under certain Indentures of Trust dated December 21, 1936 and April 1, 1937, respectively, established for my benefit by my late grandfather, Albert Bishop Chance, of the following items of personal property, viz:

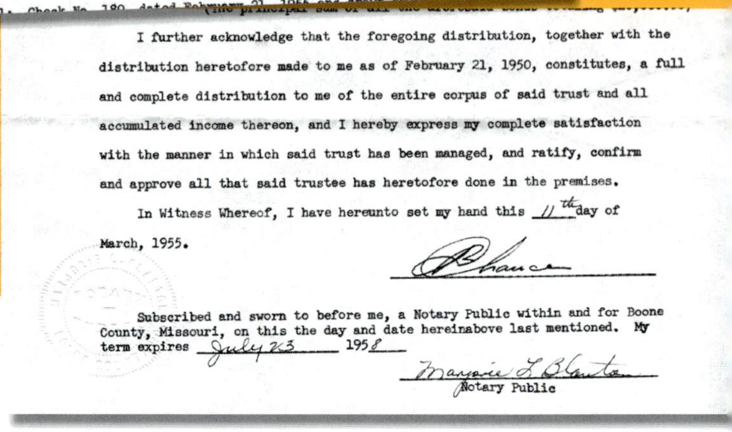
I further acknowledge that the foregoing distribution, together with the distribution heretofore made to me as of February 21, 1950, constitutes, a full and complete distribution to me of the entire corpus of said trust and all accumulated income thereon, and I hereby express my complete satisfaction with the manner in which said trust has been managed, and ratify, confirm and approve all that said trustee has heretofore done in the premises.

In Witness Whereof, I have hereunto set my hand this 11th day of March, 1955.

Subscribed and sworn to before me, a Notary Public within and for Boone County, Missouri, on this the day and date hereinabove last mentioned. My term expires July 23 1958

Notary Public

This document confirms Gano's completion of his father's wishes concerning the distribution of his brother's share of the A. Bishop Chance estate. Only the opening and closing paragraphs from this seven-item categorized legal agreement of cash, bond, stock, and note transfers, finalizing those requirements, is reproduced here. This apparently was an ongoing process from 1947 to this current "Receipt" document dated March 11, 1955. It was common practice to have one or more employees, normally secretaries, who had a notary public license. Marjorie L. Blanton was C. C. Hough's longtime secretary. The fleet area required such services frequently and this employee recalls being assisted by Billie Huffman, Carolyn Cox, Verda Estes, and others through the years. (Chance Foundation Files)

and keep "their house in order" as nearly as they can. For we do not know when, how, what is to happen. With this thought in mind I am changing some previous plans and agreements I made with you. (Late 1930)

Within these papers were A. Bishop's personal handwritten instructions and agreement as to how his estate and holdings were to be distributed. They were revised, as instead of originating in the late-1930 time frame, they originated in 1932 or later. This can be established because they are written on the back of surplus Chance Company letters, dated May of 1932. Accordingly, this is at least the third such revision, assuming that these thoughts were written prior to the late 1930 material as A. Bishop suggests:

Set forth and under terms and conditions as follows: As it was understood and has been my intention to divided the value of my properties in three parts—one part for my older son Buford, one part for my younger son Gano, and one part for myself. As Buford has passed on, I give his portion to his son A. B. Chance III with the understanding and agreement that no profits be given nor losses charged against A. B. Chance III portion until after Feb. 20, 1946 when A. B. Chance III should reach his legal age. Prior to Feb. 20, 1946, A. B. Chance III portion is to be held and operated by me and full authority and right is given me to operate his interest, and to sell or otherwise dispose of his holdings if in my judgment it is of his best interest for me to do. All profits from A. B. Chance III portion up to Feb. 20, 1946 should be given me in payment of expense of caring for and looking after in all his needs until he has reached his legal age. After that date, all holdings belonging to him shall be delivered to him and his future profits and expenses be handled by him. In the event of my death, prior to Feb. 20, 1946, when A. B. Chance III will reach his legal age, your mother Frances I. Chance shall serve as guardian and care for all the needs of A. B. Chance III until he reaches his legal age. She shall be given all the profits earned on A. B. Chance III portion or interests during this period. The handling of A. B. Chance III share or portion during this period, that Mother is acting as A. B. Chance III guardian and caring for him, shall be with full power to handle and operate. I know you will guard and protect with same interest as if your own. (1932)

Chapter 6
Reorganization and Survival
(1931–1933)

A Chance of Rain
Tough Days Ahead

Chance employees, visiting customers, or those passing through Centralia, especially the inquisitive, may ask about the history of Chance Company. Pursuit of such interests could connect them back to a colorful past

Two of the most frequently repeated Chance Company stories are the sleet storm that resulted in A. Bishop's invention of the Never-Creep® Anchor, and his son Gano's heroics in saving the company. As the son's story goes, the businesses of America were caught up in the grips of the Great Depression and a young F. Gano Chance traveled solo to Columbia in an attempt to get funds from bankers to keep Chance Company operating. Some have expanded this to say the father just turned the Company over to his only surviving son that day and said, "Here it is—if you want it to survive, then you will have to get the finances to keep it afloat!" This generated for Gano an image of stardom—"Savior of the Company"—the sort of drama that makes for a good movie!

The transitional control from father to son of Chance Company holds a great deal of intrigue. In Chapter Four, the *Forbes Magazine* article of March 1, 1931, about Chance Company and A. Bishop Chance, makes no specific reference to his son Gano. The interview occurred before the situation got economically tense at 210 North Allen Street and the *Forbes* article portrays the founder in complete control. Gano appears to have become more involved in the day-to-day operations of the company after becoming a formal partner. According to *Chanco News* of January 1947, Chance Company was operated as a proprietorship until 1929 when it became a partnership of founder and son and then incorporated in December 1932. However, other information points to an operating partnership arrangement occurring later in the 1931–32 period. The founder would pop up unexpectedly at times just when it appeared that Gano was gaining control. This points to a gradual transition that is supported by A. Bishop retaining the presidency until 1939 when he became semiretired. He then became the C.E.O. and Gano became president.

Growth and new management brings changes within a business and most small operations grow from a one-person show to a team effort. Although Gano was sometimes credited with developing the modern management team at Chance, his father was also an effective delegator and selector of fine talent. Thus it could be concluded that the father had taught

This 210 N. Allen Chance factory office scene is from August 1931. It is valuable in documetning office equipment and decor, plus more importantly, office personnel. Confirming the treasurer of the company had a private office, the lady working at the desk in the left background is Miss Ida Lou Traughber. Next is Beverley Mayes, Verna (Tucker) Palmer, Maurine Grubbs, and standing is F. Gano Chance. The picture above Miss Ida Lou's door is the "Red Snapper Piston Ring" picture reproduced on page 240. (Jack Chance Collection)

his son well. One difficulty that Gano faced was dismantling portions of what his father had organized and had in place when he came on board in 1929. Even before this, Buford, the eldest son, would face the same situation, but would choose not to confront it head on, according to his younger brother Gano:

> *My brother was unable to get much responsibility because of my father's secretary. She had done it in the past and she wanted to continue. On the other hand, business was good and so my brother decided rather than to create a revolution, he would just ride it on through. That is what he did, of course, he was killed in an automobile accident in 1928. . . . I think it was also with respect to my father's secretary, who thought that nobody could run the company but she and Bishop Chance. She came with the company in its early stages and she made many decisions for the company when my father was gone, and they were good ones. She was good! (1978)*

The core problem Gano confronted was his father was still running the company and it didn't appear he had any immediate intentions of turning his firm over to either of his sons.

It wasn't going to be an easy trip through the next few years for a young man who had just graduated from the University of Missouri in 1929 with two degrees. He would be severely tested by not only an economic climate that had gone sour, but also by those outside and within Chance Company. After joining his father's firm, it did not appear as if all doors were opened and all obstacles removed for him. "When I got out of school and came with the company, I started making decisions. I had to take it away from parties and I took some of it away from my father. My father never gave me a title." (1978) Those days weren't easily forgotten as he was working his way into the company, but this kind of status was not abnormal for young employees coming on board full time. "I was young and just out of school and these people had been in the company a good many years. It was just human nature that they would resent the loss of authority. Miss Ida Lou Traughber still treated me like her 'little boy,' however, when I began to take a few things away from her she changed a little bit." (1978) One of the most famous father and son associations within a major firm was that of Henry and Edsel Ford, which made Gano's problems seem small in comparison, especially in the eventual outcome. Gano seemed impatient and resentful of established barriers, especially since a short time before he was preparing for a technical position within the firm and not anticipating heading it.

Although A. Bishop wasn't afraid of making decisions and did so abruptly at times, there was a need for someone to maintain continuity and stability in the day-to-day operations. Miss Ida Lou fulfilled that need. The resistance that she showed and the ongoing conflict that developed between her and Gano was a natural result. Gano recognized her longstanding financial function within the company by suggesting that:

> *She was the watchdog of the treasury. A lot of growth depended upon her to see if we had the money. She didn't want to let loose of that authority. . . . When I came I realized that there should be some changes made. I thought this is what we ought to do, but authority would have to be taken from people who didn't want to give it up. I realized that parties that had authority for so many years were going to object, but I tried to be diplomatic. I did spend some time presenting the problems to the Christian minister, who was a swell guy. It was pretty tough there for a while and the minister and I used to sit out in front of his house talking about how are we going to do that. When Jack (Isbell) came along, I got him involved in my problem. He was a good salesman and could handle people. (1978)*

Gano reveals a strong will in doing things his way.

The times leading up to the Depression were relatively short for those like F. Gano Chance and Bill Hollander who came to work full time at Chance Company in the early summer of 1929. Fortunately, there would be a delay before the hard times hit Chance Company, when Gano said the Company had reached its peak of about three-quarters of a million dollars in sales. His father's original anchor, the Never-Creep®, provided a significant proportion of those sales, according to Gano, but the patent was expiring. "So business looked very good, but there was a little talk in the fall of 1929 that we might have problems before long. I was in New York, calling on our best customer, AT&T, who operated the telephone lines for the Bell Telephone System, in September." (1978) Gano recalled that their 1930 business was better than it was in 1929 because the utilities had planned about six years ahead and they just couldn't shut it off. He said business was still good in early 1931, but in the fall it began to change. "In the fall of 1931, as I began to get into the business with my father, he turned the business over to me completely. I suggested that we get incorporated because I was afraid that if the Depression continued, the bank might want their money and they would wind up by taking over our home." (1981) The incorporation was completed in late 1932. A. Bishop had suggested this in personal papers of 1930 and appears to have taken an active role in the actual process.

Although the incorporation of Chance Company, from a co-partnership structure, might have made the Chances more comfortable and secure, it had the opposite affect on the lending institutions. Just as it had limited what the Chance family could lose, it likewise restricted the banks from being able to take their home or other personal property for collateral in case of bankruptcy. This made the Columbia banker, who had made a loan to the struggling Centralia firm very jittery, to the point he called in their loan or note. Gano said, "So they contacted Dad and wanted the money." (1978) Gano described the situation leading up to this. "He had borrowed all the money he could in the local banks and two of them went broke and he lost money." (1981) This shouldn't be construed to mean that Chance Company's financial problems were to blame for the bank failures. When the banks went under prior to F.D.I.C., the Chances lost their deposits, but their notes were sold at lower, discounted rates. This is a variation of "Murphy's Law" when you lose the good and reap the bad! Thus Chance Company had lost money to help cover financial obligations that had survived in the impatient and demanding hands of third parties, which made a serious situation even more difficult.

Longtime office employee Verna Palmer recalled the drama surrounding the difficult economic times that the Chance Company endured during the Great Depression. She placed a heavy emphasis on Miss Ida Lou in her handling of the day-to-day financial requirements:

> *Talking about finances makes me think about "Boss" Chance relying on Ida Lou Traughber to take care of the finances. When the bank closed here, Ida Lou had company money in some bank in Canada, unbeknown to "Boss." Wherever she wanted to put the money was all right. A few times, if it hadn't been for her foresight, the company wouldn't have had enough money to pay the salaries or anything. When the company was smaller she could keep her fingers on each department and she wrote all of the checks.*

To further confirm Miss Ida Lou's taking care of finances, Gano said:

> *My father told Ida Lou Traughber that when my bank account got down to put some money in it. I went through college that way. (1978) He didn't care for that line of business anyway. . . . I don't think he ever wanted to run the Company from the point of finances, engineering, etc. In the past, his secretary* [Miss Ida Lou] *told him, Well "Boss," you can't get this machine yet, we don't have the money. (1981)*

There are numerous reports of hard times in the 1930s when the cash reserve wasn't adequate to continue. On at least one occasion A. Bishop asked employees to not cash their checks until the following week, but one would suspect that this occurred many times. Former employee Mildred Knowles in 1998 said that Miss Ida Lou was reported to have held her salary checks for up to four months to help the company struggle through tough periods. Gano also described how his life was changed. "I was somewhat careful about what I asked for but I usually got what I asked for. In 1932, I looked through and found out how much money I was spending. I took a 20 percent cut and that was my salary. That was the first salary I had which was about $200 per month." (1978) The company's reduced performance as the Depression intensified their financial difficulties, was discussed by Gano:

> *The year 1932 was the first year we had lost money. Our sales went from $750,000 to $196,000, which was quite a drop. We weren't quite sure whether we could get money together for the month's payroll. My father had borrowed money on his life insurance, all he could borrow. (1981)*

Gano became concerned as, "I soon realized that if I was going to have an inheritance, I would have to help build it myself." (1968) His father's borrowing on his life insurance policy was an important, frequently mentioned part of this story. It shows A. Bishop's resourcefulness and personal sacrifice toward the company and employees, as the raising of company operating capital was based on their personal as well as company assets. The insurance loan shows the desperation and reflects the harshness and powerful impact of the quickly moving Great Depression. When it hit, it did so with a vengeance and peoples' lives and businesses were shocked as if a tornado had passed through town and engulfed them, revealing the fragility of their lives and livelihoods. Being a businessman either small or large, poor or wealthy, made no difference as it became a test of stamina, faith, and resourcefulness in weathering the adverse conditions. The people living through it would never be the same as they were changed forever.

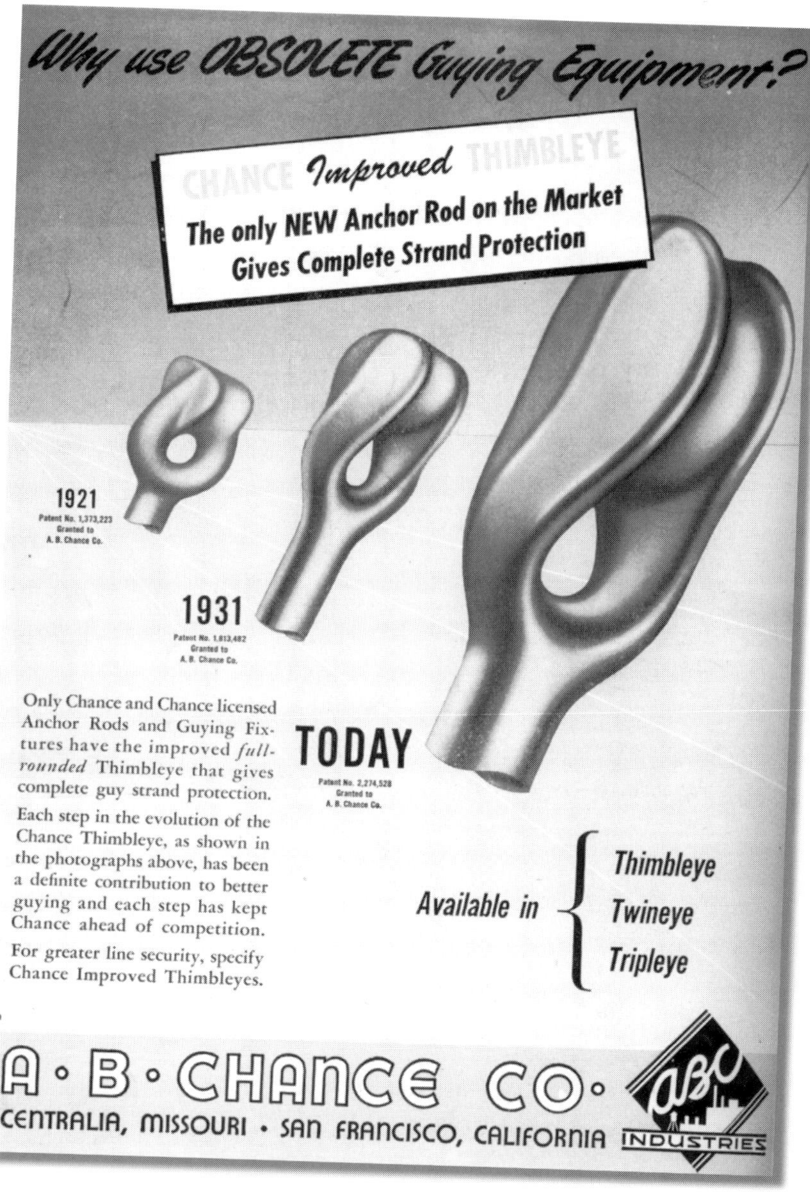

From *Chance Tips* Fall 1949. Obsolete Equipment?

Gano made a very acute observation amongst all the turmoil of this subject. He remarked, that no one seemed to notice when his father made him a partner in the business a year earlier, but when they incorporated and he wound up with about 40 percent of the stock, this opened some eyes, especially in the financial community. Apparently the Centralia loans were of smaller amounts and were dealt with in some fashion, but a certain Columbia banker was yet to be paid. The Chances didn't have the money and didn't know where they could get it. Then a surprising thing occurred; the father sent the son. Gano said, "My father sent me over to the county seat to see this bank which had loaned him money, $20,000." (1981) This is reflective of the buck deer's instinctive behavior of hanging back in the rear of his herd in a position that offered a superior vantage point. This decision reflects a hefty measure of wisdom, understanding of human psychology, possibly some self-frustration in tiring of the constant challenges, a desire to share more management responsibility, and confidence and faith in his son. It also shows the brashness of a young F. Gano Chance who didn't yet realize that he had been given an impossible mission. But this may have been a calculated and carefully organized plan that reflects neither father nor son had given up on the company. How did A. Bishop react that day awaiting word? Did he walk the floor, work in his shop, or did he have a desire to associate with and be around his friends?

As the story continued to unfold Gano said:

> *So I was still only 27 years old and my father sent me over to see those bankers who wanted the money. We did not have any money and did not know any place we could borrow it. I think my father was smart. He says, "here is a young boy full of vinegar who doesn't know that you can't borrow money. Let's let him go over there and see what he can do." (1978)*

The word *there* refers to Columbia, which generates even more tension as Centralia and Columbia were often at odds with one another. Gano spoke of that, years later, and his secretary from 1963 to 1974, Marjory Martin. "My secretary coming from Columbia to Centralia. Columbia and Centralia never did get along. It is very easy—I never thought about it so much in her case, but it was just natural." (1981) Gano was responding to Marjory's remarks about how difficult it was for her to adjust and the fact that he had actually hired someone, who worked very closely with him, from Columbia. He had not seriously viewed their relationship from that angle and in considering it, he was somewhat surprised he had not. That reinforces the idea that Gano was very objective in analyzing various things, including employees. Marjory later suggested that, "A different environment than I was used to, a little different environment. The first six months, I wasn't happy." (1981) The view fostered here was that Columbia was a city and Centralia a town, especially in the eyes of Columbians. This sparked some dissension at times, but there were also positive relationships over the years as being further established by a young Gano.

On another public occasion, Gano described the story and the challenges he faced when the note came due.

> *We had a $10,000 note coming due at one bank in a nearby town, and didn't have the money to pay it. My job was to solve the problem. Dad didn't tell me how to do it, didn't introduce me to the bankers. He just told me to get on the train and get some money. It was in the very worst part of the Depression and nobody was lending money to little companies that were already deep in debt. I didn't know that, so I just went to another bank and asked for a loan. I got it and paid off the first bank. Then I went to a third bank and borrowed enough money to meet the payroll. We did a lot of borrowing in those days, often against family insurance policies. We have always been strong believers in the value of insurance, and our Company was one of the first to provide group insurance for employees—in 1933. (1968)*

In a later interview, Gano discussed his negotiations with the Columbia bankers and the results in more detail. "Well, I went over there and introduced myself. I didn't know them. I first borrowed the money from the president of the Exchange National Bank." During Gano's sales pitch to Mr. Thurston, "He was very much interested and called in the president. Mr. Thurston was the vice president at that time and Mr. Ohling was the president and I told them we had incorporated and we would like to borrow some money." (1978) Gano said he had figures of what the company had been doing and what they were going to do. He told them that the company was a new corporation and it was going to go. Gano admitted that he really did not appreciate the difficulties his father realized. He was acknowledging the many such missions that his father had to endure.

There was electricity in the air in Centralia as a local magician had wowed Columbia bankers with optimism and raw figures. Another day was passing, but not before some celebrating that evening, as word spread and the rumors began to spill forth of what had occurred earlier in the day. How did the son break the news to his father on his return? Did he greet him like a son with all A's on his report card? Strangely this white knuckled affair had taken place in a city to Centralia's south that over the years had a local reputation of ignoring their upstart friends to the north. However, the closing of this day signaled more than the changing directions of a Depression originated storm system, as it was a new beginning for Chance Company. In some ways the old company had died that day as a crossroads had been reached, but from another perspective it had simply escaped another crisis. There would continue to be challenges facing a young F. Gano Chance and numerous trips to the banks by the Chances. But the father was watching from the sidelines, intervening when necessary as any good parent, coach, or manager would. The results reveal that the company wasn't the only thing being perfected, which stands in stark contrast to the idea of a generation gap and parents and their children not getting along, let alone excelling in their relationships. Maybe this is the most enduring part of this story, as it is applicable to any age.

The Kearney Association

High Expectations—Zero Returns

Confirming hard times had come to Chance Company, the August 7, 1931 *Guard* captured such evidence at the regular Centralia Board of Aldermen's meeting. Both A. Bishop and F. Gano Chance were in attendance. They voiced a complaint about their company's monthly one-hundred-thousand-gallon water bill being calculated from three meters at a higher rate structure than it should be if properly billed from one meter. Gano suggested there should be a special rate for factories as an encouragement for them locating there. The mayor said a lower rate did apply after three hundred thousand gallons. That being a dead end, A. Bishop then got on a common theme of excessive electrical rates, to which an official responded that Chance Company was getting it at near the city's cost from Missouri Power and Light. He stated it was still too high and his company had to "save the pennies." Using a St. Louis–based firm as an example, he said it was hard to compete as shipping costs of getting raw material from St. Louis to Centralia and back, put them at a disadvantage. This he said, was cutting into his profits and he wanted to cut all costs as close as he could in order to meet this competitor on the market.

The September city meeting announced the Chance Company water issue represented an overcharge and actions were taken to correct it. The August comments by the Chances indicate a concern in identifying and improving those areas of overhead that affected their company's bottom line and their ability to remain competitive. There was an increasing awareness by the Chances of an unnamed competitor in St. Louis, in addition to the pressures of the Depression. One growing threat to Chance business in the utility market would soon be identified as the St. Louis–based Kearney Corporation! These were tense times at 210 North Allen Street and unlike challenges from the past maybe A. Bishop wasn't able to confront them in his normal enthusiastic manner. His health, after the recent heat stroke, was now in question and it might have dictated cooperation instead of confrontation.

With the debut of the fall of 1931 came good news that A. Bishop surely enjoyed; Missouri Power and Light had announced a rate decrease from nine cents per kilowatt-hour to eight cents. While some things moved forward there was much that was moving backwards such as unemployment. The Wabash Railroad cut more employees off in the face of declining revenues and more bleakness was encountered when four more area banks failed. During that time, there were a lot of negotiations and activities occurring over at Chance Company, although it was not being publicly disclosed. A sign of this was a several-days visit by F. Gano Chance to St. Louis in the October 16 *Guard*. With what would become one of the Chances most controversial decisions, the announcement of the consolidation with the James R. Kearney Corporation of St. Louis, was to take effect on November 15, 1931. In retrospect, this was what appears to be a very drastic and trusting move toward a firm that had become an arch competitor. However, the double challenges of declining economic conditions and competition from Kearney apparently made it difficult

This Model A vintage 1928–31 Ford truck has the period company lettering with a new company phrase of "All Good Specialties." This pictures a load of painted expanding anchors in front of the new Punch Press building featuring employees from left to right: Silas Roberts, Flynt Chandler, and Ollie Chandler. This photo was originally a Chandler family keepsake. (A. B. "Buff" Chance IV Collection)

to resist. The Utility Division of Chance Company had previously distanced itself from such entanglements that had involved the Big C auto line.

An October 23, 1931 *Guard* article entitled "A Big Consolidation" explained the details of the new Chance/Kearney relationship that was to go into effect on November 15, 1931. The removal of competition in the guying specialties and anchor line by this event was considered an important move, especially in consideration of the prevailing economic climate. A. Bishop was quoted on October 20 as indicating this consolidation would increase Chance factory activities and a call for more employees. It was said the plant was one of the few that had working: "a full force and full hours in the west." Such optimistic reports of what this new relationship would grow into were a ray of sunlight in those increasingly darker economic times.

Since Kearney was said to have little or no industrial plant and outsourced their manufacturing requirements, the strong manufacturing capabilities of Chance Company was anticipated to be a benefit for both. Likewise the substantial Kearney sales organization of twenty or more sales representatives, with the addition of six from Chance would make for a union made in heaven! However, in a short time this association would reflect that it did not represent the purity of any such origin. The *Guard* article said: "The manufacturing end will be handled by the Chance Company and the selling will be done by the Kearney Corporation by an increased sales force in these lines." The Chance salesmen had been called in and given their instructions last week, it was stated. They would be working with or for Kearney, as the following details describe, which are unnerving in view of what happened in this association. "The Chance Company will be free of the selling end, as all orders and the clerical force as pertains to orders will be handled in St. Louis. All goods will be manufactured in Centralia including the new lines added from the Kearney lines, and all the orders will be shipped out of Centralia." According to discussions by key Chance officials and employees, Chance Company shut down their sales and order system and anxiously awaited the deluge of factory orders that never came. It appeared Kearney was instead strangling Chance by concentrating on their own interests!

The Kearney connection and story carries further intrigue because of the personal relationship that existed between the two families. First, the Kearneys, Jim Senior and Jim Junior, became friends with the Chances. Gano explained that he went to college with Jim Junior and they were in the same fraternity together:

> *Yes, he was an A.T.O. and I think he graduated in 1926. That's when Mr. Kearney (Senior) pulled away from Matthews, as sales manager, and set up his own company. His son went with him and a few years later the Depression hit pretty heavy and we worked out a deal. It was purely a sales agreement and actually, I never saw a contract. It was during the Depression time and we thought they were good at sales and we were good at manufacturing. (1978)*

Additional background was provided by Gano on Jim Senior, "He was sales manager for Matthews, which he later bought." Roy Hansen replied, "And he started making the triple connector," to which Gano replied, "He had one, he picked from somebody out on the road and Matthews wasn't very much with it. So that was probably the reason why he had discussed it with Matthews, and left them and set up his own company and started pushing the triple connector." (1978) This also helps explain why one of the strengths of Kearney was marketing. He likely hadn't the time, capital, or expertise to develop manufacturing plants. After he got things moving, it's clear that Kearney's operation picked up steam quickly in establishing a viable company that would become both a partner and a competitor to Chance Company.

The Kearney and Chance family similarities are interesting in the involvement of the sons with their fathers in the family businesses. According to *Family Business Initiative* the relative that most often becomes involved in a family business is spouses, sons, daughters, brothers, fathers, and sisters; in that order. "If the business becomes successful, the owner should immediately select a person to take over the firm."

Groundwork was being laid for the relationship to develop as the *Guard* of January 15, 1932, covered a "Chance-Kearney Conference" in Centralia. The consolidation took on the airs of an ongoing arrangement that was progressing. The Chances were anxious that this developed quickly as their order system was in the hands of Kearney. This conference suggests the Chances were pushing for a closer relationship and Kearney was eager to learn all they could about Chance Company. Apparently Kearney was also accumulating expenses to be charged to Chance. The sales conference was held on a Wednesday, January 13, in the Chance Club Room with "all salesmen of the company" called into Centralia for the meeting. The noon luncheon was prepared for one hundred with seventy from "distant cities" which represented major U.S. cities and even one or more Canadians in attendance. The afternoon was spent in conference, inspection of the plant, and learning facts about Chance products. There didn't appear to be much Kearney did not know about Chance Company at this point. This was followed in early February by a business trip by

A. Bishop and Gano to St. Louis, which could be presumed to be associated with the Kearney arrangement. Gano elaborated in more detail:

> *When we worked out a sales agreement with James R. Kearney, they were to sell their products and our products. We were to make as many of their products as we possibly could and be their real manufacturer. Well, it sounded pretty good but there was no money exchanged, no protected stock, or anything else. It wasn't very long until Kearney's salesmen actually were pushing Kearney products from a competitor of ours including anchors and not pushing ours. (1978)*

The Kearney/Chance business relationship, as a result, became adversarial. Jack Curtright, like other early employees remembered, "Yea, that was a bad deal. . . . I think there was some patent situation there because the Kearney Anchor was a Malleable Anchor. . . . They made it in a Two-Way and a Four-Way. But they had a Two-Way, Three-Way, and Four-Way by changing the length of the blades." (1979)

The Kearney arrangement was also remembered by Art Houston:

> *Well, Kearney had a pretty good sales organization, much bigger than Chance had of course because they had more products. They wanted to get into the anchor business too and thought they could do a better job of selling it, so they made a deal, as I recall, with Chance to handle the sales end and Chance would do the manufacturing end. But it looked like Kearney was going to promote some of the anchors that they had, I believe they had an anchor at that time too, but they seemed to put more effort in their anchors and their products than they did with Chance anchors and our products. So they* [Chance] *began to feel that this was not going to work satisfactorily. But they did have a big sales organization, I think about twenty some men, as compared to five or six at Chance. (1979)*

Gano also shed light on their relationship that suggests that the Chances, Gano in particular, were aggressively pursuing a business relationship with Kearney. The initial Kearney/Chance arrangement resulted from a meeting of Mr. Kearney and Mr. Chance in St. Louis, where the headquarters of Kearney was located. It would appear that the two sons were also in attendance, but that isn't confirmed. According to Gano, he and Kearney didn't see eye to eye on several issues but with surprising results. "Incidently, Mr. Kearney Senior and I had quite a few doubts about different things and believe it or not, he respected me. I had only been out of college for a short time and yet I was the one who did all the pushing. He treated me just like a full mature man and I doubt if I was." (1978) Gano was in St. Louis several days on business in mid-October 1931 before the agreement was announced a week later. It would appear Kearney's soft approach sucked the Chances into this partnership even deeper.

In another interview Gano continued to explain the deteriorating conditions:

> *When I got out on the road I found that Kearney was promoting his anchor instead of ours.* [Note, the November 20, 1931 Guard told of a business trip to Atlanta by Gano.] *Since all of the anchor sales went through them, I could see that it would be a question of time when we wouldn't have the anchor business we had. I began to complain to my father about it and we had several discussions with Jim Kearney Senior. I went to Mexico and hired an attorney by the name of Barnes, to advise us on how we could pull away without getting into a legal fight. My father didn't know anything about it and when letters came in from Kearney Senior to my father, Louise Kogger, the Secretary, would give them to me. I would take the letters*

to Barnes and we would decide how to answer them. He [Kearney] *was building up an expense against our company by showing the amount of sales expense, such as conventions and exhibits, that he'd spent in promoting our anchor. He had promoted, you see, for future business and for us to pull back. So I wanted the lawyer to advise how we should answer these letters, that were liable to get us in trouble, and he did. "Mr. Kearney, your responsibility is the sales and how you decide to spend the money in your own company is your own business." And he never did it. (1981)*

Another angle of the Kearney affair was the agreement to utilize the Chance field sales force within the Kearney Company. "Mr Kearney never did hire Jack Isbell and he had agreed to hire all the salesmen. They were supposed to employ all of our salesmen! The thing is, he couldn't get together with Jack and so Jack went to the Wej-Lock Company." (1978) It is revealing that Jack Isbell is the only salesman of this unhappy group that is specifically mentioned. This may be because he was not hired and was considered a most important salesman. A week-and-a-half trip to the West Coast (San Francisco, Oakland, Los Angeles) in late November 1931 by A. B. Chance and James R. Kearney may have involved concerns over Jack Isbell. It is not said which salesman was complaining the loudest, although Isbell's situation was familiar to Gano as he explained:

Kearney was slow in getting out there [West Coast]. *I don't know whether it was his intention or whether he did not fly out there in those days. Anyway, there were letters back and forth with Jack but he could not get together with Jackson on the sales deal. Kearney had a territory out there that Mabel Hartzel was assigned which was exclusive. He would have had to move Jack to another territory and Jack would have had to work for Mabel. So it was a difficult situation and I kept pushing him, "Here you guys are supposed to take care of our salesmen." Anyway, Jack got impatient, maybe rightly so, so Jack went to work for Wej-Lock Company. (1978)*

To get Jack back, especially when Chance Company was trying to reestablish their field sales group after the Kearney fiasco, it was even suggested that they made the Wej-Lock agreement just for that purpose!

Bill Hollander also recalled the Kearney and Wej-Lock situation:

After they let Nate Husted go, things were getting bad and Chance entered into a contract with Kearney in St. Louis. Kearney was to be the sales organization and Chance was to be the manufacturer for the Kearney and Chance lines. So it sounded real good. They had a large sales organization and they jobbed out most of the manufacturing. We had a very small sales organization and it looked like a most wonderful deal. Our salesmen were to go to work for Kearney and sell a combined line and we manufacture everything. But word finally came back to us through some of our men that Kearney instructed all of the salesmen, "you will sell Kearney." "You sell only Kearney goods and if they insist on Chance goods you let them have Chance but you are selling Kearney." The result was they did not sell any Chance goods and they would not send us anything to manufacture. So we were pretty low, then as the Depression was going full blast and about this time it was bad. We worked just a skeleton force and Jack Isbell had left us. He said he wasn't going to have anything to do with Kearney and he heard about E. K. Cole in Winterset, Iowa, and his anchor which he was selling for guard rail fences. He went up to see E. K. Cole and he told him, "you make me your sales manager and I will put this anchor of yours on the

> *market," and he sold E. K. Cole. But he told him, "I don't want to be in Iowa" and he sold him on letting him headquarter in Riverside, California. He sold that Wej-Lock anchor of E. K. Cole's. (1978)*

Background material on the Tips Tool acquisition by Chance in 1937 revealed that Tips/Kearney had a 1928 agreement similar to the Kearney/Chance one. Incidentally, Tips Tool was operated by Henry W. Bodendieck. According to Gano, he was not a relative of another fraternity brother, Henry Bodendieck. Combining that material with information from the Chance/Kearney association reveals an unexpected benefit reaped by Kearney in these relationships; the pirating of key personnel from both firms. Tipsword, likely the best-known and knowledgeable pioneer tool demonstrator, went over to Kearney and didn't return to Tips when the Tips/Kearney agreement was axed. Ironically, the failure of Kearney to hire Jack Isbell caused problems, but if they had hired him, the future Chance Company might have been quite different. Possibly Kearney Senior was of the same marketing mold and they didn't need two of either on the premises, but the lack of a vacancy in the appropriate sales territory undoubtedly played a significant role. In addition, Jack may not have wanted to work for Kearney and saw greater possibilities with Wej-Lock. Although Gano spoke of the Chance Sales Force transfer to Kearney, it was surprising when Bill Hollander revealed how far this agreement had penetrated within Chance Company. "Eddie McLaughlin was advertising manager and Art [Houston] was the artist who prepared the copy and did photographic work. Ed was transferred to the Kearney organization, in the Advertising Department, and Ed never did come back to us." (1979) The obvious reason for the personnel transfers, from a Chance perspective, was to build a Kearney/Chance team that would be more successful in the marketplace. Ed's employment at Kearney, however, appears to have not taken, as a January 6, 1933 *Guard* news item stated Ed McLaughlin was now of Columbia. He made a Centralia weekend visit, guest of Glenn Eierman and Art Houston, who had worked with Ed at Chance. All three were discussed by Bill Hollander earlier in this chapter. Ed did come back to Chance in the Sales Department around 1933 and left in January 1935.

As this story continues to unfold, Gano explained what occurred:

> *The salesmen were unhappy and after almost a year of it, my father ended up disgusted too. So my father said, "I think we had better separate from Kearney." So he set up a date to see Mr. Kearney and called me and asked me to go along. I said, "Why don't we go by Mexico and pick up my lawyer?" Dad said, "We don't need a lawyer and we can handle it." I said, "I think it'll be a good idea, let me call and see if Mr. Barnes can come." So I set it up and we went by and picked up Mr. Barnes and went on to St. Louis. My father didn't know all the time that Mr. Barnes had been advising us on how to write letters to keep us from too much difficulty. (1978)*

What was the last straw that brought Mr. Chance's active involvement into axing the Kearney situation? It could have been an accumulation of complaints from his son and the salespeople as well as the loss of Isbell that sparked his decision to terminate relations. Notice that Gano was monitoring the marketing area and detected problems that were looming on the horizon. It shows that in the desperate times of the Depression, it could be beneficial to be somewhat mistrustful and suspicious of other parties in protecting one's business interests, although the apparent failure of any positive activity from the Kearney association shot up warning flags. This is reflective of A. Bishop's continued attention to their competitors and covering their products with a legal blanket of patents. Notice that Gano had wisely involved a lawyer in this sticky situation and keeping his father out of the communication link as this saga unfolded, appears secretive. Gano is exonerated to a degree because his protective and preventive actions were successful and his father eventually came to the same conclusion.

As Gano continued:

> *So we met in St. Louis and my father got pretty riled up. Fact was, Barnes and myself had to keep my father from getting too far out of line and to work the thing out. I'm glad we took Barnes along. We severed relations with Kearney that day. (1978)*

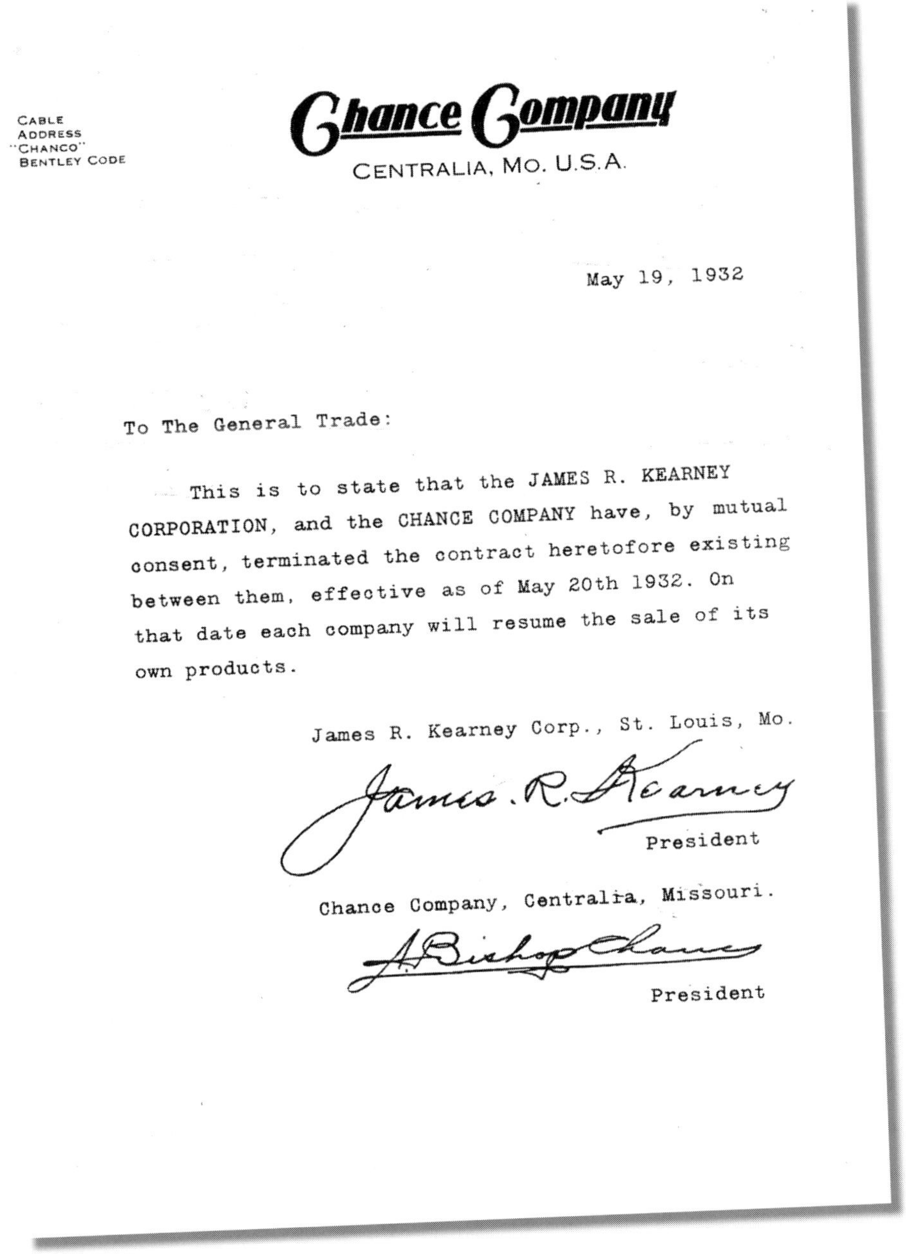

CABLE ADDRESS "CHANCO" BENTLEY CODE

Chance Company
CENTRALIA, MO. U.S.A.

May 19, 1932

To The General Trade:

This is to state that the JAMES R. KEARNEY CORPORATION, and the CHANCE COMPANY have, by mutual consent, terminated the contract heretofore existing between them, effective as of May 20th 1932. On that date each company will resume the sale of its own products.

James R. Kearney Corp., St. Louis, Mo.
James R. Kearney
President

Chance Company, Centralia, Missouri.
A. Bishop Chance
President

This letter of May 19, 1932, is a historically significant document involving two of the early giants of the utility supplier industry. This would end a relationship that hadn't materialized to a satisfactory level or expectancy of Chance Company. Note the letterhead with company name and cable address. This is even more special as most surviving correspondence is on non-letterhead carbon copies. (A. B. Chance Co.)

According to a Chance Company letter dated May 19, 1932, shown above, the Kearney/Chance relationship was terminated. The old adage of not involving family or friends in business dealings without handling it as if they were strangers, also applies. The Kearney incident was the end of an era whereby a handshake and a verbal agreement was all that was necessary, but it had actually ended long ago, although in the Midwest's rural areas, a man's word was his reputation and a reflection of his integrity. This was A. Bishop, a man who put trust in the friendships that he developed. He was a true article, not without fault, but a unique and colorful individual who was prominent not only in his material accomplishments, but in how he dealt, associated, and worked with people. In some ways he hadn't changed, as he still relied on friendships and his instincts in certain matters instead of legal council. Kearney continued to be a competitor extraordinaire over the years, whose name brought a negative connotation.

The fact the Chances pulled the plug on this Kearney deal supports the conclusion that expectations were not realized. It is fortunate that this didn't take Chance under, as the arrangement gave Kearney control of customer/product knowledge and the sales force, which they appear to have exploited. Even after it was terminated, it undoubtedly continued to damage Chance and put them at a disadvantage in facing Kearney in the marketplace. Since both father and son appear to be involved, the blame would rest squarely on their shoulders. This may be a situation in which the trusting nature of the Chances was taken advantage of and a perceived friendship instead halted at the business door. It became a warning that to survive in the current Depression climate, business people had to become wiser and colder in their dealings and relationships. However, there appears to have been a continuing relationship between the Kearneys and Chances even after their short-lived business association. So the idea that Kearney Senior was a complete villain may instead partially point to a contrast in how the two families did business. The association may have also been arranged in such a way that invited failure. For example, Kearney's role in standing sales costs would undoubtedly be supported by some commission arrangement, but it could have invited Kearney's focus on their products. Roy Hansen recalled the Kearney situation, "I remember when James Kearney died and Jim Kearney Jr. took over and he was absolutely unqualified to operate a company because his father never let him make a decision until the day he died." Then Roy asked Gano, "How did your father handle these things when you came?" Gano's response was, "That is a good point and it is a good example of the extreme. My father for some reason really had a characteristic of concentrating on whatever he was working on and if the other fellow was doing a good job, he did not get involved. He would check once in a while and see how he was getting along." Gano went on to say:

> *The Company was still pretty small and so he pretty well turned it over to me and in turn, of course, thought that was the answer, to give a fellow authority and responsibility and see if he can grow, if he can grow fine, if he can't get somebody else. The problem with Kearney, and even to the point that my father was a great friend of old Mr. Kearney, told him, "I think you ought to let your son try to run that company." What happened was he would go away in the winter and Jr. would make a lot of decisions while he was gone. He would come back and his father would change them and it wound up with two factions, a faction following the father and a faction following the son. It really was too bad and I even tried to help the situation and my father would too because we both knew Jim and Jr. But Jim* [James Sr.] *was just exactly the opposite* [of A. Bishop Chance], *he just did not let loose of anything. I remember he was very domineering but a fine fellow. He really was. I was surprised that the number of times he and I locked horns when we had a conference meeting and he never did anything but respect what I said. I thought about it afterwards, he could have probably said: "Well you little son of a gun what the devil do you know about it." (1978)*

The Kearney/Chance social relationship is puzzling at times, but it points to a hard-nosed Kearney Senior in which friendships apparently didn't mix with business. There may have also been a cooling-off period after the breakup and then their relationship was reestablished to a degree. The Jim Junior references suggest the Chances, especially Gano, may have found Jim Junior easier to deal with than Senior. The Chances may have also kept an eye on that firm for acquisition possibilities as they did for most other competing lines and firms. Notice the comment by Gano a few pages back of Kearney's purchase of Matthews, which Chance Company later acquired. In the theoretical textbook world of business associations, the chapter dedicated to the short-lived Kearney and Chance relationship would undoubtedly qualify as one of the most unusual. After generally knowing what occurred, the question then arises of how it came to be. It seems so out of character for A. Bishop, who scrutinized patent details to the nth degree, but Gano was also involved, which raises the issue if one or the other was more responsible for this association. However, this is what makes the Chance Company story so interesting; expect the unexpected!

Plant, Products, and People

THE EARLY 1930s

The dawning of 1931 brought the death from pneumonia of fifty-three-year-old H. B. Rodemyre (1877–1931), the business manager of the *Guard*. He began working in his father's newspaper office when he was eleven years old. A member of the Centralia Christian Church and a bachelor, he was survived by E. T., a brother. Retirement would also bring closure to the local Wabash agent's career, P. R. Diggs, a fifty-year veteran of that railroad. His family connections first brought Mrs. Fancy Chance to Centralia on a visit. Mrs. Bryan Fenton, now of St. Louis, visited with Miss Ida Traughber and then Miss Ida Lou traveled to Chicago in late March to visit the J. J. Milligans. Mr. Milligan was the Chicago-area Chance utility salesman. At the same time N. C. Husted traveled to New York and other Eastern points on business for Chance Manufacturing Company. The company's influence extended even to the new 1931 Centralia license plates, which carried a panoramic view of the plant. The progressively harder times brought news of the Alton dropping two passenger trains serving the Chicago, Centralia, and Kansas City market. One of their competitors, the Wabash, was also reducing their shop workforce in nearby Moberly.

The Chances kept things hopping in Centralia and it couldn't be described as a sleepy little rural town where nothing ever happened. One of A. Bishop's frequent complaints and concerns was high energy costs. A modern source for residential and industrial heating was natural gas that was supplied by Missouri Power and Light, as was local electrical power. Not yet available to Centralia, the April 3, 1931 *Guard* described the gas franchise proposition that was before Centralia voters for approval. It was an attractive opportunity in the eyes of Mr. Chance as he supported and stood expenses for promoting it. His son Gano rigged up a broadcasting station on a truck and brought it uptown to attract attention to the election, not unlike the unit in the Paris, France scene in the movie *Casablanca*, some ten years later. This suggests that natural gas was viewed by Mr. Chance as a potential benefactor to his company and the community, and his vision was on beam as it would become a vital local heating source. Its economical impact is not as certain, but it provided advantages as it was utilized by Chance Company to fire the many company furnaces. This is another early example of the forward thinking A. Bishop Chance supporting something that is still being used today. It did pass easily and was in use by July 10, 1931. Something positive always seemed to be occurring in the Prairie Queen during the A. Bishop and F. Gano Chance eras that will not soon be erased.

Confirming the versatility of the Chance Gym, located in the Anthony Building, Centralia High School basketball was played there. This was brought to light by a freshman-junior basketball game in late March 1931 and another game played there by Centralia High School in February of 1932. This was only a sample of the many such events hosted in the Anthony Gym, showing the community's need and use of this facility. In other Chance Company news, a surprise announcement in the April 10, 1931 *Guard* covered N. C. Husted's resignation, effective June 1. His frequent coverage in the paper including social events, business travels and involvement in everything from sales meetings to the A. Bishop Chance tribute suggest his prominence. A June 24, 1932 news item in the *Guard* a year later told of Mrs. Gano

Chance, Mrs. Arthur Sames, and Miss Jennie Green leaving on a Tuesday for Niles, Ohio, to visit Mr. and Mrs. N. C. Husted. Such a trip by three ladies would surely be by train; but no, it was by "motor." Mrs. Anna Lee had two weeks earlier motored to Kansas City with Miss Jane Austene, which may have been a warm-up for the much longer and more challenging journey. Mrs. Husted apparently returned to Centralia with the Anna Lee party, as the July 22 *Guard* reported she was returning to Niles after spending a week as guests of local friends. For many years there continued to be contact between the Husteds and friends in Centralia, indicating company and social relationships were sometimes at odds.

In the plant, the reality of business associations didn't always match up with the social aspects being depicted away from work. According to Paul Sturgeon there were several early plant managers, including Superintendent Bryan Fenton (1926–28) and T. B. Carpenter in the late 1920s–early 1930s. The general manager under Carpenter was Nate Husted, followed by Joe Cheetham, per Paul, who came to work under Carpenter. Husted and Cheetham, generated a lot of comment and Nate Husted was remembered by Bill Hollander in a not too complimentary way:

> *Just before I came to work* [May 1929] *here, they hired Nate Husted as Factory Manager. Carpenter was Superintendent, more or less, but Husted was kind of a promoter. He had worked his way up through the ranks, I believe with General Electric. He was very loud and he hadn't gone to school but he had a chip on his shoulder for anyone who had. The thing he used to love was to get Gano and I off to the side and give us a good cussing out. He would tell us, "you —— —— college men—you think you know everything. You don't know a —— —— thing.' He liked to dress us down and since he was the Factory Manager, he could dress us down in real royal fashion. A. B. was not too happy with Nate Husted. A. B. liked to putter around the plant and Nate would run him out. Every time he would catch him in the plant he would say: "You're the President of this Company—you get up there in the office and you take care of the presidential duties. I will run the plant." But A. B. didn't know any particular way to get rid of Nate and he kinda put Gano on the idea to act as hatchet man. So they hung on Nate's words for a long time and one day Nate made the mistake and said, "It just seems I can't get things done around here." Gano said, "yes, I notice that Nate—you don't seem to fit in." So he couldn't do anything but resign. They were very happy when they got rid of Nate. (1979)*

Another personality conflict, at about the same time, was remembered by Bill's former classmate F. Gano Chance. There is a problem, however, as Gano in interviews pronounced his name Houston, when it appears instead to be Husted, a similar-sounding name. Discussions about Chance Company invariably shift to certain high-ranking employees whose actions stood out. When staff members were brought in from the outside there was an unknown element, and not until they were around a while did it become obvious if they had horns or wings. It is more difficult for operating managers, if they are too tightly controlled, although Chance managers were routinely given enough rope to hang themselves. Part of the problem Gano may have had early on with company officials, was his subordinate position. Gano repeats on frequent occasions, especially during this discussion about Husted, his personal situation and lack of status within the company. "The Company was not incorporated. I had just been out of school about a year and I had no title or anything else." (1978) Gano seemed impatient in working himself in and proving his worth, and as the son of the owner he was pushing for position. What can be seen is the struggle of Gano's strong will in the process of taking charge in uncharted waters.

One of the reasons for special outside hires was the gaining of expertise and knowledge from their experiences and management attributes. After the initial gleaning of those positive aspects, another period set in where a person's demeanor and ability to contribute to their areas and to the overall operations was taken into account. Ironically, one who may have contributed immensely to Chance's body of knowledge may have failed miserably in the day-to-day routine. They could be let go or pressured to leave if they didn't work out. An individual's long-range performance usually dictated their permanent employment status. This was an advantage of a technical position over one who managed and had to deal effectively with people. A situation was recalled of a person coming into Centralia, interviewing for a marketing job in the 1970s from one of our competitors. Overhearing a discussion, it was contemplated that this person possessed inside information that would be beneficial to Chance and since he wanted a job it would be a purpose of the interview to gather as much information from him as possible. There may not have been any serious consideration in hiring him and he wasn't.

Gano described Mr. Husted's credentials as follows:

> *I had a little trouble with Ed moving too fast. What was interesting of course was being with Mr. Husted who had been a General Manager for a hardware company and a very hardworking man who knew pole line hardware manufacturing. He was a little fellow. He was very aggressive, but he needed a little extra finesse in salesmanship, etc. (1978)*

During this period there were other life-changing challenges that Gano faced, such as those arising from matrimonial issues. Newly wedded couples can either look forward to a shivaree or consider it a disdaining result of the marriage process. In Gano and Ann's case it became the latter because Husted was the ringleader in developing an elaborate and humiliating plan to place the Chances in cages on a truck that was to be paraded around town. When Gano found out about it he asked Ann if it would bother her. She thought it would be better after their honeymoon and so owing to Ann's and undoubtedly Gano's appraisal of this plan and Husted's involvement, Gano decided they weren't going to participate or cooperate. "Well, Mr. Husted and I had our problems anyway." (1978) Miss Ida Lou called Gano and told him they were on their way with the cage. They got away but because of some heated words that Gano made to Husted, Ann thought Gano should call back and apologize. "So I laugh and say, my first discussion with my wife was what she said I ought to do—and what I did not think I should do. So I compromised and did what she said." (1978)

As Gano and Husted got "further apart" one of the areas that generated more conflict was sales:

> *He was making our customers mad. Every time I went out on the road—I believed he was hurting our company. . . . We had a sales meeting and the salesmen were all here and he was being awfully mean to them. He was really pressuring them too much, I thought. In 1932 I know business was bad, and I was on the sidelines. I had no title, office, or anything. . . . The sales meeting was over on a Friday evening and several of the fellows were catching the train. About 2 o'clock in the morning a train was coming through here. I was down at the factory office with them and talking about the problems and some of the pressure that Mr. Husted put on. Normally you worked on Saturday mornings. I didn't get down until late that morning. He said, "You are late this morning skipper!" And I said, "Yes, I was up with the salesmen until they left about 2 o'clock." (1978)*

Most employees can relate to situations developing when salespeople "caught the ear" of a manager who would listen and take action. It appears that Gano was closely tuned into the sales area during his career. Because of the remote location of salesmen on the "front line," in direct contact with the customer, their concerns carried more weight and salespeople could naturally be very persuasive. Nevertheless, it was important to listen to employee problems, especially sales people. A capable manager could then sort out that which needed correcting, as apparently this case warranted.

Gano cited one reason that caused problems between he and Husted: an educational gap. "He resented a so-called technical man who had an engineering degree and he did not have one." (1978) There can be a cockiness of recent college graduates and it can be difficult for those with just a high school education to supervise those with college degrees. Although there were successes at Chance, there were also examples of those who were over-promoted. There were contrasting levels of educational credentials among employees, but there were many who furthered their on-the-job training/education and continued to grow. However, there is something beneficial about the sacrifice of time and money and the experiences of being a full-time student on a college campus environment that is reflective of the seriousness and commitment a person places on education. Making choices early in life of placing more importance on a job over education was usually a wrong choice, but surprisingly early birds earned promotions into management positions as a result of exemplary effort, natural talent, or just being in the right position at the opportune time. Gano was a supporter of both higher education and training programs to make better employees. His two degrees in the 1930 time frame were very uncommon and special although his educational credentials carried with it some resistance, which he never forgot or ever accepted as correct behavior.

Rural electrification was being promoted by Missouri Power and Light in early 1931, but its success would represent only isolated pockets of rural service until the federal government's R.E.A. program. A three-day business trip to Fairfield, Iowa, in late April 1931 by Gano and C. B. Taylor (Big C sales) likely involved Iowa Malleable from whom Chance was outsourcing anchors. At the same time, Mr. and Mrs. A. B. Chance were in St. Louis on a Saturday and Margaret Greenlee hosted a club meeting "T" at the Chance home. At a Commercial Club meeting during that time A. Bishop gave a talk on the clock-punching system in support of discussions for the city to install such a system to compel night watchmen to make their rounds for fire protection. Another *Guard* article of May 29, 1931, told of a surprise fifty-eighth birthday celebration for "Boss" Chance. Office personnel gathered in his office with a cake after he was called to Superintendent T. B. Carpenter's office under false pretenses, while employees made preparations. The cake bore the "A. B. C." company trademark with a "miniature Never-Creep®' Anchor. A play was made on the famous Heinz slogan concerning the number 58. His accomplishments were highlighted with a bicycle with his reversible handlebars, an orange bringing attention to his payment for six weeks of work in a local drugstore, his hotel invention, Never-Creep® Anchor, and his newly acquired hobby golf, which he played left-handed and called "Modern Shinny."

To be someone without a title, the son of the "Boss" was doing considerable traveling. The June 12, 1931 *Guard* captured a trip by Mr. and Mrs. F. Gano Chance to Atlantic City to attend an electrical convention and then on to New York and other Eastern points. Although the often-mentioned influence of the Depression is cited as the reason A. Bishop stepped aside and allowed his son Gano more control of the company, there is another element, Mr. Chance's health. All the traveling and constant drive to grow the company over his adult life, were taking its toll. There was also an invisible barrier suggesting that when you neared sixty, you were over the hill. Summer weather in central Missouri can be hot and humid, and in late June A. Bishop had a heat stroke while at the factory. Immediately, one would surmise that he was out in the hot plant working on a project, but instead he had spent several hours in Columbia on business and returned to his plant office complaining of feeling badly. After going home ice packs were placed on his head, giving him relief. The *Guard* article gave further details of how Mr. Chance hadn't felt well for the past ten days, "but like the old Indian he is, he refused to give up and kept on the job as usual."

Mental stress and fatigue are as real as physical exhaustion. Although Gano got considerable attention over his famous trip to Columbia to visit bankers, this may have been a similar trip by his father that finally took its toll. There are several occasions when employees remembered A. Bishop's trips to Columbia to get funds to make the payroll. Apparently he had a heart problem and his progressing age led to the following statement in personal papers during the previous year: "So far as I know I am in excellent health. My skipping heart gives me no anxiety, but I must not forget—60 is not many years away and it is only good judgement in making my arrangement and plans for putting my house in order!" (1930) What this episode reveals is a vulnerability that he had not faced in his adult life. It may have accelerated his part-time status, and could explain that health conditions and concerns helped drive changes at Chance Company from father to son.

Social activity represented in the country club setting would lead to support of such a Centralia facility by Chance Company families. This is why Centralia took on the airs of a larger city with the wide variety of facilities, entertainment, etc., to meet the interests and demands of a diverse makeup of area residents. The Chance payroll brought a level of sophistication to area people who advanced into the middle class. This drove positive advancements, but in 1931 residents had to travel to Columbia or Mexico to take advantage of country club facilities. The close friendship between the R. L. Hopes and the A. B. Chances, led to the Chances entertaining the Hopes at the Mexico Country Club per the July 3, 1931 *Guard*. This particular event may have derived from the inclusion of the Hopes' son Linwood Hope, wife, and daughter Mary Lee visiting from St. Petersburg, Florida. This holds future significance for both families, as this is where both would spend the winter in the later stages of their lives. The elder Hopes would permanently move there in 1941.

R. L. Hope. (Centralia Historical Society)

Just when A. Bishop appears to have retired because of health developments, he blows away such logical reasoning in late August with a trip to St. Louis. Of course, concerns about the well-being of his company drove further travel, but instead it was a business trip representing the local Commercial Club in finding a proper suitor for the overalls factory in the Bagby Building! When things get tough it is time to send the best, especially when other communities were also looking for stable firms who could provide employment to their growing number of unemployed. For example, nearby Columbia had 1,131 adults out of jobs. There was a group of Americans who had succeeded in spite of hardships and A. Bishop had faced his share of adversities. Such people gritted their teeth and confronted this recession as they had others, remaining cautiously optimistic that it would soon pass. It is not always good to know too much, because if the people of that time had known that the entire 1930s decade preceding World War II would be plagued by the Great Depression, it would have surely broken their spirit, hearts, and will. In a display of the Bible's wisdom of teaching to take each day one at a time, it is suggested those burdens were sufficient without piling all of the tomorrows onto one's back.

Little information was escaping from Chance Company although an injury accident in August resulted from material falling on employee Harry Johnston's foot. Government business seemed to be the big news around town with the new, long-discussed Post Office to become a reality in the next several months. This must have made a former postmaster happy, but Francis M. Traughber (1848–1931), eighty-three, passed away in mid-September before seeing its completion. A member of the Christian Church, Traughber was survived by seven of nine children including Misses Elizabeth and Ida Lou of Centralia. Early area pioneers were quickly passing. As the foundation work on the new Post Office progressed, it revealed where the early Christian Church was located. Interestingly it didn't face the Square to the north, but instead was located in the center of the lot facing east. However, it is unclear if this was the 1875 or the 1890 building. Per an October 9, 1931 *Guard* article, the 1875 building was still standing, but had long ago been moved to the west onto a former Bush-owned lot.

The Chance family has a long tradition of supporting historical preservation of various sorts, as revealed in 1931. The first Centralia museum may go back to the *Guard*'s prominent founder Adam Rodemyre who had a collection of local artifacts in his office many decades earlier. The *Guard* of October 23, 1931, carried a news item about A. B. Chance's "factory museum" adding another piece to its collection. It was an anchor, but it was of the expanding type and a competing model from another manufacturer. What makes this even more significant is where it was found. The new Post Office foundation work was removing remnants of what the early settlers had stuck into the ground. But this was a more recent protrusion (post 1912) by A. Bishop into what had become an empty lot after the second former Christian Church building had been torn down in 1910. This reveals that anchor testing and experimentation by Chance's first product demonstrator may have been occurring soon after his anchor was invented. The Chance move away from the south Square area to 210 North Allen in 1917 placed the factory in a landlocked kind of situation, so it is possible that the empty lot as described in the article, was used by Mr. Chance before or after 1917. The article continued:

> *Many years ago Mr. Chance placed this anchor in the ground on the lots, then vacant, and placed one of the Chance plate anchors a distance away. He made a pull, one anchor against the other, with a steel cable upon which was suspended a large barrel full of dry sand. The sand was then wet and still neither of the anchors came up out of the ground. The test demonstrated the efficiency of the modern anchor over the old "dead man" method of digging down and putting in a large timber, as the new anchors pull against the old undisturbed soil.*

This documents the extensiveness of Mr. Chance's historical early anchor testing. This

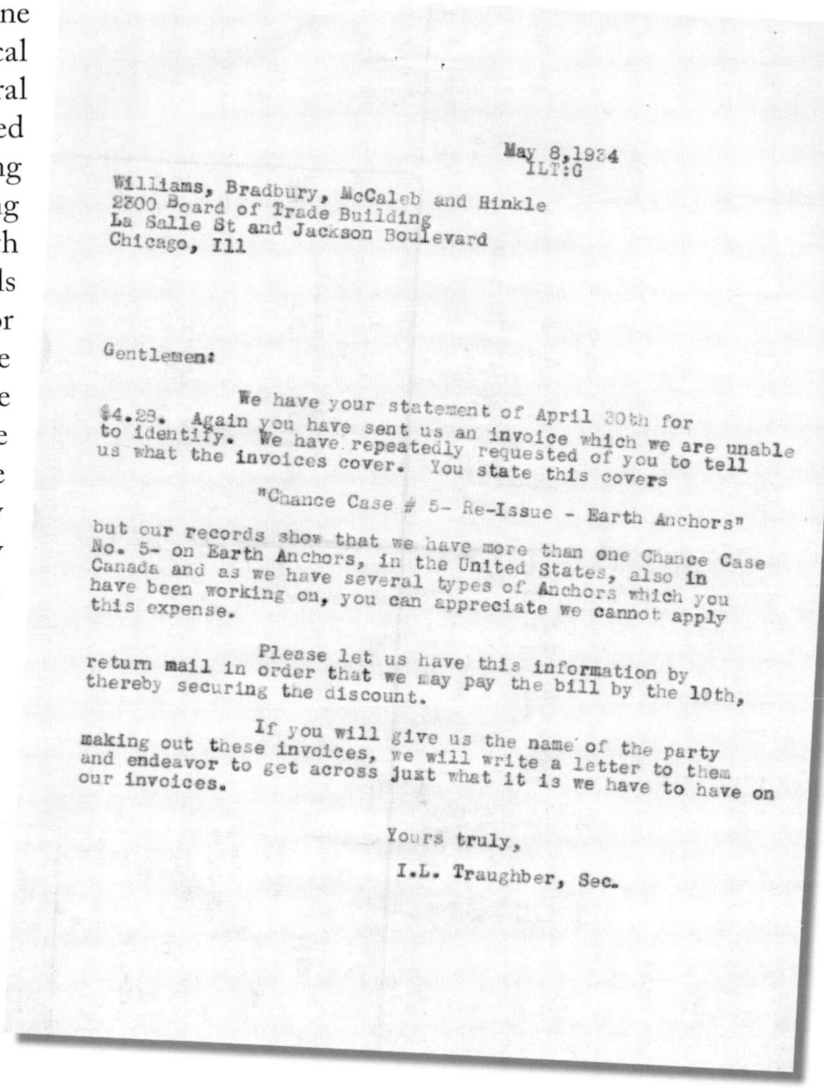

May 8, 1934
ILT:G

Williams, Bradbury, McCaleb and Hinkle
2300 Board of Trade Building
La Salle St and Jackson Boulevard
Chicago, Ill

Gentlemen:

We have your statement of April 30th for $4.28. Again you have sent us an invoice which we are unable to identify. We have repeatedly requested of you to tell us what the invoices cover. You state this covers

"Chance Case # 5- Re-Issue - Earth Anchors"

but our records show that we have more than one Chance Case No. 5- on Earth Anchors, in the United States, also in Canada and as we have several types of Anchors which you have been working on, you can appreciate we cannot apply this expense.

Please let us have this information by return mail in order that we may pay the bill by the 10th, thereby securing the discount.

If you will give us the name of the party making out these invoices, we will write a letter to them and endeavor to get across juxt what it is we have to have on our invoices.

Yours truly,

I.L. Traughber, Sec.

In a 1930-era patent file was a 1934 letter penned by Miss Ida Lou, apparently for Gano. It is firm and reflects some frustration of not being able to close out invoices in an organized manner, even though it was only for $4.28. But that was during Depression times and as the letter portrays, Miss Ida Lou was taking care of business and making sure Chance Company records were in order and company financial obligations were fulfilled. (A. B. Chance Co.)

was during a period when not everyone was convinced that the new mechanical earth anchors were superior to natural kind of dead weight objects, labeled dead man. This reveals his convincing work and salesmanship in showing the superiority of man-made earth anchoring systems and also reveals he was not overlooking other anchor types such as expanding. Chance Company would later acquire the Taylor Expanding Anchor for the purpose of broadening the Chance anchor line, which would eventually include almost every anchor variety in the rainbow! Miss Traughber's letter of May 8, 1934, reproduced above, is reflective of continuing pursuit of new anchoring systems and their patenting by A. Bishop.

As the year 1931 was coming to an end, the bachelorhood of Edgar T. Rodemyre also ended in marriage. He was the lone survivor of the Adam Rodemyre family, so long representing a prominent/influential presence in the local area. In early December, the A. Bishop and F. Gano Chances, and Reverend Burrus and Mrs. Burrus motored to St. Louis for an American theater event. A. Bishop was reported to have had some large limos and it would have taken such an extra-seated vehicle to accommodate all of them. The first news from Chance Company in 1932 was an injury accident by employee Carl Prewitt who cut his thumb in a milling machine. Some good news was a refund of $294.60 from the city to Chance Company for overcharge on current used by the plant. A friend of A. Bishop, O. B. Mayes, was a director at the lone surviving First National Bank of Centralia during January 1932, when two more banks in the state of Missouri closed. As earlier stated, Commander Byrd was using Chance equipment and the January 22, 1932 *Guard* reported Chance Company's receipt of a testimonial from Commander R. E. Byrd of the Antarctic Expedition to the Pole. It was written on a large seventeen-by-twenty-two-inch parchment and was framed and placed on a wall at the factory and stated:

> *Rear Admiral Richard E. Byrd, U.S.N. (Retired) and the Executive Committee of the Byrd Aviation Associates express to the Chance Company their enduring appreciation of the most valuable co-*

> *operation in the equipping of the Byrd Antarctic Expedition. It is our privilege to attest, and we most gratefully acknowledge, that without such assistance the accomplishments of the Expedition in the Antarctic would not have been possible. R. E. Byrd, Leader.*

This shows the growing dependence on Chance products even in diverse areas, the benefits of which could mean the difference between failure or success of a polar expedition or current flowing over a telephone or electrical power line, somewhere in the world.

The deteriorating economic conditions were accelerated when loans in now closed banks had to be paid off. This was to cause future hardships at Chance Company and farm loans were also being called in and farms were to be sold unless the loans were paid. Local businesses were closing in increasingly larger numbers and such retailers as the Hopes Drug Store, who lost funds in one of the defunct banks, had to consolidate with a partner to stay solvent. Before February closed, another accident over at the Chance plant was recorded as L. D. Coulter bruised his left hand while unloading a machine. In spite of Chance Company still operating, there were others businesses like the bottling company in Centralia that were quitting or moving on. Some firms like Ford Motor Company made drastic changes to try and outdistance competition, renew consumer faith, and drive more purchases. Their new V-8 powered 1932 Ford was the ticket. Because of the Ford V-8's quickness and speed potential, outlaws and bank robbers who had become celebrities to many of the down and out, depended on the new Ford V-8 to escape the law. Bonnie and Clyde even sent a letter to Henry Ford expressing their product satisfaction, although they had likely stolen it.

There was little good news in the spring of 1932. Car stripping may be considered a modern problem, but it existed back then when Granny Ryman's unlicensed Ford roadster was stolen, stripped, and burned. In addition, the Ashland bank closed, robberies became more frequent and suicides, theft, and injuries seemed to be on the increase. Businesses such as the Globe Hotel changed hands again, plus teachers' salaries were cut. Even the Chances weren't immune from these black days when everything seemed to be unraveling. The April 22, 1932 *Guard* told of a fight A. Bishop was involved in and you would think someone of his age and position would have known better. However, Mr. Chance's opponent was a buzz saw at the plant, which cut two of his fingers on his right hand. The news item explained that: "Ever ready to joke about his own misfortunes, Mr. Chance explained that he tried three times to grab the saw and finally got hold of it only to be compelled to let go!" Bill Hollander spoke of Mr. Chance's energetic nature leading to injuries, but one might consider the problems that he faced, such as current company economic conditions and separating from Kearney a month later. He didn't spend the day at the tavern, but a stress reliever may have been his frequent habit of getting lost in another invention or project at the plant or in his workshop.

The 1932 graduating class of Centralia High School including Paul Sturgeon, may have wondered, now what? But locals could always depend on Chance Company and consider that of the many things now shut down or dormant, one of them was not the firm at 210 North Allen in Centralia. The enduring spirit of the Chances was another thing locals could latch onto, much like going to a movie and losing oneself in a setting that distanced itself from the stark realities of the day. Gano found the dance floor in Columbia more inviting than meeting with bankers as Mr. and Mrs. Gano Chance, Lucille Fountain, Glenn Eierman, Art Houston, Bernard Wilson, and Gilmore Stone attended the A.T.O. Fraternity dance in Columbia on a Saturday night in late April 1932. The mention of longtime Chance advertising head Art Houston suggests his likely employment during this period. An imposing structure, the new Centralia Post Office was dedicated on June 29, for which Gano hooked up a public address system for the ceremonies. In other news former prominent drug merchant W. O. Baker was killed in a Kansas auto accident. Other names becoming household words around the United States were Amelia Earhart, Al Capone, and the governor of New York, Franklin D. Roosevelt. The Depression and the auto were reducing train ridership and locally the Alton cut passenger trains on Sunday and closed their station on the Lord's Day. In August, the city agreed with Chance Company that the fire call was four long blasts of the siren. September1929 information stated the siren on the Chance plant sounded at seven, twelve, and six, and five minutes before. This was for the beginning and ending of working hours and noon at the Chance plant. A contributor of local history events for the *Guard* and this book, a former Confederate and local schoolteacher Harvey H. Hulen, eighty-seven, passed away in Oklahoma on October 9.

A new commander in chief that would simply be referred to by his initials, F.D.R., was elected president in November and would quickly attempt to change people's fear to hope during his *Fireside Chats* over the radios of America. Progress was still being made in these desperate times, such as the conversion of farm tractors from steel wheels to rubber tires. An Allis-Chalmers, Firestone-equipped tractor set a speed record from Milwaukee to Chicago. A. C. was also the manufacturer of large electrical equipment and machinery. Barney Oldfield would later receive a speeding ticket passing through an Indiana town on one of these tractors, which also raced at state fairs and racetracks across the Midwest, drawing huge crowds. And during this time, an up-and-coming movie star, Clark Gable, was making his mark; while locally a Centralia landmark, the A. R. Toalson home was badly damaged by fire. To close 1932, a Chance social event with business connotations in early December found the A. B. Chances and Gano entertaining (not clear where) Mr. and Mrs. Jack Isbell of Winterset, Iowa. This residence suggests Jack's Wej-Lock association, but the Chances were not about to lose contact with him!

Bill Hollander remembered the employees and physical plant of the 1930s Chance Company in 1979. T. B. Carpenter was an early plant superintendent, which included the supervising of the machine shop. There were around a hundred employees at that time with ten to twelve in the office. The office was in the front of the first floor (Main Office Building at 210 North Allen) where A. Bishop resided at times, when he wasn't performing some function outside the office, which was a regular occurrence. Miss Ida Lou was his secretary, but Bill considered her to be the office manager although she was not called that. Beverly Mayes was the bookkeeper, a position his son Brent would also be involved in for decades at Chance. Like another future employee, Harry Donahue, the *Guard* reported in November of 1912 that Beverly was a schoolteacher in a public school. He and his family moved from Rowena to Centralia, as the paper recorded on October 8, 1915. Bill recalled that Verna Palmer was secretary to Jack Isbell, the sales manager, and there were two to four additional clerks and secretaries, some or all of which were temporaries. F. Gano Chance also had a space in the front end, as Bill described it, and as Gano acknowledged during a 1978 interview that his office was on the second floor in the time frame of 1931–32. Most employees, like Mrs. Charlie Palmer, who worked in the Piston Ring Packing Department, had the versatility to handle other jobs, such as coming into the office when an extra stenographer was needed. In retrospect, this was a team effort concept that allowed the job to get done all over the company, especially during the busy and slow cycles that seemed to dominate Chance production activities.

Bill said the Chance factory complex at that time included the 210 North Allen Street buildings and the building across Allen Street that was the Anchor/Warehouse facility. Behind the offices (Foundry Building) were the Machine Shop, Tool Room, and then the Punch Press Department. At the far end of the building was the Gray Iron Foundry. The Club Room occupied the third-story additions to the rear end of the original sixty-by-one-hundred-foot main building, as did the Print Shop, X-Ray Room, and First Aid area on the second floor. The second floor also contained the boxing and storage of piston rings in the main building. The third floor housed the manufacturing and machining of automotive-type piston rings and the east end of it was the Club Room with a pool table where parties were held. The forging operation was in the same area as it was later, except it was smaller. The Galvanizing Building was also a pole-type building on the north side of the Wabash Depot. The acid fumes from that operation had resulted in holes in the roof that Bill said, "You could throw your hat through. Every time it was raining, why you had to cover up the smelter or the rain would get in that hot smelter and make a little fire works." There were other reports of its deteriorated condition in later years from other employees. A. Bishop had envisioned replacing it in 1930, however, the Depression halted his plans for a new brick and steel structure.

Although World War II was a starting point for many women joining the work force, this time frame would have to be pushed back over several decades to cover the activities of women at Chance Company. The day after Christmas in 1927, Verna Palmer came to work in the office. Verna said the office was on the first floor, as you walked in the door. The office area wasn't very wide, possibly only twenty feet. She was hired to take the place of Maureen Chambers (Note: This appears to be Maurine Chamberlain) and recalled:

> *I did all of the typing and was also the receptionist. There were only three of us in the office, Miss Ida Lou, Mr. Beverly Mayes, me and of course "Boss." At that time we always called him boss. I was the "Girl Friday" and about a year after I came they increased the office to another girl* [Jennie Green]. *(1979)*

Verna also mentioned that Lillian Grubbs followed. A report of 1943 Chance Board minutes indicated that, "A similar resolution [in addition to Miss Ida Lou's] was passed by the Board wishing to make note of the retirement of Mr. Beverly Mayes who will retire from the company July 15, 1943." (1978) The early office bookkeeper, Beverly Mayes, was the father of Brent Mayes (1915–2002), who would work for Chance Company forty-five years and like his father he spent decades in the Chance Accounting Department. Forty-five-year employee Les Reynolds and four months shy of forty-five years Flynt Chandler, both prominent factory employees, remembered that Beverly always brought the checks. Receiving your paycheck never got old, but was always a significant occasion. Although some felt it was never quite large enough, those who maintained a long-term relationship with the company usually received one.

An early employee, Flynt Chandler, who started August 25, 1919, recalls two early women who worked at Chance, Mrs. Sires and Miss Ida Lou. Describing Ida Lou, Flynt said:

> *Oh boy, she was a good gal. Ida Lou was the only one in the office at the time other than Mr. Chance and me. She just took care of everything and saw that everything was done. (1979)*

Les Reynolds said, "Whenever you saw her she would stop and have something to say." (1979) According to Gano Chance, "My father's secretary practically started with the company. She was good—even better in finances than my father. She was the watchdog of the treasury—no question about it." (1978) According to the Board of Directors meeting of May 25, 1943, she started working for the company on

February 7, 1910, and retired in 1943. Because of her long tenure and daily work activities Miss Ida Lou Traughber may have been second in prominence only to Mr. A. Bishop Chance in the early years. The remarkable thing about this was it occurred at a time when it was considered a man's world, but don't you believe it. She was a local Centralia product and by all accounts a very capable lady.

A. Bishop Chance was a people-oriented as well as a hands-on person, which didn't make it easy for him to be confined to the office. Being drawn out of the office frequently was fueled not only by his outside interests such as traveling, the rock garden, and farming, but also by his continuous experimentations and product improvement/development within the company. Business requirements necessitated him to be out on the road for long periods of time. Selling customers, for example, that they needed the early Chance products was a large hurdle that had to be overcome. As a consequence, there were areas like finances that he didn't have time for or had little or no interest in. This is confirmed by his son F. Gano, who said, "My father never did care much for the financial end." (1978) Each of us would like to have someone to do those things that we never seem to have time for. This doesn't necessarily mean it's something we're not capable of performing, but is the kind of thing we would readily entrust to someone close to us whom we have confidence in. Mr. Chance had a foolproof method of quickly analyzing the company's financial health. If there was more money in his checking account at the end of the month than in the previous month, then they had made a profit. Gano continued, "He could get interested in most anything, like myself, but he had to drop some of these other things to do it. He couldn't handle them all." This explains why Miss Ida Lou became involved in the financial end of the business as well as in most other areas within the office and she had gained Mr. A. Bishop's trust before the Chance family world cruise of 1926–27. Gano said, "When we went on the cruise around the world my father made Ida Lou Traughber his power of attorney to buy and sell anything." (1978). Gano later discussed Miss Ida Lou's position of being hired originally as secretary who did the bookkeeping, etc.:

> *She became the Treasurer of the company and really the boss of the company. My father didn't care too much about business. He was interested in making it and getting this machine. But Miss Traughber would say to my father, "Boss," I don't think we can get this machine yet. We just don't have the money!" "OK, whenever we can, let's get it." So she was the key in the business end. (1981)*

This is confirmed as well as the opinion that Ida Lou ran the plant, so to speak, according to Verna Palmer, "while 'Boss' was out selling, etc." Nathan Toalson also recalled that, "She ran it also when they took the around the world cruise." (1979) Finance wasn't the only area Miss Ida Lou was involved in. Nathan Toalson was asked who was running the company in 1930, Gano or Boss Chance. He replied:

> *Mr. A. B. was really running it. Gano had just gotten out of school. Mr. A. B. and Miss Ida Lou Traughber were running the company at that time. Jack Isbell was there. Gano and Verna were there. The office consisted of 4 or 5 people at that time. She* [Ida Lou] *was a wonderful lady who knew the business and she knew everything that was going on and things that weren't going on. (1979)*

One important office function in those earlier years, just as it is today, was the processing of customer orders and initiating the paperwork for shipping. Paul Sturgeon remembered that, "Ida Lou opened every order each morning and tried to get it shipped before the next morning." (1979) Verna Palmer also recalled the daily office activities in detail:

I can still see Bus [Dutch] *Wisdom coming down the hall, going by my door, to see Ida Lou. She was writing up all of the packing slips for the orders that he had checked off, that had to be shipped. He was the foreman of the shipping department and he came over and picked up those orders and took them across the street to the warehouse for shipment. She went to work at 7 o'clock and she stayed until 6 o'clock if she needed to—six days a week for a long, long time. When I started to work there [1927], we worked from 8 to 6, six days a week. (1979)*

Mr. and Mrs. N. A. Toalson (Jack Chance Collection)

In the later years, from the mid-1930s on, the activities of both Ida Lou and Verna were recalled by a longtime employee Roy West:

I knew her [Ida Lou] *pretty well because I was in shipping then. She was handling the orders through the various stages and we had a lot of telephone correspondence back and forth. This was necessary because she was across the street and we were shipping out of the other building. Ida Lou was in there pitching to get the job done. Verna was too; Verna was damn good help for us, I'll tell you, she was on the ball. She knew just about all the customers Chance Company had. For a long time she and Ida Lou were their main contacts, when they called in to the plant because Jack* [Isbell] *and them would be out on the road. Granny Ryman had the shipping at one time too. He was in there before Dutch was. (1979)*

Through marriage and Chance Company positions, the Toalsons had Chance family connections and each person that was knowledgeable of their relationship had their own version. Bill Hollander, whose employment predated that of Nathan Toalson, had these remembrances: "Nate's daddy was a contractor up here and at that time he was a very wealthy man. Most of the sidewalks in this town were built by A. R. Toalson." (1979) During noonday walks in Centralia, the name A. R. Toalson would frequently appear on a slab of sidewalk concrete confirming its origin and quality of construction. Mrs. Nathan Toalson said that he didn't use sand that was contaminated with clay or dirt as it weakened the final concrete product. Such adherence to quality construction methods is admirable and a credit to his character and high standards. The positive influences of the father was passed on to the son and daughter, Nathan and Anna Lee, which had a direct influence on Chance Company. Many examples of items built in earlier times reflect a quality and construction detail that can't be found today, which supports the proposition of saving and restoring instead of madly pursuing "the everything has to be new" phobia.

The name Toalson began to emerge in the *Guard* of 1895 with an advertisement for A. R. (Alex) Toalson who was a dealer in Foundation Rock and Sand. He also was into house moving with "Raising a Specialty!" House moving was also an early occupation of Theodore Gary in

Macon, as previously covered. The February 25, 1910 *Guard* said A. R. Toalson had erected a building to build concrete blocks. Sidewalk work began to be mentioned more frequently, during those times, as a Toalson sideline. Later on June 17, the paper continued the coverage of Mr. Toalson's concrete work in the city park area (Square). "He is making the job a good one for many reasons, one of which, is that this is the sort of work he always does." Early on the Toalsons, native Boone Countians Alexander Robinson and Alice Lee (Hardin) Toalson, lived a block west of the Square and the business office was apparently located in the family home.

In communities across the county, one of the most sensational new businesses was the local auto dealer. This was due in large part to Henry Ford's Model T (1907–1927) that offered a simple, quality product that was priced so low it was affordable to the masses. Newspapers poked fun at it, such as one of its many nicknames, Tin Henry, and the *Guard* was no different in suggesting the Ford should have access to the sidewalks, along with pedestrians and bicycles. The lucky recipient of the Ford (also Overland dealer) franchise in Centralia was the Lyon Auto Company. The June 6, 1913 *Guard* said they had sold over thirty cars so far that year and three the previous week. Sales and maintenance activities outgrew their store so a large, new building (50 x 125') was built one block west of the Globe Hotel. This would be on Railroad Street and could be the location of the old bowling alley building, now occupied (2000) by a bank and printing concern. It was built by A. R. Toalson with cement blocks that he fabricated, according to the August 8, 1913 *Guard*. The September and October papers indicated Mr. Toalson was also continuing to build sidewalks as he was in May 1916. Also in March 1916, daughter Anna Lee was a new convert at First Baptist Church. A March 9, 1928 news item in the local paper told of the A. R. Toalson's living in the same house, at the Sneed/Barnes Street corner, for sixty years!

Bill continued to explain the Toalson connection. "He cast building blocks and Nathan at that time was a young man in high school. He had the job of mixing concrete by hand and pouring into these forms to make the building blocks." After Nathan had entered college in 1928, his father passed away in 1929. Bill continued:

> *Nathan had just started in school and he had to drop out and Gano and Ann got married. Gano figured that I'll find a job for Nathan. He (Nathan) came over here and that would have been about, I am guessing, 1930. So A. B. found a job for Nathan. He worked in the foundry carrying iron and when they weren't pouring he was kinda grease monkey in the punch press department. I mean he had a very menial job but he was working, he was making something. I would say he did not make much. The boys in the factory at that time, the experienced good hands made 25, 35 cents an hour, worked 5 days a week, 9 hours a day, all straight time. They worked from payday to payday. Come payday all merchants were outside the door trying to get theirs and trying to be the first one in line. About the first promotion Nathan got, I believe, they were running analysis on iron and Gano would set up carbon train and the other various tests to run analysis of the iron. Gano taught Nathan how to run these various iron tests so he would go down and make drillings of the pig iron and casts and run carbon and manganese and set up an analysis of the metal. Nathan was a shrewd fellow and a smart fellow. He did it on his own. Anything that he was ever put to do he found out what he was doing and he continued. He was a good boy. Nathan and Paul Sturgeon are a pair that had the opportunity to go on to school could have reached any heights. In all they were 2 very brilliant and very shrewd boys and neither one of them had that opportunity. But they made by reading and knowing*

> *what they were doing. They both developed to the very utmost. (1979)*

Jack Curtright recalled that he and Nathan played together as boys growing up in Centralia. They also "roomed together when we went off to school." This appears to be a reference to college as Jack noted that both of them joined fraternities. According to *The History of Missouri*, Nathan graduated from high school in 1928 and then attended the University of Missouri for one year (1928–29). As the economy worsened, jobs were hard to come by but like other fortunate Centralians, he got on at Chance Company in 1930, which was still doing well.

The discussion about Nathan and Paul is further expanded with remarks from Paul about Nathan, Gano, and himself. Paul suggested Gano first put Nathan in the lab and later placed him over the Purchasing Department. Part of Gano's revamping of the various departments within the company included Purchasing, of which Nathan became manager. Paul said, "Gano came into the Purchasing Department. . . . He finalized the purchasing system and he took Nathan. Nathan and I were working together up in the Piston Ring Department." (1979) *The History of Missouri*, in its coverage of Nathan, said he became a purchasing agent (buyer) in 1934. In 1939 he became office manager and in 1941 he became vice president and a member of the Board of Directors, plus duties as assistant general manager. Promotions continued as in 1950 he became general manager and later executive vice president. When Gano semiretired in 1960, he appointed Nathan as president of A. B. Chance Company. Like the Chances, Nathan was active in many community and civic activities, which reflects the Chance Company policy of encouraging employees to be service-minded citizens. He was involved in and had an abiding, enthusiastic interest in the Boy Scouts. His Christian faith was reflected by his longtime involvement in the First Baptist Church of Centralia and in the widespread activities of the Baptist denomination. This included serving as a trustee of the William Jewell College in Liberty, Missouri, and a member of the Missouri Baptist Hospital Association.

One insight into Nathan's executive talent was mentioned by Robert J. Tanner many years ago. Bob said that in meetings, Nathan had the uncanny ability of asking key questions that could reveal the core or heart of a matter. This shows an active, penetrating thought process that gave him the skill of analyzing and assimilating various issues coming before the Chance management group. Undoubtedly, this quality benefited A. B. Chance Company immensely. In a question about his early Chance years, Nathan confirmed he was born (January 25, 1911) and raised in Centralia and began work at Chance in:

> *1930. Of course I had known the A. B. Chance Company. I remember when they built the foundry back in 1924 or 25 which was really the extension of the offices on back. There was a building about 300' long in which they had the punch press and the factory operations making piston rings, etc. (1979)*

Nathan recalled that his first job after being hired at Chance was organizing the items the Chances had accumulated on their around the world cruise:

> *Menus from the various countries. They had them in big boxes and I had to separate them by countries. They had all of these coins they had picked up over the various countries and I had to sort out. They were all mounted and framed. They hung in what is now the International Department on the third floor of the office building. (1979)*

Not unlike many other businesses during the Depression, the local Pruett Theatre underwent change. It was owned by Joe E. Green with forty or more stockholders and was featured in the *Guard*'s issue of January 6, 1933. Movies were important during the troubling times of the Depression, as patrons generally left the theatre refreshed, uplifted, and their hopes temporarily restored. The Pruett seated 350 and in April it came under new management. E. F. Bebermeyer made several improvements and renamed it the "Vista," which does sound more Hollywood. There is a Chance Company connection with this theater as *Chanco News* pictures profitsharing meetings being held there, as Chance outgrew its plant meeting room in the postwar years. To fill their silver screen in 1933 such actors as John Barrymore, Barbara Stanwyck, Irene Dunne, and an already super-star Tom Mix were given star sketches in local newspapers. News items suggest the Chance family were regular moviegoers. Will Rogers was also becoming a household word, as he was a correspondent for nearly two hundred major newspapers. Lowell Thomas, radio broadcaster, author, lecturer, and world traveler was becoming recognized before his World War II London radio broadcasts would make him world famous. And on March 4, 1933, the most prominent person in the United States, the newly elected president, Franklin D. Roosevelt, was inaugurated. People needed role models and heroes to look up to, such as Babe Ruth and his contributions to orphanages (he was an orphan) made him even a bigger hero. Dizzy and Paul Dean, who formerly knocked squirrels out of Arkansas trees with rocks, were now setting strikeout records with the St. Louis Cardinals. To further counteract the dismal conditions, including 16 million unemployed, there was a continuing effort to maintain hope and not give in or give up.

Community duties were not always of an enjoyable nature, as A. B. Chance in late January 1933 was a pallbearer for seventy-five-year-old John R. Sames (1858–1933), a fellow Christian Church member, born northwest of Centralia. Mr. Sames was senior proprietor in Centralia with fifty-two years of grocery business service. When his career ended, Mr. Chance would also enter into the ranks of over fifty years as a local businessman, combining his telephone and Chance Company service. Over at the plant, Galvanizing employee Ralph Tucker was struck on the head by a falling galvanizing bucket, but he wasn't seriously injured. This was another danger of this operation as materials had to be hoisted up, swung over and down into the galvanizing pots. Charles P. "Cab" Atkins, crossing Allen Street, was hit by a Producers Grain truck, which resulted in a fractured arm in February 1933. He would later become a longtime employee at Chance, as the May 10, 1935 *Guard* told of his graduation from Centralia High School. A public benefit of Chance Company medical capabilities resulted in Mrs. Ralph Fast getting her fractured leg x-rayed at the Chance plant in early May.

In spite of the Depression, life continued in the face of more frequent local and national bankruptcies. Internal Revenue Service statistics revealed that 62 percent of U.S. businesses were operating at a loss during that time. There was the feeling that if a positive attitude could be sustained, it would eventually overcome these hard times and like the recessions of the past it too would go away. A birthday party for Gano and his mother in February 1933 was attended by Mrs. Margaret Greenlee, Mrs. Gertrude Gano, A. B. Chance, and A. B. Chance III. Mrs. Greenlee attended the P.E.O. convention in Hannibal in June of 1934, indicating her continuing Columbia connection. In February 1935, she visited friends in Columbia and returned to Centralia, which suggests she now was living with her mother Mrs. Gertrude Gano. In the *Guard* in March it was revealed she had sons, Bob and Russell of Kansas City. Russell was a future Chance employee. Mr. and Mrs. R. P. Greenlee and son Bobbie were entertained by Mrs. Greenlee and Mrs. Gano Chance in June 1935.

Part of the financial investments of A. Bishop was in real estate. This included homes such as one on east Singleton that he rented to the new Kroger manager W. M. Duren in February 1933. Investments in farm property would also come about, especially since a cash-short public found real-estate sales at the courthouse steps or in bank auctions. Certain legal issues voided the first auction of the Bank of Centralia in November 1938 in which A. Bishop had bought a farm. A second auction was held in which he purchased the former bank building, the farm going to a higher bidder. Gano, like his father, also purchased a property on South Jefferson for $500, known as the J. Sam Brown place. In a short time this might be the location of his new estate.

America wasn't standing still in 1933 and there was a renewed optimism that the new Democratic administration now residing in the White House would turn the economy around. The love affair with lighter-than-air crafts would blossom and quickly fade, especially after the 1937 *Hindenburg* disaster. But April of 1933 found America with the same sort of loss with the destruction of the airship *Arkron* off the New Jersey coast during a storm, with seventy-plus killed, and then the *Macon* went down in February of 1935. But back in Centralia there was a concern for the less fortunate, which was becoming an acute problem. Gano and R. E. Ausmus were on a Commercial Club committee to raise funds to buy seed potatoes for the area's needy. A late April seed potato report in the *Guard* told of many making excuses and the results were only about half of what had originally been expected. A. Bishop was also involved in a charity ball to raise funds for the needy. He offered his Chance Hall and the *Guard*'s E. T. Rodemyre gave free advertising. That same Anthony Building was also home to Centralia Motor Company, which continued to be the local Ford dealer, but now also had the Plymouth franchise and the John Deere agricultural line. Without a school gym, Centralia High School basketball games were played in the big second-story hall above the dealership. Coach Knepper, however, was having problems with rowdy fans, not so much at the games, but at practices!

According to Gano, after Chance Company was incorporated, his father had become less active in the day-to-day operations. Along with the tough economic times that continued to handicap the plant's operations, it would suggest that A. Bishop had retired to a rocking chair, but *Guard* news items like the ones in the previous paragraph indicate otherwise. A month before in early March, he made a business trip to St. Louis and H. A. Jones (Chevrolet dealer), Morris Dann, and Hugh Howell were also there on business, suggesting they were together. This was likely for the Commercial Club or other Centralia interests in trying to get aid or other support for the community. So instead of slowing down he just directed his energy into yet another area. This volunteer work away from the factory shows the Chances' strong community support and concern for the needy. During the same period A. Bishop was toastmaster at a Fire Prevention banquet held at the Christian Church, which also had Commercial Club connections. Within four days of that Friday-night event, his second grandson, Phillip Gano Chance (1933–2005), was born at Noyes Hospital in Columbia on March 3, 1933, to Mr. and Mrs. F. Gano Chance. A couple of weeks later, the March 24 *Guard* told of another A. Bishop business trip to Jefferson City, again with H. A. Jones and S. T. Dawson. Federal headquarters for the new recovery programs were apparently located in both cities, although it was early for New Deal programs to have begun. Also that week Plant Manager J. H. Cheetham had become a new member of the Commercial Club.

A new era after the Kearney association was ushered in with Mr. and Mrs. Jack Isbell's return to Centralia in mid-April 1933. They were dinner guests of the A. B. Chances and a six o'clock supper by the Gano Chances on East Singleton. Mr. Isbell was remaining, it was said for a time on business with Chance

Manufacturing Company. This clearly suggests if he wasn't already a Chance employee, there was a Chance offer on the table to rehire him. An unexplained situation was Isbell's Kansas City residence, but his employment status and the changes at Wej-Lock would help explain that. Fast-forwarding a couple weeks, the May 5 *Guard* told of Chance employees Jack Curtright and Art Houston spending the weekend in Kansas City, accompanied there by Jack Isbell. Mrs. Isbell came back with them on April 30 and would spend the week with her husband in Centralia. Jack was obviously now an employee of Chance Company, but what occurred on Tuesday, May 2, suggests Mrs. Isbell was overly optimistic. Mr. and Mrs. A. B. Chance, and Jack Isbell left on Tuesday on a business trip through the East. Mrs. Chance, upon their arrival in Niles, Ohio, would visit with the N. C. Husteds while Mr. Chance and Mr. Isbell headed further east. The wives of Chance Company employees who were frequent travelers may have never accepted their husbands absences, but Mrs. Chance and Mrs. Isbell in this instance, coped with it differently. In early July 1933, Jack Isbell was home on a Monday morning from Springfield, Missouri, where he had spent several days on business. Not only did Chance sales personnel depart Centralia, eventually for most parts of the world, but Chance salesmen such as veteran Chicagoan Jack Milligan came to Centralia. In this case as the *Guard* recorded on July 21, 1933, he and his wife were guests of the A. B. and the Gano Chances, which along with other recorded visits suggest the Milligans had close social ties with the Chances and others within the company. On November 10 the *Guard* covered an uncommon departure, but one that was reflective of the future. Office employee Miss Jennie Green and the Isbells departed Kansas City for California. Miss Green was apparently on vacation and was going to visit her brother on the West Coast. The Isbells on the other hand were going to stay over for about three months. It is assumed that Jack would be performing Chance Company business such as visiting customers and/or lining up Chance sales representatives.

In early June 1933, the P.E.O. Chapter and B.I.L. members were entertained at the A. B. Chance home with a travelogue of moving pictures of their around-the-world trip of a few years previous. During this time there were frequent bank robberies taking place and other incidents of stealing being reported. Chance Manufacturing Company wasn't immune from such threats and night watchman John Botkin was there to prevent any such occurrences. In early July, while searching for two men suspected of breaking the seal on a boxcar loaded with Chance materials, John fell off the boxcar knocking himself unconscious for a short period. Remarkably, showing the toughness and endurance of the people of that time, he continued his shift, undoubtedly with a substantial headache! A week later, A. Bishop and his frequent traveling companion H. A. Jones were in Columbia on a Sunday on business. The frequent and substantial *Guard* advertisements of Chevrolet vehicles during 1933–34 was an indicator of the success Jones was having, especially against his Ford and Plymouth competitor, Centralia Motors. The irony of this was A. Bishop liked to be around successful people, but Jones was competing against Mr. Chance's lessee in his Anthony Building, Centralia Motors. Mr. Jones' business also included the Oldsmobile line, first noted in late 1934. This business still exists today (2000), less the Olds line, which lasted nearly seventy years! Bill Forrest is the current owner of the original Jones Motors, the son-in-law of Jack Curtright mentioned in the previous paragraph. But things turned dark again for the Chance family as other prominent people had found, that such success could bring threats to them and their families. The purpose was usually to extort money as it was in this case of anonymous and threatening letters that A. Bishop had begun receiving prior to August 23, 1933, when he received another. It was turned over to authorities and since it involved the

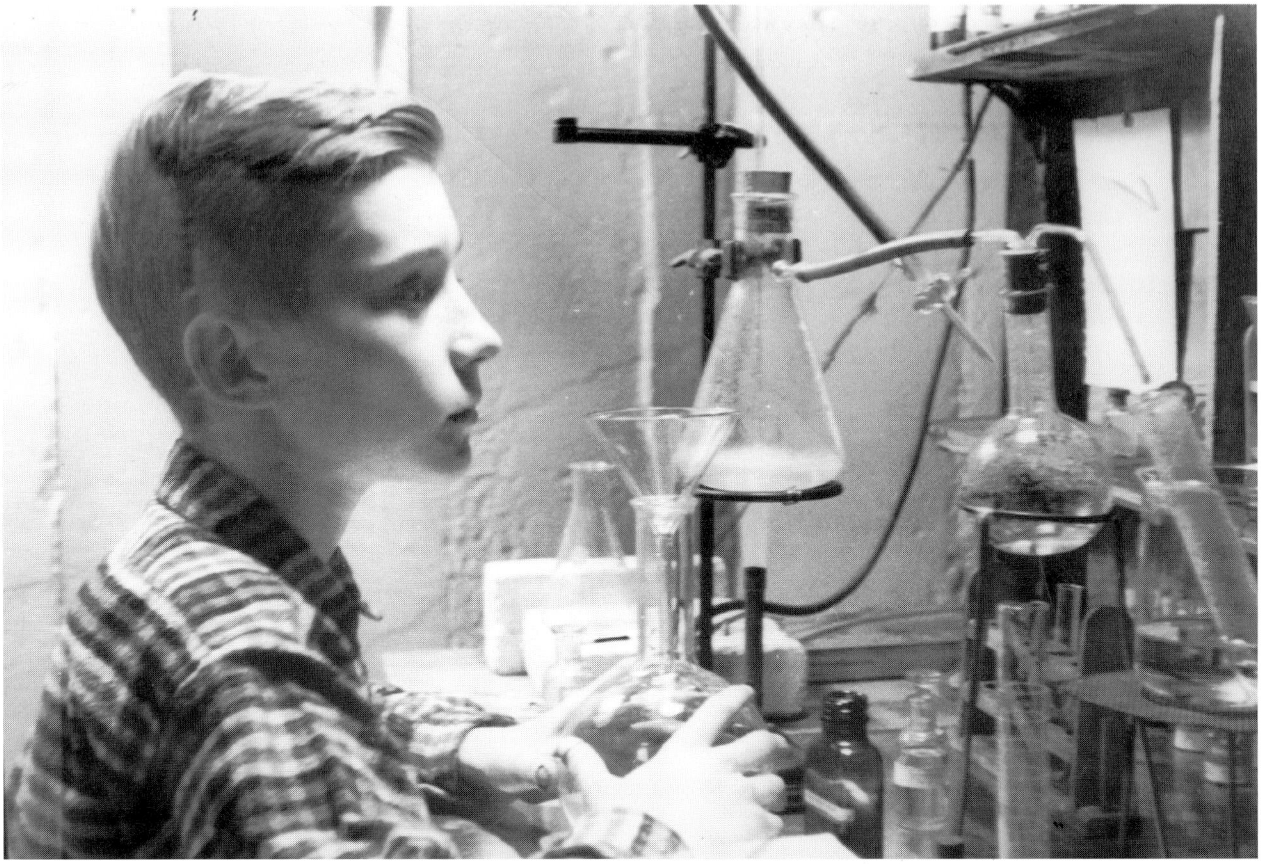

"Little" A. B. III in his laboratory where he was in his element. A former college professor suggested he was the most outstanding chemical student he had taught. The location of this photo was apparently at either his grandfather's basement lab or at the Borns, his parents. (A. B. "Buff" Chance IV Collection)

mail it was a federal matter. To halt gossip and rumors about Mrs. Grant Irvine, who had been questioned by authorities, Mr. Chance issued a statement that he didn't believe Mrs. Irvine was connected in any way, but was instead just routinely questioned.

One of the most striking, popular, and successful attempts to counteract the Depression was the Century of Progress Exposition in Chicago. All roads, railroads, shipping lanes, and air routes led to Chicago as it opened June 1, 1933, and was extended another year into 1934. Many attendees from the Centralia area were reported in the *Guard* as the summer progressed. Two examples in the July 28 coverage were G. V. Proctor and son Roger and A. B. Chance III leaving on a Friday night for Chicago. "Little A.B." however, was going to visit his mother Mrs. Allan Born, which may have included a side trip to the big fair. The sensational technical aspects of the Century of Progress had to have whetted A. Bishop Chance's appetite to attend. The September 29 *Guard* confirmed that Mr. and Mrs. A. B. Chance and Mrs. Margaret Greenlee did attend the event that was continuing to grow in popularity and recognition, worldwide.

During the early 1930s another form of professional entertainment that had a Century of Progress connection was the "Big Bands." Guy Lombardo (1902–1977) and His Royal Canadians playing "the sweetest music this side of heaven" was an early example. The Lombardos, originally from London, Ontario, were one of the longest-surviving dance bands and one, if not the most popular, with over 100 million record sales. Their first record was a 1924 cut in Cleveland and then they moved to the Granada Café in Chicago during 1927 where they later had difficulties leaving because it was a gangster-controlled nightclub. They were able to relocate to the Roosevelt Hotel in New York City in 1929 and became nationally famous for their New Year's Eve show and *Auld Lang Syne* to which Steve Allen said: "What Santa Claus is to Christmas, Guy Lombardo is to New Year." Among the many entertaining events taking place at the opening of the 1933 Century of Progress in Chicago were four orchestras chosen to provide dance music at the Pabst Blue Ribbon Casino. Featured were Ben Bernie, Buddy Rogers, Tom Gerun, and "newly returned from his conquest of New York," Guy Lombardo. It is not known if the A. B. Chance party, later in 1933, stopped by to listen or to dance.

By the 1930s the Chance Advertising Department had in-house artistic and photographic skills plus printing capabilities for internal and external needs. These benefits would bring prominence and recognition to Chance Company and also influence and benefit the electrical and communication utility industry. Bill Hollander commented about this department and the longtime Advertising Department head-to-be, Art Houston: "When he came here we had an artist, Harold Hund, an old bachelor out of St. Louis. . . . He was a good artist and photographer." (1979) When Gano Chance suggested that the company do their own advertising at that time, Bill continued with details of their many activities:

> *Anything in the magazine, we prepared the advertising copy and the pictorial part. . . . We printed our own letterheads—our own boxes that the piston rings were in. The artwork and the printing was all done in our own print shop. The foreman in the print shop then was Emmit McBride. . . . Besides Emmit, P. D. Botkins ran one press and Lena Welch set type. Bark Ledro also ran a little shop press that was about the print shop. (1979*

When the name of Sherman Jones was mentioned, Bill said:

> *Sherman came after Art came. Hund was here about two years and he went back to St. Louis. Then Ben Crump was doing the artwork. Ben did the artwork for a short time . . . then Art Houston came. Art and Ed McLaughlin came about the same time. Ed would have come here first and he was the one that hired Art. He was preparing the advertising layout and what not. Art at the time was painting posters for the Pickler shows over in Columbia. He hired Art to come on up here and handle the artwork and camera equipment that we had up here. I don't think Art had ever taken a picture with anything but a box Kodak. The camera they had up here was an old Bellows type. You cut film and loaded it into a plate holder. Glenn Eierman was one of Gano's schoolmates at Columbia and he was working with a photo service down around Rolla, a civil engineer. He urged Glenn to come on up and join the Engineering Department. Glenn and Art were pretty good buddies. They wanted to take some pictures and so they went down to load some plate holders. They took the pictures and they developed them—oh they were fuzzy—you couldn't make anything out of them. So they decided the time wasn't right. So they loaded some more and they were no good. Art asked me what the trouble was. I said, "You got the film upside down!" You could tell by where the notch was. . . . They finally turned it over. Art became a real good photographer, but when he first started with a box camera, that was the limit of his photographic work. (1979)*

During this early 1930 period, Ed McLaughlin was the manager of the Advertising Department and Art Houston was the artist who also prepared copy and did the photography work, as Hollander recalled. When Chance became involved with Kearney, McLaughlin was transferred to Kearney. As Chance Company reestablished their operating areas after the Kearney separation, a new Advertising employee was recruited. Sherman Jones was said to be Houston's brother-in-law and was working with a newspaper in northwest Missouri when he became the new Chance printing shop operator. Before the 1930s would close, the Chance Advertising Department activities would culminate with a trendsetting company publication, *Chance Tips*, directed toward the electrical utility industry. During the war, *Chanco News*, an internal organ, kept employees and military inductees informed of company activities and re-enforcing the feeling of family within Chance Company.

Crime and crime figures were prevalent commodities in the first half of the 1930s. One by one, law enforcement, with the aid of pictures sent over the wire to the nation's newspapers and two-way radios, closed in on the most prominent and ruthless of this group. It was the age of the "G" men and Public Enemy Number 1, as the F.B.I. became prominent under J. Edgar Hoover's leadership. The license stolen off the Chance car in December 1933, however, didn't qualify for a nationwide manhunt. The capture of John Dillinger in January 1934 and his jailing in Indiana led to his daring escape a month later with a wooden pistol. In May 1934, Bonnie and Clyde were gunned down in their 1934 Ford sedan down south. In November two Public Enemies, "Pretty Boy" Floyd and "Baby Face" Nelson were killed by lawmen. Soon after the Lindbergh baby was kidnapped with the trial and a conviction early in 1935. Mr. and Mrs. E. F. Bebermeyer, managers of the Vista Theatre in Centralia were also kidnapped upon the closing of the theatre one night in August 1935, but later released after the movie receipts weren't found. Bank holdups were as frequent as bank failures. Just as Prohibition had been blamed for the growth in organized crime, now it was hard times that were pointed to. It was a violent period in certain aspects, and a time when drastic actions were the rule of the day. Even in the Centralia Township, law-abiding citizens had reached a boiling point in their belief that Boone County officials in Columbia didn't acknowledge their existence. Their solution to the Centralia/Columbia feud was to secede from Boone County and join Audrain County! The October 13, 1933 *Guard* carried the story of the meeting held and no matter how hot the water there was one prominent Centralian who was always willing to jump in, if he felt a situation had merit. A. Bishop Chance presided over the meeting and pointed to one-half the distance to the county seat in Mexico from Centralia versus Columbia. He said lack of Boone County attention and perks or "the wheel that squeaks loudest gets the most grease." After discussions on both sides of the issue, it was tabled for the time being. Opposing views suggested it was legally impossible to secede because of outstanding bonds and some felt it wouldn't be a good move.

The behind the scenes coverage of what would begin to be called the A. B. Chance Manufacturing Plant, was subdued to an extent during 1932–33. The stark nature of the times was not always publicly obvious from the pages of the *Guard*, as there was no dwelling on editorial materials setting such tones. Centralia was a special place and perhaps did not represent the level of difficulties others endured, although there were many around town hurting. Chance family travels and activities continued, influencing others in the community to not fail and keep the faith. It also led to the Chances becoming a target as success did breed jealousy and those who tried to take advantage of them. The many people involved in the company also led to those who had bad experiences. In November 1933 Donald Wood, a nine-year die-maker, left for a similar position in Alton, Illinois. In a more obvious example of problems, Carlyn B. Taylor sued Chance Company for eighteen shares of stock valued at $1,800, per the December 8, 1933 *Guard*. He stated that written evidence was held as to how this transfer was to have taken place. There doesn't appear to be any connection between Carlyn and Herbert M. Taylor of Canadian Taylor anchor fame. A January 1930 news piece described C. B. Taylor as the "Big C Sales Manager." Incorporation details stated C. B. Taylor in April 1933 was made a director of Chance Company. His status within the company appears to have dramatically changed during one of the most difficult years in the company's history. Although suicides weren't uncommon, the news in the April 6, 1934 *Guard* of thirty-nine-year-old Carlyn Taylor committing suicide was made more significant because of the December 1933 lawsuit. It stated he suffered a serious stomach disorder, was a veteran of World War I and had come to Centralia about six years earlier, accepting employment at Chance Manufacturing Company, selling piston rings. He had left the company about four months previously, which coincides with his lawsuit. He was buried in Carthage (Missouri?) and his mother was then living in Centralia. Mr. Taylor was described as a well-liked, friendly, and witty person.

N.R.A. stickers were appearing in the paper and in store windows and there were those who disagreed with its intrusion into their business practices. F. Gano Chance, after a time, also became frustrated with the "red tape" per his following comments. Although Gano was not an advocate of bigger government, he did try to comply with N.R.A. (National Recovery Act) in 1933. Gano said, "We thought it had possibilities. We had the N.R.A. insignia on our stationary." (1981) However, Gano did disagree with the setting and raising of prices under the program. Chance Company was making so many different things that they were assigned five different industry codes. This generated a problem in transferring employees from one area to another (Drop Forge Department to Machine Shop, Foundry, etc.), which was a common practice at Chances, to maintain steady employment and keep overhead down in slow areas:

> *I went to St. Louis to try and get us under one code. I found that the State Code Administrator, because the organization was only a few months old, just did not know what the procedure was or what you could do. The Associated Industry, we had just joined, and the staff did not know any more what the government was going to do or try to do than the government did itself. Well, I wound up making a trip to Washington to follow through after I did not hear for about three months. I was not sending reports in as I was a little disgusted. It took money and time to fill our those forms and I said to myself, "well they can throw me in jail but I just am not going to spend anymore time filling out these forms." So I did go to Washington and saw all of the different code administrators and gave them my song and dance. They said, "Yes, we will get you under one code." I said, "when can you make it effective?" "Well you just go back and forget about it." So I went back and stopped sending in the reports and then it was declared unconstitutional. Therefore I did not get thrown in jail. (1981)*

In addition to coverage of the Arthur Brutons' Centralia visits to see friends, a July 1933 news item told of Judge Bruton's ill health over the last several months. This is out of character for his usual rock-steadiness; a person the people of Centralia had leaned on and now Boone Countians were depending on. Just as unexpected, the November 10, 1933 *Guard* told of Judge Arthur Bruton's death on the fifth of cancer of the stomach. Arthur Bruton (1871–1933) was born in Centralia and graduated from the University of Missouri in 1902 with a law degree. He married Miss Susan Gale Henry (1873–1934), from which one daughter, Frances, a Mrs. C. P. Greene of Chicago, survived. He was affiliated with the Christian Church and was said to have had an office there at one time. A. Bishop Chance was an honorary pallbearer with burial in the Centralia City Cemetery. Judge Bruton served nearly twenty years in the mayor's post, from 1911 to 1929. The only way Centralia citizens would allow him to vacate the position was his new Boone County judgeship and Columbia residency forced the issue. He was a man who positively influenced the Centralia area. It doesn't appear that he and A. Bishop were close friends, but later the accomplishments and differences of both these men came to be respected by each other, although they represented opposing political parties.

The three-year period of 1931 through 1933 appeared to be the most crucial in Chance Company's history. Surprisingly, after tasting success crises did develop that in some ways were more threatening to the company's survival than earlier challenges encountered when it was a small and struggling firm. It was initially punctuated with the highs from the 1920s boom times, which quickly evaporated. This ushered in a period with new lows, developed from a mixture of the Depression and a bout of management indirection. Yet out of this came decisions that would prove of value to future success. The Wej-Lock association, incorporation, and the rehiring of Jack Isbell were positive reminders that the company would soon be heading back up.

Wej–Lock

Anchors and Guardrails

The Wej-Lock Company was a going concern and when the Chance Company association was established, the products didn't require the immediate attention as did the Taylor anchor. Mr. E. K. Cole of Winterset, Iowa, had control of his operation, especially on the West Coast where a new salesman, Jack Isbell, was making inroads into Chance Earth Anchor sales. This wasn't something that went unnoticed in Centralia for a couple of reasons. First, competition was always a threat that required constant monitoring and secondly, Chance Company was all too familiar with Cole's flashy West Coast salesman. In fact, they knew his selling strengths as he had been on the Chance team beginning in 1928, according to his retirement coverage in early 1959. Jack Isbell's separation and loss in sales support of Chance Company products shouldn't have occurred, as a special arrangement had been agreed to with Kearney. A part of this agreement was that Chance salesmen were to be absorbed into the Kearney sales force, but this didn't occur as expected. The year 1932 would see the eventual end of this problem as Mr. Cole brought all or a portion of his Wej-Lock operation southeast a couple of hundred miles to Centralia, after a mutual agreement was made with Chance Company.

Chance Marketing executive L. C. Hansen recalled that Jack Isbell had told him early in his career that after quitting Chances he began representing the Wej-Lock Company. In addition, Jack had said he "took a great part of the anchor business on the West Coast." (1978) F. Gano Chance had further remembrances about Jack Isbell:

> *So Jack went to work for Wej-Lock, the name of the anchor they had, and he was an excellent salesman. He went out and started to take the business away from us. He tells the story, he would call on a customer back in California and the customer would kid him, "Listen Jack, you told me the Never-Creep® anchor was the finest anchor made, now you are back here with this anchor!" He says, "that is right, it still is but I am selling this anchor." And he sold it. That is the way he answered, we used to kid him about it. Jack wanted to get back. Jack came to our Sales Department and became the Sales Manager. (1978)*

Bill Hollander explained, from his point of view, how Jack Isbell regained Chance employment again in the early 1930s.

> *The way Jack happened to come back to Chance Company is when we dissolved our relationship with Kearney, why we had to start building up the Sales Department again. So A. B. wanted Jack to come back but he had a good thing going with this Wej-Lock anchor that he was selling. He really had California solid with that Wej-Lock anchor and A. B. went up and he bought out E. K. Cole and moved Cole down to Centralia in order to get Jack back with the Chance Company. (1978)*

Jack became a Chance employee again no later than April/May 1933, as covered in the previous section. Incorporation notes suggest the Chance/Wej-Lock association had been set by December 1932. The sequence of events would appear to be the Kearney separation, Wej-Lock association and then the Jack Isbell hiring as this interview confirms. Andy Anderson, in recalling a conversation with Jack Isbell said, "Well, they got Jack back when they busted up with Kearney." (1979)

There is some confusion as to when Jack Isbell came back to Chance Company, after a short stint with Wej-Lock. It could be assumed this occurred when the association with Mr. Cole's firm began, but Jack appears to have had to personally work his job situation out in the spring of 1933. There may have been downtime while all or a part of the plant was relocated to Centralia from Iowa. As to Jack's employer, it would logically be Chance Company in October 1931 (Kearney/Chance consolidation) and Wej-Lock by May 1932 (Kearney separation). His Chance Company reemployment would surely follow no sooner than the Wej-Lock association in late 1932. Jack's employment with Wej-Lock appears to have influenced the establishment of the Chance and Wej-Lock relationship.

A sketch of Jack Isbell's career and the details surrounding the Wej-Lock and Chance Company association, was presented in the January 1946 *Chanco News*. This article was also describing the Bowie Switch purchase and the appointment of Jack to a new position on the West Coast an area it said was not unfamiliar to him. Jack was sent earlier to the Pacific Coast in February 1928 as a Chance sales representative and that lasted until October 1931. He then left Chance employment to help form the Wej-Lock Company of Winterset, Iowa, with E. K. Cole. Jack became vice president and other sources confirm Mr. Cole was president. In October 1932, a partnership was formed between Chance Company and Wej-Lock and Jack then became the Chance sales manager and later vice president in charge of sales in 1939. This information ties down some important dates and also raises the issue of the type of arrangement existing between Chance and Wej-Lock. According to that article, it was a partnership arrangement. In reference to Isbell, it may have been cleaner to suggest Jack immediately became Chance sales manager and the agreement between the two firms may have spelled that out, but if not, it appears to have taken place a few months later in early 1933. In a discussion with Bill Meador (September 2005), he was asked about the Coe family, which led to a discussion about the closely similar name of Cole. Bill knew both families and vividly described the location of the Isbell farm that the Coes purchased. He also said something else that was very interesting. When Mr. Cole passed away in 1956, Bill transported the body to Winterset, Iowa, for burial. He said that "Cowboy Town" (John Wayne) and while there he witnessed a plant operation making anchors!

Chance Company's dependence upon Kearney for the marketing of Chance products and the resultant loss of their small sales group generated a crisis of sorts in Centralia when this relationship fell apart. Jack Isbell was more than just a super salesman he was also a capable manager who could organize a new sales department, which is just what Chances needed at that time. There was a problem however, as Bill Hollander pointed out; Jack

From *Chance Tips* of June 22, 1939. Jack Isbell

who had developed a relationship with the Wej-Lock Company. In some ways, this is predictable as he had discovered and hired Jack initially and was still in charge of Chance Company in the early 1930s, although his son Gano was also involved in the Kearney/Wej-Lock affair as his remarks indicate. However, it appears Bill saw A. Bishop as the boss.

F. Gano Chance said that:

E. K. Cole was the owner and inventor of the Wej-Lock anchor. I was up in Iowa and met Mr. E. K. Cole. I went up and spent some time getting acquainted with Mr. Cole and he was getting along in years. He had quite a few items, an anchor and also a guardrail that they were selling to the highway departments. So we worked out a deal whereby we took over his guardrail business and made a contract with him on anchors. We paid him a royalty on anchors a good many years. We acquired our first acquisition which was the Wej-Lock Company. (1978)

was doing well in selling the Wej-Lock anchors and he was in California, which he loved. (Note a December 1932 news item stated the Isbells were residing in Winterset and then a spring 1933 news item suggested Jack was then living in Kansas City.) Although all information comes short of indicating he may have refused offers to come back to Chance, it is a possibility. It is also interesting that Bill referred to A. Bishop as the main character in attempting to get and wanting Jack Isbell back and the one

This reference to the "first acquisition" reflects the earlier suggestions made about the variations in recognition of Chance acquisitions. Of course, no acquisition was ever the same as some were more substantial, even including real estate and manufacturing facilities. There were also immediate and long-range benefits that were different and the Wej-Lock agreement would appear to be the first one Gano was involved in. However, the most important revelation of Gano's remarks is the accord didn't fully cover the Cole anchor, as it instead was a contractual arrangement, possibly similar to the royalty agreement Chance had on anchor rods with other firms.

Within handwritten notes about important highlights in A. Bishop's life, covered in the next chapter, was a note that stated: "Our first acquisition—Guardrail business—also Wej-Lock anchors." Story can be told by FGC [F. Gano Chance]." The recorded heading, "Return to common sense"—association with Wej-Lock sprouts with numerous possibilities. The Wej-Lock purchase signaled the regathering of their marketing capabilities, especially in bringing Jack Isbell back into the fold. The Wej-Lock products also appeared to be well engineered and they could hit the ground running instead of having to reengineer inadequate products. All of this pleased A. Bishop and he was temporarily happy about the direction of the company, especially after getting off the beaten path with Kearney and the thorny issues that developed.

The Wej-Lock story was remembered with additional product details by Bill Hollander. His knowledge of what occurred fills in some important informational gaps plus there is another very illuminating aspect of his involvement with E. K. Cole. Bill and Mr. Cole worked together at Chance even after Mr. Cole's retirement in 1940. What Bill revealed was that he worked with Mr. Cole in the development of the 8-Way® Expanding Anchor and other products. Professional engineers and supporting personnel are not always recognized as inventors, but this was a significant part of their jobs as well as the improvement of existing products, which became a trademark for the Chance Engineering group. Bill was asked how long the Wej-Lock anchor was made to which he replied:

> *We made it for quite a while but we changed it. Originally, the anchor was expanded by screwing the rod down and causing two nuts to pull together to force the blades out and turned it over towards the expanding with a tamping bar. They were still in malleable iron and the cost of malleable iron was becoming extremely high. It was after E. K. Cole left Chance Company that the idea that he could make the thing out of steel rather than cast iron. He called me to come on over and see what idea he had and of evenings I would go over and work around. We gradually worked on that, how he could make it out of steel, and come up with the 8-way anchor. His first idea was just a lit, make 4 blades out of a square piece and to shape it on up but the 4 blades just did not work out. We kept working until we could make 8 blades out of it so that you could fold it up. (1978)*

Jack Curtright also recalled the Wej-Lock and the 8-Way® Expanding Anchor:

> *I first went to work for Chance Company when they were selling piston rings and then I left and went to work for Chevrolet. Then I came back and sat across the desk from Mr. E. K. Cole. You know he was in the anchor business up in Iowa and he was an inventor too, just like Mr. A. B. Chance. He invented this Wej-Lock Anchor, which was made of malleable iron. We made some progress in that as it used to have two fingers that wrapped around and the boys over in the shop would have to hammer them down and they used to break them off every now and then. We developed a little heavier stuff and then we put a little wedge we drove in there to complete the hinge. Then later on we made a steel expanding anchor patterned after the old Wej-Lock and we called it an 8-Way® Anchor, stamped out over there in the shop and formed. That was an intricate process as they pressed that thing down the blades would fold. There were eight different blades, four short blades and four long blades. As the anchor was expanded, it would form a cone type and you remember Pierce Cone in the Cone anchors in St.*

Mr. Cole & Gano with the new 8-Way® Expanding Anchor. (*Chanco News* August 1948)

> *Louis. So we had the combination of an expanding anchor and a cone anchor, you know it's a tremendous project. You can't hardly pull them out of the ground. But they will creep, as they hold, they wedge themselves. (1979)*

Jack Curtright in a December 6, 1935 *Guard* item, had resigned his position as secretary to J. T. Isbell at Chance. His employment began as early as the spring of 1933. His new position was a traveling salesman for Chevrolet Motor Company with the territory of central and southern Illinois. A few months earlier, Centralian H. A. Jones had taken over a Chevy dealership in Robinson, Illinois, but it's not clear if he may have influenced Jack to join General Motors. Jack worked out of the St. Louis headquarters and was in the G. M. Parts branch for Chevrolet in January 1937, but he had accepted a position with the Pontiac Division and would be transferred to Jacksonville, Florida. However, Jack would come back to Chance before Mr. Cole retired in December 1940, which suggests the time when they both were located in the same office area.

In explaining that Mr. Cole had developed the guardrail equipment and the Wej-Lock Malleable Anchor, Andy Anderson concluded that: "He had an inventive mind, Mr. E. K. did. As a matter of fact, in later years, he developed the 8-Way® Anchor which was very similar in some ways to the Wej-Lock except the Wej-Lock was a two- or three-blade anchor made of malleable iron, but it also folded up much like the 8-Way® he developed in later years." (1979) According to Board minutes in 1932, Wej-Lock also had a water sprinkling system.

Bill Hollander recalled that Mr. Cole:

> *More or less worked up any ideas that he could. E. K. used to come up and spent most of his time with me. One of the things he was working with was anchors and some hardware items for the highway guardrails and he tried to expand it. When he came one time, why he was the head of the highway guardrail division and he would have some ideas, but E. K. could not make much of a drawing either. He would get an idea, either beat something out or try to explain to me what he had in mind and try and work it out and we would make a model. Sometimes we would make the model together and sometimes the other boys in the shop would make the models. (1978)*

It isn't easy to find a person that was remotely close to being like "Boss" Chance in developing products and how they operated in the engineering and manufacturing realms. Bill Hollander's remarks, made in the prior paragraph about E. K. Cole, in certain ways mirrors what Bill said earlier about A. Bishop. One obvious example is the difficulty both had in reading engineering drawings. Bill felt strongly about having engineering drawings in the factory for all products, to insure their dimensional integrity for tooling and production purposes. Bill brought this out in such a way that even the most nonengineering types can easily understand why drawings were necessary to maintain dimension control. In the following story of the relationship between pattern maker Donald Woods and A. Bishop, Bill explains in

J. S. CURTRIGHT

Jack (million dollar smile) Curtright was born on a farm in the outskirts of Centralia, Missouri, home of the A. B. Company. He was traveling in Florida and thereabouts for General Motors Corporation when he decided to come back to the A. B. Chance Company, where he had previously assisted the sales manager. Now he's on the west coast covering the territory indicated and making a lot of friends -- for you can't help liking this optimistic chap with the sunny side up.

Jack Curtright. (*Chance Tips* Septmeber 1940)

his colorful manner the development of a new product and what occurred after it was put into production:

> *A. B. had probably been out of town for two weeks and the first thing he would do when he got back to town is he would go through the plant and greet everyone, all up and down the plant. He would come through the machine shop and say "hi" to everybody and Don would just hold this out and give it to him. "Yes, that is very pretty Don, what is it?" "Oh that is that thing-a-ma-jig you wanted." "Oh yes, I remember. No Don, you didn't quite get the idea. Over here, this ought to be rounded at the right. Oh, this is not big enough. It ought to be so much longer and this groove ought to go on through." So Don would cut it down and do it all over again and maybe three or four weeks later hand it back to A. B. and go through the same process. He would finally say, "well that still isn't quite right but it is close enough. Let's go and make it." So he would finish up the two dies and they would go into production beating out these forgings. Of course, you get to beating on these dies, they keep changing, they cave in and they wear. They polish them out and they get sticky and so the forgings just keep changing. Every one you made was slightly different from the last one. The dies would get to where they were sticking so bad they would just take a light cut off of the top to square them again and polish out the impression. They eventually reached the point where they just did not make any more forgings at all. By that time the forging had changed very considerably from the way they started. So they would take one of the late forging as a model and they would sink out a new die. So they would start with a new die like the other one was when it was worn out and*

pretty soon you couldn't even tell what it looked like to start with. (1979)

Gano continued to speak about some of his Wej-Lock experiences:

Mr. Cole moved here and we made guardrails and sold them to the Highway Department. An interesting experience took place one time, I had Ernst and Ernst set up a cost system to find the ratio of dollar labor to the indirect and all the cost expense you know and arrive at a percentage. Well, over in the warehouse we had very little direct labor, it was practically all indirect labor and the dipping of anchors, etc. was not direct labor so that the ratio was so far out of line that we had about a 2000% overhead. Well, we started painting Mr. Cole's guardrail in the warehouse there and with direct labor about 1000%, it would come out rather high. Mr. Cole spoke to me about it and said, "Look here Gano, this is awfully high!" I said, "well that is the way it figures." He said, "if you don't mind, I'll have them move the guardrails out on the street and we will paint them out there!" I got the point that our ratio was out of line. We continued with the guardrail business for several years but it was a difficult type of business to deal with. Number 1, the steel companies got into the business themselves and they could fabricate it and sell it and eliminate us. Then again, we had to deal with the governmen,t which I wasn't too happy with. So we got out of the guardrail business, quit it but continued with the Wej-Lock anchors. (1978)

Bill Hollander continued with additional remarks about the guardrails. Pictures of early highways show these cable-dominated guardrails being utilized:

We developed a bunch of different items for highway guardrails, offsets and the socket they used to gauge the end of the cable. We had been using U bolts and clamps to hold the cable, spread it, drive a wedge into the cable to hold it, but mainly worked with guardrail items. But he spent more time up with me. Get some idea and we would explore it, sketch and work around trying to work something that we thought might produce a product. Make some models and tested them. Spent an awfully lot of time on these wedge grips for the cable. Tried different systems and pulled them and tried to develop the strength all of the highway departments called for. Whatever means you had to deadend the cable, to develop the strength. They were up to about 100,000 pounds in size. It took some doing trying to develop that kind of holding power and some of them did and some of them did not. We thought we could hold pretty well with the things. (1978)

The guardrail business was also remembered by Paul Sturgeon, who confirmed what Gano had earlier said about that part of the business:

And we acquired with the Wej-Lock anchors the guardrail equipment, which was a cable-type guardrail and all the fittings, jay bolts, screw bolts, nuts, brackets, and everything that went with it. Of course, the steel companies put us out of business on that finally when they came out with the rolled steel section guardrail. You know, it just fastened on to the posts. (1979)

The Chance Board dealt with Wej-Lock and Mr. Cole on at least two occasions. "At the Chance Board of Directors meeting of December 1936, $3000 of A. Bishop's salary was charged to the Wej-Lock Division of the Company." At the December 1940 Board meeting:

A discussion was held in regard to the termination of the Wej-Lock Guard Rail contract. Mr. Chance [F. Gano] *explained that the Board had made a lump sum settlement with Mr. E. K. Cole and had*

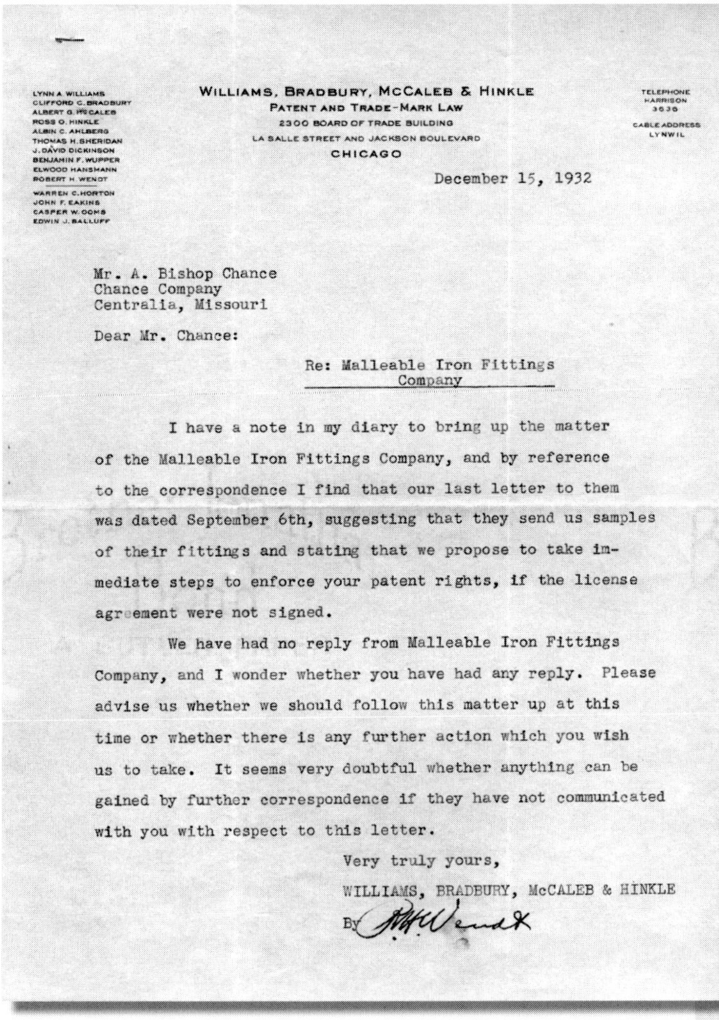

LYNN A. WILLIAMS
CLIFFORD C. BRADBURY
ALBERT G. McCALEB
ROSS O. HINKLE
ALBIN C. AHLBERG
THOMAS H. SHERIDAN
J. DAVID DICKINSON
BENJAMIN F. WUPPER
ELWOOD HANSMANN
ROBERT H. WENDT

WARREN C. HORTON
JOHN F. EAKINS
CASPER W. OOMS
EDWIN J. BALLUFF

WILLIAMS, BRADBURY, McCALEB & HINKLE
PATENT AND TRADE-MARK LAW
2300 BOARD OF TRADE BUILDING
LA SALLE STREET AND JACKSON BOULEVARD
CHICAGO

TELEPHONE
HARRISON
3630

CABLE ADDRESS
LYNWIL

December 15, 1932

Mr. A. Bishop Chance
Chance Company
Centralia, Missouri

Dear Mr. Chance:

Re: Malleable Iron Fittings Company

I have a note in my diary to bring up the matter of the Malleable Iron Fittings Company, and by reference to the correspondence I find that our last letter to them was dated September 6th, suggesting that they send us samples of their fittings and stating that we propose to take immediate steps to enforce your patent rights, if the license agreement were not signed.

We have had no reply from Malleable Iron Fittings Company, and I wonder whether you have had any reply. Please advise us whether we should follow this matter up at this time or whether there is any further action which you wish us to take. It seems very doubtful whether anything can be gained by further correspondence if they have not communicated with you with respect to this letter.

Very truly yours,

WILLIAMS, BRADBURY, McCALEB & HINKLE

By RHWendt

The Pole Guy Fixtures Patent No. 1813482 of July 7, 1931, had a series of "infringement" correspondence including these last two letters reproduced here. The December 15 letter is significant because it reveals the Chicago-based patent attorney firm who represented Chance Company during this time, which was just prior to incorporation proceedings. The reply letter is most valuable in its reference to "the company that we have taken over in Iowa. . . !" The third paragraph identifies that concern as Wej-Lock. (A.B. Chance Co.)

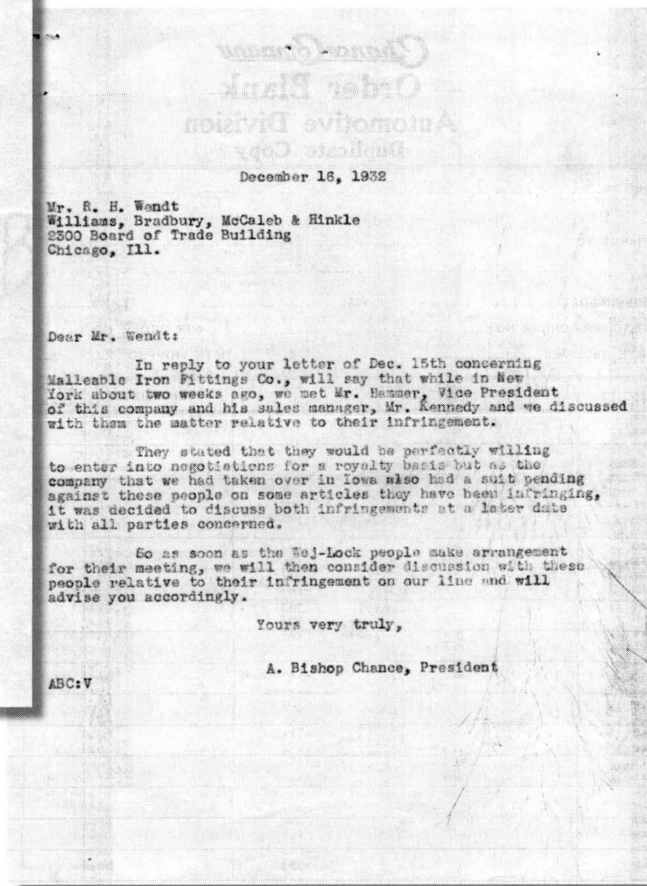

December 16, 1932

Mr. R. H. Wendt
Williams, Bradbury, McCaleb & Hinkle
2300 Board of Trade Building
Chicago, Ill.

Dear Mr. Wendt:

In reply to your letter of Dec. 15th concerning Malleable Iron Fittings Co., will say that while in New York about two weeks ago, we met Mr. Hammer, Vice President of this company and his sales manager, Mr. Kennedy and we discussed with them the matter relative to their infringement.

They stated that they would be perfectly willing to enter into negotiations for a royalty basis but as the company that we had taken over in Iowa also had a suit pending against these people on some articles they have been infringing, it was decided to discuss both infringements at a later date with all parties concerned.

So as soon as the Wej-Lock people make arrangement for their meeting, we will then consider discussion with these people relative to their infringement on our line and will advise you accordingly.

Yours very truly,

A. Bishop Chance, President

ABC:V

purchased his stock in said company. A letter of appreciation was also written to Mr. E. K. Cole who had resigned. Rather it would be more appropriate to say that he retired. (Board Minutes, December 1940, L. C. Hansen interpretation)

Notice again the language used, "contract," applying to the guardrails as Gano earlier suggested applied to the Wej-Lock Anchor. The buyout of stock also suggests the Wej-Lock relationship had apparently not been of the common, outright acquisition variety.

Mr. Cole remained active and loyal to Chance Company, even after his official retirement. Like Colonel Sanders, of food chain notoriety, one of Mr. Cole's most significant accomplishments was in his late years. This can serve as an inspiration, that we can be productive our entire lives. Not unlike the abiding spirit that co-worker Mr. Chance possessed, Mr. Cole saw the future as a place of continuing opportunity. It is complimentary that as a transplanted person, he adjusted to Centralia, Chance Company, and to a position in which he no longer was entirely his own boss. A young sales correspondent, Jack Curtright, remembered Mr. Cole as he sat across the desk from him. "I learned a lot from this gentleman." (1979) Notice the use of "gentleman" and the time Mr. Cole gave to helping this young man learn the ropes. His relationship with Bill Hollander also reflects the talent of working with other people in developing new items, which isn't always the case with inventors. Utilizing professional assistance from the Chance Engineering group speaks of a level of wisdom that served Mr. Cole well, especially later in a life already filled with business successes and significant accomplishments. According to Jack Chance, Mr. Cole was very tall, possibly six-foot six-inches and had reaped substantial financial success from his 8-Way® patent.

According to *Chanco News* of November 1956, E. K. Cole passed away in Centralia on October 23, 1956. He was seventy-nine. He spent several years in the Chance Engineering Department and designed the 8-Way® Expanding Anchor after he retired. "It required less steel and assembly time and provided greater holding power than previous expanding anchors. Placed on the market in 1948, the anchor has been a major factor in maintaining Chance Company's position as the leading producer of earth anchors."

Incorporation

In personal papers originating in 1930, Mr. A. Bishop's handwritten notes refer to the need to incorporate the company to reward salesmen and others in official capacity. "In order to make such arrangements, it will be necessary for us to incorporate so that such interest can be carried by shares of stock." One of the traditions handed down within the company was encouraging employees to offer suggestions to improve company operations. Acceptance of new ideas often requires the planting of seeds in the minds of those in power positions. It might require a year or more to germinate, but it did result in their personal adoption by management and frequently it became "their idea." It appears this concept may have originated and was practiced by the founder as several of his earlier suggestions were later adopted, including incorporation, as seen here. How this transpired was explained by Gano:

> *For many years, Dad operated the Company as a proprietorship. Not long after I got out of college, he made me a partner. In 1932, we incorporated and sold a little of the stock outside the family. Dad was the first president; I was the vice president. My first assignment from the president was to face the bankers. (1968)*

In a few hours the old year would be rung out and the new one rung in, which mirrored the restructuring taking place within the company. The actual incorporation of Chance Manufacturing Company took place on December 30, 1932, per a reference by R. E. Ausmus in a legal resolution concerning the Chance farm on December 30, 1944. Additional organizational details were worked out over at least a two-day period. At the incorporation Board meeting of December 31, 1932, the president and vice president were elected as well as a secretary/treasurer. The first two positions were filled as could be expected by the founder and his son respectively, but the third position of secretary/treasurer was filled by a lady, Miss Ida Lou Traughber. When the company was incorporated in 1932, according to Gano, "Yes, I was given the title of VP and general manger and a salary of $300 per month, as I remember." (1968)

An important point in Miss Ida Lou's selection from the ranks was she represented one of the worker bees. Those who have labored within such an environment understand there are those who talk a good talk and those who walk a good walk in getting the day-to-day work done. Miss Ida Lou was a known quantity, whose past performance strongly dictated that she fill this important, functioning position. A. Bishop was a practical man who at times didn't much care if he was going against the flow, if it was the reasonable, fair, and beneficial thing to do. Gano's feeling about this isn't clear although he was getting what he personally wanted in a position and he too was a practical person who went along with his father in certain instances. In the 1935 Board meeting, salaries are mentioned for the ensuing year for Ida Lou Traughber of $2,295.00. In 1930 dollars that was a substantial salary. In 1936 her salary was confirmed to be $2,297.50. She not only served as secretary/treasurer of Chance Company, but also was on the Board of the Chance Development and Investment Company in the same position.

The stock in this new corporation was issued to four Chance family members, Mr. and Mrs. A. Bishop Chance and Mr. and Mrs. F. Gano Chance, and one longtime employee, Ida Lou Traughber. It clearly shows the respect and the importance that was garnished on Miss Ida Lou's performance and position within the company and it also shows A. Bishop's generosity in taking care of his people. Although she and the other two Chance ladies only received ten shares each (2 percent each) of the five hundred total shares issues initially, there were three ladies on the five-member Board of Directors seventy years ago! In addition to Miss Ida Lou, there were at least four other nonfamily employees who Gano remembered that also received shares of stock after the company was incorporated. They were Mr. Milligan, Mr. Clark, Mr. Isbell, and Ms. Irene Traughber. There is some confusion as to who Irene Traughber was. Miss Ida Lou's mother was Agnes (1848–1927) and her two sisters were Miss Elizabeth (Lizzie, 1875–1943) and Mrs. Myrtle Bryson, although there were other family members. Myrtle went to California in 1912 and Elizabeth lived in Centralia with Ida Lou until her death. Gano couldn't recall exactly how many shares they received (three sales personel) but he believed it may have been associated with the number of years they had been with the company. This is confirmed by Roy Hansen's analysis of Board minutes. "At a stockholders meeting on April 10, 1933, the Board of Directors was changed to consist of eleven members. In addition to the five on the Board, J. J. Milligan, C. L. Clark, E. K. Cole, J. T. Isbell, C. B. Taylor, and Alfred C. Wilson were elected to the Board." (1978) Milligan, Clark, Isbell, and Wilson were in Chance Sales and Marketing. E. K. Cole was the former head of Wej-Lock, which became associated with Chance Company prior to the incorporation and C. B. Taylor was the Big C auto line sales manager. According to Gano, his father also made a deal with this group to buy stock in the company until it equaled what had been given them. In a year or two only Miss Ida Lou continued to buy stock, ending up with a nice retirement fund in stock, especially after the stock split according to Gano:

> *In fact, she and her sister were the only out-of-family stockholders that had any stock in the company to speak of, when we started to sell over the counter. . . . So we were already started with a little bit of the stock out of the family. Most of it was in the Chance family. We had little or no stockholders, other than the few I mentioned. We got interested in the Bowie Switch Company right at the close of the war. Still we had made no effort to sell stocks. (1978)*

Adjustments did occur as it became clear that certain initial ideas didn't blossom as expected. It appears that one of these casualties was the eleven-person Board of 1933 as reported by Roy Hansen: "At the annual stockholders meeting in December of 1936 the number of directors was reduced from eleven to five and the following directors were nominated: F. Gano Chance, A. Bishop Chance, J. T. Isbell, C. C. Hough, and Ida Lou Traughber." (1978) This group appears to be a working Board who were aware of the day-to-day activities within the company, because they held positions there.

In examining Board minutes of Chance Company and the Chance Development Company, Roy Hansen reported details of the first meeting held in the office of Clark, –––, –––, (apparently two names were lost when the interview material was recorded) and Peterson on December 31, 1932. This was apparently a reference to a law firm retained for legal matters. It was reported that those present were A. Bishop Chance, owning and representing 250 shares of common stock, Gano Chance (220 shares of common stock), and Miss Ida Lou Traughber (10 shares of common stock). A motion was made to elect A. Bishop Chance as president, Gano Chance vice president, and Ida Lou Traughber as secretary and treasurer:

> *They have been operating as a co-partnership previously and owned a certain amount of property in Centralia. This property was assigned to the A. B. Chance Company and included not only the buildings but land in Centralia. Pardon, I doubt that there was land, just buildings plus some leases for a 99-year period of time* (land between the tracks where main plant was located?). *In addition there was a Blackburn royalty contract, several other royalty contracts as well as the water sprinkling system from the Wej-Lock Anchor Company, and the royalty contract by C. L. Clark on the lineman's wrench. All of the above they agreed to sell, transfer, and assign to the A. B. Chance Company for 3,000 shares of stock, same to be issued 985 shares to F. Gano Chance and 1,515 shares to A. Bishop Chance. This was signed on the seventh day of January 1933. (1978)*

As 985 and 1,515 only add up to 2,500, it's unsure if other parties received a portion or the net was instead 2,500. C. L. Clark may have been a recipient. A paper entitled "History of Chance Stock" clears up the increased shares of stock, but ignores the initial 500 shares authorized when the incorporation occurred:

> *The Chance Company was incorporated in 1932 with authorized capital of 3,000 shares of common and 1,000 shares of preferred stock, all without par value. The authorized common stock was increased to 6,000 shares in 1938. In 1946 the authorized capital was changed to 2,000 shares of preferred and 106,000 shares of common stock with the 6,000 shares of common then outstanding being exchanged on a 10 for 1 basis.*

The Board minutes just covered in the last two quotes, suggest the change to 3,000 common and 1,000 preferred took place after incorporation in January 1933.

On December 31, 1932, a Board meeting was also held to form the Chance Development and Investment Company, according to Jack Chance. "Articles of Agreement, Certificate of Incorporation, and By-Laws were also passed." (1978) In addition two other stockholders with 10 shares each were identified: Frances I. (Fancy) Chance and Anna Lee (Ann) Chance, the spouses respectively of A. Bishop and F. Gano Chance. As Roy Hansen continued to analyze Board minutes, he found:

> *In addition the authorized capital stock also includes 1000 shares of preferred stock. . . . The Articles of Agreement of the Corporation stated that the corporation was founded for the following purposes: Engage in the business of manufacturing, jobbing, distributing, and selling at wholesale, retail, or both, all kinds of machinery, equipment, parts, accessories, etc. to carry on the trade or business of engineers, founders, smiths, machinists, manufacturers, and patent fees. (1978)*

This covered a wide variety of industrial activities, which would be stretched to the limit before the decade of the challenging 1930s transpired. In fact, when A. Bishop became involved in agriculture, which later came under

Chance Company control, these elastic appearing boundaries would have to be expanded even further to include the farm operations.

During the 1930s it was a fortunate person who could consistently bring home a paycheck. Even those having jobs were most often in what would be described today as poverty, although they didn't necessarily consider themselves in a hopeless situation. This writer's father spoke of long, hard days spent in the agricultural sector where a salary of one dollar per day was welcome in part-time employment. Dee Daugherty is quoted later in this chapter of making twenty-four cents an hour in the Chance Plant during the mid-1930s. That translates to about $10.00 for a forty-hour workweek or some $500.00 annually. Many movies of the 1930s carry a common theme, portraying the lifestyle of the rich and famous. Movies gave people hope during a period when they could only touch prosperity briefly, in a world of make-believe. In Centralia, like on a movie set, people could focus on the Chance family and conclude that America was still the land of opportunity. A. Bishop had become a local hero, as he was an example of one who had thrown off the shackles of poverty and business failure, and was living a life that was enviable by the populace. Those having, what could be considered substantial salaries, could exercise considerable influence in their areas of interest and could yield a good measure of economical clout and purchasing power in those Depression days. As the company had stabilized somewhat, the minutes of the Board meetings indicate that A. Bishop had become what locals could describe as wealthy. This perspective is more impressive when confined to a small, out-state area and remotely separated from other more capital-centered areas of the country. It was quite surprising to find how well the Chances were faring after getting the company turned around. To support this, in examining Board minutes Roy Hansen reported:

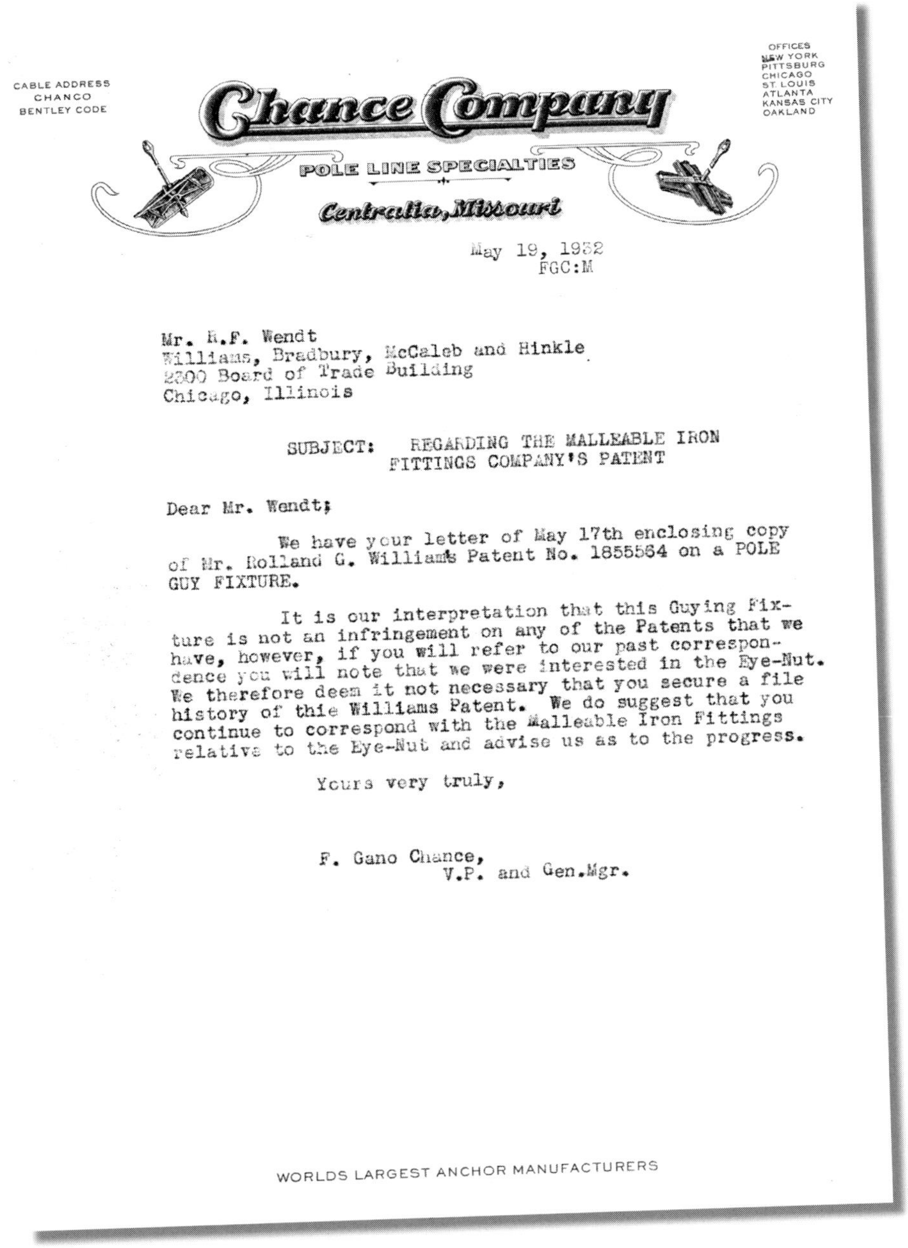
CABLE ADDRESS
CHANCO
BENTLEY CODE

Chance Company

OFFICES
NEW YORK
PITTSBURG
CHICAGO
ST. LOUIS
ATLANTA
KANSAS CITY
OAKLAND

POLE LINE SPECIALTIES

Centralia, Missouri

May 19, 1932
FGC:M

Mr. H.F. Wendt
Williams, Bradbury, McCaleb and Hinkle
2300 Board of Trade Building
Chicago, Illinois

SUBJECT: REGARDING THE MALLEABLE IRON FITTINGS COMPANY'S PATENT

Dear Mr. Wendt;

We have your letter of May 17th enclosing copy of Mr. Rolland G. Williams Patent No. 1855564 on a POLE GUY FIXTURE.

It is our interpretation that this Guying Fixture is not an infringement on any of the Patents that we have, however, if you will refer to our past correspondence you will note that we were interested in the Eye-Nut. We therefore deem it not necessary that you secure a file history of thie Williams Patent. We do suggest that you continue to correspond with the Malleable Iron Fittings relative to the Eye-Nut and advise us as to the progress.

Yours very truly,

F. Gano Chance,
V.P. and Gen.Mgr.

WORLDS LARGEST ANCHOR MANUFACTURERS

One of few letterhead stocks surviving. Note this 1932 period letterhead differs from the Kearney letter (see page 381) which may have represented executive stock. Note "Pole Line Specialties" is a variation of that represented on the Model A truck on page 376. Chance Company at that time considered itself the "World's Largest Anchor Manufacturers. (A.B. Chance Co.)

At a subsequent Board of Directors meeting in April 1933, Mr. Chance's salary was set at $6,000.00 per year. It was also made a motion and carried that Chance Company enter into a contract with A. B. Chance whereby the said corporation negotiates a royalty contract with A. B. Chance extending royalties of 5% of the net sales price on all articles manufactured by the Chance Company under the patent now belonging to A. Bishop Chance." The 1935 salary for Mr. A. Bishop was $10,625 and $16,200 in 1936. Gano's salary in 1936 was $7,320. "At a December 24, 1937 Board of Directors meeting the salaries were A. Bishop Chance $29,275 and F. Gano Chance $12,900. (1978)

So evidently business in 1936 and 1937 was good. The Board of Directors met in December 1939 and A. B. Chance's salary was set at $24,000, and F. Gano Chance, president, at $8,400 reflecting the down year of 1938. In December of 1940, Nathan A. Toalson was elected a vice president of the company and Gano's salary as president of the company was $18,000 a year and $34,000 for A. Bishop as chairman of the Board.

Continuing to interpret the Board minutes, especially in reference to the special property and investment company that was set-up, Roy said:

The Chance Development and Investment Company was established in which it received as assets some 22 different properties and land locations in and about the City of Centralia, for which A. Bishop Chance received 300 shares of stock in the corporation. The corporation authorized capital stock up to 450 shares all within part value divided into 350 shares of common stock and 100 shares of preferred stock. The amount of capital with which the corporation will commence business is $100,000.00. . . . In June 1935 at a special Board of Director's meeting of the Chance Company, it was noted that the company is indebted to banks in the sum of $30,000.00—in order for it to be paid promptly to the credit sum withstanding and for these purposes borrow $40,000 to $45,000. The money was raised by selling to the Chance Development and Investment Company $45,000 of its preferred stock and accepting therefore nine $5000 6% demand notes of said Chance Development and Investment Company. Also at the meeting on December 28, 1935, in the Directors Room of the Company, Mr. Bishop Chance was elected President and F. Gano Chance was elected Vice-President and General Manager. Salaries were also set for the ensuing year, Bishop Chance $10,625, Gano Chance up to $100 [sic] *and Ida Lou Traughber $2,295. . . . The Board of Directors met in December of 1936. The company caught up the arrears on preferred dividend payment for the year 1935 and also declared the regular 6% dividend for 1936. Gano Chance moved that a 5% dividend be declared on common stock outstanding and salaries were set at $16,200 for Bishop, $3,000 to be charged to the Wej-Lock Division of the Company, Gano Chance $7,320 and Ida Lou Traughber $2,297.50. (1978)*

Jack Chance also reported on the Chance Development and Investment Company:

The minutes of the special meeting of the Board of Directors of Chance Development and Investment Company held Thursday, December 17, 1937, it was explained by F. Gano Chance that the purpose of the meeting was to accept the resignation of A. Bishop Chance as Director and President of said corporation because Mr. Chance had transferred all of his stock of said company and was no longer a stockholder thus not qualifying to be Director and President. . . . At the Chance Development

> *and Investment meeting held December 17, 1937, there was an election of officers consisting of F. Gano Chance president, Anna Lee Chance vice-president, C. C. Hough secretary, and Ida Lou Traughber treasurer. (1978)*

An insertion by Roy Hansen set the stage for further change:

> *On the 25th day of November 1940, there was a special meeting of the Stockholders. The purpose of the meeting was a proposed plan for reorganization of the A. B. Chance Company and the Chance Development Company, The authorized capitalization of A. B. Chance at that time was 1000 shares of 6% accumulated preferred stock and 6000 shares of common stock. The authorized capitalization of Chance Development and Investment Corporation (was) 450 shares, divided into 350 shares of common stock, at a stated value of $400.00 a share and 100 shares of preferred stock. Under the plan of reorganization, the shares of the Development and Investment Company would cease to exist and surrender its corporate charter and all investment in the development corporation to be assigned to the A. B. Chance Company. It was moved and passed unanimously that the present Secretary of the A. B. Chance Company be authorized and directed to issue 1400 shares of unissued common capital stock of said company to the Chance Development Company. (1978)*

Jack continued:

> *In a meeting of the Chance Development and Investment Company held on the 25th day of November, 1940, Chairman F. Gano Chance stated that the purpose of the said meeting was to carry into effect the instruction of the stockholders of said corporation with reference to reorganization of the A. B. Chance Company and the Chance Development and Investment Company as per resolution adopted by special stockholders meeting held at 9:00 o'clock on this date and for the further purpose of declaring an additional cash dividend at this time on stock of said company. A resolution was made and passed that the Chance Development and Investment Company will cease business on the 30th day of November, 1940. All of the foregoing being done in pursuant to the resolution of the stockholders. The company ceased doing business on November 30, 1940. (1978)*

What appears to be a result of the November 25, 1940, meeting was a reorganization of the Board of Directors membership per Roy: "At the annual meeting of the stockholders held the 28th day of December, 1940, the number on the Board was increased from 5 to 10 when the following directors were elected: Bishop Chance, Gano Chance, Frances Chance, Anna Lee Chance, Nathan Toalson, Ida Lou Traughber, Jackson T. Isbell, O. E. Walker, C. C. Hough and R. E. Ausmus." (1978) As was covered earlier, the original five-member Board was increased to eleven in 1933, five in 1936, and ten in December 1940. One reason for the 1940 addition of Toalson, Isbell, and Walker as directors, was their promotion to vice presidents.

Gano spoke of how decisions were often made outside of the Boardroom:

> *So when we incorporated, he practically turned the business over to me. Our Board of Directors consisted of my father and mother, my wife, myself, and my father's secretary. Well here was 1933 and business was tough and if a problem came up, I would call a meeting of the Board of Directors and my mother was unfamiliar with business, and my wife was young and unfamiliar with business. So it was my father's secretary, my father, and myself. I*

wound up deciding there wasn't much use of calling a Board of Directors meeting as the same group was working every day. If I wanted advice and council, I would ask my father." (1981)

Gano would get frustrated at times as this example shows but keep in mind that Mr. Chance was semi-retired in the mid-1930s and early 1940s when this situation likely occurred. "There was no use to have a Board of Directors meeting because I couldn't get Dad to sit down to even talk about it." (1978) Gano felt comfortable with a more complex organization, but one in which he could maintain stricter controls over what was occurring. In a personal record of his likes and dislikes, Gano said, "Like efficient systems and records." (1941)

Although Gano suggested that it was difficult to get his father into Board meetings, there was one occasion in which Paul Sturgeon regretted that he had. It isn't recorded who was in the meeting, but one could assume that Mr. Chance was in attendance, as it took place in his office. It all started simply enough as Paul recollected that he:

> *Nathan (Toalson) and Les Moore were going to odd man for a Coke. Well, we flipped about three times and we all had the same side of the coin. So about the fourth flip they still had the same side of the coin and I didn't. So I lost and I started getting my money out of my pocket to buy the Cokes when I suddenly realized I had been had. I had been hoodwinked. H——, they had been together and I let out a squall you could hear over in town. I didn't know that just across the hall in Mr. A. B.'s office, they were holding a Board meeting. I don't know exactly what happened but on Monday morning I was advised to move from the front end of the second floor to the back of the third floor. I got the job of moving my own desk and everything else. I can tell you it was about fifteen years before I ever got back in the Main Office! (1979)*

Paul, in focusing on the highlights of his career, considered the first prominent thing "was managing to get a job in the middle of the Depression. The second thing, I knew the best job in Centralia was running that plant, short of owning it. I had no idea of ever owning it." (1979) But a third and fourth significant development during his time would not only make employees feel like an owner, but in fact would allow them to share in the company ownership. Profit Sharing would connect employees to the performance of the company and would give them additional incentives to insure its success. But a surprising remark was made by Paul about a fourth rewarding consequence of employment at Chance Company, even surpassing Profit Sharing:

> *Even more important, to me, than Profit Sharing was the day Gano let us start buying company stock . . . I don't remember what year.* [Note: Postwar material reveals company stock sales] *I remember we could buy it for book value. I will never forget, I borrowed $1,950, I believe to buy 100 shares of company stock. I thought, Oh Boy, how will I ever get that paid off! (1979)*

Another positive result of the incorporation of Chance Company to Paul was, "The day I became an officer and director of the company. So those were the outstanding things and in between them a lot of fun, a lot of work, and a lot of long hours. But there is one thing about it, nobody ever in the world enjoyed their work as much as I did. You are lucky if you can really enjoy what you are doing. I enjoyed every minute of it." (1979)

Jack Curtright also thought highly of the sale of company stock to employees:

> *That was a good thing for the Chance Company employees. . . . I don't remember what year it was but I remember when it went on the market . . . Chance Company*

allowed the employees to, as a matter of fact, they could make a deduction if the person wanted to buy stock and take it out of their check, do it little by little. Then after they paid for it, why they got the stock. (1979)

Granny Ryman purchased Chance stock when it first went on the market. "I bought maybe fifty shares on time payments. It was kinda hard to do that at that time too." (1979) Lower salaries in earlier times made it difficult to set aside a portion of their paycheck for investments in the future. But those who sacrificed had no or few regrets when looking back on their decisions to do so and the idea that they were part owners forever changed their perspectives. A. Bishop and F. Gano realized the benefits of such programs to the company, besides the idea of rewarding and providing those opportunities to their employees.

Business continued at Chance Company, but hardly any firm could long survive without making a profit. Gano continued to explain:

So 1933 came along and we started to get a little increase in sales. I think we ended the year just about breaking even. In 1934 we made a little money and we have made a profit ever since. Our first vice president in charge of sales (Isbell) was a fellow that I started out with when I got out of college, going out on the road to sell. He was a typical salesman and he was terrific! He could sell anything. He told me years later, "you know back there in the 30s when I came to your company, I did not like some of the decisions you were making and thought that we might even go broke." But he said, "I decided you were not going to break the company and it looked like you were going to make it. It really wasn't my place to tell you that you were deciding wrong so I did keep still but I did not want to." So our company was most interested in changing from a one-man organization to a private family corporation. We first went to selling our stock over the counter. A brokerage firm in St. Louis was pretty well sold on us and helped us to acquire one or two companies and we kept growing from there. (1981)

One of the reasons Gano later started to get outside professional businessmen on the Chance Board of Directors was to gain credibility with investors as he further explained:

They in turn strengthen the positions of the company so that outside investors would be interested. So it was all a part together, so we were in contact with Stifel Nichlaus in St. Louis and they were very good brokers. They got interested in our company and began to sell some of the stock over the counter. (1978)

This of course eventually led to A. B. Chance Company being listed on the New York Stock Exchange. More coverage of this subject will be provided in Volume II. To find people who could contribute to the growth and development of A. B. Chance Company, Gano became active in several statewide organizations where talented and high-ranking officials tended to migrate. Having this caliber of person on the Chance Board was a boost in having the benefit of their counsel as Gano spoke to:

To do that and get some men who were more experienced than I, I became active in several associations. I was on the board of the Missouri State Chamber and the board of the Associated Industries, the two major organizations in the state of Missouri on business, and on the board of a third organization known as Missouri Public Expenditures Survey that analyzed government expenditures. Through that contact I found fellows that I thought knew finances better than I. . . . We wound up with some pretty good board members. We had a lawyer that was my type of lawyer who instead of telling you what you can't do tells you what you can do.

He was always in a position to figure out how to do it. Had one on the board who was treasurer of a big corporation in St. Louis, who was treasurer of the Associated Industries of Missouri, and recognized by most businessmen in St. Louis as being one of the top financiers. So I began to have good support from the outside board and about 1970 to go on the big board (New York Stock Exchange). Prior to that time we were over the counter but went on the board (NYSE) about 1970. Of course that is a real thrill in which you build up a family company that starts out with six people in the office and Board of Directors consisting of your mother, father, wife, father's secretary and yourself, to being accepted as an investment by the investment field and go on the New York Stock Exchange. To go to New York for the opening, I walked up to the front of the New York Stock Exchange and see in lights "Welcome the A. B. Chance Company" is quite a thrill. We were in prior to 10 o'clock by a few minutes and we had the president of the New York Stock Exchange, as they do, announce here is a new company coming on the exchange and here are the executive officers, etc. here down on the floor. I bought the first share of stock that was sold over the stock exchange that morning just after ten o'clock. We went to Chicago and were on the board there as a member selling stock, but in a few years we just concentrated on the New York Stock Exchange. (1981)

Returning to 1939, the importance of the 1930s is not only reflected in the progress and changes in the corporate organization of A. B. Chance Company but also in the changes in key company officials. Roy Hansen stated, from Board minutes that:

On May 12, 1939, a $.50 per share dividend was declared immediately on all outstanding common stock of the company. In August of 1939, a further dividend of $1.50 was declared immediately on the outstanding common stock of the company. At a meeting on August 31, 1939, a special meeting of the stockholders, Mr. A. B. Chance explained at the meeting that it was his desire to retire from active management of the company and to resign from the position of President of said company. Thereupon Mr. Gano Chance stated at the meeting that it was his desire and the desire of other officers of the company that Mr. A. B. Chance continue in the capacity as chairman of the Board. Consideration was then given to the selection of a successor of the president and to recognize by title the duties actually performed by this member of personnel. After the discussion and voting the following officers were chosen for the balance of the calendar year, 1939. Gano Chance president and general manager, Nathan Toalson vice president, J. T. Isbell vice president in charge of sales, O. E. Walker vice president in charge of production, Ida Lou Traughber treasurer and C. C. Hough secretary. . . . On December 15, 1939, the Board of Directors declared a dividend of $30.00 per share on common stock. (1978)

Chapter 7
A Depression Rebirth
(1934–1936)

The Mid-1930s

N.R.A., C.W.A., C.C.C., P.W.A., F.D.I.C., and R.E.A.

A. Bishop Chance's contribution to his local community during the early years of the Depression is impressive. One result was the success of his manufacturing firm, although the Chance plant had only operated at about 60 percent capacity in November 1932 through February 1933, but activity for the balance of the year had been adequate for Chance employees not to seek help.

However, it was anticipated the factory would begin closing a number of days per month in 1934, requiring relief help for employees. His work in 1933 for the needy and unemployed generated another mid-November trip to Jefferson City to try and obtain a federal work camp for Centralia under F.D.R.'s "New Deal" programs. Later in mid-December, forty men were employed under Civil Works Administration (C.W.A.). A Depression news article in the December 1, 1933 *Guard* told of relief requirements for eighty-two local people. A new group of women C.W.A. workers were given jobs in the old Overall Factory to gain experience and potential future employment. A. Bishop had purchased the idled equipment and donated it for the use of the C.W.A. group of fifty-five women, who began work in mid-December. The program closed in mid-February and there didn't appear to be an extension of this much-needed work program in spite of local efforts in support of it. The Civilian Conservation Corps (C.C.C.), another employer of idled local men, had a quota of five Centralians. This New Deal–generated group would construct many buildings, roads, parks, and planted three billion trees that are still evident today.

The Civilian Conservation Corps

The C. C. C.
(*American Profile Magazine*, Publishing Group of America)

In late January 1934, A. Bishop accompanied Mr. and Mrs. Ed Whitmere to Mt. Vernon, Missouri, to the State Sanitarium, which treated tuberculosis sufferers. In a less dramatic but generous act, A. Bishop loaned his car to Nathan Toalson. Returning from Paris, a tire

This four-generation portrait is a late 1933 or early 1934 setting. A young Philip Gano Chance was born March 7, 1933, and his Great-grandfather John, holding Phil, passed away on March 3, 1934. F. Gano is standing on the left beside his father A. Bishop Chance. (Jack Chance Collection)

blew out and the ensuing wreck injured Nathan and his two companions. Nineteen thirty-four was a tough mid-Depression year, which was intensified by a drought in the Midwest. Assistance was welcomed as a resourceful and much-needed Federal Deposit Insurance Corporation (F.D.I.C.) brought money out of mattresses and back into banks. This would help stabilize the banking mess and regenerate faith in banking institutions with deposits protected up to $2,500. Another benefactor to local area individuals, even those who began second careers was Chance Company. Charles F. Hiilzer, sixty-nine (1864–1933), had retired from farming in 1925 and was employed at the A. B. Chance Manufacturing Plant, working even the Saturday before his death on Monday, January 1, 1934. He had worked in the "big garage" which was Mr. Chance's Anthony Building at Railroad and Collier Streets. Older adults were welcome at Chance Company as wisely management realized the steadiness, skills, and experiences of such people were needed inside the firm. Men and women leaving the farm on a full-time basis and seeking employment at Chance became a common practice around the Centralia area.

On March 3, 1934, another Boone County pioneer, John A. Chance (1849–1934), an important link in the Chance family tree of A. Bishop Chance, passed away at the home of his daughter, Mrs. W. T. Fountain. Surviving were his wife, the former Miss Adaline Sexton, whom he married on October 5, 1871, a daughter Sallie, a son Albert Bishop Chance, one grandson F. Gano Chance, and a granddaughter Mrs. Gilmore Stone of St. Louis. He left three sisters: Mrs. Elisha Sexton, Mrs. Tom Sexton of Tulsa, and Miss Afton Chance and five living brothers: D. B. [*sic*] of Canon City, Colorado; B. R. and Charlie of Los Angeles; Grover of Tulsa; and William, whose home was unknown. John Chance was of the Christian faith and a longtime Lodge member. He had begun as a farmer and had several different business interests, including yet another new name for a Centralia hotel, the Central. His most prominent position and one long associated with his name was becoming manager of Home Telephone. He was a strong family man, even tempered, a noted gardener, and was a longtime solid and supportive citizen of the Centralia community. He and his wife's most noted contribution was being the successful parents of two offspring, one of whose influences would spread worldwide.

A. Bishop continued his business trips to Columbia, as recorded on a Monday in early March 1934. Although community service work had occupied a lot of his energy in recent years, he occasionally took time to rally support for his firm. In mid-March at the Commercial Club meeting, he asked for better cooperation as there had been charges that the A. B. Chance Manufacturing Plant had a low scale of wages.

John A. Chance's residence, west end of Singleton Street. (Centralia Historical Soceity)

He didn't feel this was conducive to good feelings or helpful to the plant. He said they were living up to federal codes (hours/wages) but they weren't making adequate return on investment. He appeared to still be leading his firm with assistance from his son Gano. Early April found another injury accident at the plant when Clarence Edelman fractured his right arm. The *Guard* was now consistently using the initials of A. B. Chance in referring to the firm. This may have changed at the time of incorporation and would officially become the company name in a couple of years. His other business interests in property was going forward as the former Bank of Centralia building (South Allen and West Singleton corner) was being prepared for the C. C. Jennings and Son Insurance and Real Estate business and an upstairs stairway was being put in on the east front wall for apartments on the second floor.

A big news item in the May 4, 1934 *Guard*, was a Chance sales conference of five days beginning April 30. It revealed the strides made in reestablishing the field sales force, since the Kearney detour. The organization depicted appeared to be the largest group yet fielded by the Chance firm. The Wednesday dinner in the company's Club Room was directed by Sales Manager J. T. Isbell, acting chairman of the conference. This was a change in which high-ranking Centralia plant officials had previously chaired the meetings, but Gano had not been happy with the tone and direction of certain portions of those sales meetings. The attendance of Mr. and Mrs. J. J. Milligan of Chicago and Mr. and Mrs. C. C. Musgrove of Dallas, suggests wives were also visiting in Centralia. M. P. Barber of Boston; C. L. Clark of Pittsburgh; E. K. Cole of Winterset, Iowa (Note: Later in the year he would relocate to Centralia); W. D. Howze of Los Angeles; T. B. Dally and Louis M. Meckler of New York; Arthur E. Bacon and John J. Cooper of Denver; Alfred C. Wilson of Atlanta; J. I. Ellmann of Washington, D.C., rounded out the Chance out-of-town participants. Certain members of this group may have been Chance agents. Chance Company distributors were also represented during Wednesday's agenda. A. J. McCall, sales manager, and Bud Arthur of Graybar Electric Company from Kansas City were present. Graybar was a large, if not the largest, distributor of electrical products. Distributors like Graybar had salesmen just as did Chance and warehouses in strategic locations to serve their customers with a full line of products from various manufacturers. This was normally high volume, discounted business for industrial firms like Chance Company. Also present were Frank Dansboe of Hubbard and Company from Omaha; Harvey Nehr of Graybar Electric Company of St. Louis; and Earl Nelson of Hubbard and Company of Kansas City. This group represented nearby regional sales offices and warehouses of Graybar and Hubbard. The Chance headquarters group in addition to Isbell, included A. Bishop Chance, president, and F. Gano Chance, vice-president; Miss Ida Lou Traughber, secretary; Ed McLaughlin, assistant sales manager, and J. H. Cheetham, factory manager.

The 1934 sales meeting identified certain key positions and personnel within Chance Company. The head sales job was now Jack Isbell's, which was no surprise. The assistant sales manager, Ed McLaughlin (sometimes spelled McLaughton), appears to be the former advertising employee who had transferred to Kearney. Besides, an earlier visit by Ed from Columbia, an April 27, 1934 news piece covered Mrs. Ed McLaughlin of St. Louis visiting her husband in Centralia over the weekend. It stated that Ed was employed at the A. B. Chance Manufacturing Plant. This helps define roughly when he returned to Chance. In August of 1934, the McLaughlin's residence was now in Centralia per their relocation to Toalson Apartments on South Hickman. A news piece in the January 11, 1935 *Guard* told of Ed McLaughlin's resignation of his position in the Sales Department of the A. B. Chance Manufacturing Plant. Further details suggested he had come to Chance Company four years ago, but had resigned to take a position as advertising manager with a newspaper in Tokyo, Japan. After a year, he returned to the United States and came back to Chance in sales. No future plans were revealed.

One of the most colorful Chance employees, William (Bill) Hollander, had been with Chances since 1929, but no public confirmation was found in the *Guard* until the May 18, 1934 issue. He motored to St. Louis over the previous weekend visiting his parents Mr. and Mrs. W. B. Hollander. A repeat was also found in June. As the midpoint of 1934 approached, the effects of the drought were being felt, especially in Iowa and Missouri. Without plentiful mechanized equipment there were few large ponds and lakes to provide a water source for people or livestock. Crops and gardens were total failures and temperatures reached levels more normally associated with southwestern regions of the United States. Without air conditioning and fans many people slept outside under the stars. Unemployment was made worse by the 1934 drought and by the year's end there were 1,155 Boone Countians on partial or full relief. One diversion from such conditions could be found in the local movie theatre and even some offered relief from the heat. Mid-1930s movies such as Fred Astaire and Ginger Rogers' *Top Hat* and *Gay Divorcee* were advertised in the *Guard* and were playing at Centralia's own Vista Theatre. Another Academy Award–winning movie, *It Happened One Night* with Claudette Colbert and Clark Gable, came to Centralia and Mexico's "Cooled By Refrigerated Air" Liberty Theatre in March of 1934 and had a return engagement in July.

The Century of Progress in Chicago had proven such a success and a morale booster that it was carried over into 1934. A Chicago legal conflict had soured Henry Ford on the area and he refused to have Ford Motor Company participate in 1933. However, the overwhelming success of the

E. G. (little man with big stuff) Ryman, our assistant sales manager, is another prodigal son who returned to his native haunts, after several years in various parts of the country, and joined the A. B. Chance Company sales force. He has been with the company about five years now, and we're beginning to consider him a pretty swell guy. You'll probably hear from him again.

E. G. Ryman. *(Chance Tips* September 1940)

event and the substantial presence of competitors led Ford to go all out in 1934. The timing couldn't have been better to also introduce the Burlington Zephyr to the world with its record-breaking Denver to Chicago speed run under its belt. It was then displayed at the exposition so attendees could tour this new streamlined, diesel-powered, lightweight, stainless steel passenger train. There were eventually several copies of this train set produced, surprisingly one called the Ozark Zephyr that ran through Centralia on the Alton between St. Louis and Kansas City! An ad introducing it appeared in the January 15, 1937 *Guard*. The Alton and Burlington had track agreements whereas the Alton could use the C.B.&Q. shortcut at Mexico to Old Monroe and the Burlington could use Alton's Mexico to Kansas City line. In both instances there was a fifty-plus-mile advantage to share trackage.

The 12:55 a.m. departure of the Alton Hummer and its 9:00 a.m. arrival in Chicago made it an attractive choice in attending to Chicago business or the Century of Progress. The Alton Night Hawk, departing Centralia at 3:30 a.m. and arriving in St. Louis at 7:15 a.m., likely had much more competition from the Wabash. Like the Zephyr the Alton also introduced in August 1934 a new gas-electric train from Bloomington, Illinois, to Kansas City through Centralia. As a consequence, Centralia saw some of the latest advancements of the day. Besides the many train tickets purchased by Chance Company travelers, the Wabash's widespread destinations had a romantic connection as already seen with A. Bishop. The Omaha (Council Bluffs) Wabash line angled across the northwest section of Missouri passing through another town served by two railroads, aptly named Burlington Junction, clearly a railroad-originating label. An August 1934 news item told of former Burlington Junction resident, now Chance employee, Art Houston's marriage to Miss Thelma Roberts in St. Louis on July 28. However, he reportedly had hitchhiked all or a portion of the way when he first landed a job in Centralia.

The N. C. Husteds and Centralians surprisingly continued contact as a September 1933 and a mid-January 1934 visit by N. C. Husted to Centralia indicates. A former Chance Company employee, W. H. Kinkead (1856–1934), passed away, per the June 8, 1934 *Guard*. A Baptist and retired farmer, his family included three sons and one daughter. Politically speaking, it was advantageous for Missouri Power and Light, a Chance customer and power supplier, to move their offices from Kansas City to the state capitol in Jefferson City. This did not help Centralia's population, however, which had reached a peak of 3,318 in 1926, but had dropped to a 1934 level of some 2, 220. The frequent loss of the original pioneers didn't benefit this cause as William H. Carpenter, ninety-seven (1836–1934), who was one of the signers of the Petition for Incorporation of Centralia in 1857, died. Of the Christian faith, his parents homesteaded in southern Boone County in 1825. A. Bishop Chance and close friend R. L. Hope were two of the pallbearers. Marcus Alvin Silver, eighty-four (1849–1934), also passed away in late June 1934. He was another pioneer of northern Boone County and had faithfully served as clerk and director of the Dunbar School for fifty-eight consecutive years! He was a Baptist. As the usually hot month of July was entered, the drought continued as did the state of Missouri's testing of guardrail designs to provide more highway safety. The Wej-Lock guardrail with a unique anchoring system was a Chance product.

Even during the Depression there were new hires such as Granville "Granny" Ryman who became a prominent Chance Marketing employee, but not before a couple of aborted starts. Born in 1907 on a farm adjacent (east) of the Centralia Country Club, he recalled winding wafers or coils for the Chance lightning arresters. The January 11, 1924 *Guard* stated that Granville Ryman was in town, which could have been a home visit while in college. Nothing was found again until December 23, 1927: "Granville Ryman of Central College, Fayette is here spending the holidays with his mother, Mrs. Emma Ryman." The November 9, 1928 *Guard* wrote a news piece about a F. G. Ryman who after thirty years of local residency was relocating to Bakersfield, California. Since Granny (E. G.) was only about twenty-one at that time, it would appear to be a relative. F. G. was leaving a large number of friends who regretted to see him relocate. Returning to Granny, after attending Central College he became a time study engineer for Swift and Company in Texas. From there, he went to California to work for Standard Oil and in 1930 (Great Depression) he returned jobless to Centralia. Very quickly he got a highway inspector job with the Missouri State Highway Commission, which lasted about a year. He then got a factory job at Chance Company, but in a short time he was laid off, as was everyone with less than five years seniority. This would have been in 1931 or 1932. He then began working for the Wabash Railroad at the Centralia Depot. A news item of July 1933 found Granville Ryman spending his vacation in Kirksville with friends. Notice this was during the summer school break. His mother Mrs. Emma Ryman, whom Granny stayed with, was still living in Centralia at the Chance apartments as of December 1934. In May of 1934 there was a Mrs. Granville Ryman! Her parents were the Frank Bowlings of Kirksville, which was convenient as Kirksville was easily reached by Wabash passenger trains. A July 1935 news item said Mrs. Granville Ryman, former Centralia schoolteacher, had now become the new Hallsville Grade School principal.

Granny's job at the Wabash Depot included handling heavy shipments, which included

Chance Company materials. A November 17, 1933 *Guard* local news item told of Wabash employee Granville Ryman dislocating a vertebra in his back. He was x-rayed at the Wabash Hospital facilities in nearby Moberly and sent on to Decatur, Illinois, for treatment. Although Wabash headquarters was in St. Louis, the center of the railroad, promoted as "Serving the Heart of America" was Decatur where their most advanced and capable facilities were located. According to Granny, in 1934, while working at the Wabash Railroad Depot in Centralia, "I went over and talked to Gano and he said I could go to work in the factory as a manager." According to Verna Palmer, he was working for the railroad and Jack Isbell saw possibilities in him and decided he would hire him." (1979) Verna is referring to Isbell latching onto Ryman and making him his assistant in marketing after he was already working in the Chance Plant. The main customer at the Depot was the Chance Company, and Granny became acutely interested in how Chance customer shipments should be classified. He found that by reclassifying Chance shipments to a lower freight classification, a substantial savings could be realized in shipment costs. The tariff list didn't include such special Chance items as Guy Anchors and Pole Line Hardware, which gave Granny a lot of freedom in setting new classification for these items as he remarked:

> *I never said anything to Gano about it except when he went for the job interview. I told Gano, "I would go to work for him and it wouldn't cost him anything." He said, "how do you figure that," and I said, "well here's a record that I have saved for the last year in freight rates and I figure if you pay me this amount you will still be ahead." Gano said, "you've got a job." (1979)*

Jack Curtright remembered Ryman, especially his handling of the heavy baggage for the Columbia Branch traffic of Columbia College students:

> *Yeah, Ryman was working for the railroad. He hurt his back some way lifting some of those heavy trunks. You used to have girls and boys taking their trunks over to Columbia. He was a feisty little rascal. (1979)*

In spite of piston rings and hammers seemingly occupying center stage in the mid-1930s, the old reliable earth anchor, was still alive and well at Chance Company. Granny Ryman gave an overview of what was going on in the plant, especially in reference to anchors and other products when he came with the Company:

> *We had first started making Never-Creep® anchors. We had the cast iron foundry and were making swamp anchors, cast iron cone anchors, and we had just started making the no-wrench screw anchors. We bought a malleable Helix, pinned it on to the rod and we developed that. We had bought the Wej-Lock Company, which had the Wej-Lock anchor and guardrail and all those were malleable iron castings. Another one of the jobs I had, when I first started work for the Company, was traffic manager. I was over in the warehouse directing the shipments and that sort of stuff. Ollie Chandler, Dutch Wisdom, Silas Roberts, and myself and our job consisted of assembling anchors. We would bring over the steel expanding anchors and load those on to the dipping racks and we would paint them. Then we bundled them and put them in stock. We were also the testing department. We did pole guying and conducted those tests on it. (1979)*

As Verna Palmer recalled, "For a while Ryman worked in shipping handling bill-of lading, etc. and then he moved into the office and did correspondence and selling over the telephone." Granny will become a prominent marketing/sales executive at Chance as revealed in Volume II.

In a World War II–era railroad movie by a St. Louisan, one scene covers incoming passenger trains, including an old Negro gentleman waving and welcoming them into Union Station. Not until a Centralia story was told of a similar individual did this remembrance of the St. Louis person connect up. How many among us are dedicated and unselfish enough to spend our time welcoming strangers into town? Then a special individual in Moberly also came to mind, known by all the train crews as he often can be seen at the Coates Street crossing welcoming Decatur and St. Louis trains into town. Kenny, with his Big Chief tablet in hand, also takes notes at Moberly Area Community College Greyhound basketball games. What could he be thinking or writing? It's serious stuff to him, a concentrated level of concern and feelings few of us could ever muster on such occasions.

One of the ageless aspects of small communities and towns, not unlike a Broadway play or movie, is there are those religiously performing certain roles. Centralia was no different as there were community-leader types represented by Judge/Mayor Bruton and C. E. and A. Bishop Chance. There was also a town drunk and those held in low esteem and others of perceived high standings. Edgar T. Rodemyre, in his richly harvested, investigative and compassionate view toward his fellowman, also suggested there usually was what he described as the "Town Character." One common trait of this individual was he or she always seemed to be drawn to where people were. Albert "Snaps" Kelly wanted to be everyone's friend. He grew up and remained in adulthood near the Centralia Wabash Station and seldom missed a train stopping or coming through town. He was known and accepted by railroad people and traveling men alike. Travelers coming through Centralia on a regular basis became familiar with this big figure of a man, greeting them on their arrival and waving goodbye on their departure. Although he didn't have the title, he was truly Centralia's longtime ambassador. Usually we don't sufficiently appreciate such people until they are gone away to be rewarded their crown of diamonds for being a good and faithful servant. Albert "Snaps" Kelly, fifty-two, would be missed. This beloved Negro man, who lived a life of childhood innocence, was always concerned about the sick and some felt he had an uncanny ability to sense impending death. He was also a heroic figure in that he was credited with saving a number of lives by stopping and warning those walking and driving up and down Allen Street of approaching trains. He had an aversion for paper money and big coins, but would accept five-cent pieces, which he labeled "shiners." He felt he could budget his expenditures with shiners, but bigger stuff would complicate his life. How many of us don't want more money and aren't lustfully corrupted in some way by it? Albert "Snaps" Kelly's funeral was held in the Second Baptist Church, per the December 14, 1934 *Guard*. Many of his friends filled the church along with the white pastor from First Baptist Church, who preached the funeral service. Burial was in the Centralia City Cemetery. He was one who had faithfully fulfilled his duties and job, a position he personally had conceived and had permanently employed himself in for his entire lifetime. Most of us can only hope for such security and fulfillment!

Dr. Zebulin Monroe Hampton, seventy-seven (1857–1934), had been a Centralia physician for forty-five years. He had attended the University of Missouri and St. Louis University (Marion Sims), was of Baptist faith, born in Illinois, and buried in the Centralia City Cemetery. Dr. Hampton's obituary in the October 5, 1934 *Guard* indicated he was kind to the poorer classes and didn't "accumulate much of this world's goods," but was rich in honor for those he served. One of the special things about Centralia was the people living and working there always had those they could look up to and depend on.

In more pleasing news, A. Bishop Chance was unknowingly going to become reacquainted with one of his former telephone contacts of twenty-five years ago. Fred Coulson of Abilene, Kansas, was invited to speak at the local Rotary Banquet and Mr. Chance, after being happily surprised in seeing Mr. Coulson again, introduced the speaker. The October 12, 1934 *Guard* suggested they were Independents (telephone exchanges), who battled the giant telephone corporations. As the fall progressed, off-year political races were heating up with the Stark candidacy for governor and another newcomer Harry S. Truman, the Democratic candidate for United States Senate. It also was a time when Chance officials were vacationing. Mr. and Mrs. F. Gano Chance spent a week in late October in the popular resort of Hot Springs, Arkansas. Although Mr. and Mrs. Jack Isbell's return to Centralia after several months in Tampa, Florida, appears like a vacation, it was instead stated that it had been on business. Headquarters personnel were always joking about business trips south in the winter and north during the summer. Jack liked the more exotic lifestyle of a Tampa or the West Coast even during the summer when he was busy trying to establish Chance business, new sales territories, or hiring sales personnel that he felt could do the job.

Entertainment, like manufactured goods, could be of the self-generated "in-house" variety. The radio was a powerful medium in 1934 even as it is today. It was a fortunate family that had a radio to tune into F.D.R.'s *Fireside Chats*, the news, or one of the many popular programs offered. Movies, like the radio, were faith restorers. Noticing all the new movies coming out each week, it was astounding the number produced, many including actors and titles not recognized. Its growing popularity was developing many stars including Laurel and Hardy, Ginger Rogers, Fred Astaire, Dick Powell, Boris Karloff, Cary Grant, Mae West, Jackie Cooper, Clark Gable, Carol Lombard, Gracie Allen, Bing Crosby, and radio comedian Jack Benny.

Mrs. Edmonia L. (Sneed) Cupp, seventy-seven (1857–1934), passed away in late November 1934. Her father, T. S. Sneed, was the first North Missouri Railroad station agent in Centralia, a hotel operator, and president of the early Centralia Board of Trustees. The Sneeds were a prominent Centralia family and one descendent, Sam Sneed, was one of those people who could generate no-cost variety "in house" entertainment. The pay scale in the mid-1930s reflected the hard times, and jobs were not plentiful according to Dee Daugherty:

> *People today won't believe it, but 24 cents an hour is what we started for at that time. Forty hours was only $9.60 and we were getting all the overtime we wanted, straight time for overtime. One pay, I'd do $27.60 and boy I'll tell you, that was a big check. It takes a lot of overtime to make $27.60 when you're only making $9.60 for forty (40) hours. The guys today can't believe that anybody ever worked for 24 cents an hour. (1979)*

Roy West recalled an incident in which the idea of being down to one's last dime generated an unexpected response from a Chance employee. "I remember one time, back in the piston ring days, it was payday and Sam Sneed had one dime left. He just walked up there and gave her a fling across the railroad tracks. He paid his bills." (1979) Sam was working in the packing area of the Piston Ring Department and this employee was privileged to have known and worked with Sam many years later. Sam was one of those independent spirits who had a talent for life and people. He was a courageous World War II veteran, wounded in action high over Germany, and he was one of those individuals who you meet in a lifetime that forever captures a firm remembrance.

Travel and social events have proven valuable in revealing much about Chance Company. Mr. and Mrs. E. K. Cole, Iowa transplants, returned home to Centralia in early December 1934, after visiting relatives in Des Moines and Fairfield, Iowa. They had a daughter, Miss Elizabeth

Mrs. Gertrude (Chanslor) Gano (1851–1941) came to Centralia in the 1910–20 time frame to be near her daughter and son-in-law, Mr. and Mrs. A. Bishop Chance. Her daughter, Mrs. Margaret Greenlee, would later reside with this dignified lady in a brick home near the Catholic Church. (Jack Chance Collection)

Cole of Des Moines. Mrs. Gertrude Gano, who made her home in Centralia, celebrated her eighty-fourth birthday. The December 14, 1934 *Guard* news item described her as "active in all social and religious activities and is a pretty and attractive lady." The A. B. Chances kept family matters somewhat personal and although the family circle had grown smaller on this occasion there were some new additions: Mr. and Mrs. W. T. Fountain, Mrs. J. A. Chance, Mrs. A. R. Toalson and Nathan, and Mrs. Margaret Greenlee. Nathan Toalson was already involved in company operations and A. Bishop and Gano saw potential in him. In February 1935 Mrs. J. A. Chance, seventy-eight, fractured her hip and entered Audrain Hospital in Mexico to recuperate. There were also other family relationships being maintained, as Merritt Beck, a University of Missouri student from Sikeston, visited his sister Mrs. R. E. Ausmus in Centralia. Chance Company business travels took J. H. Cheetham to St. Louis on a Saturday according to the January 18, 1935 *Guard*. A week later found A. Bishop in Kansas City on business for a week. Suggesting the close social relationship with his sales manager, Mr. and Mrs. J. Isbell accompanied by Mrs. A. B. Chance motored to Kansas City on Saturday to meet up with Mr. Chance. A frequent Centralia visitor, Chance sales representative J. J. Milligan of Chicago was entertained at a Chance dinner in early February.

H. A. Jones, a local boy made good, had become a successful Chevrolet dealer in Centralia and accepted management of a General Motors dealership in Robinson, Illinois. Although the automotive industry was helping maintain industrial activity during the Depression, the bicycle was again gaining popularity. In the late 1800s, the bicycle had exploded upon the American scene and now because of the hard times, its low initial cost and very low operating expense, it became popular again. It was also becoming a fad as the newspaper pictured Hollywood starlets with their new bikes. Business casualties were still occurring as R. L. Hope sold his 1885-established drugstore to L. M. Haynie of Nelson, Missouri. His son later said an early Depression bank failure had impacted his father's sellout. Centralia wasn't long without some commotion to stir up controversy and reenergize local gossip. The March 1, 1935 *Guard* ran a front-page item on the Vista Theatre's stockholders revolt and lawsuit against the Pruetts who in 1932 had obtained the financial support of the Chances, R. L. Hope, O. B. Mayes, and others to convert to "talkie" movies. The Centralia Commercial Club, the predecessor of the Centralia Chamber of Commerce, had solicited this venture. First leased after the sound conversion, the theatre was recently sold with charges that sales funds hadn't been properly accounted for.

Infantile paralysis was becoming a threat and F.D.R., a most famous victim, established the Georgia Warm Springs Foundation facility where he also built a home, the "Little White

House." His revolutionary work for polio victims and for all the disabled also included the founding of the "March of Dimes" program in 1938. The idea and name came from the famous entertainer Eddie Cantor. When the program was failing to sustain itself from large wealthy sourced donations, they took it successfully to the common people and even schoolchildren. So much mail was received at the White House that it temporarily shut down the government. The February 6, 1942 *Guard* told of $51 being donated by Centralians during the March of Dimes national drive. In March of 1935, frequent coverage of movie star Ida Lupino was partially due to her recent bout with infantile paralysis, from which she was recovering. She would later star in such movies as the 1941 classic *High Sierra* with Humphrey Bogart. This writer's mother Willa Mae (Johnston) Darst (1920–) also had polio during this time, postponing her eighth grade studies until a year later when she recovered, partially due to her determination.

A. Bishop wasn't immune from injury and the dangers of constant travels finally caught up with him on Thursday, March 14, 1935, on a return trip from Kansas City. He suffered severe bruises about the body sustained when the steering locked on his car and it turned halfway over on Highway No. 40 near Boonville. Gano motored to Columbia and picked his father up and took him to the Audrain County Hospital in Mexico. His mother was also a patient there as earlier stated. The injuries to Mr. Chance were more serious than earlier thought as two weeks later the March 29 *Guard* told of Mrs. A. B. Chance traveling to Mexico on Sunday, March 24, along with Miss Ida Lou, Miss Juanita Bryson, and Mrs. W. T. Fountain to visit the Chances at Audrain Hospital. After three weeks, the April 25 *Guard* told of Mr. Chance's release and triumphant return to Centralia. The factory siren sounded as the car carrying the president of Chance Company ran its full passage down Allen Street. It wasn't a warning of a fire or tornado, but to a man many likened to a human tornado! Mrs. J. A. Chance was dismissed a week later on April 7.

Because of increasing employment at the Chance factory, former locals such as George Robinson of Fort Collins, Colorado, moved back and had accounting and much later data processing duties. Mr. and Mrs. Jack Milligan of Chicago in April 1935 escorted A. B. Chance III to Centralia to visit the A. B. Chances and his other grandparents, the M. H. Pembertons. The school year was also ending locally as fifty students were graduating from Centralia High School including Cab Atkins, Helen Cheetham, and Marjorie Forrest. There were also twenty-seven eighth grade graduates including Ruth J. Cheetham, Carl McManama, and Majorie Blanton. Downtown, Attorney R. E. Ausmus was establishing an impressive law library with the procurement of the late Judge Bruton's collection and additional law materials from a Mr. Chinn of Higginsville.

A. Bishop and F. Gano Chance believed in promotion schemes and salesmanship. A May 10, 1935 *Guard* article carried such a Chance event that was performed to capture attention. The Chance Company, on Monday May 6, paid their $500 (*sic*) payroll in silver dollars. These new coins were said to be rare and this event was staged to show how much money was being placed into circulation by the company. Since 1929, silver dollars had disappeared from circulation due in part to the U.S. government setting a stable price of $1.29 an ounce for silver. The December 1950 issue of *Chanco News* carried a picture of the silver laid out on a table (shown on the following page), stacked for payment. Along with Chance officials, the photo included Mr. Edwards and Mr. Palmer of the First National Bank and Mr. Renie of the Post Office. It was revealed that at this time in the spring of 1935, there were 275 local employees at A. B. Chance Company.

Following the silver dollar payroll in May, the June 1935 Chamber of Commerce meeting

Payroll in silver dollars. From left to right: Beverly Mayes, C. C. Hough, Julius Edwards, "Boss" Chance, Mayson Palmer, Pat Renie, and Gano Chance. (Jack Chance Collection)

was held in Chance Hall at the factory and covered by the June 14, 1935 *Guard*. After the regular meeting, F. Gano Chance addressed the assembly and told of the accomplishments of the A. B. Chance Company. He said there were now three hundred employees, an all-time high. Gano said the average $22,000 per month payroll had virtually eliminated unemployment in Centralia. How many communities were so blessed in the middle of the Great Depression? Following Gano's talk, Jack Isbell explained the various Chance products and their positive points. Various Chance products were spread over a large table. The Wej-Lock device was bringing national notoriety as it was the guardrail used along state and federal highways across the United States. There were various devices that made up the Wej-Lock package. The discussion about the hammers, plow points/blades, and axes will be covered later in this chapter. Showing a stack of customer orders for such products as pliers, it was pointed out the lack of capital to purchase new equipment to manufacture such items. This could have been another Buffum hand-tool item that could be supplied to the big catalog houses. J. H. Cheetham also spoke on the need for plant expansion and was noted as the expert on cast steel, especially cast-edged tools that would hold their edge and compared favorably with more expensive processes of manufacture. The acceptance of those cast-steel Chance products could either spell success or doom for J. H. Cheetham at Chance Company. He also suggested that it was possible that the factory workforce could be doubled if money could be secured for plant operations. Because most Chance employees lived in or near town, this could generate a housing crisis in Centralia.

The three most important management figures at Chance Company had spoken, except one–"The Boss!" A. Bishop concluded the speaking part of the program with some new ideas and some old ones. First, capital needs were highlighted for daily operations and to fill large orders. This had become a problem with the new regulations limiting one bank from loaning more than $6,000. Earlier there had been three banks in Centralia and some suggested Chance Company owed them all. This may have been why Miss Ida Lou kept money in several banks so those accounts could be borrowed from, although prior to F.D.I.C. there was a fear of keeping all the eggs in one basket with banks going under right and left. Capital shortages were said to have resulted in lost business as orders wouldn't be placed where the: "terms spot cash" was applied. Again, Mr. Chance raised the issue of moving the factory where money could be secured. He said this shouldn't be taken as a threat, "as he wanted to continue to stay in his home town." Undoubtedly he had everyone's attention, and in a receptive mood to hear his bond proposal. Bonds would be issued against the real estate of the company, which according to a recent insurance appraisal was valued at $157,000. The bonds would have a ten-year term and a 6 percent rate, but could be paid off at any time. Mr. Chance thought this proposal was a way out of the difficulties the company was facing and offered the public more interest than government bonds. The only reason these bonds wouldn't be safe was the failure of the company, which he explained that such a contingency was not impossible. A statement revealing A. Bishop Chance's unwavering faith in his enterprise, however, followed in the *Guard* article: "but one he had not considered possible at any time." It could be suggested that at this particular time when business was good, it was easier to make such a statement. There had been talk over the last several years that the Chance firm would go under, but now A. Bishop proudly silenced any remaining doubters. He didn't accept failure and when it did occur in his life, he worked over or around it. Although there would be fall-out from some of the current business activity evaporating, there doesn't seem to be the potential for crisis as was present earlier in the 1930s.

A plant tour followed the bond proposal presentation including watching plow points being made. Piston rings were also being fed into a machine at the rate of 1,500 per hour, automatically roughing both sides of the rings. The Chance plant was said to be one of the country's largest producers of piston rings. The largest order for 580,000 rings was placed with the Chance factory in the spring of 1935. The Piston Ring Department was crowded in every corner because of the high level of activity and the recreation room had become an assembling room. In the mid-1930s, Dee Daugherty remembered yet to become executives and managers who were working in the Piston Ring Department just as he was. "He [Paul Sturgeon], at that time, ran the rent-all machine and he swept the floor just like the rest of us. Brent Mayes, he was the timekeeper . . . Les Reynolds was there. Cab Atkins was running the rent-all machine." (1979) Further information about Mr. Chance told of his early schooling in the Burnham District School and in the Centralia public schools. Details about the factory said the three-story 210 North Allen building housed the offices and machine shops behind the offices on the first floor and the second and third floors also include machine shops. The long structure directly behind the main building was the iron foundry and part of the punch press and machines for roughing. Several other buildings mentioned had other purposes. As the Chamber group departed the factory at 11:35, the plant was still in operation. In fact there were three shifts in operation day and night to fill the rush of customer orders that had been received. A. Bishop had further comments about a wide variety of products now being embraced at the Chance plant:

> *In other days we thought it was best to specialize on certain products, but today we are expanding to make any saleable article we can get, so the plant can be operated all year around. If one product has reached the point of saturation, we will not have to shut down, but can take up some other article that can be sold in its season and so keep the plant busy all the time.*

A. Bishop's remarks are significant about the direction of the company into becoming involved in a wide variety of products. With a small plant and limited capabilities, specializing in a few products was all that was possible for the early company. After the many new processes and machinery were put on line, it became easier to branch out into other areas with no relationships to the utility market. The first of these was piston rings and the auto line, which would confront the utility seasonality problem with new items having year-around demand. In 1935, A. B. Chance Company was now into at least four distinct markets: utility, automotive, small tool, and agricultural items. However, Chance was often unknown in the overall markets, so name recognition and sales-promotion were challenges. It's like the saying: "Jack of all trades—Master of none." When the bubble burst on the small tool and agricultural plow items, this quickly removed Chance's participation in those markets. Then the piston ring area became highly contested and the company was back to its old standby, the utility market. Thus the multiple industries concept would die on its own and then Gano could concentrate more heavily on the surviving utility market. This was a fortunate situation as the future World War II and postwar years would prove the advantages of becoming a bigger player in one market, especially when the electrical utility area showed tremendous growth potential with the new R.E.A. program. Although Gano is frequently credited with that move, A. Bishop had already captured a portion of that utility market with over three decades of product development in the anchor and hardware areas. Gano had again just built upon what his father had already established.

The wide variety of known products at Chance increased by two, as identified from Paul Sturgeon's remembrances:

> *When I went to work for Chance Company, they were making piston rings. I went to work in the Piston Ring Department. Prior to that time, they made other things that went into the automobile line such as tire sealer; the stuff you put inside of the tire. You remember when your bicycle tire would go flat, you would go down and buy a tube of it and squirt it in. Well, we made that. I don't know where in the world the hog rings ever came from, but we had a very elaborate supply of hog rings at one time that were pretty fancy. (1979)*

June 1935 brought good news for those seventy years or older as they might be eligible for an old-age pension. Travel continued for all ages during the Depression, especially for Centralians and Chance employees. Miss Jennie Green visited her brother in Sedalia and the Joe Cheethams visited friends in St. Louis. William (Bill) Hollander, Bernard Wilson, and Mr. and Mrs. Luke Horack left on August 2 and motored to California to visit Hollywood, Long Beach, and other locations. A week later relief from the hot weather may have been consideration for a several-day business trip to Minneapolis by Mr. and Mrs. F. Gano Chance and Mr. and Mrs. Jack Isbell. The Clifford Houghs and daughter visited his parents and Mrs. Hough's mother in Garden City. In mid-August, A. Bishop motored to Kansas City with brother-in-law W. T. Fountain, who visited with his mother. Mr. Chance remained several days on business. Mr. and Mrs. Perry Roberts of Rocheport were Centralia guests of her sister Mrs. Roy McManama in early September. In mid-September, Rotarians A. B. Chance and J. H. Cheetham gave short five-minute talks to the Rotary and Brent Mayes, employed as timekeeper at Chance since March 1935, was marrying Miss Edith Leach. If there was any thought that the president of the Chance firm was retired, another business trip in early October to Kansas suggests otherwise.

On the national scene, one of the Christian giants of the day, former major league baseball player turned evangelist, Reverend Billy Sunday (1863–1935), passed away. In mid-October 1935, President F. Gano Chance of the Rotary Club of Centralia gave a three-minute talk on his hobby. The main speaker, Odie Walker, discussed his activities in the Boy Scouts. Accidents within and outside the Chance plant continued as employee Ray Crigler's injuries weren't serious from a car accident and he was soon back at work. Keeping up on current trends and technological advancements, Chance Company had already begun the long tradition of attending trade shows and educational programs. An example was Art Houston's attendance in October at a Kansas City direct-mail advertising convention. All those October activities, however, were overshadowed by a bomb blast over at the company. The November 1 *Guard* announced that J. H. Cheetham was leaving employment at A. B. Chance Company! Mr. Cheetham's eventual replacement as factory manager, O. E. (Odie) Walker, head of the Cost Department, was in Chicago on business in early November. Headed in the same direction on a Chicago and Wisconsin business trip were A. Bishop, Gano, and J. T. Isbell. A later news item suggested this trip continued into Canada where A. B. Chance Company was doing business and in a couple of decades would establish a facility there.

As the end of 1935 was in sight, Chance employee Miss Ruth Brown of the Printing Department was marrying Virgil Stanford. Besides the A. B. Chance Company payroll directly benefiting employees, it also fostered the growth of small businesses. The Chance-owned Anthony Building, home of the Ford dealership, was vacated in late November. Mr. Leader, owner of Centralia Motors moved to his new filling station and garage at North Allen and Highway No. 22 (Switzer). There may have been issues of a better retail location or Chance Company may have needed the Anthony space for expansion. The Anthony Building was destined to lose another patron when a construction firm from St. Louis was poised to begin construction on the new High School gym in 1936. Commercial entertainment for all the working people around Centralia was boosted in December 1935 with a new bowling alley on West Singleton. To help power such advancements, Bagnell Dam in the Missouri Ozarks, had increased electrical power generation in mid-Missouri. However, the transmission line near Jefferson City had an outage when an unknown party dynamited two of the steel towers. Those displaced from land now under water, generated strong feelings as it did in other states where huge dams were being constructed.

The capabilities of the Chance Foundry, in spite of problems encountered, was also increasing as 1935 was closing. One of the advantages of having your own foundry is personal items could be cast, like the Rotary Wheel symbol. Rotarian President F. Gano Chance gave each member a "wheel" made and finished in the Chance factory. That was a thoughtful gift for Christmas and it would be an interesting project to try and locate surviving Chance Rotary Wheels some seventy years later. The Chances were also on the receiving end as A. Bishop received a Christmas card from a representative in Japan. During this time a lot of piston rings were shipped to Japan, many of which may have been used in the buildup of their war machine. However, not everyone understood the extent of the threat in the 1930s anymore than we did six years later in 1941.

Reflecting the safety of brick and steel construction, an early 1936 fire in the paint shop of the Anchor Department was extinguished, with damage only to the equipment and roof. An hour later another fire broke out in the same place and city firemen were called out a second time. Minor burns were suffered by I. S. Kuhne a Chance employee. Noting other Chance plant items, Mrs. Irene Oldham was working in the Packing Department of the factory and the C. C. Hough family was now Centralia residents. A promoter of the Centralia area and Chance Company, Edgar T. Rodemyre, sixty-seven (1869–1936) died of pneumonia on April 9. Mr. Rodemyre was of Christian faith, a Rotarian, and the last member of the Rodemyre family that had operated the Centralia *Fireside Guard* for sixty-seven years! He and his brother had bought out their father's interests in the paper in 1903. Ed was a bachelor most of his life with no children, but in 1931 he married Mrs. Edna

E. T. Rodemyre. (Centralia Historical Society)

Miller who had four children. A stepson George Harold Miller, an M.U. graduate, would become the new editor. As a new era is about to begin, it's significant that Mr. Rodemyre's death signals the close of the earlier one. Early editors held an important position in their communities and Centralia couldn't have been more fortunate in having the Rodemyres, who were recognized as having one of the better weekly newspapers in the state. A friend, Clarence L. Northcutt said Ed continued to grow in mental stature. They never had the advantage of wealth or education, but this didn't alter the Rodemyres compassion for their fellow man, the revealing of rights and wrongs, the eulogization of service, promoting advancements of the day, or the successful and influential positions they held for so long.

In mid-June 1936, a trip to Bagnell Dam reveals some close Chance Company and family relationships. The J. T. Isbells, Mr. and Mrs. F. Gano and Mrs. and Mrs. A. B. Chance, Miss Jennie Green, Mrs. Gertrude Gano, and Mrs. Margaret Greenlee motored to the lake. Another transportation alternative, passenger trains, was getting faster which made business travels more efficient. The Wabash No. 9 and 12 were now on a six-hour St. Louis to Kansas City schedule with only four out-city stops: Mexico, Centralia, Moberly, and Carrollton. It is significant that Centralia, being close to its larger neighbors Moberly and Mexico was included which indicates it was generating significant ridership from the Columbia Branch, Centralia, and Chance Company. The new *Guard* editor, George Harold Miller, married Miss Ruth Lillian Quigley of Cameron per the July 3, 1936 *Guard*. Miss Quigley attended St. Mary's in Leavenworth, Kansas, and graduated from the University of Missouri with a bachelor of science in education in 1931. In October 1937, they became the parents of a fellow Kiwanian and company associate, William Harold "Bill" Miller, a longtime Chance employee. Rotarian and Chance employee E. K. Cole spoke to the club about transportation. Mr. Cole was the inventor of the Wej-Lock guardrail that Chance Company was now producing and he was said to be much-traveled. The latest Chance employee-trio traveling was O. E. Walker, Paul Sturgeon, and Russell Smith. They motored to Springfield, Illinois, to attend the Illinois State Fair on a Saturday and a baseball game in St. Louis in late August 1936.

In late June 1936 the Golden Gate Bridge was under construction as was the Highway No. 36 bridge at Hannibal, named the Mark Twain Bridge, which was to be dedicated by our president on September 4, 1936. It was a P.W.A. project that this writer's Grandfather Denver Johnston (1896–1981) said had to be accelerated to make the president's scheduled appearance. This generated added local drama as the president's train, en route on the Wabash from Des Moines, had a water/ice stop at dawn in Moberly before proceeding on to Hannibal for the dedication ceremonies. A month later, as the 1936 campaign heated up in anticipation of the nearing November 1936 general election, President Roosevelt began a nine-day swing through the Midwest. As the stumping in Kansas ended, his opponent Alf Landon's home state, the president's ten-car train was run again on the Wabash from Kansas City into Moberly, arriving at 11:03 p.m. according to the October 14, 1936 *Moberly Monitor Index*. But the crowds were disappointed again in not seeing the president, although both U.S. Senators Clark and Truman, plus Governor Park tried unsuccessfully to appease the large crowd, estimated at five thousand. As the train headed east toward Centralia and St. Louis, the *Guard* reported it parked on the Sturgeon siding and remained there until 4:10 a.m. The Sturgeon location suggests the train was on the Wabash and later passed through Centralia to make St. Louis by early morning. The *Guard* stated that news reporters made the Centralia Wabash station headquarters for a few hours while the President of the United States slept over in Sturgeon.

For well over half of Chance Company's existence, the train was used extensively for business travels. F.D.R. also primarily traveled by rail and during his twelve years as president he recorded over 544,000 miles by train and in 1944 alone he traveled more than 50,000 miles! He rode in a special car with three of four additional cars for Secret Service, reporters, and communication gear plus a freight car for his automobile. The car allowed him a platform to speak from as he did at the Hannibal bridge dedication, view troops and other activities that couldn't be handled without revealing his paralyzed condition to the public. It can only be roughly estimated how many miles A. Bishop Chance traveled by rail, but beginning at near the turn of the century, he likely traveled several thousand miles a year over a fifty-year period, which could have also generated some impressive numbers.

One of the fondest remembrances of early employees was the annual picnic, which began on Saturday August 19, 1936, per a front-page article in the *Guard*. However, the previous chapter noted a company picnic in July 1927 and another in 1928, although it apparently hadn't become an annual event. The 1936 picnic was an all-day event with a flag golf tournament at the Mexico Country Club beginning at 7:30 a.m. At 1:30 attendees gathered at the "Fountain Woods" for the main picnic. This reveals that a portion of the Sunnydale farm was formerly Fountain-owned. The picnic area had a stream and wooded area and was located northeast of the farm's headquarters on old Highway No. 22. These grounds would be improved by A. Bishop and would later become Buford Park, named in memory of his eldest son. There were various activities such as a scavenger hunt, chicken chase, blindfold boxing, horseshoe pitching, croquet, ping-pong, ball games, and numerous other games and activities. Mr. Chance dubbed the occasion the "First Annual Chance Company Picnic." One of the benefits of the picnic coverage in the *Guard* was the inclusion of the names of many Chance Company employees and families who won prizes and/or participated. They included: Charles Atkins, J. C. Blanton, Virgil Blanton, A. Bradley, Velma Bradley, Mrs. Orville Brown, Franklin Brownell, Mrs. Victor Bugg, Virginia Carey, A. B. Chance, Brick Chase, Fred Chase, Flynt Chandler, Ollie Chandler, Lois Chandler, Harold Cox, Mrs. Tom Cox, Fred Crawford, Velma Curry, Dee Daughtery, Ruth Dawson, Clarence Edelman, Buddy Fordyce, Virginia Fordyce, Bill Gallup, Mr. and Mrs. Wilbur Gholson, Leon Groves, Mildred Hofter, W. L. Hollander, C. C. Hough, Art Houston, Paul Hulen, Paul Hutchinson, Neville Hutchinson, Jack Isbell, Bertha Johnson, Billie Jones, Dick Jones, Margaret Jones, Wagner Jones, Mrs. Verdell Kanatzer, J. B. Kaufman, C. Kettrey, Ralph Koewing, Tom Marshall, Brent Mayes, Francis McDaniel, Maxine McDaniel, Doc McDonald, Junior McKinzie, Lex More, J. J. Mundy, Russell Mundy, Claude Owings, Mildred Palmer, Betty Parker, Howard Parrish, Madge Parrish, Gilbert Phelps, Mrs. Luther Powell, Melvin Pruett, Leland Purdy, James Reams, Perry Reams, George Robinson, Granville Ryman, Esther Smith, Russell

Smith, Sam Sneed, Howard Sturgeon, Paul Sturgeon, Nathan Toalson, Ernie Townsend, Joe Townsend, O. E. Walker, Roy West, L. B. Wilkes, Pete Wilkes, B. L. "Dutch" Wisdom, and Granville Wright. It wasn't said how many employees were in attendance, but it was a good turnout.

As fall came again there were some positive economic signs with a number of business changes, start-ups, and expansions such as the local Maytag dealer, McManama and Farmer. The new school gym was opened on September 5, 1936, with Harry O. Donnahue becoming the new high school principal. Another drought year similar to 1934 was ended in mid-September when rains returned. The silver screen was being lit up in 1936 with such performers as Randolph Scott, Gary Cooper, Jean Harlow, John Wayne, Maureen O'Sullivan, Mae West, Don Ameche, Myrna Loy, William Powell, Spencer Tracy, Bing Crosby, Jeanette McDonald, Loretta Young, Fred MacMurray, Shirley Temple, and Will Rogers. There were also considered to be key performers at A. B. Chance Company, according to an O. E. Walker proposal before the Foremen's meeting on Saturday, September 12, 1936. A club consisting of officials and "Key Men" of the company, according to the September 25 *Guard* was to be called the "Chance Keys" Club. The Walker proposal did include the Club's purpose, which was "to bring about a better understanding of the problems of management and foremanship." This sounds familiar as recorded in the late 1920s and reflects a Chance philosophy of plant operations. Sure enough one of the speakers at the first meeting held September 23, 1936, at the Chance Gym was the president of A. B. Chance Company, A. Bishop Chance. Club officials elected were Harold Cox, president; Wilbur Gholson, vice-president; and Brent Mayes, secretary/treasurer. O. E. Walker appears to have brought a changed atmosphere to the plant that A. Bishop was in tune with. Mr. Isbell also spoke of the operating practice established at Chances of supporting employees who improved themselves for promotions. Mr. Isbell explained the opportunities within the company and the importance of study to better qualify for advancements. One of the results of plant growth was the need for increased numbers of capable management and a fertile field was local talent who were a more known quantity. The company policy of promoting from within developed, and this directly benefited employees and was a reflection of the American principle with a Chance twist, that if you work harder and smarter you will grow more successful. This was a grass roots concept that flourished all across the United States during over a century of industrial growth.

The railroad and the prominent colleges in Columbia brought many VIPs through Centralia. Now taking up lecturing, Admiral Byrd was to stop briefly in Centralia on Wednesday, October 21, 1936, on his way to Columbia. A prominent user of Chance anchors, he was surely welcomed to "Anchor City." A longtime fixture four miles south of Centralia, a new pipeline pump station was near completion, while in Centralia a young Phil Chance suffered a fractured right wrist in a fall at his home. There was also unwelcome news of Mrs. C. E. Chance's death on October 27 at the age of eighty-four in St. Louis, at the home of her daughter Mrs. Maud Bates. On the national scene Franklin D. Roosevelt was reelected to a second term as president and locally F. Gano Chance became a director of the National Association of Manufacturers. Also trying to gain the spotlight was the annual new car introduction, which was becoming a big fall event. Taking a page out of one of Henry Ford's daring initiatives, the $5.00 per day wage, A. B. Chance Company announced their version. The November 20, 1936 *Guard* reported Chance employees on December 1 would receive a wage increase of 10 percent. During the Depression when employment alone was important, such an announcement was significant, although there were rumblings of labor union activities, suggesting the need for fringe benefits such as

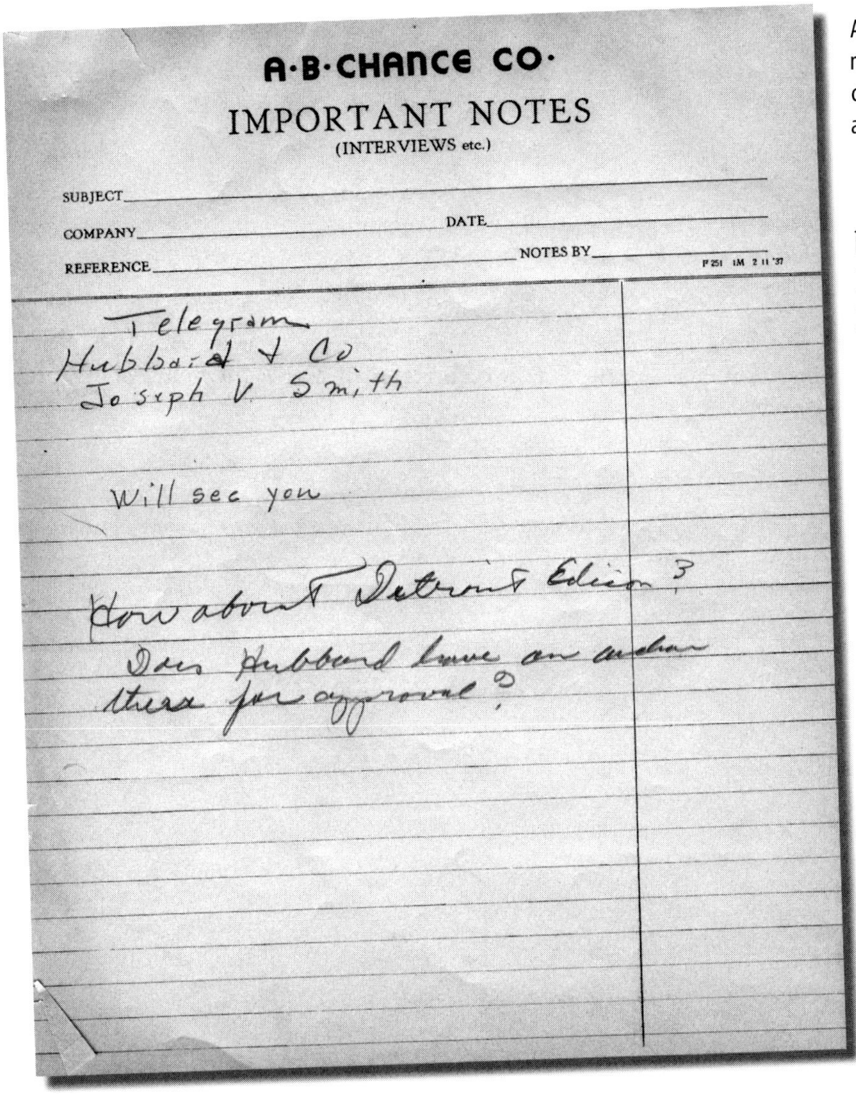
A·B·CHANCE CO·

IMPORTANT NOTES

(INTERVIEWS etc.)

SUBJECT

COMPANY DATE

REFERENCE NOTES BY

F 251 1M 2 11 '37

Telegram
Hubbard & Co
Joseph V. Smith

Will see you

How about Detroit Edison?
Does Hubbard have an order
there for approval?

A notepad of the post–February 1937 variety with notes from possibly the founder. The letterhead had changed to the familiar A. B. Chance Company name and appearance. (Chance Foundation Files)

vacations. The wage announcement was made by the vice president and general manager of A. B. Chance Company, F. Gano Chance. He said that better business conditions and the outlook for continued improvements made such a wage increase possible. He suggested this had also been occurring at many other firms over the last several weeks. So there was renewed optimism that the Depression was weakening, but in spite of the courageous leadership of F.D.R. and business leaders like the Chances, hard times continued to linger on. The strength and faith of the people of that day, however, brightly shined through in an optimistic spirit that denied defeat.

As Thanksgiving 1936 approached, there was much to be thankful for around Centralia. The employment brought by A. B. Chance Company sheltered many people from the hardships endured by surrounding neighbors. Good jobs are so important in our free enterprise system, adding to the dignity and independence of our citizens and gluing the fabric of our nation together. But this season also brought remembrances of a mother for her son, although his death had occurred eight years previously. Mrs. Fancy Chance was seriously ill, according to the paper, possibly with a recurring stomach ailment. Maybe she was able to regain her health and spirits as the Christmas season approached, and news of the planned visit of her grandson A. B. Chance III of Chicago, was anticipated.

Claw Hammers and Cultivator Shovels

A claw hammer is defined by Webster as a hammer with one end of the head forked for pulling out nails. If there ever was a hammer that spoke with a "forked tongue," it surely was the one made by Chance Company during the Great Depression. When Flynt Chandler was asked if he remembered the hammer and ax business, he stated: "Yes sir, I sure do, that was about as near nothing as anything I know. But they sold a lot of them and they did haul a

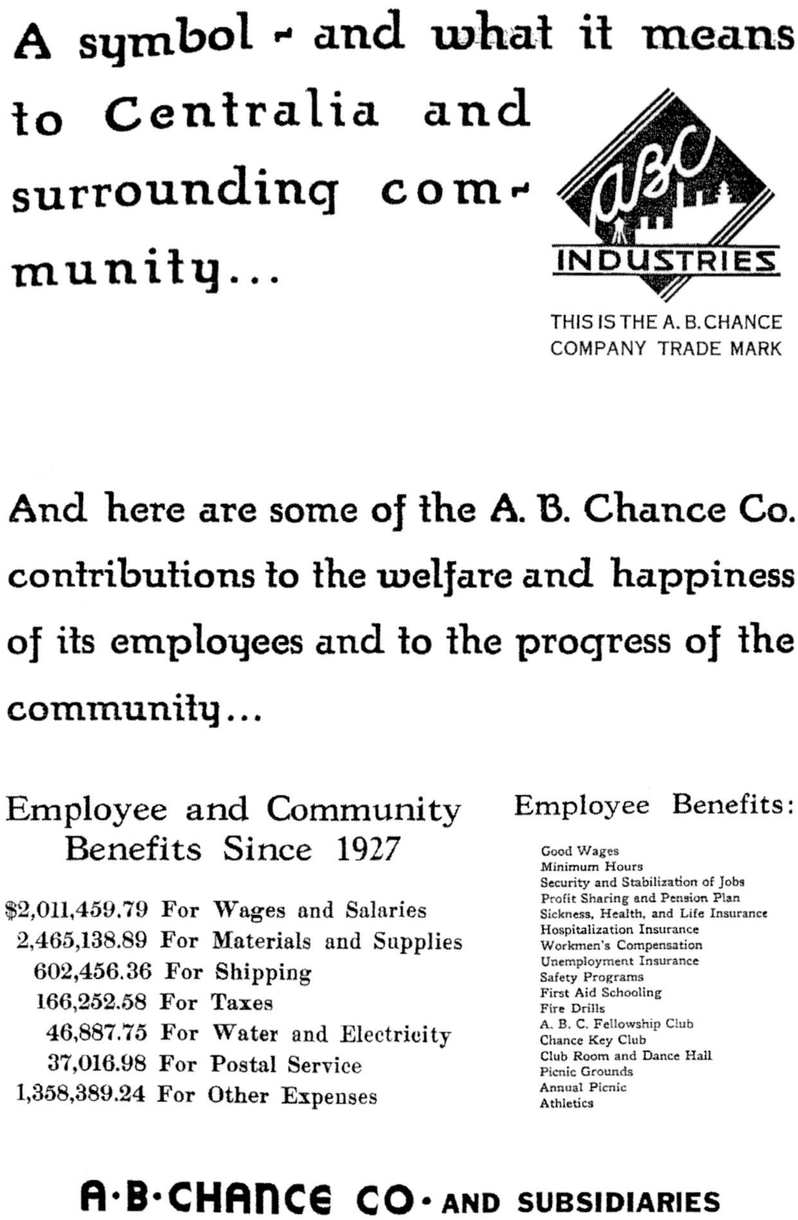

The classic trademark symbol in the ad suggests the Chance public concern for their community and employees and the contributions and benefits that each had received and could expect. (*Guard* of July 26, 1940)

lot of them out of here." (1979) Les Reynolds remembered that he had one somewhere in his garage and said they were "no account." Well, if you hit them just right, they would fly to pieces!" (1979) Granville Ryman discussing the hammer said: "It would break once in a while," (1979) a somewhat guarded statement that you would expect from an experienced marketing man or more assuredly from a public relations person.

In the early 1930s, the iron grip of the Depression was being felt everywhere. The Centralia area was no exception as the Chance Company "was the only company in town" according to F. Gano Chance. "I thought the main job of our company was to hold the employment and not have to lay anybody off." Area residents had come to depend upon the jobs that the Chance Company had generated over the several decades of its existence. "I felt a deep responsibility for the welfare of the town," Gano continued, "so the best thing I could do was to try to hold the employment even with sales that had gone from $740,000 down to $196,000." (1981) A young F. Gano Chance continued to be tested and christened with fire, so to speak. The first half of the 1930s was challenge enough for an experienced veteran, but this didn't deter Gano. An explanation of why he was driven like a man with a mission was he felt the burden and responsibility for the welfare of the family business, plus he appeared to enjoy the challenges of directing the company and wanted it to continue.

On several occasions certain actions that Chance Company took can be traced back to a Chance philosophy of how they wanted the company operated. The emphasis upon keeping the plant doors open and people working is a focus during the very difficult Great Depression period, because of its threat to the stability of employment. It was a surprise for a former executive to speak of an incident on this subject. This is especially significant as he hadn't had the opportunity to see this book material and it could be assumed he was hired too late in the company's life, in 1970, to experience this first hand. J. G. "Jim" Everhart, employed as vice president of manufacturing, had become general manager a year later in 1971. Although F. Gano Chance was supposedly retired by this time, Jim said, "Gano was normally in the office each day and attended the Monday executive meeting." (2001) This was a weekly, Monday, noonhour meeting at the Chance Garden House. A slump in business activity had led Jim to lay off about eighty production workers in the Chance Plants. Gano hit the ceiling when Jim reported his actions. Even after the bitter 1961 strike, Gano still felt a strong responsibility toward his people. Looking back over his six decades of being on Chance property, it could be suggested that everything had changed—but on the other hand nothing had changed in the Chance philosophy of plant operations! That is a remarkable aspect of A. B. Chance Company, as its people not only embraced the values of the past, but also made bold, innovative changes in confronting the future.

The dramatic drop in company sales and profits in the early 1930s that Gano alluded to threatened the survival of Chance Company. Like his father, Gano also faced the threat head-on when he had secured an important loan, helping to insure the company's survival. These hard times had swiftly followed his college graduation and had not given him much of a chance to grow into the company. He was spurred on by a remembrance of his older brother who was the natural heir to lead the company, but now the family and company needed his leadership in providing hope for the future. The exuberance of a recent college graduate must have played a role in giving Gano a confidence that young people often have of feeling they are invincible and nothing is impossible. One thing Gano felt his father owed his success to was:

> *I think it was primarily his enthusiasm and his energy. If it doesn't work the first time, try it again. I think you have to have that enthusiasm, interest, and dedication to succeed. I think he was an excellent salesman and with that enthusiasm he could get people to follow him and believe in him. He was a man with a lot of enthusiasm and was therefore a good salesman. (1981)*

Although it would appear his father had deserted him, this wasn't true. Even Gano realized that his father was watching company activities carefully, allowing his son to experience successes as well as failure that would christen him into the sometimes lonely and difficult job of guiding the company. These experiences would help shape and mold F. Gano Chance for future challenges just as his father had dealt with the struggles of the start-up process and directing the expansion and reorganization of the company during its phenomenal growth period of the 1920s.

The goings-on within the company during this period undoubtedly brought the usual dosage of skepticism, especially out in the plant. Hardly any good factory person ever thinks management knows what they are doing. This is an attitude that commonly prevails out where words and orders are transferred into sweat and toil in generating products. The Lord spoke his creation into being over an evolutionary time frame, but people are much less efficient, requiring several layers of development before something becomes a reality and can be sold to a customer. Certainly management needs to keep connected with the plant, as did A. Bishop. But doubt about a young inexperienced F. Gano Chance was also felt within management from one who had gained experience and business savvy from the school of hard knocks. One could conclude that this was an obvious situation, as the Chances didn't always do things in a manner that an experienced and more traditional businessman understood, especially in the sink-or-swim atmosphere of the early 1930s. Gano admitted his aggressiveness during this time and burdened himself with the resulting failures as his father allowed him to develop a personal conscience and be responsible for his actions. "My father never did say to me, 'you did a good job or you did a lousy job'." It often is a lot easier if you get a chewing out instead of carrying around personal burdens in silence. "Jack Isbell was with our company, who was up at the sales end. He later told me he thought a lot of things you put in and did weren't right, but he said he wasn't in the management end, so he kept still. He said it turned out you were pretty right and I am glad I kept still." (1978) Of course Jack knew how to butter the bread on both sides without ruffling any feathers.

At least three different methods were used to keep work in the Chance plant. First, Gano said, "We did that by making things that we formerly bought as parts that were used in our anchors, hardware, etc." Along the same line, A. Bishop, to cut expenditures, would have new machines built from pieces and parts of old machines. Bill Hollander said, "We built all kinds of machines. Some of them were wonderful and some of them terrible." A direct result was, of course, giving more work to those in the plant. Another method was the manufacturing of other items not associated with the telephone and electrical utilities industry or auto line. This included axes, cultivator shovels, guardrails, hog rings, and of course hammers. According to Dee Daugherty, "We made just about anything that you could get a buck out of." Gano's thoughts were:

> *I thought I had a way to make some money, having worked out a deal with this company to make cast iron hammers. They were not much good, and they did not come through at the time in paying their bills. Buffin* [Buffum] *started selling the hammers in big quantities and getting the money for them but he wasn't paying us. That was when I began to seek the advice of lawyers, etc. . . . and I think the company lost about $10,000 back in 1932 and that was a lot of money. (1978)*

This was a time when boldness could be utilized in attempting to slow or stop the economic slide backwards, but yet caution was also needed to determine what was expedient to get involved in. Gano honestly admitted that not all of his special projects worked out:

> *After losing my shirt once or twice in deals like the one in making hammers for a company, my father never said a word such as, "Son you got messed up with the guy and he is going to cheat you out of money, or this or that." But he didn't need to as I realized that I had really muffed the job and the fact is in 1932, losing two or three thousand dollars was bad. (1981)*

In another response Gano said:

> *When I got out of school I pushed my way in and when I saw something that had to be done, I did it—made a lot of mistakes popping all the buttons on my vest—lost the company some money. The first one I think was about 1932 when we got a contract for hammers made out of cast iron—most of them were made out of steel—for Buffet* [sic Buffum] *from Louisiana* [Missouri], *a young fellow. (1978)*

This description of Buffum as a "young fellow" by an under-thirty Gano suggests this may be Frank Buffum's son or relative and not the turn-of-the-century lumber yard/telephone entrepreneur as originally thought. It also could be a different family as two other spellings of the name (Buffet, Buffin) could suggest in interview material. However, from this point forward it is assumed Buffum is correct, as the *Guard* used in earlier material concerning Frank Buffum and per the 1911 advertisement for Buffum Tool Company. Frank Buffum will be remembered earlier as associated with LaCrosse Lumber and then operating several competitive telephone exchanges in central Missouri when it was spun off of the lumber yard business. After the sale of the exchanges to

Bell Telephone in April 1906 his area presence evaporated. But five years later an ad was noted being run in the *Guard* of November 24, 1911, about a Buffum Tool Company, who had the capabilities of doing nickel plating. This may be the firm that became associated with Chance Company in the early 1930s.

The Chance Company involvement in the axes, plowshares, and hammers began during the reign of Plant Manager Joe Cheetham, according to Bill Hollander. Joe's Chance employment was in 1931 as revealed by his later separation, which agrees with the departure of Husted in April 1931. Bill, in his forthright manner said:

> *He was a salesman and he sold A. B. on making him factory manager. . . . Somehow he heard about a fellow over in Louisiana, Mo., who had a little shop. Buffin* [sic] *had all kinds of ideas and he was a promoter. He had a tiny little forge shop in which he made screwdrivers and tools. They made a few tools for working on a car, etc., and he had made some cast-iron axes hatchets, and claw hammers. Being a real good promoter, he had sold these things to Montgomery Ward, Sears, and several big retailing outfits, but he wasn't able to supply them. Joe brought Buffin over and he sold the deal of giving Chance Company some stock in the Buffin Company to manufacture the stuff. Joe also got a bunch of stock from Buffin for promoting the whole deal.*

Combining Bill and Gano's remembrances of how the hammer business was initiated, suggests Buffum presented the hammer proposal to Gano and he approved it. Unlike the Kearney agreement, Gano stated it was a contractual arrangement. Affixing a date to the Buffum association is more difficult, although the interview material suggests 1932. This appears too early and with the Kearney situation festering during this time and the incorporation not taking place until late 1932, the hammer era surely began no earlier than 1933 but more likely in 1934. The first reference to hammers at Chance found in the *Guard* was its November 9 and December 14, 1934 issues. There is also the question of how Joe Cheetham got involved and what role he played. His involvement with Buffum in production questions after the relationship was established would not be all that unusual. However, Bill's remarks about preproduction contact does raise questions. The 1933–34 years were desperate ones for Chance Company, when drastic actions would not be unexpected such as the involvement in diverse new ventures and product lines.

In the November 9, 1934 article there is no reference to when these products were first contracted or built in the Chance plant, although it's obvious that the description of the hammer was of an existing item. They were made out of a "special alloy and are polished up to look like 'seven hundred dollars'. . ." but were designed to retail at just twenty-five cents. However, the most impressive revelation of this article was anticipated production levels of some two million hammerheads and assembling them on handles! This is why Nathan Toalson was sent to the Ozarks to procure hardwood handles. Just as astounding were the plans to manufacture: "250,000 forged plow points of a modern design." There seemingly was enough optimism to go around several times. The article did confirm that an order "for several thousand to be put out at once . . ." had been received for the hammers. This was a boon for the local area as "Additional help is being trained for these lines." So there was cause for celebration as the Depression had dampened hope and confidence in most everything that had previously been economically successful, and in those dark days of 1934 it was deservingly met with enthusiasm.

The November article headlined: "Making Two Million Hammers" was an immediate attention-getter. The burden of fulfilling such a prediction may have led to a less flamboyant news item in the December 14, 1934 *Guard* with no title, but carrying a no less-significant message.

Employment of two hundred at Chance Company was cited as an all-time high. Near destruction in 1933 forward, this transformation was on the order of a miracle. What a Christmas present to celebrate the upcoming holiday. The immediacy of this rebirth is very suspicious in regards to the work that must have preceded all this current activity, although preparatory work isn't always newsworthy. New products could be brought on-line quickly as the Depression did provide certain advantages of positive work attitudes, no extensive current-day regulations, and no shortage of manpower. A month earlier it was suggested additional workers would be required and would have to be trained. Prior to Christmas a record number were then on site!

The mid-December news item continued with "has made this increase possible." As already revealed, the new Chance cast hammers and plow points were listed plus a new product, axes that were credited with the record employment levels. Later employee interview material reveals axes were part of this new line of products normally found at the local hardware store. Hardware stores carried agricultural equipment and parts before tractor dealers split off when servicing and repair work required a separate facility. The axes were even described as of cast one-piece construction and were of both poll (one blade) and double-bit styles. The double-bladed style is preferred by the novice because if a rock is hit, the ax can simply be turned over to utilize the other bit or blade, thus eliminating a trip back to the shop to sharpen it. This was said to be a new innovation, but it didn't reveal who had invented this new twin-bladed ax. Granny Ryman said Chance Company was also making both plow points and cultivator shovels and Mr. Chance would send him out to farmers who were using or testing them. Paul Sturgeon said the plowshares or replaceable blades were a Sioux Falls Plowshares Company item, which surely meant they had developed the original designs. These don't appear to be Buffum items, but another product line that developed from a Chance outreach to another firm during that time to generate new business and industrial activity at 210 North Allen. Incidentally, the December article said: "The factory still makes the old lines of pole line hardware, anchors, piston rings, pulleys and other articles." Attractive and more glamorous products would come and go while the less-colorful utility hardware items would survive and continue to pay the bills!

It's amazing that by the close of 1934, Chance Company had such a high level of plant activity and had made a rags-to-riches recovery. But the hammer's publicity in the *Guard* ceased, so the situation may have quickly changed from wonderful to something less, or nothing newsworthy was occurring. The April 5, 1935 *Guard*, reminiscent of the late 1920s, titled a news item: "Chance Company is Busy." It said that this past week a lot more employees were added raising the weekly payroll to 250, which was a new peak in employment at Chance. But the article indicated that a portion of this activity was due to the utility and piston ring business. Noting the April date, spring had come and the surge in new construction projects and auto use was driving increased business activity. The article stated: "Practically every department of the factory has its full quota and the foundry seems to be in full blast."

According to employee material concerning the hammer and axes, there were at least two distinct sets of problems, the first in the choice of castings instead of forgings. A hammer's metal-to-metal strikes make it much more prone to breakage than an ax against the dampening effects of wood. Did this design come packaged so to speak from Buffum or his customers and was simply being duplicated in the Chance manufacturing process? The issue of the design being to blame for quality issues comes into focus versus the integrity of the Chance manufacturing process. There are two *Guard* articles in June of 1935 that helps explain that Joe Cheetham was receiving praise for these new processes, but more importantly also becoming responsible for them. The first was on June 14, 1935, during the open tour of the Chance plant when the proposed bond issue was introduced. Among the Chance officials presenting programs was J. H. Cheetham, who the paper described in a built up fashion:

> *Mr. Cheetham is the man whose knowledge of metallurgy and casting, is making possible the development of these new lines. He grew up the atmosphere of the foundry and factory and had added to his practical knowledge of these things by an intensive study of metals, the application of heat to metals and the chemistry of the factory products. He is a forceful talker as well as being the man who knows how. Mr. Cheetham is considered as a sort of wizard in the handling of steel and other metals as he has been able to do the impossible with cast steel especially in the making of an all cast edged tool that will hold its edge and compare, favorably with tools of more costly manufacture.*

Rotarian Cheetham also gave a talk on foundry operations, moldings, and metallurgy, wowing his Rotary audience, per the June 28, 1935 *Guard*. When the shattered remains of Chance hammers began returning back home, it raised doubts about the sophisticated, scientific formula for making hammers at Chance. However, most new processes do take a "teething" period to perfect. It is not clear if Joe's metallurgy knowledge or his plant activities came under fire and led to his separation. The *Guard* of November 15, 1935, reported his departure from A. B. Chance Company—which indicates something must have taken place over a short, few months—although public coverage is not always in tune with internal employee status.

The difficulty with the Chance hammer, according to Gano, was "they broke too easily, particularly for a claw hammer. . . . You try to pull something and it was going to fail. This product should have been made out of malleable or steel." (1978) Dee Daugherty said:

> *All you had to do was drop one and you had two! We had to be very careful when we bagged them up. They went to Frank Lattimore* [not Buffum] *over in Louisiana, Missouri. They were cast iron deals, you can imagine how they were. They forged* [sic] *them at the foundry. (1979)*

Most of us have at least one claw hammer in our homes. A favorite is a quality cushioned steel-handled one that belonged to Mr. Ira Mozingo, my grandfather-in-law who was a talented farmer/carpenter. Also possessed are several wooden handled models, some of which no longer have handles in them. Since the handles were broken on a couple of them, it was natural to assume that is what occurred to the rest of them. However, there was a comment made my Nathan Toalson about the infamous Chance made hammer that now raises grave doubts on that theory:

> *I remember going down into the Ozarks, southern Missouri and northern Arkansas, buying carloads of hammer handles. We would get them back to Centralia and would put hammerheads on them. Because they would shrink, when they finally got to the customer they would be in two parts, a handle and head.*

Possibly some of the Chance-made hammerheads can still be found. Hammers aren't easy to throw away as they can regularly be found in those "mystery boxes" at estate sales. Remember when examining these treasures, look closely because if a claw hammer is found with no handle and one ear on its fork broken off, you may have relocated an oddity, something that was made at 210 North Allen Street that wouldn't always work!

A *Guard* news item of September 9, 1927, raised interest in recalling the later hammer problems at Chance, although there doesn't appear to be a direct connection between this earlier incident and the Chance-produced hammers. It does point out the dangers of a faulty hammer, the quality of which generally doesn't come into question today. William B. Laforce lost sight in one eye after an "improperly tempered and defective" hammer exploded. His suit was against his employer Donelson Motor Company under the management of L. D. Donelson and M. W. Smith, who provided the hammer. Such incidents surely led to the idea of mechanics and factory machinists providing their own tools. Apparently Chance officials did not recall the incident when they got into the hammer business, especially when the quality problems developed. Along with hammers, there was another association with an Iowa firm that Paul Sturgeon recalled:

> *At one time we had a relationship with Sioux Falls Plowshares Company in making plowshares with replaceable blades, which is common now you know. In the old days you didn't replace the blades but had the* [local] *blacksmith work on them. That is when we went into spot welding and we liked to have knocked the city's electrical system out because every time we would sock that spot welder on everything in town went dead! (1979)*

Bill Hollander said the plowshares were a product that were designed to allow only the small cutting portion to be replaced instead of the entire plow blade:

> *The idea that rather than having a one piece plowshare or cultivator shovel that as it wore down you either threw it away or you had a rather major job to build it up and be able to use it some more. His idea was to make plow shares and cultivator shovels like a safety razor. You had a thin blade that did the work and when it would wear down you would just replace the blade and the back part were reuseable. So we started making cast-iron hammers and axes and hatchets and plowshares and the whole trouble with the plowshares were they worked better in theory than in practice. (1979)*

The plowshares and cultivator blades depended on spot-welding little bolts on their backside, which sometimes resulted in the welder "blowing" a hole in the thin material. Another problem cited was the breaking off of the attaching bolts in service, not unlike the problem with the early Taylor expanding anchor that A. Bishop corrected. Further details of the cultivator point and the large plowshare were revealed in the June 14, 1935 *Guard*. This information was gleaned during the Chance open house promoting the bond issue:

> *. . . the new plow or cultivator point made by A. B. Chance Company. This device is self-sharpening and is made so that after it has worn down—if it ever wears down—it can be reset by a boy with a wrench to the second position on the holder and will wear a like number of years. One of these plow points was later placed at the end of an iron rod similar to the position it would have on a cultivator, the rod placed in a vice and seven blows struck on the point by a powerful workman with an 18-pound sledge, without an injury to the point upon which the blows fell. This was to demonstrate the ability of this bit of steel, hardly larger on its face than a man's hand to withstand the shocks it might receive in stump land or among rocks. J. H. Cheetham later took a newly broken file and cut on a pane of factory glass and then tried to scratch the case-hardened surface of the plow point, but could not make a scratch on it.*

This is quite an impressive demonstration and positive description of a much-maligned family of Chance manufactured products. Later that June evening the group was taken for a plant tour:

> *Especially interesting was the manufacture of the new plowpoints, which were being cut out of sheet steel and carried through all the processes that go to make them into the finished product. While the crowd was*

> *in the Punch Press Department three wheelbarrow loads of these plowpoints were fed into the furnace and on through the processes that include the tempering and case hardening of the points. Two bottle-neck electric spot welders were placing the studs in the points to be later threaded for the holder.*

The metal-hardening process appeared to be a perfected art, but in practice variations in various factors involved did cause problems, per employees who were there. In the early 1990s, Chance was trying to recapture the art of casting anchors. Even after lengthy research and experimentation in developing the operation, production glitches continued to slow production. This reveals that even with the aid of modern technologies, how difficult it is to perfect certain processes. Many cases of ulcers, long hours, disappointments, and broken dreams preceded successes and were inherited by those attempting to develop methods that resulted in better products. A concern today should be the exporting of and the loss of such manufacturing capabilities that were developed with the sweat and toil of previous generations. It is reassuring to be able to take a positive reading back to 1935 when these special, short-lived Chance products, if given more time to develop and perfect, may have proved more successful. However, trying to enter new lines and competitive areas, or depending upon a third party can present some very difficult challenges.

A short paragraph in the June 1935 *Guard* article focused on the large plowshare, which Mr. Isbell exhibited to the bond entourage visiting the Chance plant. It was said "to be a future product of the plant." This reveals there was a developing stage in these special items just as Bill Hollander later alluded to. The large plowpoints appear to be near the end of this chain of developments at Chance. Although Joe Cheetham was in the plant when these items were born, it was undoubtedly a company-wide involvement and credit or blame could be broadly spread. The Chance association with Sioux Falls Plowshares Company apparently had brought these products into Centralia that were being developed and tooling made to produce them. Chance Company was able to add new innovations due to their more-sophisticated manufacturing capabilities. Employee information confirms what appears to be unique features, suggesting Chance may have been first to apply them to these products. Further confirming Chance's deep involvement in these products, the June article continued: "It will be of the removable type and will be of material similar to the cultivator points. Dies and other equipment to make them are in process of manufacture." There is historical significance attached to putting form to these new ideas, even though newer methods of soil tilling and no-till methods now dominate the agricultural scene. The modern processes used, such as heat-treating, would serve Chance well into the future, with the experience/expertise gained. A prominent example would be in World War II for aircraft component production, just a few years removed. Even failures drove the body of factory knowledge at Chance, sometimes more so than successes as it brought closer analysis of mistakes made and intensified efforts to overcome them. Complicating the understanding of this unique era of Chance Company history was employees tended to group all these products together, when there appears to be a distinct division between Buffum and the plow items. Nathan Toalson, who was in Purchasing at that time, also recalled his involvement in these products:

> *The cast iron axes we made here, as tough as they were—they might be 1000 pieces* [break into]*! That is when we needed quality control on our iron, if we did not have it. We really came out with a fancy plowshare. Cultivator shovel is really what we made and it was a replaceable head for the cultivator shovels. The cultivator shovels were just one piece and when they wore out, they threw them away. What we made was*

> *one made of 1/16" high central steel, heat treated, and when they wore out they were thrown away and replaced. It was a good idea, but it never worked, partly because we couldn't heat treat the replaceable part very well. As I recall it was made out of a nickel steel, 3140, or something like that. That was when Cheetham was here and he really had some times with the steel companies. I was the Purchasing Agent and we got a carload of the material in and we didn't get it heat treated right. Cheetham was intent on making the steel company take all of the steel back because it wasn't right, Well, the steel was right, we just couldn't do the job right. I think they finally gave us about half credit on it but they didn't completely take it back because it represented somewhere around $2,000. That was big money in those days. (1979)*

Bill Hollander continued:

> *I worked quite a bit with plowshares and anchors* [sic] *and they looked real pretty when we got them finished, but the first time you tried to chop or hammer them you lost a good part of the casting. We made a whole slough of those outfits and the back up part of the plowshare was a form stamping. It was a pretty tricky outfit and it took some drawing to shape them up where it would fit on the plowshare or cultivator blade that supported it so it wouldn't break down. The blade was about 1/16" thick and tempered so it was kind of a brittle piece. It took a lot of trial and error modifications on the blade until you come up with a piece that shaped up the way you wanted it. . . . We made the eyes, etc, and they didn't quite come out. They were long some places and short in others. We tried to alter the form die to use up the back stock, which wasn't a real good process either. Joe Cheetham was getting kind of impatient. He wanted to be selling plowshares because that was the only way he was going to get anything out of his stock. He would say, that is a waste of time to develop a form and make a guess of what it will be." We will just go into production. They sold just as many axes and hammers as they could make but the trouble was in a short time they all came back because they broke in service. The plowshares were the same thing, in use the little spot welded lugs on the replaceable blade would break lose and couldn't be used. (1979)*

Advertising head Art Houston recalled the plowshares and said: "I didn't think it was too successful. We did do a bulletin or two on it and it moved along reasonably well for awhile but it finally faded out." Granny Ryman also got involved in the plowshare and cultivator shovels. For those without an agricultural background, the plowshare fitted on the bottom of a plow and was the cutting edge of the plow that required a somewhat sharp edge to cut through the soil. A plow was a tool for penetrating into the soil a few inches and turning it over exposing fresh soil with no stubble or debris for developing a seed bed for corn, beans, etc. The cultivator in earlier times and up through roughly the 1960s, was used to control weeds and to break any soil crust. The cultivator was a multi-bladed tool used in the softer seedbed, between the rows of growing crops. Granny Ryman explained the whys, hows, and the problems encountered with these products:

> *At this time we were still looking for things to do to make money and we got into the disposable cultivator shovel and disposable plowshare business. Mr. Chance would send me out to the farm and the farmers would allow me to put them on. They would plow corn with them a half day, bring them back in and check and see how they were working, and to see if they were holding up. One of the problems we had was the electric weld of the bolt that went on it, kept breaking loose when it hit the hard ground. So we had to develop that,*

> *but for some reason or other the disposable cultivator shovel business didn't go over as good as the safety razor blade. (1979)*

Granny was likely referring to the hard compacted soil at the end of the fields where the implement was dropped after turning, generating a heavy shock that broke the retaining spot weld attaching studs on the shovel or blades.

Nathan Toalson recalled how Chance Company disposed of part of these products after a comment about the Chance Board approving such sales:

> *I think that was the hammer and ax that we got involved in and we finally sold it. As I recall, to a fellow by the name of Buckham* [Buffum?], *who was a brother to the head of the LaCrosse Lumber Company. And I don't think we got very much out of it—we weren't getting much out of it by being in it yet, of course! (1979)*

The LaCrosse connection with Centralia goes back to at least 1890. The *Guard* reported that on October 18, 1890, the LaCrosse Lumber Company of Louisiana (Missouri) bought out the E. J. Sears Lumber Yard of Centralia. The remarks by Nathan are a little puzzling about Board approval for the sale of the Buffum items. Possibly, it refers to the tooking and or remaining stock.

Although the technical aspects of these Chance-produced consumer-type items were a focus of employee remembrances, their promotion/marketing is little discussed. Apparently this was in the hands of the originator Buffum and the Sioux Falls firm and was their major role. This is similar to the role-playing of the earlier Kearney/Chance relationship and is logical as Chance was redeveloping their sales force. However, Chance marketing involvement would have been of limited value because these were non-utility/auto-type markets that didn't fit the usual Chance coverage and sales expertise. There was the additional problem of selling anything during the Depression, so Buffum must have been successful in his promotion efforts. There are always two sides of every story and the non-Chance perspective is missing and would likely generate some interesting client/customer feedback. It's not clear if the cultivator shovels were a variation of the plowshares or a Chance generated item. There also is the question if these items were being made to certain "specs" which were faulty or was it the Chance manufacturing process? A. Bishop's involvement per Ryman's comments suggests he became interested in the agricultural items, as his farm was receiving a lot of his attention at that time. There is a question, however, if he applied his problem solving skills to the farm shovel's technical problems as he had in the past to telephone and utility items. His level of involvement and success isn't stated and it may have ended before he had time to resolve the problem issues.

The desire to know more about this chapter of Chance Company history has been inspired by the importance of the subject and the material that was provided. It is an episode in the Chance story that holds a great deal of intrigue, and will close with the hope that additional information may be forthcoming.

Fairways and Fireworks

There were several employees over the years who came with a friend when they applied for a job at Chance, such as Columbians Dee Daugherty and Roy West, who became veteran Chance factory foremen. Their meeting of Chance officials F. Gano Chance and Joe Cheetham on

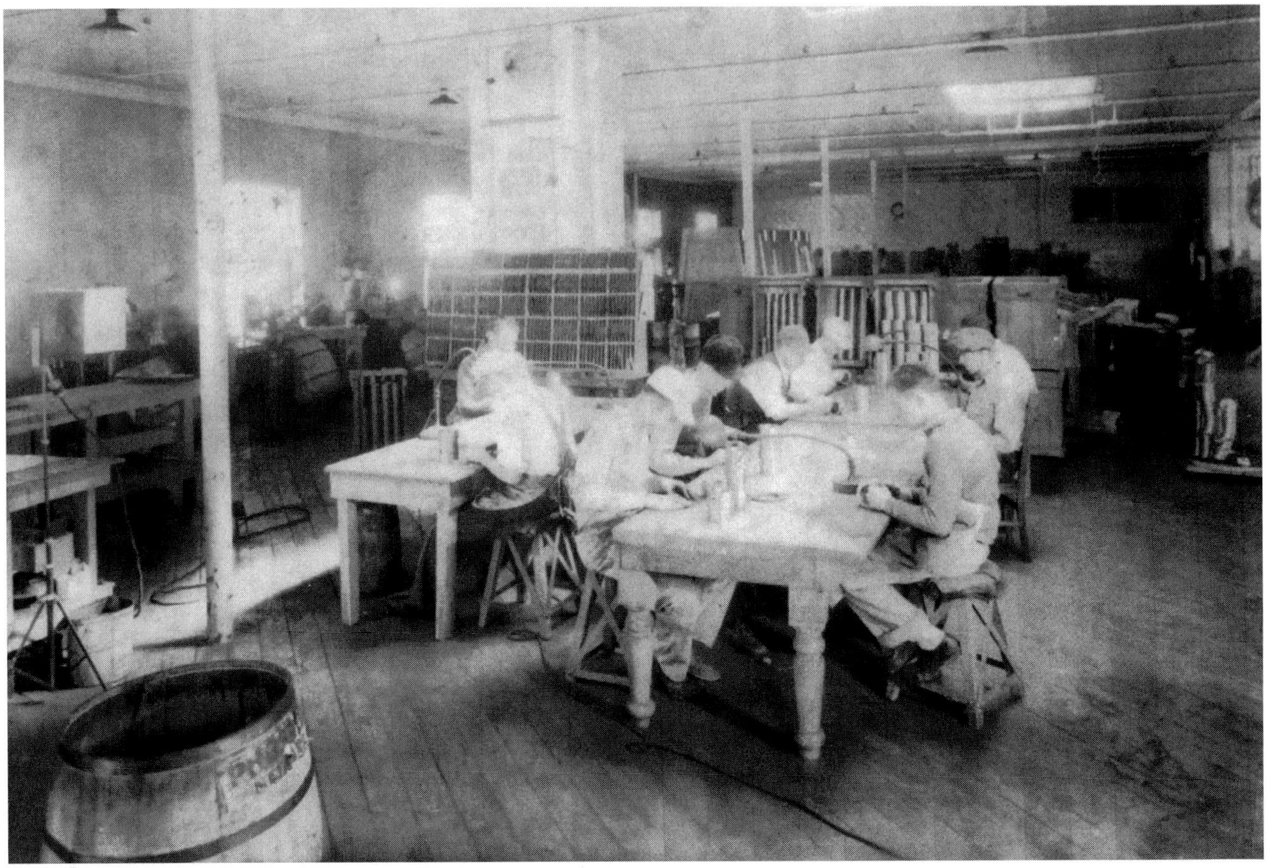

No. 29 Operation 13: These Big C employees were kept busy gauging and inspecting piston rings before they could be approved and boxed. (Centralia Historical Society)

Pictured are Dee Daughtery and Jack Chance. Jack, the grandson of the founder, was a key figure in the 1978–79 Chance employee interviews that brought forth valuable details about the company. They also bring to light the golf story as told by Dee and Roy West, incorporated into "Fairways and Fireworks." (*Chanco News* December 1963)

a Columbia golf course, led to their eventual employment on May 5, 1935. Dee commented about he and Roy's friendship:

> *We had known each other for some time. We were in school together. We happened to be caddying for Gano and Cheetham and about the time school was out, we hit them up for a job. Joe said: "You'll have to come to Centralia." So school was out on Friday and we hitch-hiked up on Saturday. And Joe wanted to put us to work on Saturday but we didn't have our clothes or anything. So we had to go back to Columbia and came back on Monday. We went to work in the Piston Ring Department, I was a gauger and Roy was an inspector. (1979)*

There are several amusing sidelights to the Columbia golf course experiences that Dee and Roy remembered. They came to the conclusion that Gano and Joe were the guys they wanted to caddy for. Their resourcefulness is revealed in the unorthodox and clever way in which they developed a foolproof method of making sure this occurred. The reason why they went to all of this trouble was a financial one, especially since caddy tips during the Depression era were likely on the light side. Another angle was the competitive confrontations on the golf course by Gano and Joe. Undoubtedly, this was entertaining to young men who were in the late stages of their spirited high school days. Dee explained how they worked this plan to make sure they got the twosome from Centralia:

> *Well, we would hide out and wait for them. You drew numbers to go out* [caddy] *and whoever came you went out according to your number. So in place of drawing numbers, we waited outside the road and they'd come by and pick us up. Then we'd come in with them. (1979)*

4 CHANCO NEWS, CENTRALIA, MISSOURI

Some of the golfing contestants disproved the theory that golfing requires a lot of practice. Some who had never swung a club before made the old timers step right up and do their best in order to remain competitive. Of course, there was a little matter of handicaps which gave the beginners some advantage. The forms displayed in these pictures, however, fail to distinguish between the experienced, the dubbers, and the beginners.

Warren Vance, of the Engineering Department, made the longest drive to win the driving contest.

Roy West had the best score, but didn't win. See picnic article.

Harry Donnohue tees off.

Top picture — Soap Markwell takes a hefty baseball swing. Dee Daugherty, second picture, hits out a long one for admirers. Ye editor, third picture, follows through, stomach and all. Lower picture, Dee Daugherty, Country Club champion golfer, ushered a gang of beginners from the machine shop around the course for their first game of golf which took them all morning to complete. Some balls and some tempers were lost. Left to right, lower picture — Dee, "Tiny" Quisenberry, Ray Adams, Shannon Adams, "Doc" McDonald, tournament winner.

Golfing pictures and Roy West (*Chanco News* October 1946)

After they got on the golf course, Roy told of the antics that Gano and Joe got involved in. This reflects the competitive nature of both these men and the enjoyment of games that Gano relished:

> *They played this mingle, mangle, mongle thing with the longest drive, closest approach, longest putt, and what not. They created quite a bit of comment at the Columbia Country Club because they spent quite a bit of time on the greens arguing about who was the closest, and what not. Some times there were some doubts about some of the scores that were turned in. (1979)*

Dee remembered that what Joe lacked in golf skills, he made up with slick little tricks to make up for any handicap he might have:

> *Joe told me: "The more I win, the more you get." Gano is going to appreciate this when he hears this. Joe would come out the winner. Gano accused me of hitting the balls, dropping balls for him, and everything else. He had a hole in his pocket. Cheetham would get to the distance and*

say, "Here it is!" I think the second time I came back up they had caught up with Joe and he was gone. (1979)

It's surprising that over all the years they worked together that they didn't tell Gano about Cheetham's antics. However, Gano took winning seriously and was the sleight-of-hand artist, the magical-trick sort of fellow, which likely led him to suspect Cheetham. It is amusing to think Joe took advantage of Gano at his own game, although those who knew Gano would confirm that he was not one who would cheat. Gano's magical skills, which he described as, "Pocket Tricks and Club Tricks" (1941) were learned early in his life. "My father taught me magic at the age of twelve and it became my hobby." (1941) Later Chance engineer Dave Hubbard would tell of visiting in his home and playing games that Gano had collected from across the country. When Gano would lose, which apparently occurred against Hubbard, he would try to find some explanation for his loss. He would even apply some of his mathematical skills and get out his slide rule. This strong desire to win and analyzing problems with his math skills would serve Chance Company well.

Joe Cheetham was another one of those controversial yet colorful figures that became factory manager after Nate Husted left. Bill Hollander discussed the Foundry problems during this time and mentioned how Cheetham tried to maintain a positive image. "He said that never in all of the time he had been in industry, admitted that anything he was asked that he did not know." (1979) The development of the earlier Chance Foundry that primarily catered to iron for anchor and piston ring production had many challenges along the way. Bill suggested that someone during the World War I period had come up with a way of predetermining the properties of gray iron castings by analyzing the raw materials that went into it and Cheetham suggested he was a pupil of this mentor. Sometimes the end casting:

was light iron, sometimes gray iron and sometimes full of chill spots. The end product was quite different, sometimes you could tap in it, sometimes it was harder than the tap. So he started out Gano, of course a chemical engineer, and had Gano running chemical analysis on all the pig iron we had around. Cheetham would say, "take so many pounds of this" and eventually we learned how to control our castings but it was trial and error and not what we learned from Joe. (1979)

Another prominent Chance executive, Nathan A. Toalson, also had a connection back to the old Foundry operation and explained his understanding of the evolution of getting into production on items first out-sourced:

The Foundry was started about 1925 or 26 and I don't know how they procured the products prior to then. Much of it, I guess, was done on the outside. Of course the Never-Creep® was always purchased on the outside because it was malleable iron until they finally changed and made it out of some other cast iron. That was back in Cheetham's time when there was an awful lot of trouble with cast iron but I think primarily they were purchased on the outside. Joslyn made the rods for us and Hubbard made some of them before we actually set up the forge shop for the actual forging of the Thimbleye. (1979)

One of Joe's favorite sayings, according to Bill Hollander, was "Figures don't lie but liars can figure." Bill continued to explain how Joe was expanding his empire out in the plant: "Everything had to go through him as he didn't let anyone in the factory talk to anyone in the office." Not only was Joe in charge of production, but he would become the plant accountant extraordinaire in which his ingenious piston ring calculations would lead to a crisis of sorts at inventory time when his figures were found grossly inaccurate according to Bill:

Piston rings were a large line, there wasn't any money in it but it took quite a few people. They were all separate castings and required quite a bit of machine work. It was up to Joe to see that the factory was making a profit. They would make a bunch of castings and the foundry would get credit. [Each department that touched the rings from machining to boxing got involved in the count].

So finally inventory time came around and I think there were about a million less piston rings in stock than all the books showed. Cliff Hough was with an accounting company in Columbia. They used to come up here about two days a week and they audited. They decided they needed someone up here all of the time to watch and oversee. They had installed an accounting system that had everything that General Motors had except the volume of business. So it was decided to have Cliff Hough here as a full-time employee to keep up with this accounting system. (1979)

Gano recalled that Hough was an accountant for a firm in Columbia. "He came with us shortly after I came out of school. He was our auditor. . . . In 1932 our company was losing money and at my insistence we called in Ernst and Ernst. They set up a budget and accounting system." (1978) Flynt Chandler also remembered that Hough came from an accounting "outfit" in Columbia who were spending one or two days a week at Chance Company. "That is when I got acquainted with Hough. He was a pleasurable guy too. I always liked him." (1979) Les Reynolds, who lived for many years across the street from Hough, considered him a good neighbor and "as good a fellow as I was ever with." (1979) C. C. Hough is pictured earlier in this chapter in the "Silver Dollar Payroll" photograph. Joe now had someone looking over his shoulder, which according to Bill Hollander, didn't generate a very peaceful plant atmosphere when Cliff Hough came calling:

Cliff got to work trying to find out where this extra million piston rings were. He would find something that looked like it was kinda messy and he would say, "Joe why did you do this?" Joe would say, "Because that is the way it is supposed to be." The two of them would go tearing through the plant, Cliff trying to find out why he had done something and Joe insisting that he did not know what he was talking about and it was none of his business. But Cliff finally found out where all of these extra rings were that Joe was getting a bonus for. Every time they went through the rework why the factory was credited as if they had made them from scratch. Joe was a capable boy but he got into lots of things that he had better have left alone and he colored them enough to make it appear that he had done the job. They finally decided they could get along without Joe. (1979)

Although most Chance employees got involved in all sorts of activities, the more promising ones were moved around to gain experience in the various areas of the company. Nathan Toalson moved from the lab or foundry testing area to: "Well, I went into the Standards Department. That was time study and working up costs on various products. That was in about 1932 and as I recall George F. Mayes came in and made a study. Then he actually set up the Standards Cost System for the entire company. That was after Husted. About that time Cheetham came in to run the factory." (1979) Nathan also mentioned that before he went into Purchasing, in about 1938, that he worked in the "Epoxi Department, Punch Press, Galvanizing, Drop Forge, and the Piston Ring Department. Made rings day after day. I worked in the factory close to a couple of years and then went into the Standards Department." (1979)

It is not unusual to have the quality assurance and manufacturing people at odds with one another. This couldn't be any more true when you had two fiery guys like Granny Ryman and

Joe Cheetham working together, as fireworks were sure to occur. As Granny remembered:

> *Gano appointed me as quality supervisor and I had to be an inspector at times. We got into some pretty heavy conflicts with Cheetham on the quality of products that would come over. I would sent it back. Gano said, "I want you to do it," so that is what I did. Then he would have a foreman's meeting and spend a half-hour talking about me indirectly to the other guys, about what I was doing! (1979)*

It's impressive that Chance Company had a formal quality program in place in the early to mid-1930s. In a way, the confrontations and brashness of the manufacturing manager drove progress. This was beneficial from the standpoint that Chance Company now had professional auditing and a quality program in place prior to World War II when growth and government requirements would require it. Gano's leadership was positively reflected in the controls and departmentalization within the company, as he relates. "As our company began to grow and I set up different departments, I wound up with the Public Relations and Employment Relations Department." (1978)

It seems as if everyone had a story about Joe. Flynt Chandler said he was a "feisty little guy and was always sneaking around trying to catch somebody doing something." (1979) Les Reynolds added, "That Cheetham, I was fired and hired and never did get out of the building!" (1979) Les had inherited some new process or machine attachment that Cheetham said:

Les Reynolds (left) shown making a purchase. (*Chanco News* November 1950)

> *"They had to work!" Les responded, "Cheetham, they will never work like you have them in there!" We messed with them for a month and Mr. Chance came up there one evening and said, "Les those things don't work do they?" I said, "not very good." He said, "Well you shut them down." That was on a Thursday and he told me to go down to the house and help Dick Cole (plumbing project). He told Cheetham that he wanted them fixed by Monday morning. So I went in Monday morning and there was Mr. Chance. He came over and looked at it and said, "Les has anything been done to his?" What were you going to say. "Mr. Chance, I don't think there has ever been a wrench on it." He says, "That is just what I say." He called Cheetham and told him that he wanted those things tore out of here, right now. So Cheetham looked around at me, that cross-eyed thing, and says, "Tear them out. You never wanted them in here anyway." I took them out and about 10 o'clock Milt Beaman came over and said, "Les, I am going to have to let you go." That is when you had to knock on Cheetham's door to get to Mr. Chance's office. I walked in and Cheetham said, "You know you are supposed to knock before you come in. Don't you know that?" I said that it doesn't make any difference after you are fired. He asked where I was going and I said to see Mr. Chance. I thought he was in his office, but I happened to look out and Mr. Chance was crossing the road coming over to the factory. So I just went on in his office. He came in and said, "What happened?" I just up and told him. He said, "We will see! I have to be gone for about thirty minutes. You just sit down there." So I just sat down and lit up a cigarette, but Cheetham opened the door and said, "Come in here a minute." I*

said, "He told me to stay in here and this is where I'm staying." He said, "Now I'll tell you what to do. You just come on back up there and Milt will put you back to work." I said, "I didn't tell Milt to fire me and I'm not going to tell Milt to hire me back. Somebody else is going to tell Milton, I am not." He said, "Well come on," and I said, "No, I told Mr. Chance I would be here for thirty minutes and I'm going to be sitting right here." So Mr. Chance came back and said, "Let's go in and see Cheetham." Mr. A. Bishop was told by Joe that he had already asked Les back to work. "Mr. Chance said, "Why don't you go on back and go to work." And I said, "Mr. Chance, Milton told me he was going to have to let me go and somebody is going to have to put me back because I am not going up there!" So he went up and told Milton, "That boy is going back to work and he isn't going to lose anytime this morning." So I went back to work. . . . (1979)

The Cheetham story continues with further remarks by Bill Hollander on the workplace relationship between Joe and Gano. Bill had an inside view as he reported directly to Cheetham. The golf story indicated there was some association between the two away from work but how close or how long it lasted is uncertain. The most important findings gleaned from Bill's remarks is a glimpse of how the Chances made company decisions at that time. Gano was developing his management legs and it was quite natural for him to be hesitant on occasion in making decisions without first consulting with his father. It reveals the value of A. Bishop's experience in not only handling business matters but a deeper insight into evaluating people and personnel. According to Bill Hollander:

It was shortly after Gano got married that A. B. began to spend more of his time fooling around with his farm and letting Gano take over. At that time, Gano was reluctant in trusting his own gray matter on anything that came up because he would always go and ask Dad. A. B. made the split decision right then and there and that is all there was to it. Gano would muddle over it a long time before he decided to go ahead with anything. Joe Cheetham was here then and Joe had done a pretty good job of selling Gano that he was a fair-haired boy, that what there was to know, he knew. He would want to do or try something and Gano would say, "Well we have to see what Dad says!" A. B. was not so well sold on Joe Cheetham as Gano was. He would say, "Well, I don't know about that, maybe we had better wait a bit." So Joe Cheetham set out to build up Gano's confidence in himself where he would make a decision rather than go ask A. B. whether they should do it. Why he would say, " Yes, let's do it." He would hang on Gano's heels and anything that Gano did, he would run up and say, "Boy, that is a good job Gano, you're really getting the job done. You are really up there and you can get things done. You don't have to rely on anybody." (1979)

Although the focus on Joe Cheetham has revealed a fiery view of plant activities, he would bring in an acquaintance, Odie Walker, who would receive praise for his operation of the Chance plants. According to Bill Hollander, Joe brought Walker into Chance employment from their former employer at Decatur, Illinois:

They . . . were having trouble in their foundry in Chattanooga. They transferred O. D. Walker down to Chattanooga where they made large valves for power plants and things of that nature. The manager down there was in quite poor health and the manager just turned his files and everything over to Odie. Odie for all practical purposes was manager of the plant and assumed that he was the main manager. I don't know how long he was

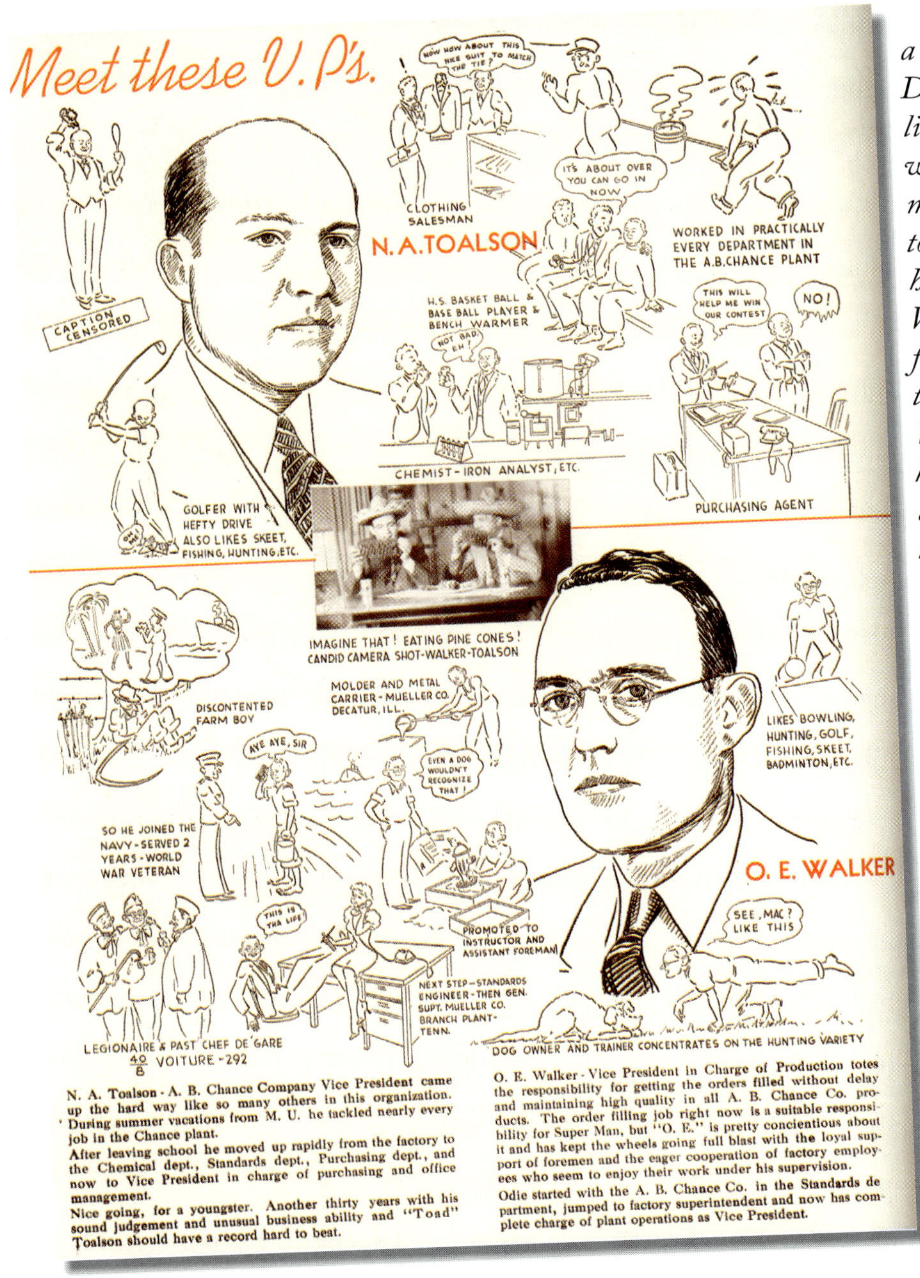

N. A. Toalson - A. B. Chance Company Vice President came up the hard way like so many others in this organization. During summer vacations from M. U. he tackled nearly every job in the Chance plant.

After leaving school he moved up rapidly from the factory to the Chemical dept., Standards dept., Purchasing dept., and now to Vice President in charge of purchasing and office management.

Nice going, for a youngster. Another thirty years with his sound judgement and unusual business ability and "Toad" Toalson should have a record hard to beat.

O. E. Walker - Vice President in Charge of Production totes the responsibility for getting the orders filled without delay and maintaining high quality in all A. B. Chance Co. products. The order filling job right now is a suitable responsibility for Super Man, but "O. E." is pretty concientious about it and has kept the wheels going full blast with the loyal support of foremen and the eager cooperation of factory employees who seem to enjoy their work under his supervision.

Odie started with the A. B. Chance Co. in the Standards department, jumped to factory superintendent and now has complete charge of plant operations as Vice President.

From *Chance Tips* of January 1941. Toalson and Walker.

at that plant, but one day a boy comes and introduces himself as the new manager. Walker kinda blew his top and went on a big drunk. He finally sobered up and in pretty poor health and hungry so he got to looking for a job. He found out that Joe Cheatham who he knew in Decatur with Mueller, was with Chance and he corresponded with Joe. Joe told him to come on up, why he could possibly find a job for him. So he hired O. D. to be kinda a hatchet man, figuring that O. D. was really down to the bottom line. He was broke, hungry, he wasn't too well, and very low morally and he was really down to the bottom of the heap. He hired Walker and when he told Walker to do something, why Joe felt that there was no question that he owed everything to Joe. Walker would do just what he said and would act as a go between so that Joe could keep things going the way he wanted them. But Walker thought a lot of the things Joe asked him to do were a little less than honorable and he did not much like it. Joe would say, "Well that is the way it is going to be." So Walker got to going to Gano and telling him that he did not think it was the thing to do and yet he was supposed to because he was working for Joe. Gano would say, "What is wrong, what is it all about?' Well, I don't think I ought to tell you because Joe told me to do it. So Gano would say, "You are right, that is the way it ought to be." So when they let Joe go, that was after Cliff Hough checked around and found out that Joe had been feathering his nest at the expense of the Chance Company, why then they made Walker factory manager. Walker was here for a good long time. (1979)

As Paul Sturgeon recalled:

Cheetham brought Walker in. Walker came from the Columbian Iron Works but his home was in southeast Missouri. [Note: A visit to his ill father in October 1936 suggests it was Cape Girardeau.] *He worked in Decatur and then went*

> *to Chattanooga and came to us from Columbian Iron Works. He left and went back to Decatur and then went up to a Canadian operation. (1979)*

Nathan Toalson was another employee under Cheetham who "had a little problem with that" (Cheetham making his records look good). "I started out working in there for Glenn Eierman, who was here at that time. Then Glenn left and I worked directly for Cheatham." (1979) Glenn was mentioned again by Paul Sturgeon who said he was in the same class (College at the University of Missouri) as Bill Hollander and F. Gano Chance. This was the class of 1929. The spelling of the last name was Armin, but it's obvious it is the same person. However, neither name was found in the 1926 yearbook and both will be referred to as Glenn Eireman in the indexing, although Eireman is not listed either in the 1926 University of Missouri yearbook. Per *Chanco News* of July 1949, Glenn Eireman had recently resigned as head of the Fabrication Division. An additional comment was made by Bill concerning A. Bishop's increased absence at the plant: "He had been spending a lot of his time working of his farm, back in the thirties, because while Joe Cheatham was here, why he spent an awful lot of his time out there at the farm." (1979)

The departure of Joe Cheetham was remembered by Paul Sturgeon as he was present at the Saturday morning meeting when it occurred:

> *His office was right in front on the second floor and every Saturday morning he had a supervisors meeting. By this time, I was in the office doing time study work and I attended the meetings. Gano walked in and sat down on the file case beside Joe's desk. He turned around and he said, "Mr. Chance, do you have something you would like to say this week?" Gano said, "Yes, I have, gentlemen, you are no longer with us," and stood up and walked out. Joe sat down for about a minute and then said, "Well, I guess that will conclude the meeting!" (1979)*

The separation of J. H. Cheetham from A. B. Chance Company took place in late October 1935 as the November 1, 1935 *Guard* carried the news story. Mr. Cheetham was described as the superintendent of the company since 1931 and "has severed his connection with the company and will take up a similar position elsewhere." The official version suggests Joe had left on his own and even Gano in the article suggested Joe "has been a very valuable man since entering the Chance plant." The paper regretted their plans to leave Centralia although they thought the family might remain as the children were in school here and in Columbia. Two weeks later the November 15 *Guard* covered a party honoring Mr. Cheetham given by employees. A silver shield mounted on hardwood was engraved with: "Presented to J. H. Cheetham as a token of esteem by A. B. Chance Co. Factory Personnel, Foremen and Assistants, Oct. 31, 1935." This was thought to be the first such testimonial given to any former employee.

Without employee interview material we would be no wiser to some of the problems that occurred out in the plant. It is routine to put a pretty face on departures as not only would such a positive public stance save face for the Cheetham family, but it would also be beneficial to the company from a public relations standpoint. What pushed the decision to terminate Mr. Cheetham, which apparently was supported by both Chances? Plant turmoil, credibility issues, and the turnaround in the integrity of the new products developed under Joe's watch, were reasons to change directions in plant management, especially with Odie Walker available to take over his job. Most former employees would likely concur that the company would put up with certain circumstances if they were getting the job done and putting out a good product. There were always aggressive managers around in this employee's forty-year experiences and sometimes they were purposely brought on board to change directions, especially where there was resistance. Such people were also easier to get rid of than someone universally liked. Certainly, these circumstances were not unique to Chance. However, Mr. Cheetham clearly brought positive change during the Depression era when drastic measures were often needed. And, of course his side of the story isn't recorded.

The Cheethams were still in Centralia in early February 1936. The June 18, 1937 *Guard* said the Cheethams had been in Toledo, Ohio, the last fifteen months, but were now moving to Philadelphia where Mr. Cheetham was employed as superintendent of an unnamed manufacturing company. The January 28, 1938 *Guard* carried the news of J. H. Cheetham's death (age fifty) on January 13 from cancer. He would later be buried in the spring beside a daughter in the family lot in Fairmount Cemetery of Buffalo, New York. It was stated that he loved to meet people and talk with them.

Highlights of a Life

ALBERT BISHOP CHANCE

It is doubtful if many people take the time and effort to list the important events and achievements in their life. How many personal highlight lists have you seen? Since they aren't that common, this tends to make their structure and contents loosely up to the individual. Such information can be very private and personal, like that found in a diary. Most of us will allow someone else to pen the story of our lives and we even come to expect that. Examples are our family Bibles which may record two highlights—our birth and death, but with nothing in between. An author would develop such material about an important individual they are writing about, but most of us won't have a book written about us. Much more probable is a preacher preparing a funeral message or someone doing our eulogy or obituary.

Most of us don't feel we are important enough to take the time to do such a thing. It doesn't necessarily have to include our highest office or position, how many holes in one we attained, or how much money we made in the stock market. It could be about an important thing we did for someone or they did for us: our favorite grade school teacher; when we were baptized; or the day we donated our fifteenth gallon of blood. We might be pleasantly surprised to find what we see as the important things might not agree with someone else's assessment. Why should we do it? One reason is because A. Bishop Chance thought it was important enough to do. The impact of his rendering and its existence and value today makes for a convincing argument.

The influence of our highlights can be very strong and real to family. In the F. Gano Chance interviews, certain questions about early company/family history would bring a frequent response, "Find the little black book" or "Find my father's listing of important events." Gano could have been instrumental in suggesting his father produce such a record, but did he himself make up such a list? Or did Gano fall prey to the same reasoning or excuses most of us would point to? Aren't we richer with A. Bishop's and poorer without Gano's? However, the early company interviews were supported and likely initiated by Gano, which have proven invaluable.

One inescapable revelation about A. Bishop was his very fertile mind. Not only did he generate ideas, but he also made a strong attempt to transfer these ideas to his workbench and into a working model. If he had spent his life making

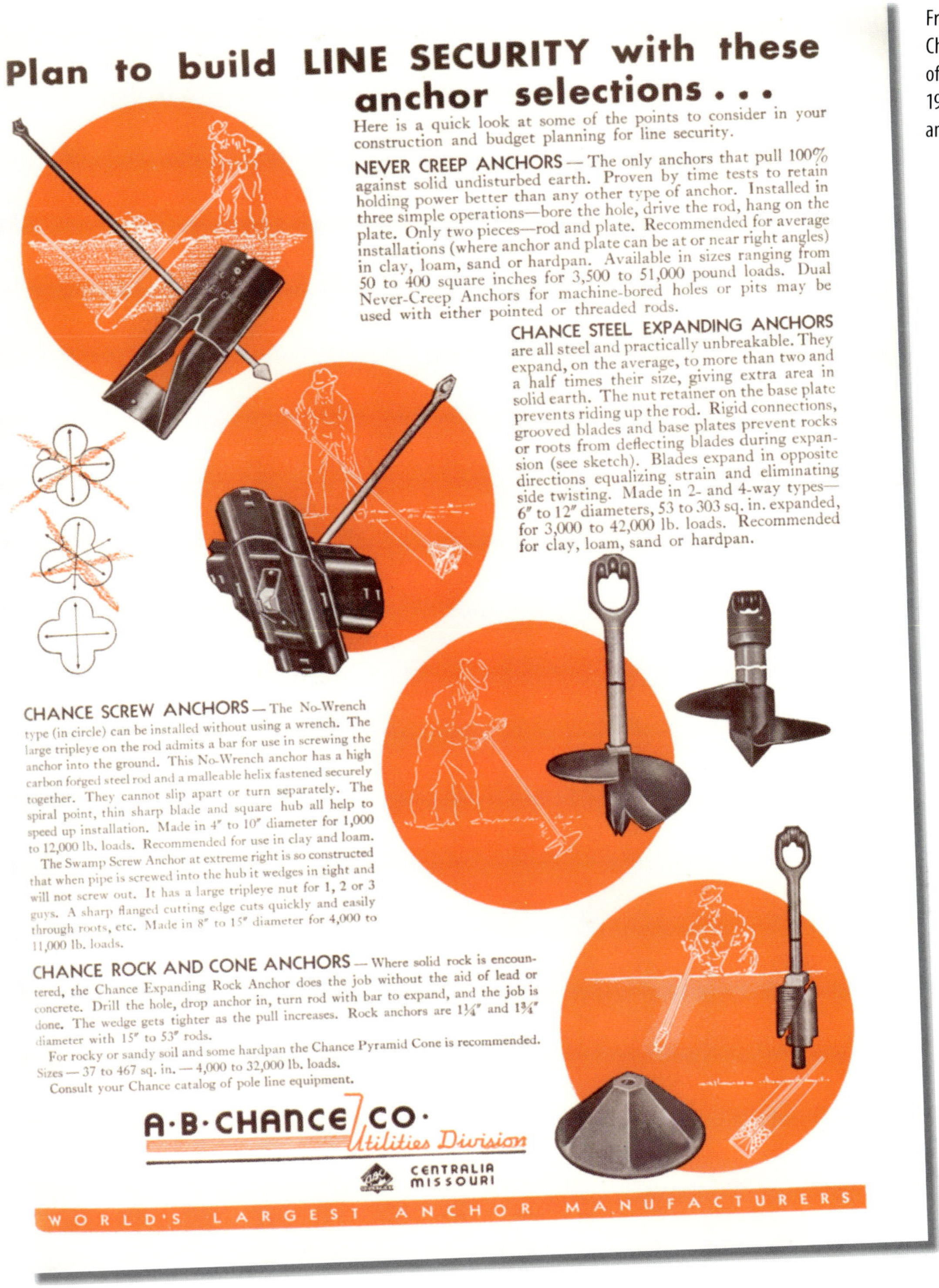

From Chance *Tips* of January 1941. Five-anchor ad.

electric toasters, he would undoubtedly have found it difficult not to plug each one in to see if and how it worked. Maybe he had more time for special projects as he could more freely come and go than the majority of his employees, but consider his many obligations and that some of his significant activities were in his retirement years.

A. Bishop labeled his highlights, "A. Bishop Chance—History—Record of Facts." It is composed of an interesting combination of personal and business subjects. Obviously, it's primarily oriented toward his business activities, which makes it valuable to Chance Company history. A. Bishop had a sense that what he did was important, but one would suspect those around him prodded him into writing his highlights. Certainly, such a list is more in demand for someone considered a success than a failure, but his highlights includes some glaring failures that helped shape and generate future successes, especially for Gano and Chance Company. This document appears to have originated in 1936 or 1937, so even it is incomplete as some of his more prominent personal activities were yet to be completed. Another reason he did his highlights could have been health concerns, first initiated in the 1930 time frame, when he got the idea of "getting his house in order." His highlights shouldn't be construed to represent his personal side although even when his chief motivator was business issues, he couldn't fully separate it from other people or personal items. The information on the original 8½-by-11-inch typed page is reproduced, except for the omission of penciled-in information, etc.

A. Bishop Chance—History—Record of Facts

Age	Year	Highlight
17	1890	Saw my first telephone.
18	1891	Built a telephone and telephone line.
19	1892	Decided to become a pharmacist.
20	1893	Designed the reversible handle bar.
21	1894	Quit drug business—took job clerking in bankrupt store; got job with E. Kahn at $20.00.
22	1895	Hotel clerk—started development of hotel apparatus.
23	1896	Started work on building telephone equipment. Filed patent on hotel signaling apparatus and pushed my work as rapidly as my savings from $20.00 per month would permit.
24	1897	Assumed the title of "Telephone Engineer." Sold the idea of future possibilities of telephony and got my first job in telephone exchange construction at Macon, Mo. And sold the plant to my old friend, Theo. Gary.
25	1898	Married December 17, 1898.
36	1909	Ten years of general telephone exchange construction in Iowa, Kansas, and Missouri.
37	1910	I decided to direct my endeavors in the direction of manufacturing and Miss Traughber came with me in Feb. or that year.
39	1912	Big sleet storm—birth of Never-Creep® idea.
40	1913	Original patents on Never-Creep® filed—and our first sales representative came with us.
41	1914	Take back of stock sold in 1913.
42	1915	Business vacation.
43	1916	Jack Milligan—second salesman added.
44	1917	Built new factory.
45	1918	Moved out of old church building to new.
46	1919	Chas. Clark.
47	1920–25	Good steady growth and addition of products.
52	1925	Plant enlarged—foundry rebuilt.
53	1926–7	Trip around the world—greatest profit year.
54	1928	Big business year.
55	1929	Big Warehouse fire—net fire loss $66,000.00.
56	1930–1	Depression arrived in Utility industry.
59	1932	Kearney episode.
60	1933	Return to common sense—association with Wej-Lock.
	1934–35–36	Going up.

(Span of 46 years)

Undoubtedly, the reader has enjoyed personally pondering the various aspects of Mr. Chance's highlights. How many of us ever get our priorities in order? There have been some convincing studies that suggest people have burnouts. That's one reason why vacation time

A. B. Chance III and his wife Carol (Tusler) Chance with Mr. and Mrs. A. Bishop Chance. (A. B. "Buff" Chance IV Collection)

Mr. and Mrs. A. Bishop Chance and Mr. and Mrs. F. Gano Chance. Pictured in 1948 at the elder Chance's home on the occasion of their fiftieth wedding anniversary celebration. (Jack Chance Collection)

on the job was developed. A. Bishop was a man who possessed wisdom. Notice that his highlight in 1915 indicates a "business vacation." Did he say enough is enough? Turning this around could suggest that he simply took a vacation from business. A mere decade later he did the same thing but in a more sensational way. He and his family took a world cruise. The gall of this man, a businessman no less. Notice that he remarked that it was, "greatest profit year." It's unlikely that he is suggesting that he profited from the vacation, although he surely did. In a reassuring way he is noting the greatest profit year, 1927, which portrays the idea everything was under control, especially upon his return in early 1927. An interesting sidebar might be any affects this might have had on Chance employee vacation policy.

There is information contained in these highlights that can only be found there. It gives a panoramic view of his early influences/interests and shows the development of his life and Chance Company. The telephone influences over a twenty-year period would propel him into the manufacturing business.

What does "take back of stock sold in 1913" mean? During the many Chance employee interviews that L. C. Hansen did, he would ask that question. Roy had become privy to the contents of the highlights document. No one seemed to have a clue as to the meaning of this statement. Roy's persistence eventually led to the discovering of information about this short-lived Arizona incorporated entity. It was covered in an earlier chapter and referred to the dissolving of a corporation that had been formed with a Kansas City interest and a couple of Centralians.

Most of the material in the highlight is self-explanatory, except the last, "Going up." Was this a literal remark or is it to be taken in a figurative sense? It could be construed to represent an attitude developed from poor health. Since he was an active Christian who appeared to have a strong spiritual faith, he could be referring to his ultimate destination. Instead of a personal-type statement, however, it could be an evaluation of business conditions and due to his optimistic attitude he saw business as looking better even in the grips of the Great Depression. The mid-1930s described earlier in this chapter at Chance Company, did produce record employment and plant activity that kept the doors open. Indications suggest this turnaround occurred in 1934 and accelerated like a spacecraft, "going up!"

The Rock Garden

A Chance Legacy

A. Bishop Chance wasn't someone who did things in a small way and even his backyard reflects the dynamics of this man. Although the Chance Rock Garden was not a company project, it reflects one in which he gained a high level of personal delight and satisfaction. The spark that was said to have ignited the movement of three hundred tons of rocks and boulders to his backyard and to Centralia where only rich prairie soil could previously be found, was a world cruise by steamship. According to some personal papers of F. Gano Chance, this was the "Ryndan University World Cruise–1926 and 1927." Taking nearly a year to steam around the world and visiting thirty-five countries was no little thing in 1926–27. For a Centralia family in that time frame to do it was on the order of going to the moon, or that was what Grandmother Chance thought. Her grandson, F. Gano recalled she was quite upset when they departed in the fall of 1926 as she felt she would never see them again. "My Grandmother Chance told us that she would not get to see us as we would never make it back." (1981)

Gano added:

> *It was something you just wouldn't think of. So it was somewhat unusual. . . . Today there are so many people traveling around the world, particularly with an airplane, it isn't any problem—anything much to glorify. However, a couple of years after I was on this cruise I was introduced to a little girl, about 10 years old, by her mother. She said, "Well, Helen, Mr. Chance has been around the world." The little girl looked up at me and she said, "Have you been to Akron, Ohio?" I said, "No." "Well, you have not been around the world!" That stuck with me over the years and it kinda cooled me off. I wasn't as good as I thought I was. (1981)*

This would appear to be in reference to the Joe Cheethams' eldest daughter Helen, as they had Akron connections. One common comment of those traveling abroad is how fortunate they felt in being an American. The best part of those trips was often the return to their native soil. Gano made such a comment about being away from home and their triumphant return to the U.S.A. after their nine-month world tour:

> *The first time away from home at Christmas was an experience of course, but as we got closer to England, our last stop, we began to want to go home. I remember as we pulled into New York Harbor, one of the boys was standing at the rail with me looking at the Statue of Liberty and the Statue of Liberty is facing the ocean, the face. He said, "Listen lady if you want to see me again then you are going to have to turn around!" (1981)*

Like the Chance farm, the Rock Garden and Chance home eventually came under the control and management of A. B. Chance Company. The Mr. and Mrs. A. Bishop Chance home was used for many years as the "Garden House" where company visitors could stay. It provided services normally offered by motels/hotels, but the smallness and remote location of Centralia deemed this to be a practical solution in entertaining and housing overnight customers and guests of A. B. Chance Company. This employee recalls an evening spent there in the late 1960s when some field sales employees were in town. This tradition of a guesthouse was carried on a few blocks south in Mr. and Mrs. F. Gano Chance's former home, the beautiful Art Deco–styled mansion Gano designed and had built in 1939. But it did not have an elaborate garden, like his father's, for evening strolls. Headquarters changes and less

This beautiful home at 319 East Sneed became the Mr. and Mrs. A. Bishop Chance residence after World War I. It would also be described as the "Garden House" because of the Rock Garden constructed at the rear. For a time it housed overnight guests of A. B. Chance Company before it became the site of the Centralia Historical Society. (Jack Chance Collection)

use brought its sale to a private party in 2006 that are related to Sallie (Chance) Fountain.

The Rock Garden is a significant piece of Chance history due in part to its survival over all these years. This may be so because the Rock Garden and the former home of Mr. and Mrs. A. Bishop Chance hold the special distinction of still being under local control and the influence of the Chance family. They remain much as they were in earlier times. In contrast, the Chance Company, Chance farm, and Buford Park either no longer exist or do so in a progressively difficult to recognize fashion.

Gano spoke of the Chance Foundation and how it was established to serve. "That foundation contributes and encourages a lot of community activities. We wound up by taking the old home place and making a museum out of it." Gano said the home that used to house customers overnight was turned into a library and his home became the new depository for Chance Company guests. "We began to separate the Chance Foundation from the old Chance Company. The Chance Foundation is continuing to do what was done before." (1981) The Foundation would carry on the work the Chance family had been involved in during the past, showing a strong family desire that this work continue even though there was no longer a family connection with Chance Company. The former A. Bishop Chance home and garden are on the National Register of Historic Places supported by the United States Department of the Interior, per a plaque in the Chance Rock Garden.

It was June 9, 2000, a sunny and breezy kind of day, some sixty-three years after A. Bishop

Chance's Rock Garden was completed and opened in 1937. This employee's noonday walk along the ash-lined Centralia streets now included a side tour to and through the Rock Garden. Daily visits would hopefully generate the transmission of a message and when the Garden's turn came, words and the story about it would flow easily. However, even the garden this spring, although touching with its physical beauty of new colors and fragrances appearing and changing daily, didn't generate the desired effect as something was missing. Then it finally sunk in and became obvious that people make the difference. A personal infatuation with the Rocky Mountains and the Estes Park area in particular, led to an invitation of fellow employee and Kiwanian Bill Hollander, to give a Kiwanis program. He graciously agreed, but like the initial response to the Gettysburg Address, it was momentarily stunning to hear someone who had a second home in Estes Park, around all of the Rocky Mountain's natural beauty, dare to suggest that the scenery would soon lose its edge without people to share it with. Man, like a flower, has a way of bringing things to life. The spirit of man, inspired from above, is the mover and shaker in the world, although there are those who suggest nature would be better served without man's interference. But natural things like rocks and flowers are to be used and enjoyed by man, of course in a responsible manner, because people are very important. A. Bishop believed that because he used the natural beauty of rocks, flowers, exotic structures, and flowing water to touch and reach people. That was what was missing, the spirit of the man who once roamed the garden greeting and talking to strangers in the same manner he did with local friends. To prolong the day, he even turned to electric lights to enhance the garden's beauty and to attract and draw people in the same way an airport beacon reaches out into the darkness of the night to reveal its source. Of course it's been fifty-one-plus years since A. Bishop last greeted those that pass through his garden, a privilege many of us have missed. But he and his son Gano, in 1947, assured future guests that the gates would remain open and the garden immaculately maintained through the auspices of the Chance Foundation. So in an indirect manner, guests are still welcomed and Mr. Chance's spirit still lingers there, in view of what he left behind; it just requires a little deeper thought to discover it and to recognize his charitable and uniquely original contributions to his fellow man.

One of the advantages or disadvantages of a small town is there are few secrets. Our extracurricular activities, and most of us have one or more, can come under the scrutiny of those around us, especially if they are of an unorthodox nature. The building of the Chance Garden was undoubtedly a public thing, just as it would become after its completion. So along with town's people, Chance Company employees were eyeing the activity out behind A. Bishop's house. Hardly any employee is ever satisfied with their wages and benefits, which can result in a union coming in as it did at Chance's. Gano recalled:

> *I was out of town when it* [CIO proposal] *was presented to the company but they had a majority of our employees and wanted to negotiate. My father had pretty well gotten away from business, however, my father was building his rock garden. That caused the union to make statements, "that is where the money is going, in the rock garden, instead of paid to us!" It really made my father mad and he fired two or three employees. I came home and had a little more experience with the government and said, "You can't do that. You will have to take them back." I suggested to go and get them and so they went back to work and that was about as much friction that we had. (1981)*

Gardening and flowers apparently weren't something that A. Bishop had always been involved in, in fact it was to the contrary. When asked of such an interest, Gano replied:

No, not very much. He got to traveling around the U.S. and seeing some rock gardens and flower gardens. But he wasn't originally that I know anything about as a kid. He wasn't a guy that liked to work in the yard. But he could shift from one thing to another, just like that. So I think he decided, well Gano is going along pretty good, let him alone. I will put my time in the garden. (1981)

Replying to a question if Fancy Chance appreciated A. Bishop being around the house more, Gano said: "Well, I don't know. I expect she did appreciate him being there more. She was a wonderful lady, of course, and she and dad got along wonderfully well, as far as I know. I don't know that I ever heard them argue among themselves so I expect she liked it." (1981)

In the mid-1930s when the Rock Garden was constructed, Centralia was a much smaller town and hadn't yet extended very far south. But by that time the Chance's owned a sizable amount of real estate in Centralia including a house on the lot where the Rock Garden is built. Roy Hansen, in examining Chance Company Board meeting minutes noted, "about the same time the Chance Company was incorporated, the Chance Development and Investment Company was incorporated." (1978) F. Gano Chance, in responding to why it was established:

The tax angle was to separate the patents of my father, particularly since he was getting a royalty from other companies besides the Chance Company. So it was pulled out separate because of contracts with other manufacturers and of course it was the result of the tax advantage. As it turned out it was a personal holding company and we had to dissolve it. (1978)

Roy further added that the minutes of a 1940 Board meeting show actions taken to dissolve it. Gano and Roy both agreed that one of its purposes was a holding company for Chance family real estate. Gano responded that, "It was a corporation under which all the land was held. There was quite a lot of land. I don't think there was any of it that the factory was on." (1978) Roy added, "About seventeen tracts in town. I was amazed. I told Jack [Chance], look at the land, it was three-fourths of Centralia!" (1978) One of the early employees, Flynt Chandler, recalled what occurred when A. Bishop decided to build the Rock Garden where the Chandlers were living:

When they got ready to put that rock garden in he was going to move my house. So he fixed me up a house out where Harold Cox's [Chance Purchasing Agent] *house is now and I moved out there* [north of Gano's estate]. *It had three or four acres of ground then. All that where Gano's big house was; that was grass out there and I would buy me a couple of pigs in the fall and fatten them up and around the first of the year butcher them. I had a pony out there and we had a regular little farm. There were three houses between where that little store and where the school office is now. That was about the end of the houses. Kinda like living in the country. (1969)*

Mr. Chance's involvement in the Rock Garden project wasn't something surprising or unusual as Roy Hansen considered him as sort of an "inventive guy." Roy West said: "He was creative, yes he was. He worked on a lot of things that never did hit the market. Some of them, I'm sure, he could of patented, but it was something to occupy his mind, keep him busy." (1979) As to the Rock Garden, there was mutual agreement that it occupied a lot of his time. West said, "Oh yes, he did. That was quite a project. Once he got in that Rock Garden he stuck with it until he got it completed." (1979) News items from the *Guard* tend to tie down when this construction began. Since

This Rock Garden scene is not long after it was officially opened in 1937. The mid-1930s Chrysler Air-Flow sedan and the more basic garden layout suggest its newness. (Jack Chance Collection)

it consisted of only rocks from Missouri, a vacation trip by Mr. and Mrs. A. B. Chance to the Ozarks where beautiful rocks can be plentifully found, was noted in September of 1935. A few weeks later a November 1, 1935 *Guard* item told of A. B. Chance having a large fishpond constructed at his Braford and Singleton Streets home. In November 1936, a Fulton, Missouri garden group visited his "rock garden" that reflects further progress in its development.

The Chance Garden House, as remembered by Granny Ryman, was first the R. L. Hopehour home and then the Charles Early property in which the Hopes lived for a time. A. Bishop purchased if from Mr. Early and remodeled it. Ryman went on to say that:

> *He* [A. Bishop] *built the Rock Garden in memory of a trip around the world when Gano was going to the university and took some valuable leave. This lady, a garden expert/architect and he built all of those canals and he built a lot of those grottos, etc. in the Rock Garden. He went down in the Ozarks and brought back all those crystals, etc. He spent a great part of his time at that. When he started something he put a lot of effort in it. (1979)*

Granny Ryman's point associating the Chance's around-the-world cruise with the Rock Garden's genesis is also confirmed by other sources. The impressions and contrasts of the many nations visited had a deep and lasting affect on the Chances, politically speaking, from a business perspective, moral and human rights issues, compassion for the underprivileged masses, and it even affected their private home life with an elaborate public garden in their backyard. Gano described its educational benefits:

> *That first trip was very educational because it was composed of a lot of students. We were together and we discussed the problems with students of foreign countries. We had discussions aboard ship of what we had seen, etc. and our professors pretty well steered us if we analyzed anything a little bit wrong, etc. (1981)*

One of the lasting legacies of F. Gano Chance was his support of education and educational institutions. His political views and stances didn't bring embraces from that community later in life, but he would have liked a closer walk and acceptance while exercising his right to the freedoms Americans enjoy. One very wise observation of Gano:

> *The Japanese are wonderful people. The Russians are wonderful people. All the people are wonderful. There are only a few people that get control of a country that gets the country in difficulty. It is not the people and I learned one very important thing, always remember the difference between government and people. (1981)*

According to Gano, a highlight of a 1968 cruise was to be a visit to Japan. Between Honolulu and Tokyo a storm pitched the ship violently causing Gano to lose his balance when he had gotten up during the night. Even the ship's doctor was sick and a naval surgeon instead sewed up a twenty-three-stitch head wound and set a broken collarbone. Gano's condition resulted in he and Ann missing the Japan visit they had planned. Gano did mention his former trip, forty years earlier:

> *Japan was so much different than America. My mother said to my father, "Look at those foreigners over there on the corner." My father said, "Those are not foreigners We are the foreigners!" We left Japan thinking it was a wonderful country, primarily because of the people. It was very unusual too because the temples were so different than anything we had seen. They were very clean people—took a lot of baths. We had the experience of taking some Japanese baths. (1981)*

Although Gano didn't mention the beautiful gardens in Japan, they must have caught A. Bishop's eye as his Rock Garden had Japanese connections. Examples in the garden influenced by their world tour is a replica of a Japanese pagoda, according to *Chanco News* of March 1949. A June 1957 issue also speaks of the rock grotto and other garden features. "There is a rock grotto with alabaster replica of the Taj Mahal, a wishing well, a fountain of youth and a pool with miniature stone mill, water wheel and lighthouse."

An editorial in the June 18, 1937 *Guard* spoke of the new Chance Rock Garden. The garden's guest book had been in use for about two months, which would put the official opening around mid-April of 1937. However, a group from Fulton toured the Chance Garden in November of 1936 and this surely was only one of many other visits prior to 1937. For several years the Rock Garden would be a work in progress and wasn't something that Mr. Chance saw in its final form immediately. This of course runs counter to a commonly held belief that it was long inspired in a form seen in the spring of 1937. *Guard* editor Miller's detailed story of the garden fills many other information voids such as the garden designer/artist is identified as Mrs. Maude Dawson Taylor of Kansas City. She was involved early in the project and worked with her hands and others to bring form and shape to the garden's features. Missouri surface rocks were featured such as a limestone type known as cathedral rock. All the pottery was made from Missouri clay, most by the Kansas City Pottery Company. Many of the black wrought-iron rails, stands, and other fixtures were made in the Chance plant. The garden featured many beautiful flowers including 330 varieties of roses, said to be Mr. Chance's favorite flower. The watering system of pipes and valves was very elaborate allowing all or any part of the garden to be watered. The water in the pool was kept fresh by the aerating process as the water went over the waterfalls and was pumped back to begin its journey again, so the water wasn't wasted. It is special to view the garden today in the spring when the water and flowers come to life again as it did in its first official full season in the spring of 1937, just as A. Bishop Chance would have wanted.

Getting the Rock Garden on Missouri state road maps and in other publications led many people over the years into Centralia to tour it. A letter dated November 24, 1941, from a Mr. J. H. Dean of Shelbina to Mr. A. B. Chance, confirms visitor enjoyment of the garden. Mr. Dean said he was attracted to it, "by the pictures, in the *St. Louis Globe Democrat*." That newspaper had done an article, apparently in that time frame, of the Chance Garden. The garden also impacted Centralia residents as the January 14, 1938 *Guard* told of Mrs. Eliza Cleeton composing a song last summer, "Centralia's Rock Garden!" A July 15, 1938 *Guard* article stated that over three thousand people had registered their visit to the Rock Garden since Easter Sunday. It was estimated many visitors didn't register. One Sunday there were visitors from nine different states and each Sunday there appeared to be at least two carloads from Kansas City, likely influenced by Mrs. Taylor's role in the garden. Apparently nothing else in Centralia continually drew the crowds as did the Chance Gardens in the early years. *Chance Tips*, Volume 1, No. 1 of 1939, indicated about eight thousand people visited it in 1938. It's surprising the adventurousness of Americans and their extensive travels during the Depression to visit notable attractions.

It is not a certainty that A. Bishop envisioned the Rock Garden existing into the postmillennial period, but it has stood the test of time and it and his home have become personal reminders of the sustaining influence of this man. There is a plaque in the garden from a letter A. Bishop wrote that captures his feelings in a splendid manner. "I am . . . just one who enjoys and appreciates the geological wonders of our grand old state of Missouri and gains pleasure and happiness in their display—in the assembly of our garden, that the ones passing this way may stop, rest, reflect and enjoy." (Letter to a friend, 1948, A. Bishop Chance) His son, F. Gano, also made a comment to a Mr. Petrakis (author) while they were sitting in the Rock Garden one evening in the 1980 time frame. "I am sure there are other significant memorials, but I think these gardens are special because they provide so many people so much pleasure." (1981) The thought that a son is like a father in certain ways rings true here as both father and son were builders who were ultimately concerned about the welfare of their fellow man, nation, and world. How could each of us excel more than to leave behind a reminder that we truly cared?

Epilogue
A Trip to the Farm

In 1943 the world was stuck right in the middle of World War II. On the "Home Front" there was a lot of work to be done in support of the war effort and Americans responded in a responsible and courageous manner. Employers were begging for people to fill positions left by men being recruited into the services and from the tremendous increases in productivity needed to fuel the needs of a two-front war.

On the farms and in the factories of America, many women were pressed into service as well as retirees and many young people. Opportunities for jobs were excellent at that time for young men who were just short of eighteen. The catch was no long-range commitments could be promised, but then a Sturgeon youth of seventeen had the whole summer to enjoy before facing new and difficult challenges that would come that winter. He decided to accept employment at a nearby farm and little did he know that fifty-five years later he would recall these events as if it was yesterday.

On the A. Bishop Chance farm between Sturgeon and Centralia, John Stone was grinding feed for the livestock, which consisted of hogs, dairy cattle, and a herd of Aberdeen Angus cattle. As he fed more corn into the grinder a big Cadillac automobile made its way into the farm driveway. As it came to a stop, the door opened and a tall, lanky, Charles Lindbergh–type of individual exited. It was the summer of 1943 and although this approaching gentleman was seventy years old, there was still a gleam and sparkle in his eyes and a broad smile on his face. Located several miles west of Centralia on Highway No. 22 (old), this 800 + 80-acre farm was the apple of this man's eye. It reflected the years of his attention and interest and with the dark clouds of war hanging over our nation its productivity of food, milk and fiber was important as it was on millions of other farms across America.

As the man approached he greeted John by his first name. A story told over and over again was that A. Bishop knew all of his employees by their first names and took an active interest in the lives of his "guys and gals." A longstanding joke around his Centralia business was if you couldn't find a tool, it was likely at the farm. According to *Chanco News* of March 1943, lost and misplaced tools had become a problem at the plant. Many exploratory missions and retrieving operations were run to the farm in search of a missing tool. After completing his tour, he approached John and asked if he would fill his car up with gas. This visit, like hundreds of others, reiterated the fact that it was a mistake to believe this man was predictable and was satisfied with everything the way it was. Ideas still flowed generously from his active mind and he could often be found in a machine shop or his shop at home trying to work out a problem and bringing form to an idea with his hands.

Drawing ideas on paper wasn't his forte, but shaping and molding something until it looked and performed correctly were strengths he was said to possess.

Heading east toward Centralia he crossed a small stream and approached a crossroads. The road to the north was just a farm lane but it took one to a place where many memories and happy times were had and were still to come. The A. B. Chance Company of Centralia had picnics there. Employees played horseshoes or other games, or just talked, or rode the small train that was there at one time. The park had a further significance as it was named after Buford, the eldest son of the founder of A. B. Chance Company, who was killed in an automobile accident just east of Centralia in 1928. It was a loss such as no parent is said to ever completely recover from. It was especially difficult for Buford's wife and young son that would be addressed as "Little A. B."

Continuing east the big V-8 powered Cadillac purred effortlessly up Highway No. 22. Thinking of the farm he realized that at some point in the future he might not be able to manage it any longer. Age has a way of reducing what one can oversee and do and health problems were taking their toll. His only son, F. Gano, had his hands full operating the family business, especially during these tense times of World War II. Maybe it could be sold to a worthy group, but that would be for another day.

A mile or so to the south across the flat prairie cropland, separated only by an occasional hedge row, the smoke and steam from a hard charging Wabash steam locomotive shot high into the sky westbound toward Moberly and points north and west. Behind it was a troop train filled with new recruits destined for some far off training center and soldiers heading home on leave or headed back to the front. These troops fortunately found their brief passage through Centralia a couple of minutes earlier more hospitable than a group some eighty years earlier, but of course the threat to the free world was outside of our borders this day and not from within.

Instead of Highway No. 22 logically following the Alton Railroad straight into Centralia, it veers north, but that's how it has always been it seems. Suddenly rounding a curve, the road sharply turns south. Immediately the trees and buildings that develop the Centralia skyline come into focus, looking somewhat unnatural in its prairie setting. But that is perfectly explainable as both the trees and structures on the horizon are all transplants, a sign of modern civilization. Getting closer, a large, tall three-story building dominates the center of this town, shattering immediately any illusion that this is an ordinary rural agricultural community. Approaching the Alton Railroad crossing, this large red brick building looms in one's face and it and the surrounding complex alerts strangers to ask themselves, "I wonder what they make here?" But it's no stranger to the man returning from his farm, although there is a spark of newness and a certain amount of pride every time he enters town down Allen Street. His fresh look at things, however, should be no surprise because this man had never allowed himself to be satisfied with normalcy and the world has benefited as a result.

In the area between the two tracks one was transported into another world. The pounding of the huge hammers shook the ground like a passing freight train and the hum and chatter of presses and other machinery could be heard as they cut, shaped, and formed raw metal into finished products. Instead of the finished goods going to some domestic customer, much of it was destined to the Signal Corps to help reestablish communication lines that had been destroyed by the advancing military forces in the European or the Pacific Theaters. Certain items produced were for B-17s flying daylight missions high over Berlin. War production, even in the tranquility of this small community in the heartland of America, continually reminded one of the turmoil, death, and destruction that was being wrought by war at that time in such places as Guadacanal and Messina.

There are several references about A. Bishop "going to the farm," especially by his son. It is frustrating to be enthusiastically caught up in something and not be able to convince an important figure in that drama to come aboard. This is an early indication that retirement kind of interests were taking center stage in A. Bishop's life and the demands of travel, business confrontations and negotiations weren't as attractive to him in his later years. There would come a time when he was no longer there to guide and counsel his son and Gano would need to be capable of dealing effectively with the operations of the company. There was a strong urge among those around the "Boss" to desire his company as his presence lent an air of confidence and credibility. Although the evidence wasn't as clear at that time, Gano would also rise to a position in which his presence, guidance, and leadership would be in demand by those around him. But there is one tradition that will strongly be upheld in closing this book, the sacristy of the position of "Father." It's only fitting that the elder father, represented here by A. Bishop, have the last word!

During the Chances' world cruise in the late 1920s, Gano relished in the story about his father being reprimanded by one of the scholastic leaders because of frequent class absences. Of course A. Bishop found the tremendous opportunity for socializing a more important priority, as he did various personal activities later in life. Albert Bishop Chance was surely the best-known and most-beloved resident that Centralia ever produced. Fifty years after his death there are still reminders of the affection for him in the plaques and other items dedicated to him. The declining number who remember him do so with affection and speak of him as if it was only yesterday when he last shook their hand and called them by their first name. In spite of his prominence while living, his name, undoubtedly, has been

A 1940 picture of A. Bishop presenting Buford Memorial Park to employees. Odie Walker is on his right, N. A. Toalson on his left beside Miss Ida Lou. Ollie Chandler is to her right. (Jack Chance Collection)

mentioned and printed many more times after death than before due to the continuing success of the company he founded. So it should come as a surprise that official historical documentation and recognition of he and his son are weak to invisible. Hopefully this will be corrected.

One's life can often be seen more clearly by summarizing what they did away from their occupations. A. Bishop involved himself in three projects that are very revealing and they could be labeled: his Roots, Tribute, and Generosity. The farm was a connection back to his early roots and one that would harness his immense energy for many years. It was a challenge and a pleasurable adventure in growing and perfecting it. An important part of the farm was the development of Buford Park. This was done in tribute to his fallen son, but not in a private way. A worthy tribute to A. Bishop was one in which he could unselfishly enrich the lives of others and be in their company. About this time, another project was started that was one of his last major ones before age would slow and stop such activities. It would, however, be one of his most lasting legacies and one that would represent his unselfish generosity. It would be a rock garden that others would continue to enjoy long after he had grown his last flower. Although there are many things more physically durable than a flower or a life, hardly anything is capable of capturing a moment in time like the specialness of a beautiful flower or a rich life!

BIBLIOGRAPHY

A. B. Chance Company Official Materials (documents, correspondence, etc.), 1907–1974.

American Profile, Midwest Edition. Newspaper Magazine, May 28–June 3, 2006.

Bible. King James Version. Iowa Falls, Iowa, World Bible Publishers, 1979.

Chance Account Representative Hot Line, as appropriate, unknown–1974.

Chance Memo to Management. Centralia, Missouri, Weekly, 1961–1974.

Chance News Release, Centralia, Missouri, as appropriate, unknown–1974.

Chance News for You. Centralia, Missouri, as appropriate, unknown–1974.

Chance Tips. Publication of A. B. Chance Company, Centralia, Missouri. Five issues annually, 1939–1974.

Chanco News. Publication of A. B. Chance Company, Centralia, Missouri. Monthly, 1942–1974.

Chapman, Carl H. *Indians and Archaeology of Missouri.* Columbia: University of Missouri Press, 1983.

Gary, Theodore. *Independent Telephony.* Quincy, Illinois: Press of Cadogan-Hatcher Manufacturing Company, 1907.

-----. *Macon County Handbook of 1888.* Macon, Missouri, 1888.

Gerhard, Ralph. *Early Day History of Moberly, Missouri.* Vol. I. Moberly, Missouri: Randolph County Historical Society, 1991.

Grant, H. Roger. *Follow the Flag: A History of the Wabash Railroad Company.* Dekalb: Northern Illinois University Press, 2004.

Guard (Our Fireside Guard or *Fireside Guard).* Newspaper. Centralia, Missouri.

History of Boone County. St. Louis: Western Historical Company, 1882.

History of Missouri: Family and Personal History. Vol. VIII. New York, New York: St. Louis Historical Publication, 1967.

Hot Line Maintenance. Centralia, Missouri, A. B. Chance Company, March 20, 1952.

Interviews conducted by Donald R. Darst with A. B. Chance employees, 1961–2006.

Interviews conducted by L. C. Hansen and J. H. Chance with A. B. Chance Company officials and employees, 1978–1981.

Interview by H. A. Houston with F. Gano Chance, 1941.

Interview by H. A. Houston with Art Kalips and Andy Anderson, 1973.

Interview by Mark Petrakis with F. Gano Chance in Scottsdale, Arizona, beginning January 27, 1981.

LaDoux, Rita. *Missouri.* Minneapolis: Lerner Publications, 1991.

MacMeal, Harry B. *The Story of Independent Telephony.* Chicago: Independent Pioneer Telephone Association, 1934.

McCarthy, Thomas E. *History of G.T.E.* Stamford, Conn.: G.T.E. Corporation, 1990.

Menninger, Karl. Personal letter, October 29, 1943.

Missouri Life Magazine, January–February 1983.

Moberly Monitor Index newspaper. Moberly, Missouri.

Monroe County Appeal newspaper. Paris, Missouri.

Musick, John R. and Richard Chitwood. *Stories of Missouri.* New York: American Books, 1897.

Not By Accident. 50th Anniversary Sound-Slide Program. Centralia, Mo.: A. B. Chance Co., 1957.

Parry, Thomas W., Jr., "A Chance-Built Industry." *Forbes Magazine.* March 1, 1931.

Railroad History No. 133. Boston: Railway and Locomotive Historical Society, Fall 1975.

Rand McNally Handy Railroad Atlas of the United States. Chicago: Rand McNally and Company, 1985.

Robbins, Henry P. *Theodore Gary.* Kansas City, Mo.: Martin Publishing, 1932.

Shoemaker, F. C. *Missouri and Missourians: Land of Contrasts and People of Achievement.* Chicago: Lewis Publishing Company, 1943.

Show-Me State Story: A History of Telephony in Missouri. Jefferson City: Missouri Telephone Association, 1989.

St. Louis Post-Dispatch Sunday Magazine. December 6, 1925.

University of Missouri Yearbook for 1926. Jefferson City, Mo.; Hugh Stephens Press, 1926.

Wolferman, Kristie C. *Osage in Missouri.* Columbia: University of Missouri Press, 1997.

Chronology

1889 First Christian Church building moved from Post Office location west to M. S. Bush lot.

1896 A. Bishop Chance working on Macon Telephone Exchange. An electrical supply depot opened in Bush Building in July by A. Bishop Chance. Electric lights installed in Centralia store buildings and hotel electric call apparatus developed by A. Bishop Chance.

1897 Centralia Telephone Company began by A. Bishop Chance. Telephone line to Madison and Columbia completed by A. Bishop Chance.

1898 Centralia Telephone Company now had eighty subscribers.

1899 A. Bishop Chance sold his telephone line between Centralia and Columbia.

1901 The Centralia Telephone Company purchased a portion of the old Standards trunk line.

1902 Transfer-Board for Telephone/Telegraph Circuits patented by A. Bishop Chance, No. 693,784. Curcuit-Closing Devices, a telephone signaling apparatus patented by A. Bishop Chance, No. 707,761.

1904 The Centralia Telephone Company reportedly became a co-operative. Another Circuit-Closing Device was patented by A. Bishop Chance, No. 763,464. LaCross consolidated their switchboard into Centralia Telephone Company central.

1905 Centralia Telephone Company purchased by Home Telephone Company group. Chance plumbing shop converted into the manufacturing of telephones and supplies.

1907 Year selected as the official birth of Chance Manufacturing Company, also known as Chance Company, or A. B. Chance Company. Promotion of new Sky-Rocket® Lightning Arrester began in March.

1908 A. B. Chance Arrester Patent No. 894,723 issued in July. Trip to Texas by A. Bishop Chance to investigate starting a manufacturing concern.

1909 First confirmed Chance acquisition, an arrester from W. N. McAnge for $200 in October. A. Bishop Chance announced his successful run for the Centralia mayor's position. First export sales of Chance product confirmed. Chance Manufacturing Company reorganized into a corporation with stock offered to public. Stock Sales Department established in Kansas City by A. Bishop Chance.

1910 Canadian order for more than five thousand arresters was industries largest to date for protective devices. Miss Ida Lou Traughber employed at Chance Manufacturing Company.

1911 Confirmation Chance Manufacturing Company is located on Sneed Street in Bush Building. Electrical demonstration of various electrical items at the Chance Manufacturing Company plant held in June.

1912 Never-Creep® Anchor developed by A. Bishop Chance.

1913 New "dead man" device introduced in January to commercial salesmen and jobbers in St. Louis. Never-Creep® registered as a trademark in January. Mr. Montague of Kansas City first sales person hired by Chance Company.

1914 Stock retrieved by A. Bishop Chance in unsuccessful reorganization attempt of 1909. Thimbleye® Anchor Rod design and/or other anchor rod design activities progressing.

1916 Jack Milligan of Chicago was second salesman employed at Chance Company.

1917 Announcement in March that the production of Irl Hicks' Gas-Tite Piston Rings would be moved to the Chance Manufacturing Company, establishing the Big C Auto line. Home Telephone central destroyed by fire in April. Offices moved from Baker Building to Farmers and Merchants Bank Building. City granted a lease to A. Bishop Chance in May for a 50 by 150 feet lot between the tracks in the Allen and Rollins Streets area. A two-story factory building at 210 North Allen was completed by year's end. Building was originally sixty by one hundred feet.

1918 A.T.&T. developed specification numbers for Chance anchor rods. Centralia Manufacturing Company leased the second floor of the new Chance plant. Crump and Humphrey relocated to the Anthony Building on Railroad and Rollins.

1919 Chance Company request to the city for fifty feet of land to allow extension of the Chance plant. A one-hundred-foot plant extension and a third story was planned for the Chance factory. Charley Clark of Pittsburgh was the third Chance Company salesman employed. A Cross Arm Back Truss and Clamp assembly designed by A. Bishop Chance in 1917 resulted in Patent No. 1,324,585. The seventh Chance Company employee Flynt Chandler was employed.

1920 "Anchor Rod Eye," a Thimbleye® design, patented by A. Bishop Chance, No. 56,273. Chance Manufacturing Company announced intentions in September of constructing a large Foundry east of their current plant. Chance Manufacturing Company in October obtained a ninety-nine-year lease from the city on their current and the additional grounds approved in their September request.

1921 Thimbleye® Anchor Rod patented and trademark registered. Patent No. 1,373,223. Commonwealth Edison Company developing specifications for utilizing Chance Anchor Rods including the first reference to Hot Galvanized found. New 80-by-240-foot Chance Foundry plans being drafted by Ed Chamberlain in January. New Chance Blast Furnace, in a temporary building, was fired up in late January and some experimental casting produced.

1922 Chemical concern located in the Chance plant became bankrupt and its operations discontinued by a sheriff's sale in August. Gas-Tite Piston Rings were being produced at a pace of five thousand per day by mid-year. In May Chance Manufacturing Company had forty-seven men and women employees in the plant. Concern continues over the need for a low-cost reliable source of electricity to power the many Chance factory machines.

1923 Employees gave A. Bishop Chance a going-away banquet before his California vacation departure. Piston ring reorganization proposal suggesting Big C would become a part of the C-K Company. A local business, Drace and Chance involving F. Gano Chance sold, repaired, built, and installed radio equipment.

1924 First confirmation that the Anthony Building is now owned by A. Bishop Chance. Rebuilding of the Anthony Building began plus news that the J. Ford Honey's Ford Dealership would occupy it. Ed Chamberlain drew up building plans. New X-Ray machine purchased by Chance Company for plant and local community medical needs.

1925 The new Chemical Building on the northeast side of Collier was completed during the summer. Earliest confirmation that Chance Company had developed a movie for advertising purposes. By mid-year Honey Ford was in the new Anthony Building, which would now be extended further east, work beginning in May. The third story and power plant addition to the main factory and the 80-by-300-foot Foundry to its immediate rear were begun in July. Construction on the Wabash Depot began in July. New 40-by-120-foot Paint Shop near the Alton tracks, west of the new Chemical Building would begin mid-year. A new wareroom of the same dimensions would be constructed immediately to the west after the Paint Shop's completion.

1926 Frank H. Dukesmith employed in February as director of sales of the Piston Ring Department. Taylor/Sutcliffe expanding anchor acquired. Home Telephone Company sold. Chance Company purchased the Green Mule Barn in April. The Chance Power Plant built to provide an independent and reliable electrical power source for the Chance plant was tested in July. Mr. and Mrs. A. Bishop and F. Gano Chance left on world cruise in September. Stamping/Galvanizing Departments established at Chance Company in November. Chance Company School of Instruction was begun in November for Chance employees. Second furnace brought on-line in the Foundry in late December. Steel stamping presses in operation in December to make anchor and other parts.

1927 The A. B. Chance family returned in May from their around-the-world trip. First confirmed Chance Company picnic held in July with around 150 in attendance. The Chance Drop Forge Department began operation in September with eleven male employees. A Chance employee Social and Amusement Committee was formed. Overhead electric conveyor systems installed in various departments also included a "Dipper" line that submerged anchors into a paint vat. Verna Palmer employed in late December.

1928 Confirmation that the new "Red Snapper" piston rings had been introduced. Clifford Hough began part-time accounting duties. Record million-quantity piston ring order received by the Big C Department in July. Jack Isbell employed as salesman in Utility Sales. New Chemical Lab established in July at the Chance Plant at 210 North Allen. Agricultural professors from four states visited the Chance Plant in August where the new Post Anchor had been introduced. Chance plant had eighty-three employees in September. Chance amateur broadcasting station 9-CUQ was on the air at the plant. First confirmed Safety Meeting held in the Chance Plant in early November. Albert Buford Chance died in an automobile accident on November 24.

1929 A new Chance Big C patent No. 1,780,003 was issued in October as a "Method and Apparatus for Manufacturing of Split Rings." Construction of new Alton siding track that served the Mule Barn and the new Punch Press. Warehouse No.3, the Mule Barn, was destroyed by fire on April 9. New Warehouse No. 3, Punch Press, was completed by late summer. N. C. Husted was hired as the new vice president and general manager. Engineers F. Gano Chance and Bill Hollander were employed full time. Chance Company became a partnership of founder and son, from former proprietorship. Chance Company had some ninety employees, not including salesmen.

1930 N. A. Toalson employed. Slower business level noted during normal spring rush.

1931 Sales Conference of the Electrical Division held in January at Centralia. Sales were estimated at around $750,000. Kearney/Chance consolidation in October.

1932 Sales dropped to around $196,000. Kearney/Chance relationship terminated in May. Wej-Lock and Chance association began in December. Chance Company incorporated on December 30. At December 31 Board Meeting, A. Bishop Chance became president; F. Gano Chance vice president and general manager; Miss Ida Lou Traughber secretary/treasurer. Paul Sturgeon was employed.

1933 Jack Isbell reemployed as sales manager. Cooperation with "New Deal" programs brought the N.R.A. insignia to Chance Company stationery.

1934 Annual Chance Sales Conference held in May at Centralia headquarters. Production of hammers confirmed in November. Axes and Plow Points described as new factory items. Record late-year employment of more than two hundred. C. C. Hough employed full time.

1935 Employment now around three hundred. Largest Big C order received for 580,000 piston rings. Chance Company bonds proposal to develop greater working capital presented in June. Chance produced Wej-Lock Highway Guardrails being used across the United States. Granville "Granny" Ryman reemployed.

1936 Chance employees received a 10 percent wage increase on December 1. Official support of employee preparations for "In-House Promotions." Chance Rock Garden construction continuing. Chance Company officially became the A. B. Chance Company.

Volume I INDEX

Included are Chance Company employees, other related individuals, competitors, industry references, and specific products. Because A. Bishop Chance and F. Gano Chance are mentioned so frequently, they will be omitted. Also included are local, state, and national historical figures, and items which depict the times or relates to the Chance family or to the Chance Company Story. Various sources and time frames generate variations in indexed names. Upholding the integrity of these sources can generate multiple names of the same entity. The use of "nee" indicates the maiden name.

A

B

C

D

H

M

N

O

P

Q

R

T

U

V

W

X

Y

Z

Tour Guide

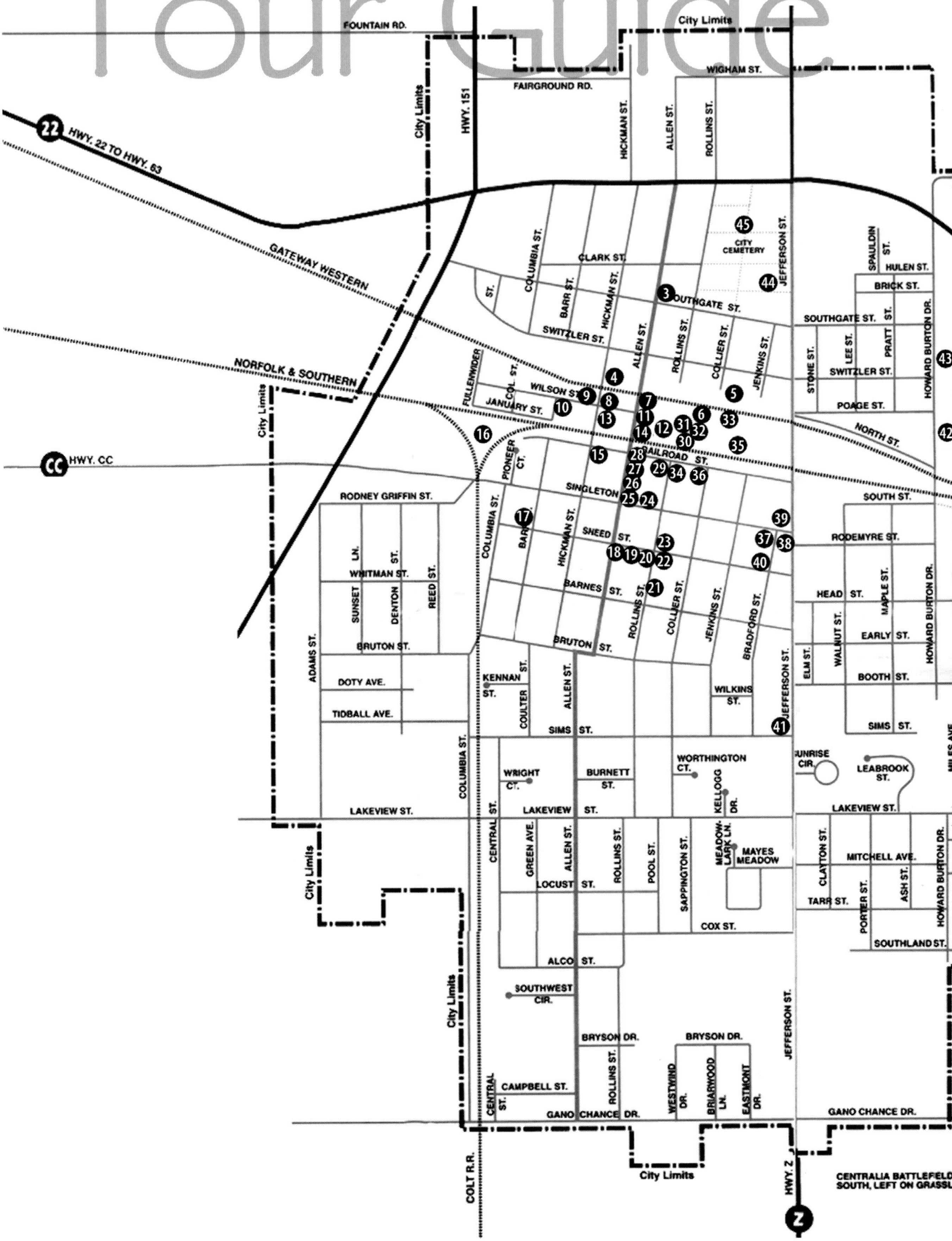

FOUNTAIN RD.
City Limits
WIGHAM ST.
FAIRGROUND RD.
HWY. 151
HICKMAN ST.
ALLEN ST.
ROLLINS ST.
22
HWY. 22 TO HWY. 63
GATEWAY WESTERN
NORFOLK & SOUTHERN
COLUMBIA ST.
CLARK ST.
CITY CEMETERY
JEFFERSON ST.
SPAULDING ST.
HULEN ST.
BRICK ST.
SOUTHGATE ST.
BARR ST.
SWITZLER ST.
COLLIER ST.
JENKINS ST.
STONE ST.
LEE ST.
PRATT ST.
HOWARD BURTON DR.
FULLENWIDER
COL. ST.
WILSON ST.
JANUARY ST.
POAGE ST.
NORTH ST.
CC
HWY. CC
PIONEER CT.
RAILROAD ST.
SINGLETON
RODNEY GRIFFIN ST.
SOUTH ST.
SHEED ST.
RODEMYRE ST.
LN.
WHITMAN ST.
SUNSET
DENTON
REED ST.
BARNES ST.
HEAD ST.
MAPLE ST.
BRADFORD ST.
EARLY ST.
BRUTON ST.
ADAMS ST.
ELM ST.
WALNUT ST.
BOOTH ST.
KENNAN ST.
COULTER ST.
DOTY AVE.
TIDBALL AVE.
WILKINS ST.
SIMS ST.
WORTHINGTON CT.
SUNRISE CIR.
LEABROOK ST.
WRIGHT CT.
BURNETT ST.
KELLOGG DR.
LAKEVIEW ST.
CENTRAL ST.
GREEN AVE.
POOL ST.
SAPPINGTON ST.
MEADOWLARK LN.
MAYES MEADOW
CLAYTON ST.
MITCHELL AVE.
LOCUST ST.
TARR ST.
PORTER ST.
ASH ST.
COX ST.
SOUTHLAND ST.
ALCO ST.
SOUTHWEST CIR.
BRYSON DR.
CAMPBELL ST.
WESTWIND DR.
BRIARWOOD LN.
EASTMONT DR.
GANO CHANCE DR.
COLT R.R.
HWY. Z
Z
CENTRALIA BATTLEFIELD

Tour Guide for Chance Company Story

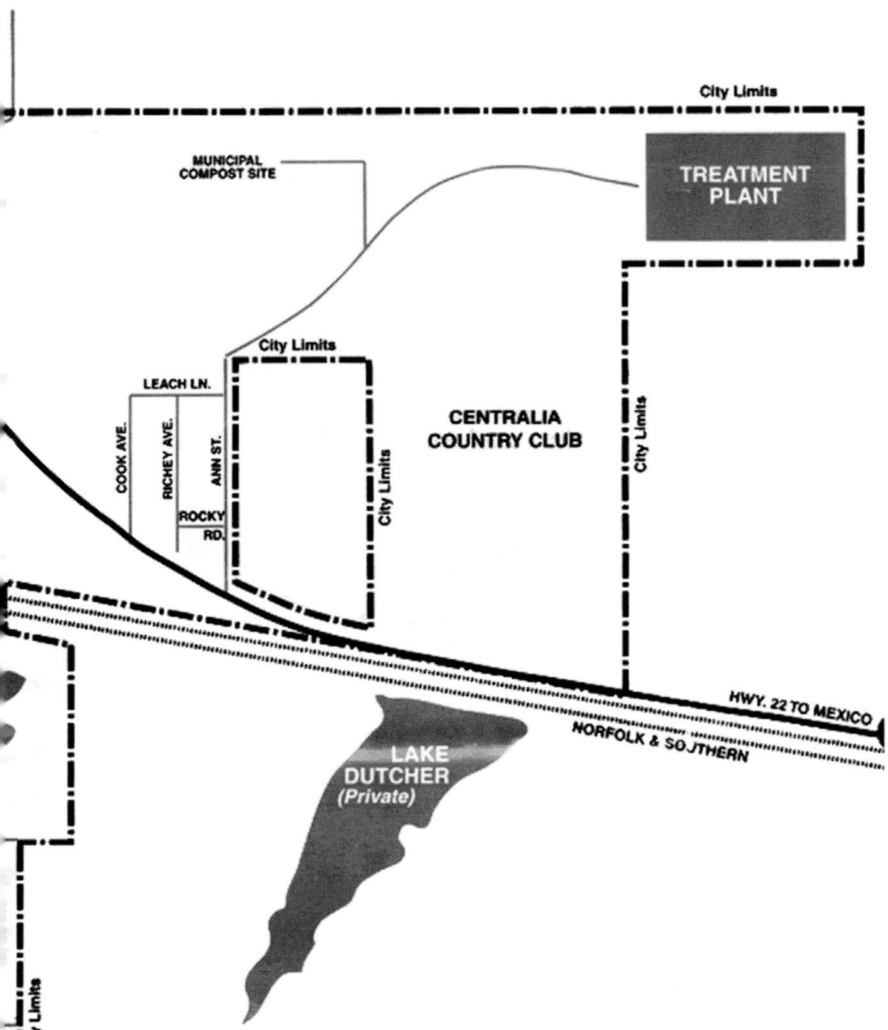

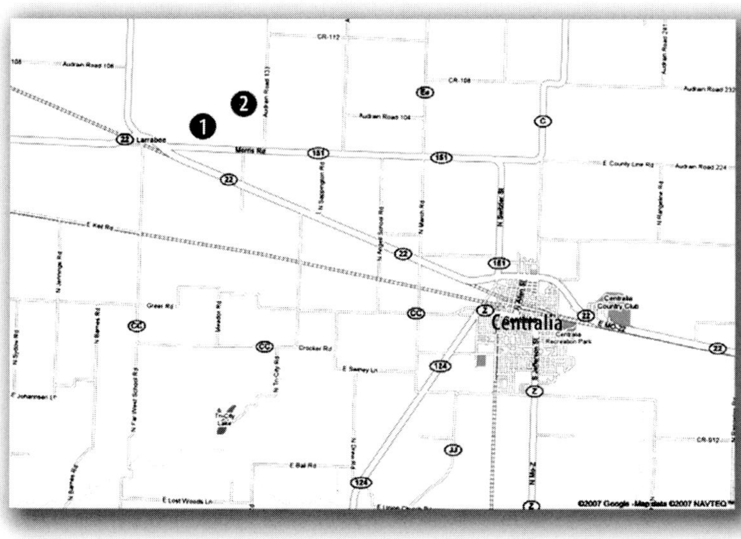

1. Chance Farm (Sunnydale)
2. Buford Chance Memorial Park
3. Wabash Station Shelter House
4. Broom Factory location
5. Thomas Lumber Yard
6. Drop Forge (Chance Company)
7. Alton Depot location
8. Producers Grain Company (M.F.A.)
9. Illinois Oil Company
10. Alton Spur Track
11. 210 North Allen (Chance Company Main Factory)
12. Galvanizing Building location
13. Anchor Assy/Punch Press (1929) (Mule Barn location)
14. Pre-1925 Wabash Depot location
15. A. R. Toalson Building
16. Columbia Branch Wye
17. A. R. Toalson home
18. Bush Building (Guard)
19. Chance Manufacturing Company first official plant location
20. Post Office
21. Miss Ida Lou Traughber home
22. Home Telephone (Meadors)
23. City Hall
24. Fireside Guard (Rodemyre/Miller)
25. Home Telephone location after fire
26. Baker Building (Chance/Home Telephone)
27. Vista Theatre
28. Globe Hotel location
29. Squires Restaurant
30. Wabash Depot location (1925) and site of Centralia Massacre of train troops
31. Chance Foundry (1925)
32. City Water Tower/Plant location
33. Chemical Building location (1925)
34. Anthony Building
35. Chance Switch/Tool Plant (1948)
36. Chance rental home (Fenton, Husted, Kalip, etc.)
37. Chance Rock Garden
38. Chance Bungalow
39. Elm Branch
40. A. Bishop Chance home
41. F. Gano Chance home (1939)
42. Chance East Street Plant (1954)
43. Chance Research Center (1961)
44. Chance Sepulcher
45. Centralia City Cemetery
46. Chance Plastics Plant (1968)

About the Author

Born in Paris, Missouri, and raised on a working farm, Don Darst attended Evans and Ash rural elementary schools. The summer of his junior year at Madison High School was spent working for a painting company in Elkhart, IN. He continued to work his way through Moberly Junior College and was employed at Magers Standard Service Station in Moberly before getting a job in the A. B. Chance Company factory in the fall of 1961. Don's Chance tour of duty was very brief as on Halloween night the plant went on strike. Don returned to Magers until being called back to Chance in 1962.

Service in the National Guard in 1962 included advanced Military Police training at Fort Gordon in Augusta, GA. Later the Officer's Candidate School program was successfully completed. Back in Centralia, an inventory-posting job came open in the office in 1963 under Moe Zoellers and later Cab Atkins. The threat of job loss due to computerization helped Don decide to enter Northeast Missouri State University in the fall of 1964. The summer break of 1965 was spent back in his old Chance job as a student worker.

Upon obtaining his degree in 1966, no openings were available at Chance so a management trainee position was accepted with J. C. Penney Co. at their new Parkade Plaza store in Columbia, MO. Personnel Manager Blair Smith knew of Don's desire to go back to Chance, so when a position came open at Grandview, MO, in the Pitman Division, it was accepted in late 1966. After two years, an opening became available in Centralia in Marketing Services under Bob Tanner. Don notes that his "last decade was spent in Manufacturing/Purchasing ending in 2001, representing working continuously for forty years at Chance! Extracurricular activities were always important, including the most difficult challenge and longest sustained project of my life, this book."